PHYSICAL GEOGRAPHY
FOURTH EDITION

PHYSICAL GEOGRAPHY

FOURTH EDITION

MICHAEL P. McINTYRE

SAN JOSÉ STATE UNIVERSITY

JOHN WILEY & SONS
NEW YORK   CHICHESTER
BRISBANE   TORONTO
SINGAPORE

Cover photo by Duilio Peruzzi
Cover and text design by Karin Gerdes Kincheloe
Production supervised by Jan M. Lavin
Photo researched by Elyse Rieder
Copyedited by Frank Menchaca

Library of Congress Cataloging in Publication Data:

McIntyre, Michael P.
  Physical geography.

  Bibliography: p.
  Includes index.
  1. Physical geography. I. Title.
GB55.M22 1985   910′.02   84-15158
ISBN 0-471-87915-0

Printed in the United States of America

10 9 8 7 6 5 4 3 2

PREFACE

Most American students, in nostalgic retrospect, surely recall the dubious thrill of locating places and coloring maps in long-ago grade school courses labeled Geography. But is this really the essence of geography? The answer is both yes and no: yes, of course, with the world as our arena of activity, place location is inevitably a part of that understanding; no, there is much more to the study of geography than these simple exercises. All of the earth and all of the human activity on it are within the realm of the geographer. Perhaps the definition of *geography as human ecology* is not a bad one because ecology means *mutual interaction*, and the real heart of the geographic equation lies in a reciprocal relationship between the inhabitants of this earth and their physical environment. Each influences the other in a classic symbiosis.

But now there arises the practical pedagogical problem of offering an introductory course that addressses adequately this huge mass of material in one quarter/semester—the general consensus is that it can't be done and still maintain the quality of a college-level course. Long-established practice dictates that we pry apart the physical and human elements of the geographic whole and treat each separately. This book, entitled *Physical Geography*, then becomes a study of the environment in its totality. Other texts and other courses deal primarily with cultural and societal matters to round out the beginning student's introduction to geography. Such a pedagogical expedient does indeed defile true symbiotic purity, but it is more a shift of emphasis than a stark disassembly of the basic structure—all through this physical geography volume there is con-stant reference to and awareness of the human factor.

This edition is for typical lower-division students who, more often than not, suffer from a paucity of organized scientific training. Basic principles and perspicacious generalizations are emphasized. When the entire physical world, as a range of operation, is coupled with beginning students lacking in geographic background, the gross image becomes of primary significance. Moreover, a conscious effort has been made to deal in clarity, directness, relevent uncomplicated illustration, and careful, nontechnical definition of terms. The method employed is systematic and the discussion is analytical/descriptive with an emphasis on the concept of nature's inherent ecological unity.

A new edition is routinely brought up to date with new photographs and diagrams, an expanded bibliography, fresh statistical data, and metric equivalents. But this edition also deals with a host of modern topics, frequently through the use of page box vignettes: El Niño, fusion energy, seabottom warm water jets, nuclear power for seawater desalinization, Mount St. Helens, and Holland's Rhine Delta Project. All of these are physical geographical matters with far-reaching human implications.

This book should present no major difficulties for lower-division or community college students, who will find that it makes considerable sense (perhaps even generating interest and excitement) without bogging down in the swampy ground of formulas and jargon. The simple treatment, concentrating on a fundamental skeleton of principles and processes, should make life in the classroom more pleasant for the professor. Secure in the knowledge that students understand the textbook, each teacher is freed to embroider those sections where additional depth seems helpful. With the conceptual framework firmly established, it becomes a simple matter to add provocative marginalia. Also revised for this edition are the student workbook/lab manual, and instructor's manual, which are both significant parts of this edition's total educational package.

It is my pleasant duty to acknowledge the aid, comfort, and constructive criticism cheerfully given by so many: my students, my colleagues, my wife, the always-helpful Wiley staff, and especially my hardworking typist, Judie Geiss. Dan Epstein, San Jose City College, and Don Reasons, West Valley College, have functioned as my valued community college consultants and reviewers; Herb Combs, North Adams State College, developed the workbook/lab manual, and Bill Takizawa, San José State University, the instructor's manual. The job could not have been done without this band of selfless accomplices.

Michael P. McIntyre

CONTENTS

PART ONE

THE EARTH AS A GLOBE

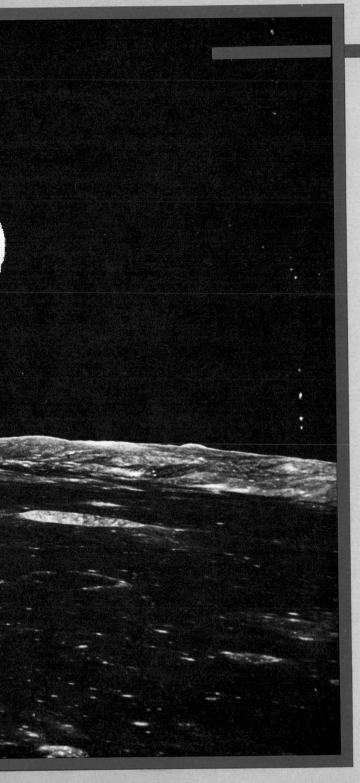

CHAPTER 1

THE EARTH IN SPACE

Viewed from the moon our earth is a startlingly handsome planet in space. The white tops of swirling clouds tend to obscure many familiar landmarks, but here and there peeping through are the darker landmasses and striking azure seas.

THE SHAPE OF THE EARTH

Let us suppose, just for the sake of reasonable generalization, that the earth is a perfect sphere. Of course it really is not. For one thing, there are bumps all over its surface and some of them are of substantial size. Mt. Everest is 29,000 feet (8839 m) above sea level and the Marianas Trench 36,000 feet (10,972 m) below. This is a total crustal relief on the order of 12 miles (19 km), seemingly difficult to ignore. But if we put it in proper perspective and measure this 12 miles (19 km) against the earth's radius of approximately 4000 miles (6437 km), it secomes minuscule. For example, on a large classroom globe, where the earth's 25,000 mile (40,234 km) circumference is commonly represented by 20 inches (51 cm), the paper that covers the globe is thicker than the total surface relief drawn to proper scale.

Further, we know that a rapidly rotating earth responds to centrifugal force and bulges at the equator while flattening at the poles. But this amounts to a maximum radius difference of only about 13 or 14 miles (21 to 23 km), again a tiny fraction of 4000 miles (6437 km). The people concerned with finite measurement—the geodesist, surveyor, and navigator—are very much aware of these relationships, but we are not subjecting the truth to any real violence when we choose to generalize.

Actually, *nobody knows exactly the shape of the earth.* Certain photos from orbiting satellites suggest a pear shape, gravity studies show wide variations from place to place, and terrestrial magnetism is constantly shifting. It is unfortunate that the geodesists have appropriated the term *geoid* for a separate and hypothetical shape because geoid, quite literally, means *earth shaped* and that is certainly the most accurate way to describe the unique shape of the earth.

It is unfortunate that the geodosists have appropriated the term geoid *for a separate and hypothetical shape because* geoid, *quite literally, means* earth shaped *and that is certainly the most accurate way to describe the unique shape of the earth.*

The key point, of course, is magnitude of variation. Admitting myriad deviations from a true sphere, *none of* them, known or speculated, are of sufficient magnitude to disturb our spherical image.

The concept of the earth as a globe goes far back into early Greek times when the classic logic of the Greek mind appears to have been tempered by a bit of inspired imagination. It was reasoned that the globe was the perfect shape, its surface equidistant from a central point and without beginning or end. Thus since the earth was the handiwork of the gods, its shape must be spherical. Once postulated, however guileless the process, it could be easily proved by the progressive "sinking" of a ship sailing over the horizon, the distinctive shadow on the moon during successive eclipses, and the predictable change of altitude of fixed heavenly bodies with change of latitude, etc. . . . . Columbus knew the earth was round even if his simple seamen did not. His error of knowledge was in the assumption of a globe of a smaller size.

So for our purposes we will assume the earth to be the *perfect* sphere that all the globes show it to be. Life is much pleasanter for the student at problem-solving time if this kind of generalization can be accepted and thereby doing away with the need for constant minor corrections. And to avoid Columbus' error of too small a globe, we will adopt 25,000 miles (40,234 km) as a workable round number for the earth's circumference.

EARTH MOTIONS

The earth is constantly in motion in two separate and distinct circulations at the same time. We as inhabitants are not immediately aware of this, since we do not stagger about on our moving platform but are held securely through gravity; nor does the wind whistle through our ears, since the atmosphere, also held to the earth by gravity, moves with us. There are, however, a number of readily observable phenomena that are the direct result of these motions. Alternating day and night, the progression of the seasons, and the differing length of day during the year are among them. The rising and setting sun and the movements of many of the heavenly bodies are *apparent* motions rather than actual; these stationary objects appear to migrate as they are observed from a moving planet.

In common parlance, the words *rotation* and *revolution* are exact synonyms, but when used to describe earth motions, they have separate meanings and must be applied properly. Rotation is the earth's movement about its axis; revolution is its movement about the sun.

Rotation

Day and night are the result of rotation, as any location is first turned toward the sun and then away as the earth spins on its axis. This is a constant and steady movement of about 1000 miles (1609 km) per hour at the equator. The movement is so regular that our system of time is geared to it and reckoned on a 24 hour basis (one rotation) (Fig. 1–1). The sun, appearing to wheel around the earth once in every 24 hours, rising in the east and setting in the west, is actually not moving at all. This is an apparent motion viewed from a moving earth and therefore the earth must rotate in the *opposite* direction from the sun's apparent motion, or from *west* to *east*.

Revolution

Revolution is a different matter—variations are the rule. The earth's orbit describes an ellipse about the sun, not a perfect circle, and thus the earth is constantly at different distances from the sun, ranging from 92 to 95 million miles (147 to 152 million km).[1] The speed at which the earth follows its orbit varies also, the most rapid advance being in January and the minimal speed in July. Despite these variations, the earth manages to go around the sun in about the same length of time each circuit, but this period does not come out even with our 24 hour days (determined by rotation). It actually measures about 365.25 days and necessitates the familiar leap year every fourth year when we add up the accumulated quarters and splice them onto the tag end of February.

An important resultant of revolution is the seasons; however, revolution is not the complete explanation. There is another factor that, in conjunction with revolution, causes the progression of seasons—the inclination of the earth's

---

[1]The earth is nearest the sun in January (perihelion) and farthest away in July (aphelion). Obviously, a couple of million miles more or less have no appreciable effect on the temperature of the earth, or our Northern Hemisphere summer would be in January.

Fig. 1–1 The sundial does not keep the same time as our familiar mechanical clock, geared simply to repeat 24-hour intervals endlessly. They would be identical if only rotation were involved, but the sundial also reflects the varying speeds and elliptical orbit of revolution, and in so doing records all sorts of minor permutations. We have chosen to ignore these.

axis. The standard statement is that the earth's axis is inclined very close to 23.5° from the vertical. This is true, but the following questions may well arise: "How can the vertical be determined when we are dealing with a turning sphere in space?" "Vertical to what?" Obviously, we need a plane of reference from which to determine the vertical before a 23.5° variation can be reckoned. There is such a plane. It is called the *ecliptic* and is the plane of the earth's orbit about the sun. The axis, then, is inclined 23.5° from the vertical to this plane and maintains its inclination, always pointing in the same direction (*parallelism*) throughout each revolution (Fig. 1–2).

Every year about June 21 the earth is in a position where the inclined axis points directly toward the sun and the Northern Hemisphere is presented so as to receive a maximum of sunlight, while the Southern Hemisphere is tilted away. This results in the shortest and most direct rays of the sun striking a point well north of the equator, but the Southern Hemisphere experiences only low-angle rays that are capable of much less efficient heating. This is our longest day of the year and the hemispheric summer. It is called the *summer solstice*

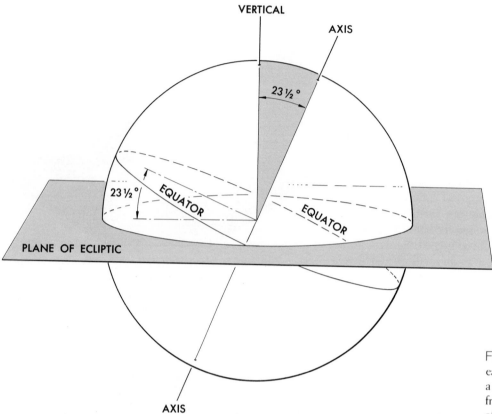

Fig. 1–2   Inclination of the earth's axis. The axis maintains a constant inclination of 23.5° from a line vertical to the plane of the ecliptic.

Three months later, September 21, revolution has carried the earth through a quarter of its orbit, and although the axis continues to be inclined the same amount, this inclination is now at right angles to the sun and both hemispheres receive equal sunlight. In such a position the most direct rays strike the equator, concentrating the greatest heating midway between the poles. This day is the first day of fall and is called the *fall (autumnal) equinox.*

After three more months of revolution, December 21, the earth has reversed its June position. The Southern Hemisphere is now tilted directly toward the sun, and summer prevails as the direct rays strike at a point south of the equator, but only weak, low-angle rays affect the Northern Hemisphere. Now we have arrived at the shortest day of the year, the *winter solstice,* and the Northern Hemisphere winter commences.

On March 21, after still another quarter-revolution, the earth reaches a point where the axis is once again lined up at right angles to the sun's rays, and both hemispheres receive the same amount of sun as they did in the fall. This is the first day of spring and we call it the *spring (vernal) equinox.*

Finally, after three more months, the earth returns to its original position of June 21[2] and it is summer again (Fig. 1–3).

Revolution alone could not accomplish this familiar seasonal sequence; revolution plus the earth's axial inclination are required.

---

[2]These exact dates, the 21st of June, September, December, and March, may vary from the 20th to the 23rd. However, for our purposes here, we will be consistent and use the 21st.

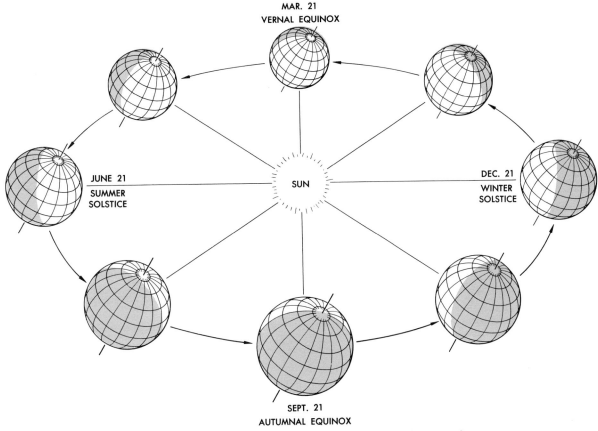

MAR. 21
VERNAL EQUINOX

JUNE 21
SUMMER
SOLSTICE

SUN

DEC. 21
WINTER
SOLSTICE

SEPT. 21
AUTUMNAL EQUINOX

Fig. 1—3 Both revolution and axial inclination are required to complete this familiar annual sequence.

THE MOON

Closely related to the earth both in physical propinquity and patterns of motion is its near neighbor or satellite, the moon. However, if the measure of the significance of an element of the earthly environment is its degree of direct influence on humans, the moon rates as relatively unimportant. For aside from its obvious role in controlling the oceanic tides (see Chapter 29) and such minor phenomena as eclipses, there appears to be little evidence that the moon affects the earth at all. But suddenly, with the advent of astronauts and international competition in space, we are confronted with casual references in the press to such things as *libration* and *synodic months*. Obviously, some under-

Suddenly, with the advent of astronauts and international competition in space, we are confronted with casual references in the press to such things as liberation *and* synodic months.

standing of the basics of moon behavior becomes imperative.

The moon is a true earth satellite approximately 240,000 miles (386,243 km) distant, slightly less than 1/100 the earth's mass, and like the earth, engaging in simultaneous rotation and revolution. It rotates about its axis and revolves about the earth in an elliptical orbit whose plane

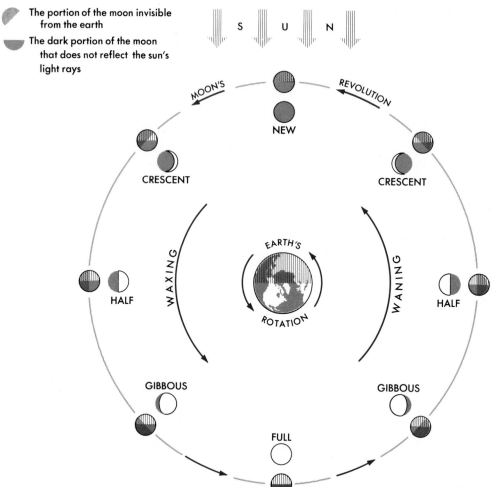

Fig. 1-4   The phases of the moon.

(or ecliptic) is almost parallel to that of the earth. When viewed from the moon's north pole these motions are counterclockwise, precisely the same as the earth's, and this correlation is in turn remarkably similar to the general pattern of elliptical orbits and counterclockwise circulation of the total solar system. However, there is one major deviation in the behavior of the moon when compared to that of the earth: both rotation and revolution are accomplished in exactly the same time—1 month. This means that essentially only half of the moon is ever seen from the earth. The actual figure is 59% because the moon travels an elliptical orbit at varying speeds and its ecliptic is not quite identical to the earth's. This apparent slight wobbling of

the moon, exposing to our view an added fraction of marginal area, is called *libration*

One element of our system of time comes down to us from the past, based on the visible repetitious behavior of the moon. Although our fundamental clock is the sun, which controls the 24 hour day and the 365.25-day year, the subdivision of the year into months (*moonths*) and their further breakdown into seven day weeks (each a quarter of the moon, beginning on Monday, or *moonday*) are only slight modifications of observable moon phenomena.

Observation of the moon on a daily basis reveals a characteristic and predictable pattern. Each 24 hours as the earth's rotation discloses the moon to us for substantial

Fig. 1–5   There are not many nights in the year when the moon looks this clear with all of its surface acne nakedly exposed. The moon is apparent to us because it is reflective of light from the sun as it revolves about the earth. Once every 29.5 days it works its way to a point exactly opposite the sun and for 24 hours is a perfectly round full moon migrating majestically across the night sky.

lengths of time, we see that in its path across the sky it lags a bit *relative to the sun*. This is a result of both the earth and the moon revolving simultaneously in separate orbits. But the apparent motion of the sun is affected only by the earth's motion, so although sun and moon appear to rise in the east and set in the west, the moon runs progressively behind, about 12° every 24 hours. At this rate it requires approximately 29.5 days before it is back at its original starting point with reference to the sun. This is called a *synodic month*. It varies slightly from an exact 29.5 days because the moon's speed of revolution is not constant, but it is nonetheless the rough basis for our calendar month.

The synodic monthly cycle becomes simple to observe and record because of the reflective quality of the moon's surface to sunlight *(albedo)*. This cycle makes at least a part of the moon visible so that we can follow a series of distinctive shapes as it waxes from new moon (no moon) to full moon (perfectly round) and wanes back to new moon (Fig. 1–4). It is simply a matter of the location of the moon and sun as viewed from the earth. When both the moon and the sun rise simultaneously from the east, we view the dark side of the moon because it is between us and the sun, but as it lags progressively behind on ensuing days, a thin sliver of *crescent* moon appears. After seven days the crescent has broadened to become a *half* moon, which progressively deforms into a larger *gibbous* moon. The maximum *full* moon is reached by about 14 days (Fig. 1–5) and then decreases in size through another *gibbous* by 18.5 days, *half* by 22 days, and *crescent* by 26 days, to *new* moon again in 29.5 days.

THE TROPICS—CANCER AND CAPRICORN

On every world map and globe there appears a pair of dashed lines, one in each hemisphere, parallel to the equator and not far from it. Everyone has carefully memorized at one time or another that the northern line is called the Tropic of Cancer and the southern one the Tropic of Capricorn. It is well to know the names, but of more importance is the recognition of why these lines exist in their particular positions. A clue is found in their latitude of 23.5° north and south. (Latitude is numbered in both directions from the equator through a quarter of a 360° circle, with the equator being 0° and each pole 90°. See "Latitude and Longitude," later in the chapter.) This points to a direct relationship between the Tropics and the axial inclination, and in turn the seasons.

The *Tropic of Cancer* may be defined as *the northernmost point on the earth that ever experiences the sun directly overhead at noon*. This only occurs on one day each year—June 21. Conversely, the *Tropic of Capricorn* is *the southernmost point on the earth that ever experiences the sun directly overhead at noon*, on December 21. These Tropics are the outer limits of the particular phenomenon of the noon overhead sun at least once each year. Only between Cancer and Capricorn does this occur. Although

Calendars

Pope Gregory XIII.

The calendar is a timekeeping device as surely as is the clock; both utilize regularly recurring phenomena in nature to regulate and organize our lives. Day and night (rotation) is the basis for our 24-hour clock, a relatively simple and straightforward arrangement. But the calendar attempts to combine this with the seasons/year (revolution) and the week/month (phases of the moon), none of which are exactly comparable. One year, from winter solstice to winter solstice, is measured in fractional days, 365.25; and one lunar month averages 29.5 days which when multiplied by 12 equals 354 days, a far cry from one full year.

How to mesh these disparate elements into a functional whole? This is the question that has vexed generations of would-be calendar makers. There is no easy or obvious answer.

The first modern western calendar (excluding Mayan, etc.) to achieve any real success was the *Julian*, executed by order of Julius Caesar in 46 B.C. It addressed the rapid cumulative error of an extra quarter day each year by introducing the concept of leap year, and solved the short lunar year problem by arbitrarily lengthening each month except February. Our familiar months, their names, and varying lengths, have come down to us intact from the Julian calendar.

This remarkable calendar served very well for over 1500 years even though, as it turned out, the Julian year was 11 minutes longer than the astronomical year. By A.D. 1582 this small problem had magnified into a 10-day error, so that on October 5th of that year Pope Gregory XIII felt it necessary to suppress 10 days by ordaining that the date should be October 15 instantaneously. An additional bit of minor tinkering with the formula took care of the congenital cumulative error difficulty. The resulting calendar became known as the *Gregorian*, and was adopted by Great Britain and the English colonies in America in 1752; it is the one in almost universal use today.

It isn't pretty but it works. We still have to go through the "30 days hath September, April, June and November" routine; and it is difficult to forecast what day Christmas will fall on several years hence—day and date are never the same from year to year. Calendar reform is always in the air; there are whole societies dedicated to it. But let's not hold our breaths. As long as the Gregorian calendar functions without error, newer, improved, slightly polished models are not likely to be adopted. Progress is not rapid in calendar circles. It took the British 170 years to make the move from Julian to Gregorian.

these Tropic lines experience the overhead sun only once each year, the entire region between them finds the sun overhead on two dates during the year. The equator, halfway between Cancer and Capricorn, has these two dates equally spaced between June 21 and December 21—on the equinoxes, September 21 and March 21—but the overhead-sun dates of all other locations are unevenly spaced, depending on how close they are to the equator. To permanent earthly residents it *appears* that the overhead sun migrates from Cancer to Capricorn and back in one year, thus passing over each spot in between twice—once as it moves south and again as it moves north (Fig. 1–6).

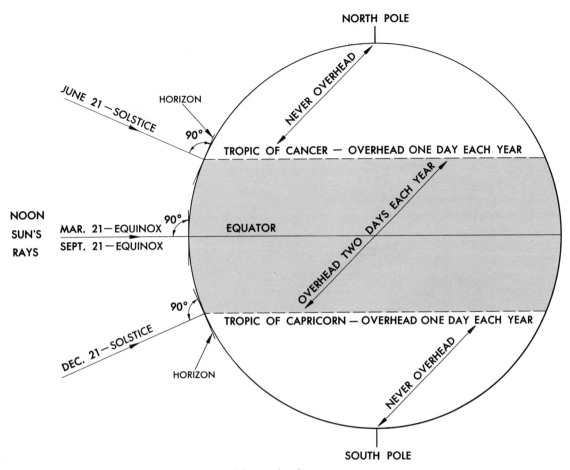

Fig. 1–6 The apparent annual migration of the overhead noon sun.

THE ARCTIC AND ANTARCTIC CIRCLES

Also seen on world maps and globes are the dashed lines in each hemisphere some little distance away from the poles. These are the *Arctic* and *Antarctic Circles* and, like the Tropics, they mark the outer limits of a special phenomenon peculiar to the polar regions. Again, we find a relationship with axial inclination when we discover that the Arctic and Antarctic Circles are 23.5° from their respective poles or 66.5° from the equator (66.5° north or south latitude). These are *the farthest points from each pole that experience at least one day when the sun never sets and at least one day when it never rises.*

At the Arctic Circle, the day when the sun fails to set is June 21, the summer solstice when the noon sun is overhead at the Tropic of Cancer; the one day when the sun does not rise is December 21, the winter solstice when the noon sun is overhead at the Tropic of Capricorn. The situation is identical at the Antarctic Circle except that the dates are reversed. Within the polar regions enclosed by the Arctic and Antarctic Circles, more than one 24 hour period of light and of dark during each year is the rule. The number of these days increases with the distance traveled toward the poles, culminating in 6 months of light and 6 months of dark at the poles themselves.

These peculiarities are brought about by the alternating presentation of each hemisphere, first toward the sun

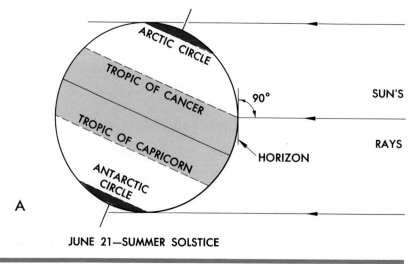

JUNE 21—SUMMER SOLSTICE

A

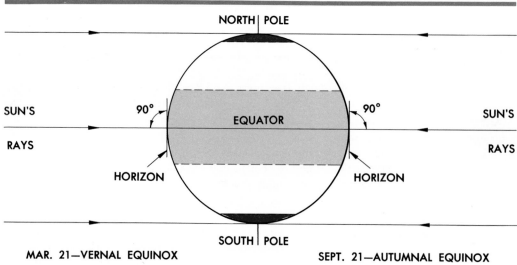

DEC. 21—WINTER SOLSTICE

B

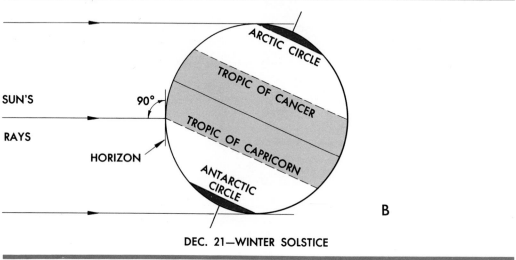

MAR. 21—VERNAL EQUINOX

SEPT. 21—AUTUMNAL EQUINOX

C

Fig. 1–7 Earth and sun at the solstices and equinoxes.

and then away as the earth travels its orbit. During the Northern Hemisphere summer, when this hemisphere is tilted toward the sun, the sun's rays strike beyond the North Pole and, at the same time, fall short of the South Pole. Rotation which causes day and night over the earth is ineffective in the polar areas during this period for, regardless of time of day, the north polar regions remain exposed to continuous sunlight while those areas near the South Pole receive none at all. The distance that the sun's rays reach beyond the North Pole and fall short of the South Pole is at its maximum on June 21 and measures 23.5°. On December 21 the sun's rays fall short of the North Pole and strike beyond the South Pole by this same amount (Fig. 1–7).

Only at the equator are day and night of equal length every day of the year.

The middle latitudes, that Northern Hemisphere zone between the Arctic Circle and the Tropic of Cancer and a comparable Southern Hemisphere zone, experience neither of these special phenomena. Those who live near the Arctic or Antarctic Circle will be familiar with extremely long summer days and winter nights. But in each 24-hour period of the year, the sun will set and rise if only for a short time. And even on the margins of the Tropics, although the sun will appear very high in the sky at noon in the summer, it will never be directly overhead.

THE LENGTH OF DAY

Only at the equator are day and night of equal length every day of the year. Elsewhere on the earth, day and night are always of unequal duration, *except for the equinoxes*. On these two critical dates, day and night are each 12 hours long everywhere. Near the equator, the variations of length of day are so small that they are not readily apparent, but as one moves poleward, the differences become striking over a period of a year, until at the poles the extreme of 6 months of light and 6 months of dark is reached. Yet even here on September 21, the sun, which has been above the horizon

at the North Pole for 6 months, finally sets. During the first 12 hours of that equinoctial day, the sun has been up, and for the last 12 hours, it has been down, and day and night are of equal length. And on March 21 at the North Pole, the sun, which has not risen for 6 months, suddenly appears, and the day is equally divided into dark and light. So, even at the poles, day and night are of equal length on the equinoxes, as they are everywhere else on the earth (Fig. 1–8).

LATITUDE AND LONGITUDE

Measurements on a map or globe might be considered logically in the next chapter, except that the precise determination of these measurements is inextricably linked to the apparent motions of certain heavenly bodies and therefore to rotation and revolution—thus the introduction of latitude and longitude at this particular juncture.

The problem facing the ancients, once they had proven the earth spherical, was how to establish artificial coordinates on its globular surface. The idea of a coordinate measuring system is, of course, very simple on a flat piece of graph paper where all of the X lines are parallel, all of the Y lines are parallel, and X and Y inevitably cross at right angles. But how to do this on a sphere? The ingenious system they devised utilizes two bases: (1) the axis of the earth with two surface fixed points at the poles and (2) the great circle.

A *great circle* is commonly defined as the *line described on the surface of the earth by a plane passed through its center.* Less awkwardly, we can say simply that such a line (circle) will divide the earth in half. This imaginary plane can be passed through the center of the earth in an infinity of ways, making possible an infinity of *great circles* (Fig. 1–9).

Latitude
The base line from which all latitude is reckoned is one of these great circles—the only one of its kind, for it is defined as *a great circle whose plane is at right angles to the axis*. There can be only a single great circle meeting this requirement, and it is the *equator* (equating the earth into a Northern and a Southern Hemisphere). All other lines

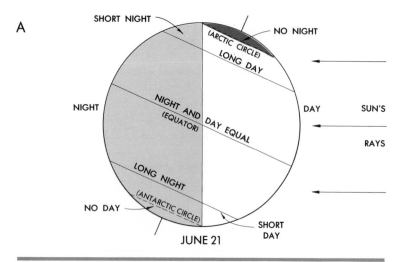

A

SHORT NIGHT
NO NIGHT
(ARCTIC CIRCLE)
LONG DAY
NIGHT
NIGHT AND DAY EQUAL
(EQUATOR)
DAY
SUN'S
RAYS
LONG NIGHT
(ANTARCTIC CIRCLE)
NO DAY
SHORT DAY

JUNE 21

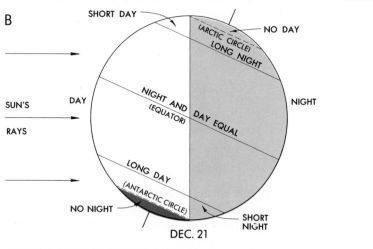

B

SHORT DAY
NO DAY
(ARCTIC CIRCLE)
LONG NIGHT
NIGHT AND DAY EQUAL
(EQUATOR)
NIGHT
SUN'S
RAYS
DAY
LONG DAY
(ANTARCTIC CIRCLE)
NO NIGHT
SHORT NIGHT

DEC. 21

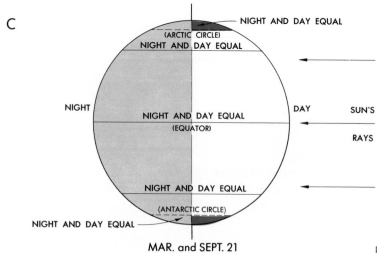

C

NIGHT AND DAY EQUAL
(ARCTIC CIRCLE)
NIGHT AND DAY EQUAL
NIGHT
NIGHT AND DAY EQUAL
(EQUATOR)
DAY
SUN'S
RAYS
NIGHT AND DAY EQUAL
(ANTARCTIC CIRCLE)
NIGHT AND DAY EQUAL

MAR. and SEPT. 21

Fig. 1–8 Length of day.

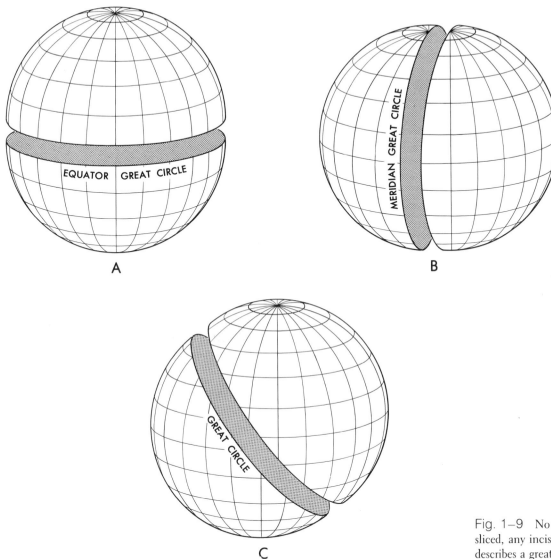

EQUATOR GREAT CIRCLE

A

MERIDIAN GREAT CIRCLE

B

GREAT CIRCLE

C

Fig. 1–9   No matter how it is sliced, any incision halving the earth describes a great circle on its surface.

of latitude are parallel to the equator just as on the graph paper, and they are circles too with their planes at right angles to the axis. But they cannot be great circles since they are north or south of the equator which itself cuts the earth in half (Fig. 1–10).

Every place on the earth has a particular latitude. An imaginary circle parallel to the equator (these circles are properly called *parallels* as well as lines of latitude) passes through any given location, and the measurement in de-

grees from there to the equator is its *latitude*. Latitude represents nothing more than distance north or south of the equator, described by a numbering system that designates the equator 0° and the poles 90°. Note, however, that in numbering in both directions from the equator, each number except 0° appears twice, once in each hemisphere, and the suffix N or S must always be added for clarification.

Why are we measuring linear distance in degrees in-

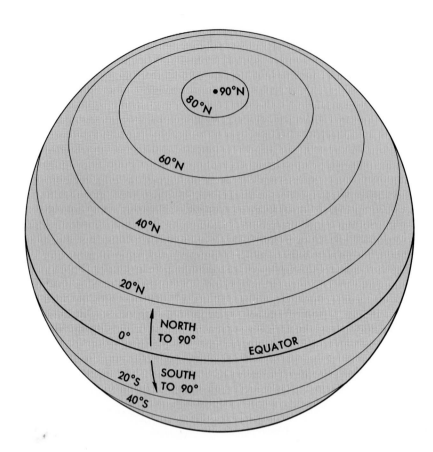

Fig. 1–10   Although they appear to be straight lines on a map, lines of latitude are actually circles, their planes parallel to that of the equator.

stead of in miles or kilometers? This is because the earth is round and the circumference of a circle is 360°; we are in essence working through one fourth of this circumference (0° to 90°) as we number the curved surface of the earth from equator to pole, both north and south. But, if it is more comfortable to work with distance in miles or kilometers, latitude can be translated: 1° of arc on a great circle whose circumference is 25,000 miles (40,134 km) comes very close to 69 miles (111 km). A location at 10° N latitude is somewhere on a circle 690 miles (1110 km) north of the equator (Fig. 1–11).

World maps and globes commonly show the equator, poles, Arctic and Antarctic Circles, Tropics of Cancer and Capricorn, and every 10° of latitude. These are meant to be utilized as references lines in map reading, and do not in any way imply that they are the only latitudinal circles. If we drew in black on a globe the latitude line for each degree, we would obliterate everything else and the lines

would still represent distances of 69 miles (111 km). If every place on earth has its own measurement north or south of the equator, there must be an infinite number of latitude lines. And if we are going to establish a more precise tolerance, with a possible error of less than 68+ miles (109 km), we are going to have to subdivide degrees. One degree of latitude breaks down into 60 minutes (1.15 miles or 1.7 km) and a minute, in turn, into 60 seconds (0.019 miles or 0.03 km). This seems pretty fine, but when extreme accuracy is required, seconds can be further cut into tenths, hundredths, or thousandths. At Meade's Ranch in east central Kansas there is a bronze marker that is the base point for mapping the whole of North America. Its latitude has been as precisely measured as any location in the world and reads 39° 13' 26.686"N. That takes us right down to fractions of inches (centimeters).

But now we come to the practical problem of determination of latitude—how does the navigator ascertain a

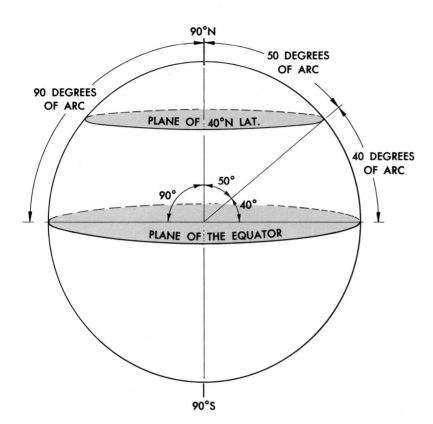

Fig. 1–11 Measuring latitude. A north latitude of 40° represents 40° of arc measured northward from the equator along the curving surface of the earth, and also a 40° angle at the earth's center.

precise distance from the equator when this is an unknown? Nobody has drawn those critical circles right out there on the ground. Somehow the location of the equator must be determined as a point of departure for reckoning. Strangely, or so it might seem, the navigator looks into the sky for help and observes the known behavior of any of a galaxy of heavenly bodies. An awkward approach? Maybe, but let us select the simplest of those heavenly bodies, one that cannot be mistaken for any of the others and one whose apparent motion follows a familiar pattern—the sun. On March and September 21 at noon the sun will be directly overhead at the equator, and since it is the equator we are trying to locate, a critical relationship has now been established. The equator is marked unerringly with a brightly shining beacon at noon on the equinoxes.

To proceed from this point one simply measures, with a handy sextant, the altitude of the sun above the horizon, that is, the number of degrees in the angle between the sun and the horizontal (Fig. 1–12). Assume for the mo-

The equator is marked unerringly with a brightly shining beacon at noon on the equinoxes.

ment that the observer is standing on the equator. The sun would be directly overhead and its altitude would be 90° above the horizon. But this 90° is obviously not the correct latitude. Because of the arbitrary numbering of latitude, with 0° at the equator and 90° at the poles, one more step is required after measuring the sun's altitude. To determine latitude, subtract this figure from 90°. Thus 90 minus 90 equals 0°, which is the latitude of the equator. If the observer steps off the equator, either north or south, the sun is no longer overhead and its altitude would be something less than 90°. And the farther one goes toward the poles, the lower the sun would appear at noon, until finally at the poles, the sun would appear resting on the horizon with no altitude at all. So it is not difficult to see that the alti-

Fig. 1–12 The navigator lines up his horizon in the sextant eyepiece, then by turning the knob with his left hand he brings the mirrored image of the noonday sun down to the horizon. This maneuver will record the sun's elevation on a vernier scale at the base of the instrument.

Old-time mariners could always be sure of reasonably accurate latitudinal measurement even though their longitude was sometimes very questionable.

tude of the sun at noon on the equinoxes leads directly to a determination of distance from the equator, or latitude. Once again, the altitude must be subtracted from 90° or, stated differently, *the latitude of any given location and the altitude of the sun at noon on the equinoxes are complementary (when added, they equal* 90°). For example, if the sun's altitude measures 65°, latitude equals 90° minus 65°, or 25° north or south depending on whether the observer is looking north or south.

The above, of course, is the simplest possible example. A navigator does not have to wait until noon on an

equinox to determine latitude. Since the apparent movement of the sun and other heavenly bodies follows regular patterns, their positions relative to the equator on any date and at any time can be reckoned and corrections applied to compensate for variation. This principle is relatively simple and has been understood and applied for centuries. Old-time mariners could always be sure of reasonably accurate latitudinal measurement even though their longitude was sometimes very questionable.

There is one other easy guide that can lead to rapid determination of latitude, at least in the Northern Hemisphere, and that is *Polaris*, or the North Star. It is permanently so close to being directly over the North Pole that the slight deviation can be ignored generally. Here the latitudinal numbering works in our favor, for the measured altitude of Polaris requires no processing but *is the correct latitude*. If one stood at the pole, Polaris would be overhead all night long, and its altitude would measure 90°. Latitude is 90° N. As an observer moves away from the

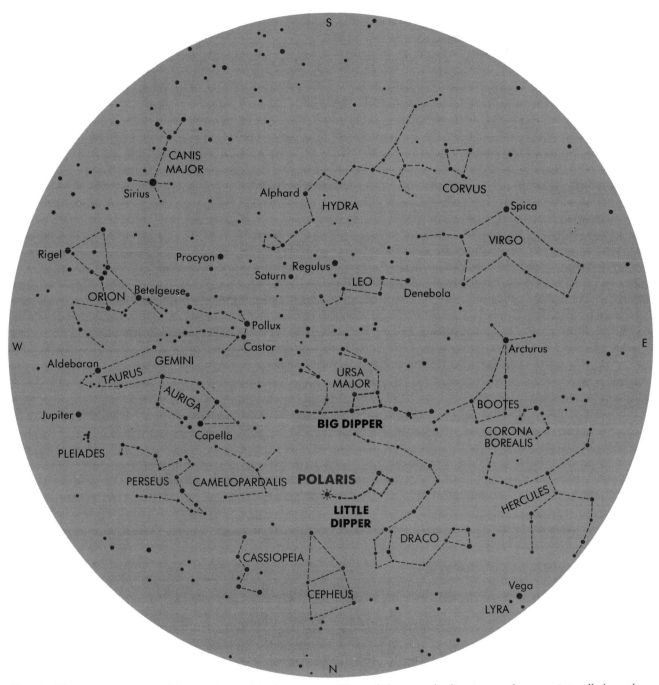

Fig. 1–13   Polaris. A series of due north monthly observations will reveal the stars wheeling in arcs about a point well above the horizon. That fixed point is the North Star and around it the Little Dipper describes a tight circle, the Big Dipper a slightly larger one. The heavenly pattern in this illustration is as it appears in March.

pole, latitude decreases and the altitude of Polaris decreases at the same rate. There is no need to wait for a particular date or time, and one never needs a correction factor. As long as the North Star can be seen (Northern Hemisphere only), latitude determination is no problem. But be sure Polaris can be isolated from its several million companions. That can be a difficult trick on a partially cloudy night (Fig. 1–13).

Longitude

To establish the other element in the coordinate grid, a series of lines crossing the parallels at right angles, we go back to the great circle, this time to an infinity of great circles passing through both poles and, obviously then, with their planes in the same plane as the axis. So if the plane of the equatorial great circle was normal to the axis and the planes of all the parallels were parallel to it, then all meridians (the longitudinal great circles) must cross all parallels at right angles. We have, after a fashion, duplicated the graph paper. Of course, any two meridians converge as they approach the poles, introducing problems not present on the simple graph; but this is the best that can be accomplished on a globe (Fig. 1–14).

A word of caution on the term *meridian*. True, it is properly used to designate the total longitudinal great circle, but its more specific and frequent use is to refer to half a great circle, from pole to pole, or a *limb*. This is because of the numbering system which, as we shall see shortly, designates the two limbs of the total great circle by two different numbers 180° apart. So we talk about the *prime meridian*, a single limb numbered 0°, and the 180th meridian, the other limb of the same great circle literally half a world away.

These lines of *longitude*, then, measure distance east and west of a base line. But arriving at a base line from which to begin the numbering was not as simple as it was

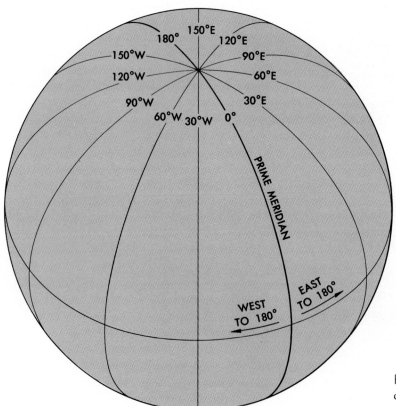

Fig. 1–14 All lines of longitude are great circles passing through both poles, their planes parallel to the axis.

with latitude. There is no natural line corresponding to the equator, and every country thought that the meridian of its capital should merit the honor. However, after many years of international bickering, it became apparent that one meridian would have to receive worldwide approval, and the meridian of the Royal Observatory at Greenwich, England, was officially adopted. The Royal Observatory has since been removed to the quiet countryside as suburban London engulfed Greenwich, but the prime or 0° meridian is still there in the form of a bronze marker (Fig. 1–15).

Fig. 1–15 Here at the old Royal Observatory, Greenwich, the building which houses the large telescope is rent, north to south, to allow observation of the 0° meridian. Sturdily imbedded in its brick and mortar matrix, the official bronze marker on the ground ignores the modern geologic notion that continents are adrift.

This line, then, from pole to pole, became the prime meridian or 0° longitude; all longitude is reckoned either east or west of it. The other half of this great circle running through Greenwich is athwart the central Pacific, and being halfway around the world from the prime meridian, is numbered 180°. Only these two meridians, 0° and 180°, require no directional suffix; all others must show a W or E to indicate which side of Greenwich they are. Thus, unlike latitude, longitude is measured through a half-circle, or 180° (Fig. 1–16). Degrees are subdivided into minutes and minutes into seconds just as in latitude, and every place on the earth has a line of longitude running through it, even Meade's Ranch where the super precise measurement reads 98° 32' 30.506"W. [3]

It becomes apparent that the latitude and longitude of any place on the earth's surface are peculiar to that spot alone. Any number of other sites may share a given latitude *or* longitude but not both.

Fundamentally, longitude is a function of time, and the most important requirement in the reckoning has been an accurate clock.

The problem of determining longitude when it is not known is no more difficult than latitude; if anything it is easier. And yet all through history, longitude has been almost impossible to fix accurately because of the lack of reliable instruments. Fundamentally, longitude is a function of time, and the most important requirement in its reckoning has been an accurate clock. These have been produced for years, but they were pendulum clocks whose perfect functioning depended on a stationary mounting, and keeping time at sea was impossible. It was not until the development of alloy steels for the manufacture of clock springs that an accurate timepiece could be relied on under all conditions. Today, with radio and other time checks to supplement the clock, longitude determination has become a very simple matter.

---

[3] A degree of longitude does not have a fixed distance that can be represented in miles. Only at the equator is a degree of longitude equal to 69 miles (111 km) for meridians converge toward the poles, and the closer the poles are approached, the shorter a degree of longitude becomes.

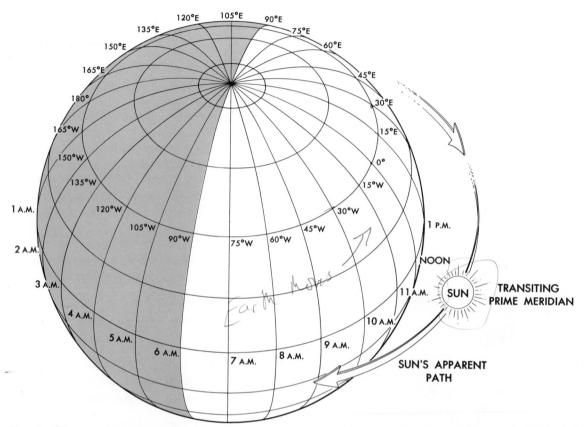

Fig. 1–16   Longitude and sun time.

The basic concept in longitude is the 24 hour day—the sun appearing to make a complete circuit of the earth (360°) in 24 hours. Therefore its apparent movement in 1 hour is 360° divided by 24, or 15°. This means that by observing the sun, we have a way of relating sun time to longitude. Now we come back to the accurate clock or chronometer. It differs from a standard clock in two ways: first, it has a 24 hour face so that there is never any question as to whether the indicated time is A.M. or P.M..; and second, *it always keeps Greenwich time.*

Assume the longitude is unknown; the first step is to determine local time. One's normal wrist watch is complicated by zone time, daylight saving time, etc., but the sun is accurate. Noon is the easiest time to check. When the sun is at its highest point, it is transiting the local meridian and will appear due south (Northern Hemisphere). Now it is noon. A glance at the chronometer will give Greenwich time, and a comparison of the two in terms of hours difference will reveal the longitude.

Working with noon and even hours will simplify the problem and at the same time illustrate the principle. For example, it is the observer's noon and the chronometer reads 17 o'clock (the equivalent of 5 P.M.). What is the longitude? P.M. has an important meaning—post meridiem. Post stands for "after," and P.M. then means *after* the sun has transited the Greenwich meridian (noon), or *afternoon.* The sun appears to move from east to west, so if it has already crossed the prime meridian, it is now west of Greenwich. The sun has just been observed transiting the local meridian (noon), so the longitude must be west of Greenwich. Now it is merely a matter of deciding how far west. At the prime meridian, 5 P.M., means that it was noon there five hours ago, so if the sun appears to move 15° each hour, the longitude is 5 × 15 or 75° W (Fig. 1–16).

If after observing local noon the chronometer reads A.M., then the locations must be east of Greenwich, for A.M. means ante (before) meridiem, or *before* noon. If the sun has not transited the prime meridian yet, it must be coming in from the east at a rate of 15° per hour, but at the moment it is just crossing the local meridian. How far this is from Greenwich is again simply the number of hours between Greenwich time and noon multiplied by 15.

TIME

Time Zones

The chances are excellent that your watch is not keeping correct time, that is, correct sun time. It is extremely unlikely that even the most expensive timepiece will show 12 o'clock when the sun is due south and at its highest point of the day. Instead, it will indicate local *zone time*, which for most of us is not the correct time at all.

This has not always been the case. In the not too distant past the attempt was made to keep one's local time without a great deal of success. For the individual who spent a lifetime on one meridian, this was possible, but those who walked east or west, constantly changing meridians, would continually have to adjust their watches to conform to the time of the immediate meridian. To most, such a situation was inconvenient but scarcely serious. However, to the railroads who pioneered zone time in this country, it became crucial, as several trains, each attempting to keep its own local time, frequently arrived at the same intersection simultaneously. So with the railroads leading the way, zone time was established, and everyone agreed to keep the *wrong* time but the *same* time for convenience and safety.

The entire world is now divided into 24 time zones each 15° wide; the time is based on the correct time of a central meridian. In actual practice, the margins of these zones, except at sea and in unpopulated areas, deviate slightly to conform to state and national boundaries and, as in the United States, are frequently given descriptive names. Here we have four such zones: Eastern Standard Time Zone, which keeps the time of the 75° W meridian; Central Standard Time Zone, based on the 90° W meridian; Mountain Standard Time Zone, centering on the 105° W meridian; and Pacific Standard Time Zone, which keeps the time of 120° W longitude (Fig. 1–17). So all clocks in the Eastern Time Zone read noon when the sun transits 75° W even though those near the eastern and western boundaries are almost a half-hour off.

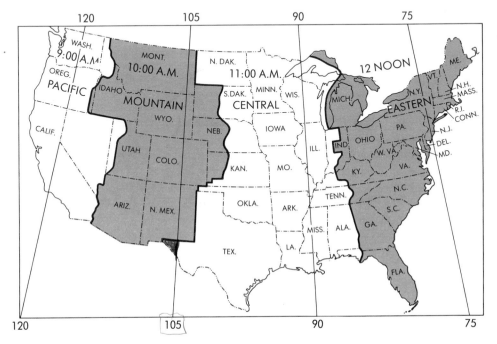

Fig. 1–17 Standard time zones in the United States.

With the railroads leading the way, everyone agreed to keep the wrong time but the same time for convenience and safety.

This arrangement is, of course, an aid to all business transacted within one time zone but brings about the problem of sudden hour changes in time as a person leaves one zone and enters another. Suppose we were driving from Georgia to Alabama, the border of which is the changeover point from Eastern to Central Standard Time, and arrived at the Georgia side of the border at noon EST. This means that the sun is transiting the 75° W meridian on which EST is based. We have already spent the hour 11 to 12 on the road and progressed 50 miles (81 km), but as we cross the line into Alabama and CST, the time suddenly becomes 11 A.M. Central Standard Time is based on the 90° W meridian and it will be another hour before the sun gets there. So we have the hour 11 to 12 all over again and can progress another 50 miles (81 km). We have gained an hour and do not have to give it back unless we return and the whole process is reversed. Those who travel all the way to the West Coast pick up a bonus of three hours. The general rule, then, anywhere on the earth (with the exception of the International Date Line, which will be discussed later), is that *as you go west you gain an hour every 15°, and as you go east you lose one.*

Daylight Saving Time

Daylight Saving Time has become a widely accepted practice that allows us an extra hour of daylight in the evening during the long summer days. Actually, it is a form of self-deception wherein we get up an hour earlier than usual and go to bed an hour earlier, thereby substituting one of the early morning hours of light—which we normally waste in bed trying to sleep despite the peeping of the birds—for one of the early evening hours of dark when we are still awake but outdoor activity is restricted. This could, of course, be easily accomplished without changing the clock, but therein lies the deception. We like to think that we are getting up at the usual time and still getting the bonus of an extra hour of daylight. So we adjust our clocks to read 7 A.M. when it is really 6 A.M. Now whose true local time are we keeping? Normally, we do not keep our own local

time anyway but that of the central meridian of our time zone, so if we have changed the clock by an even hour, we must now be reading the time of a meridian 15° away from that central meridian. If we live in the Pacific Time Zone, our usual time is that of the 120° W meridian, and if that is 6 o'clock and we want the clock to read 7 o'clock, we must turn it one hour ahead; then we will be keeping the time of the 105° W meridian, or Mountain Time. Always, in Daylight Saving Time, *the timepieces of any given zone are keeping the time of the next zone to the east.*

For a while between 1974 and 1975 the U.S. Congress decreed that we should, by statute, endure Daylight Saving Time all year around. Such an arrangement was supposed to *save energy,* which was the political buzz phrase of the time. In the summer the manipulation of the clock was simple enough to defend on this basis since plugging a wasted early morning daylight hour into a normally dark evening slot would indeed save an hour of artificial light in the home. But in winter when we both arise in the dark and go to bed in the dark, substituting a dark morning hour for one in the evening somehow did not seem to accomplish the stated objective. We have since reverted to norm—norm being that individual communities can no longer make the decision as they once did, but whole states now opt to go either to Daylight Saving Time (summer only) or not. As of the latest reading, only Arizona and Hawaii had chosen not to go along.

The Date

In this business of time change as we travel from zone to zone, there is an inherent danger that people traveling west and gaining one hour every 15° of longitude will get carried away and pick up more free time than they are entitled to. A certain amount is legal, such as the time that is gained by going from the East Coast to the West Coast and staying there. One has achieved 3 free hours—even 23 hours is considered within bounds. But a complete circuit of the earth would give one a full day's bonus, and here is where the line is drawn—literally—the *International Date Line.* When this is crossed in a westerly direction, 24 hours is lost.

There has not always been a Date Line. It is determined by humans, and until they began to go around the world, a Date Line was unnecessary. But the first circumnavigation of the globe by Magellan's party demonstrated

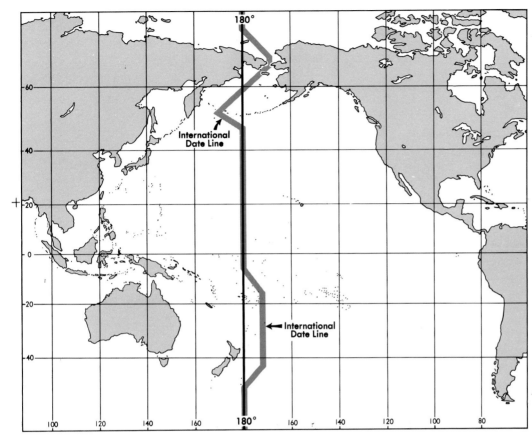

Fig. 1–18 The International Date Line.

the need for such a line, for despite keeping careful track of the date, the survivors of that voyage found that on their arrival in Spain, they were one full day in advance of the calendar. They should have had a day taken away to offset the 1 hour gained for each 15° of longitude in their westward journey. If they had gone around the world to the east, the accumulated 1 hour losses would have been rebated at a date line.

Despite keeping careful track of the date, the survivors of [Magellan's] voyage found that on their arrival in Spain they were one full day in advance of the calendar.

The International Date Line is essentially the 180° meridian running from pole to pole down the middle of the Pacific Ocean (Fig. 1–18). It is exactly halfway around the world from the prime meridian and together with that meridian describes a great circle. In actuality, the Date Line deviates a bit from the 180° meridian so that it can run up the Bering Strait to separate Asia from North America, and it detours around islands and island groups to avoid the complication of differing dates on opposite sides of a single political unit. But, for our purposes here, we can consider the two to be identical.

In this day of rapid communication we are constantly being made aware of the existence of two dates on the earth at any given time. News reports from Asia are frequently datelined a day ahead of the time of delivery. There is, however, one instant in every 24 hours when there is only

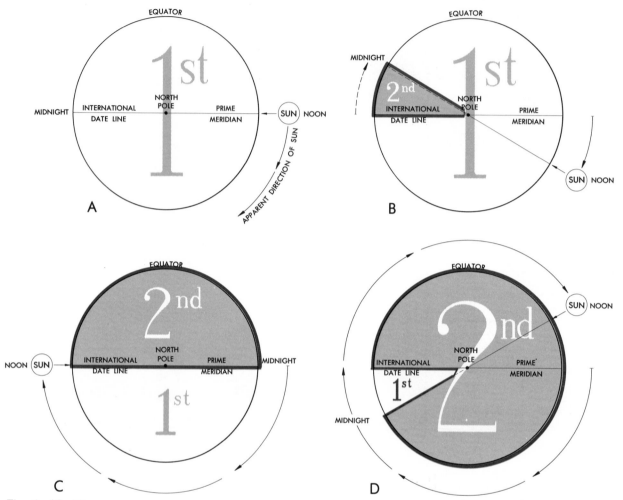

Fig. 1–19　The apparent movement of the sun around the earth from (A) to (D) illustrates the function of the Date Line. The midnight lines, sweeping around the earth opposite the noon sun, initiates a new day as it passes.

one date on the earth—noon at Greenwich (midnight at the I.D.L.). At any other time there are always two dates.

Midnight is the time when the date is advanced for reasons of convenience. Therefore if we can imagine a midnight line sweeping around the earth, as it crosses each location, the date changes. This midnight line would be 180° (12 hours) away from the sun, so that when the sun is transiting the prime meridian, it would be midnight halfway around the world at the 180° meridian (Fig. 1–19). This is the instant when the entire earth has only one date. But one hour later when the sun has moved 15° W,

the midnight line has swept a comparable 15° zone from 180° to 165° E. Each spot within this area has had its date change, so if the date everywhere one hour ago had been the 1st of the month, one little 15° zone out in the Pacific would be the 2nd. And as the sun continued to move west, the wedge of new date would become larger and larger, until 12 hours later (noon at 180°, midnight at the prime meridian), one-half of the earth would be the 2nd and one-half the 1st. Finally, when the sun had made almost a complete circuit of the earth and was again approaching Greenwich, all but a small part of the world would have experienced midnight and the date change, and only this narrow zone short of the Date Line would still be the 1st.

Now if we are crossing the Pacific headed west toward Asia, crossing the Date Line will move us ahead one day (on the calendar). A full day is lost, and the only way to recover all or most of it is to go back across the Date Line or continue west all the way around the world, picking up 23 hours one at a time for each time zone and then stopping before crossing the Date Line again. Remember, the Date Line operates in opposition to the hour lines bounding the time zones: *going west across the Date Line a day is lost, and going east a day is gained.*

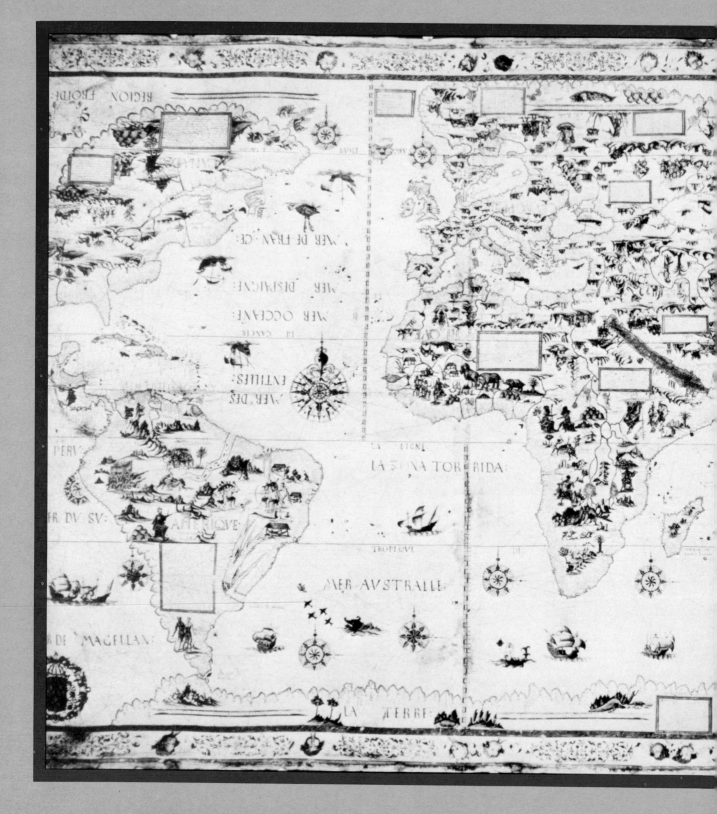

CHAPTER 2

MAPS

Cartography is an ancient art. By 1550, when Pierre Desceliers rendered this version, the renaissance mapmakers had developed an amazingly accurate conception of the world, as well as a sophisticated technical and artistic format. Meant to be displayed on a table with the map readers circulating roundabout, the lettering and sketches are placed in opposing directions—turn the book upside down to read the Northern hemisphere.

INTRODUCTION

A map is merely a symbolized representation of all or part of the earth's surface on a flat piece of paper. As such, it is a highly useful geographical tool, for in attempting any description or analysis of the earth's surface or of the human activity on it, relative location is essential. One of the problems that maps help to solve is that of the individual on the ground being unable to observe more than a very small part of the earth at any one time. With a map it is possible to visualize large segments or even the whole earth and relationships that are not immediately evident from the restricted local view. And maps are capable of much more than mere location. With the proper selection of data and symbolism, a great variety of concepts may be rendered in visual form.

The oldest known map is a fragment of clay tablet from Mesopotamia dating back to 2500 B.C., but nobody suggests that this was the first map ever made.

The urge to map is almost intuitive in all of us. How often do we grab a piece of paper to draw a crude map of road directions? How often the cave dweller must have scratched the same sort of information with a stick in the dirt, long before any means of preserving artwork for posterity were conceived. The oldest known map is a fragment of clay tablet from Mesopotamia dating back to 2500 B.C., but nobody suggests that this was the first map ever made. Maps are such obviously useful tools that they have surely been around continuously from the earliest time. We have, of course, improved on early mapping practices. The Greeks recognized the sphericity of the earth and introduced a system of coordinates for measuring. Later, with the great voyages of exploration, the distribution of the continents became known. Now *cartography*, the science of mapping, has become of necessity a world encompassing affair and is no longer restricted to the immediate local neighborhood. Today, with satellite imagery (see Chap. 19), computer assisted cartography, and highly sophisticated techniques of drawing and reproduction, maps are more widely available than at any time in history—cheaper, better, and in greater variety (Figs. 2–1 and 2–2). But the average person often does not know how to use them except in a perfunctory way. Knowledge of a number of basic practices is

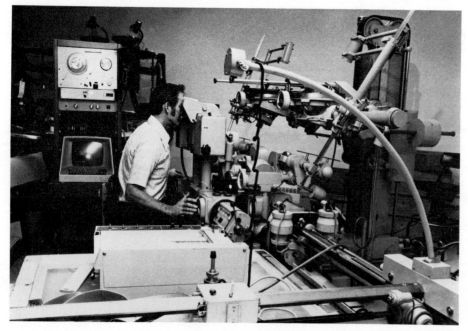

Fig. 2–1 A cartographer with pen in hand has virtually gone the way of the maker of buggy whips, shoe buttons or basketball laces. The machine and computer are the wave of the future. Here is a DPROS (Digital Profile Recording and Output System) at the U.S. Geological Survey's mapping center. The instrument digitizes and records terrain profiles from information spewed out of a diverse array of stereoplotters. Digital profile data then is stored on magnetic tape to be processed into terrain models by a computer.

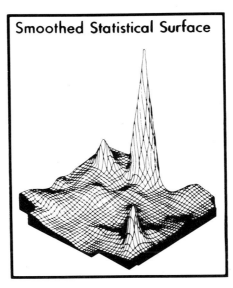

Fig. 2–2 Strictly speaking these strange looking cartographs are not maps at all, but rather statistical material presented in quasi-map form.

necessary to achieve the maximum understanding of this important geographical tool.

PROJECTIONS

If we once again assume the earth to be a perfect sphere, its only true representative is a globe, not a map, for *it is simply a physical impossibility to transfer a curved surface onto a flat one without distortion and error (Fig. 2–3)*. In other words, *there is no such thing as a perfect map*. Maps of very small areas, where the curvature of the earth is minimal, come close to perfection, but the larger the region the map is attempting to show, the greater the error involved, until it reaches its maximum in world maps. So why not forget maps and use only the globe? Cost is one factor. A good globe is many times more expensive than a map. However, of more importance is the fact that the globe

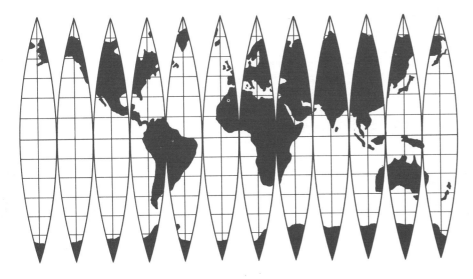

Fig. 2–3 Stripped off the globe, where they have formed a perfect picture of the earth, globe gores leave a great deal to be desired as a flat map.

just cannot do the job that a map can in certain instances. Imagine the size that a globe would have to be in order to show real detail of the country of Belgium. A map could handle this easily, but the globe would scarcely fit in the room. And a world map is a necessity in many situations, but only half the globe is visible at any one time. Add to this the ease of carrying a map around in your hip pocket, the simple printing of multiple copies from an original drawing, and the ease of storing in a flat case, and it becomes apparent that despite their distortions of the areas they purport to show, maps have many advantages over the globe. This does not mean that you should throw away your globe, for it too is useful in other ways. Only on the globe can great circles be properly understood, as well as true direction or earth/sun relationships. Maps and globes are complementary, not competitive, although they both attempt to represent the earth at a reduced scale.

There is no such thing as a perfect map.

If maps are necessary and we must accept their inherent defects, the problem of control arises. How can we keep general distortion to a minimum and limit specific error to that part of the map where it will do the least harm? This is the function of *projection*, the process of transferring the latitude/longitude grid from the earth's or globe's surface to a flat piece of paper.

The earth and globe have a number of important properties that a cartographer would hope to retain on the map. The most vital of these are:

1.   True direction.
2.   True distance (scale).
3.   True shape (conformality).
4.   Equivalence of area.

Unhappily, no map can successfully emulate the globe and have all of these properties, but depending on the method of projection, certain critical ones can be retained at the expense of others. So this matter of which projection to select relates very closely to the purpose of the map. A navigational chart must be conformal above all else, but this means that it cannot be equal area too, for no map

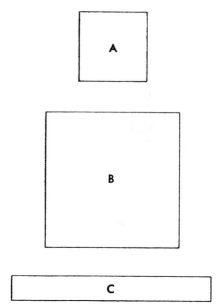

Fig. 2–4   Conformality and equivalence of area. A and B are the same shape (conformal) but obviously different in area. A and C are equal in area but far from conformal. These are the choices that a mapmaker must face, for no map can be *both* conformal and equal area.

can be both (Fig. 2–4). If true direction and distance can be maintained—and they frequently can be, at least along certain lines—so much the better, for these are useful in navigation. On the other hand, if the map is to be for classroom use to show distribution of terrain or population and the like, equivalence of area is more important than true shape and again, one must be sacrificed to gain the other. Some maps are compromises, achieving none of these qualities but coming close enough in all so as not to destroy an effective visual image. But, in every case, the map user must select the projection exhibiting the properties that are needed. Military maps, for instance, often distributed free by the government, will not always work well in the classroom any more than atlas maps can be used for navigation.

The graphic and mathematical methods by which the earthly grid may be transferred onto a plane surface are endless and limited only by the ingenuity of the mapper and the useful properties of the-end product. For example, there is a group of related projections that might be termed "true projections." Although they can be arrived at mathematically or graphically, we can also place surfaces (the plane, cone and cylinder) tangent to a model globe and actually project its grid pattern.

Azimuthal Projection: Plane

Imagine a globe made up entirely of selected latitude and longitude lines represented by wires—an open hollow wire cage. In the center of this we will put a light. Now place a flat piece of paper tangent anywhere on the globe and turn on the light. The shadow pattern of the wire grid will be *projected* onto the paper and can be traced, giving us a latitudinal/longitudinal base on which to draw in the details of continental outline. Here we have a map on our paper actually projected from the globe. Since this was done onto a plane, it is an *azimuthal* projection; the name is derived from the fact that all projections constructed in this manner have the property of true direction (azimuth) from their center or point of tangency.

If we turn the globe so that the point of tangency is the North Pole, this true direction becomes apparent. (Fig. 2–5). The grid pattern now appears as a series of concentric circles (parallels) about the pole at their center and radiating straight lines (meridians) from the pole. These meridians are azimuths or true compass headings, and if we follow any one of them out from the center we are going south. But notice the spacing of the concentric parallels. Although they were evenly spaced on the globe, on the projection they are increasingly farther apart with distance away from the pole—evidence that the projection is not a true picture of the globe. Land areas drawn on this projection will exhibit increasing distortion in size and shape toward the margin at the same rate.

Conic Projections: Cone

Now take our wire cage representing the earth and place a cone over the top of it with its peak above the North Pole. The cone will be tangent at a particular line of latitude, latitude that depends on the angle of the cone. If we construct a very low-angle cone, a coolie hat, it will be tangent at a latitude not far from the pole, whereas a high-angle cone, a dunce cap, will be tangent near the equator. Obviously, no cone could be so flattened as to be tangent at the pole or it would have become a plane, and in order for a cone to be tangent at the equator, it would have to be transformed into a cylinder. Therefore *conic* projections

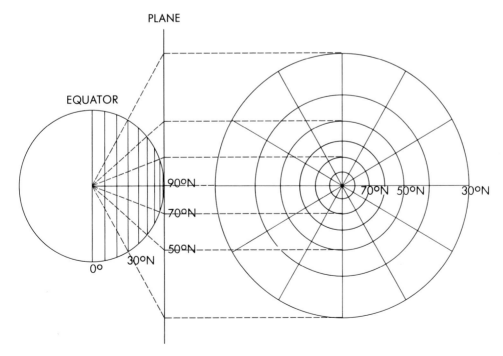

Fig. 2–5   An azimuthal projection is derived from a tangent plane.

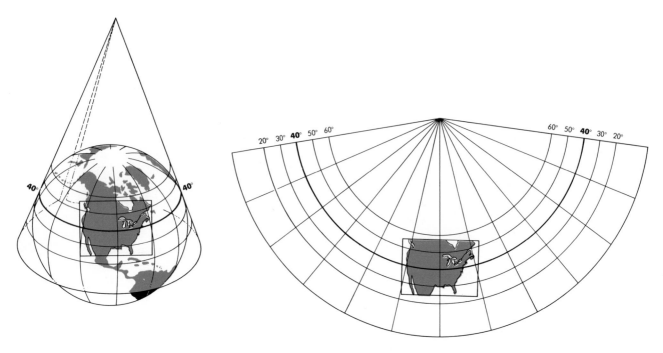

Fig. 2–6   Conic projection.

are especially suited to regions in the middle latitudes between these two extremes.

Let us construct a conic projection on which to map the United States. We would select a cone that would be tangent at a latitude running through the middle of the country, say 40° N. Turn on the light at the center and trace the shadow outline on the cone. Slit the cone up the back and lay it flat and we have a base on which to draw our map (Fig. 2–6). The lines of latitude will be curved but unequally spaced parallel lines crossed at right angles by equally spaced straight-line meridians converging toward the tip of the original cone. At the critical central 40° parallel, scale, shape, and area are all true. But since the other parallels, although representing equally spaced lines of latitude on the globe, become increasingly farther apart with distance from the 40° parallel, those portions of the map will exhibit greater and greater distortion.

✳ *Conic projections are especially suited to regions in the middle latitudes.*

Cylindrical Projections: Cylinder

Here we fit a cylinder over a globe. If the globe is upright with the North Pole at the top, the cylinder will be tangent at the equator. Then having turned on the light at the center and traced the grid, we cut the cylinder down one side and lay it out flat (Fig. 2–7). Once again, we have constructed a projection, this time of virtually the whole world. One of the weaknesses of both the azimuthal and the conic projections was their inability to reproduce a great deal more than a hemisphere, but now we have a grid that, except for the polar regions, represents the entire earth.

Offhand it looks pretty good. The parallels are nice straight lines (although spaced unequally again) instead of those disconcerting curves, and the meridians are also straight and at right angles to the parallels.

But once we draw the continental outlines onto the grid, it becomes manifest that there are inherent problems. Only at the equator do things look right. Elsewhere shapes and sizes get stretched in all directions. The meridians are primarily at fault; they are parallel to each other and do not converge at all. This means that at the higher latitudes where the meridians should be closer together, they are as

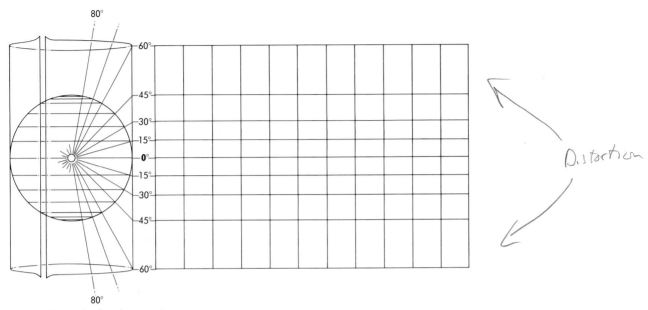

Fig. 2-7 Cylindrical projection.

far apart as at the equator and tremendous east/west distortion results.

Because the parallels are spaced increasingly farther apart, some of this meridianal stretching is compensated for, but not enough. If only the parallel spacing exactly equalled the error in meridian spacing, we would have a conformal projection. This is exactly what an early mapmaker by the name of Mercator did; he adjusted one to compensate precisely for the other. So the well-known *Mercator* map is a world map on a *cylindrical* projection, slightly doctored to achieve conformality (Fig. 2-8). As it happens, any straight line on a Mercator projection is a true compass heading, and this quality plus conformality has made it one of the most widely used of all navigational charts.

Somehow the Mercator map found its way into the classroom and the atlas and was out of its element here. Proportions were all wrong. Greenland near the pole was true shape but was so badly distorted in size that it appeared to be larger than South America, which, being astride the equator, came much closer to its true size. Obviously, a map explicitly constructed for navigational use was not

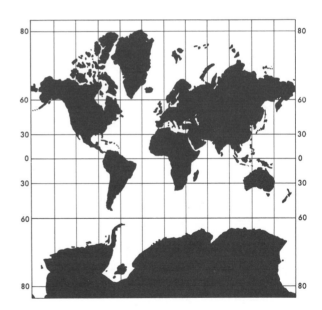

Fig. 2-8 Mercator projection. This famous conformal projection, derived from a cylindrical base, is still a standard navigational chart. But it contains the obvious flaw of gross size distortion in the high latitudes, which limits its use for other purposes.

Mercator

Mercator (1512–1584), the father of the Dutch School of Cartography as he was sometimes called, had Latinized his name from the prosaic Gerhard Kremer—this was the fashion amongst scholars, artists, and professionals in 16th century Holland, newly independent from Spain. The Dutch School of Cartography was responding to an accelerated demand in maritime Europe for practical navigational charts and accurate maps to serve both seafarers and the public with a sudden immense interest in world exploration, commercial development and colonialism.

Mercator was a cartographer, engraver, and instrument maker living in Louvain whose main interest was not particularly maritime charts. And it wasn't until he was well over 50 years of age that he addressed a longstanding navigator's dilemma—how to plot a true compass-heading on a chart that was at the same time a "steerable" line. A course could not be true on any existing chart unless it was a curve and a curve could not be steered. Mercator's projection, when finally completed in 1569, solved that problem and at the same time was conformal. Any straight line on a Mercator chart is a *rhumb line*, that is a true compass-heading crossing all latitude and longitude lines at the same angle, and this is a "steerable" line.

Unhappily the seafaring fraternity, conservative and closely knit, was not convinced immediately that a mathematician and well-to-do society cartographer could revolutionize navigational chartmaking. One of his early supporters wrote scathingly "a cloude and thicke myste of ignorance doth keep this carde (chart) concealed; and so much the more because some who were reckoned for men of good knowledge...have gone about what they could to disgrace it." After Mercator's death his projection was finally recognized for what it really was and it remains in wide use today as a practical nautical chart.

of universal utility. And this brings us back to the original proposition regarding projections—their purpose is to emphasize those qualities required for a particular use and to relegate the errors and distortions to the regions where they will do the least harm. The map is too useful an instrument to abandon simply because it cannot be perfect, so we accept it, warts and all. The choice of projection is our control mechanism.

SCALE

All maps are reductions of the region they are depicting; there are no enlargements. This means that once a portion of the earth's surface has been rendered onto a flat projection, there must be some indication of how much reduction has occurred in the process so that the map reader can measure

Scale Selection

In selecting proper scale for a map the cartographer must exercise caution, always making sure that the end-product is reduced to a size that is easily manageable. Note what can happen in this apochryphal account.

Of Exactitude in Science

. . . In that Empire, the craft of Cartography attained such Perfection that the Map of a Single province covered the space of an entire City, and the Map of the Empire itself an entire Province. In the course of Time, these Extensive maps were found somehow wanting, and so the College of Cartographers evolved a Map of the Empire that was of the *same* Scale as the Empire and that coincided with it point for point. Less attentive to the Study of Cartography, succeeding Generations came to judge a map of such Magnitude cumbersome, and, not without Irreverence, they abandoned it to the Rigors of sun and Rain. In the western Deserts, tattered Fragments of the Map are still to be found, Sheltering an occasional Beast or beggar. In the whole Nation, no other relic is left of the Discipline of Geography.

From *Travels of Praiseworthy Men* (1658)
by J. A. Suárez Miranda

---

distance and properly interpret relative size. *Scale, then, is simply the index of reduction* and is written at the bottom of each map in one of three ways:

1. *Graphic scale.*
2. *Stated scale.*
3. *Representative fraction.*

The simplest and easiest to use, especially for the general map reader, is the graphic scale. It is the common bar scale marked off in tens or hundreds of miles so that we can measure simply with a straightedge the distance between any two places.

Graphic Scale

Scale of Miles

It may also be in kilometers or feet—whatever is most convenient. Subdivisions for more finite measurements are frequently introduced at the left end of the bar. But this scale, like all others, cannot be applied with impunity all over a map, for no map is perfect, especially if it is of a large area. The general rule is: *Trust the scale only at the center of the map and regard it as a mere generalization near the margins.* The graphic scale has one advantage over all others: if the map is to be enlarged or reduced, its scale changes, but the length of the bar changes at the same rate and so remains true. Any other scale must be reworked to conform to the varied size of the map.

Stated Scale

The stated scale is a prose sentence that merely states the number of miles (or kilometers) that is the equivalent of 1 inch (or 1 centimeter), so that again, with a ruler, distances can be readily determined.

Stated Scale: 1 inch equals 8 miles

Representative Fraction

The representative fraction is in many ways the most versatile of the lot and is in very common use, but since it is merely a number, it does not appear very useful to the uninitiated. It is shown as a fraction, or more frequently a ratio, and is applicable in many situations where other scales are not by carefully avoiding any reference to inches, miles, or feet.

Representative Fraction (R.F.): 1:316,800

The ratio, alone, simply states that *1 unit on the map is equal to 316,800 of the same units on the earth.* This means that such a scale is just as applicable in a country using the metric system, or any other, as it is where we use inches and miles, and tells us how much reduction has occurred. The large number, 316,800, is the number of

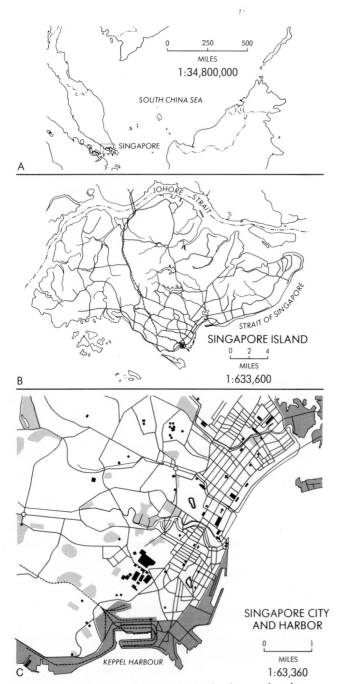

The first point of business on attempting to read any map is to check its scale.

times the map has been reduced. Obviously, if a map of the world were reduced sufficiently to fit on a piece of paper of manageable size, the number would be even larger, on the order of 1:10,000,000 (Fig. 2–9).

It is all well and good to know that a map has undergone reduction several million times, but our minds simply do not grasp exactly what this means. A snapshot of a person is a considerable reduction from the original too, but we have seen that person and know how big he or she is, and our minds take care of the reduction problem automatically. But we have never seen the United States at one glance or probably even our county, so a simple notation of how many times it has been reduced does not help much. Somehow we must make better sense out of that number. Let us come back to our original scale of 1:316,800 and use inches as our unit: one inch on the map equals 316,800 inches on the earth. Still not much help; 316,800 inches is hard to visualize. Let us change the inches to miles and then we will have something useful. If one inch on the map is equal to so many miles, measurement has suddenly become easy. The critical key to all this is 63,360, *the number of inches in a mile.* This number will make sense out of any R.F. scale when it is applied to the larger number. In this case, $316,800 \div 63,360 = 5$, or 1 inch on the map is equal to 5 miles.

Without a scale, maps are not much more than pictures, and no intelligent interpretation can be made. Therefore the first point of business on attempting to read any map is to check its scale.

DIRECTION

On a map with latitude and longitude lines the determination of direction is a simple matter, for they are by definition east/west and north/south lines. But on the occasional large-scale map without these lines and on any map that might be taken out into the field, there will appear an

Fig. 2–9   The map of S.E. Asia (A), has been reduced many more times than that of Singapore harbor (C), and thus shows much less detail.

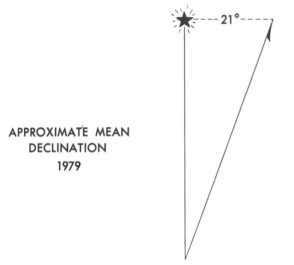

APPROXIMATE MEAN
DECLINATION
1979

Fig. 2–10 Compass variation. The star represents Polaris or true north; the arrow, the direction of the magnetic compass needle. A symbol such as this frequently appears at the margin of a map to indicate the proper correction to be made when the map is used with a compass in the field.

arrow pointing to true north and often a second arrow pointing to magnetic north (Fig. 2–10). Their direction may differ by a significant amount (termed "declination" or "variation"), for the North Magnetic Pole is over a thousand miles from the North Pole at Prince of Wales Island, Canada.

If one were to stand near the southern tip of Lake Michigan and observe a magnetic compass, it would point to true north because this particular point is on a great circle passing through both the true and magnetic poles. This is called an *agonic line.* Anywhere on the agonic line the two poles are in line and a magnetic compass will point true north; anywhere else there will be variation of some degree. Oddly, the earth's lines of magnetic force change from time to time in an unpredictable manner so that the agonic line wanders a bit and at any given moment will display waves and therefore is never quite a true great circle. Consequently, the map arrows showing the amount of compass variation are usually dated to call attention to the fact that they may be in slight error if the date is not recent.

SYMBOLISM

A map in its entirety is merely an assemblage of symbols. Even a line representing a coastline is a symbol, for it attempts to demark mean sea level, a line that does not exist in nature. Or if it shows the broken Norwegian coast on a small-scale map, only the major fiords are at all accurate, and the rest is merely generalized to indicate a ragged character. Roads, if they are to show clearly, may be actually represented as 5 miles (8 km) wide, and railroads display a single rail 3 miles (5 km) wide with 10-mile-long (16 km long) ties 50 miles (80 km) apart. These are all deliberate exaggerations to call attention to features of importance; other major features that are not deemed useful in what that particular map is attempting to show are left out. A map can show political boundaries, latitude and longitude, etc., all of which are symbols representing features invisible in nature.

It has been said that an aerial photo is essentially a map. It is not. For one thing, it shows too much. Forest foliage may hide a road, and the tremendous amount of minutiae detract from the important things. A map is selective; a map is intelligently generalized; a map is often exaggerated to illustrate a particular synthesis—and it does this through the use of selected symbols.

Through long use, most basic symbols have become standardized so that even foreign maps can be read rather easily, although the printed words may be unintelligible. To a considerable degree, symbols are pictorial, especially on *large-scale* maps, that is, those depicting a small area in great detail. Schools, churches, mines, and the like are denoted by symbols that even the novice can recognize immediately (Fig. 2–11). With the increasing use of color on maps, symbolism becomes even simpler, for here too international standardization of basic colors aids in recognition. Blue, of course, is always used for hydrographic symbols, rivers, lakes, springs, and swamps; black and red are used for cultural features such as roads, buildings, and political borders. Any symbol illustrating terrain or topography is shown in brown, while green refers to vegetation, both natural and cultivated.

But symbolism must be adjusted to conform to scale. A perfectly legitimate symbol on a world map simply will

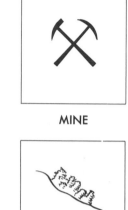

SCHOOL

CHURCH

MINE

INTERMITTENT STREAM

MARSH

CORAL REEF

Fig. 2–11   Examples of internationally standardized symbols for use on medium to large-scale maps.

not do on one of a county. Cities, as an example, are merely a dot on a *small-scale* map, and even that dot may have to be exaggerated in size in order to show up readily. But on a large-scale map, the entire street pattern of the city may be called for with major buildings, parks, and airports indicated in their exact location. The important requirement for map symbols is that within the limitations of scale, they should be readily recognizable if at all possible. Obviously, the occasion arises now and again for the use of unusual symbols, and a legend explaining their meaning is appended to the map. This is perfectly all right, and most people have no difficulty in understanding legends if their

WESTERN CANADA GROWING SEASON AND POPULATION

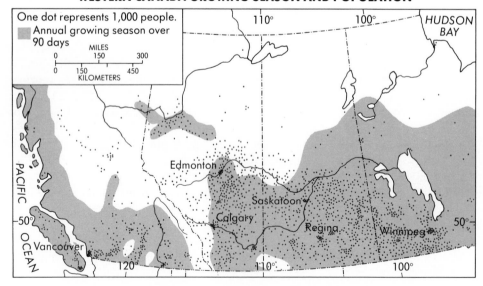

One dot represents 1,000 people.
Annual growing season over 90 days

MILES
0   150   300
0   150   450
KILOMETERS

Fig. 2–12   Value symbols—dot map with isopleths.

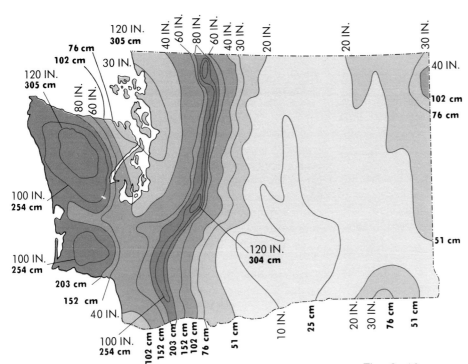

Fig. 2–13 The State of Washington value symbols—isohyets.

notations are clear and succinct. But the less explanation required, the better the map.

Thus far, we have been dealing with symbols that are indicative of location alone, and there is no question that this is the prime function of map symbolism. But if a symbol can do double duty and show something else in addition to location, so much the better. We have such symbols and they are highly useful; as a general term, let us call them *value symbols*

Value Symbols

The common dot map is an example of a value symbol. The location of the dots immediately tells us *where*, for instance, tomatoes are grown in California; but also, if we consult the legend, we find that each dot is given a value of so many acres, and such a map then tells us, essentially, *how many* tomatoes are grown in a certain locale (Fig. 2–12). Or take for example, a map locating cities with circles for those with over 1,000,000 population, and triangles, squares, or smaller circles which indicate a population of 100,000 to 500,000, etc. These are value symbols.

So are bales of cotton, stacks of dollars, or rows of little men—the variations are endless, but all show location plus.

One particular kind of value symbol that is applicable to many situations is the *isopleth* or *isarithm*, *a line connecting all points of equal value*. Within this group are many applications, some of which are undoubtedly already familiar to even the most casual map reader. Examples include:

Isotherm—a line connecting all points of equal temperature.

Isobar—a line connecting all points of equal barometric pressure.

Isohyet—a line connecting all points of equal precipitation.

Isagon—a line connecting all points of equal compass variation.

Contour—a line connecting all points of equal elevation.

There are many more, often without specific names. Corn production in a given county may be plotted by the use of these lines. All areas with over 80% of each farm planted to corn will be enclosed by one line, over 50% by another, etc. The use of color or shading, with the darkest

---

Whether it is political affiliation, soup consumption, or the percentage of men wearing derby hats that is to be mapped, the isopleth can do the job.

---

shade representing the heaviest production, will make the picture very effective. Whether it is political affiliation, soup consumption, or the percentage of men wearing derby hats that is to be mapped, the isopleth can do the job (Fig. 2–13).

Contours and the Third Dimension

One of the most difficult problems faced by mapmakers is how to show nature's third dimension on a flat map. All sorts of methods have been tried, from crude pen strokes to indicate a mountain range, to highly artistic sketches of mountains and valleys. Recently, a process called *plastic shading* has become widespread, wherein—assuming a low sun—usually in the northwest, the artist shades deeply those sides of the ridges that would be in deep shadow and highlights the sunny side. Actual construction of a plastic model to be photographed can give much the same effect. This is very striking and the third dimension is strongly suggested. However, the weakness in most of these methods, not including exorbitant cost and a high degree of artistic ability required on the part of the cartographer, is that the symbolism merely represents terrain character without allowing the map reader to pick off specific elevations anywhere on the map except perhaps certain prominent peaks that may be labeled (Fig. 2–14).

All things considered, the map using *contour lines* is the most accurate from the point of view of depicting actual elevations throughout. These maps have their drawbacks too in that they are costly to construct, are most applicable only as relatively large-scale maps, and are not easy to read without some practice and training. But the general map reader is likely to come across them, especially via the widely distributed U.S. Geologic Survey sheets that cover a large part of the United States and should know a few basic rules so that such maps can be meaningful.

All things considered, the map using contour lines is the most accurate from the point of view of depicting actual elevations throughout.

Atlases and wall maps of continents very frequently represent topographic variation by the use of graduated colors—greens grading through yellows and browns indicating lowland versus highland. These are contour maps, albeit highly generalized. The legend tells us that everything colored green is under 500 feet (152 m). The outer margin of green then is a 500-foot (152 m) above sea level. [The green color does not mean flat, merely under 500 feet (152 m)]. Similarly, the limit of each of the other colors is a contour line. But the usual contour map is of a much smaller area than a continent, and much more detail is shown.

The secret to reading contours (usually drawn in brown) is to recognize that each line represents a certain vertical distance; and therefore the closer they are together, the steeper the slope. Always at the bottom of the map a *contour interval* will be indicated in feet, the elevation differential between any two contour lines. As an illustration, let us examine a round but asymmetrical hill rising from sea level with a contour interval of 100 feet (31 m) (Fig. 2–15). Since elevations are virtually always reckoned from mean sea level, we measure a series of points all the way around the hill 100 feet (31 m) above the sea. When these are connected by a line, we will have drawn a contour line completely encircling the hill 100 feet (31 m) above sea level. Now since the contour interval specifies that contour lines must represent 100-foot (31 m) differences of elevation, our second line will be constructed above the first in the same manner at 200 feet (60m) above sea level—a smaller circle since the hill tapers to a peak, but parallel to the sea and the contour line below it. Similarly, the 300 foot (90 m) and 400 foot (122 m) contours are drawn on the hill. Since the top of the hill is above 400 feet (122 m) but under 500 feet (152 m), its exact elevation is often written in at, say, 450 feet (137 m).

If we were to fly over this hill and look down on it from the top, we would see these contour lines as a series of concentric circles, but they would not be evenly spaced. On

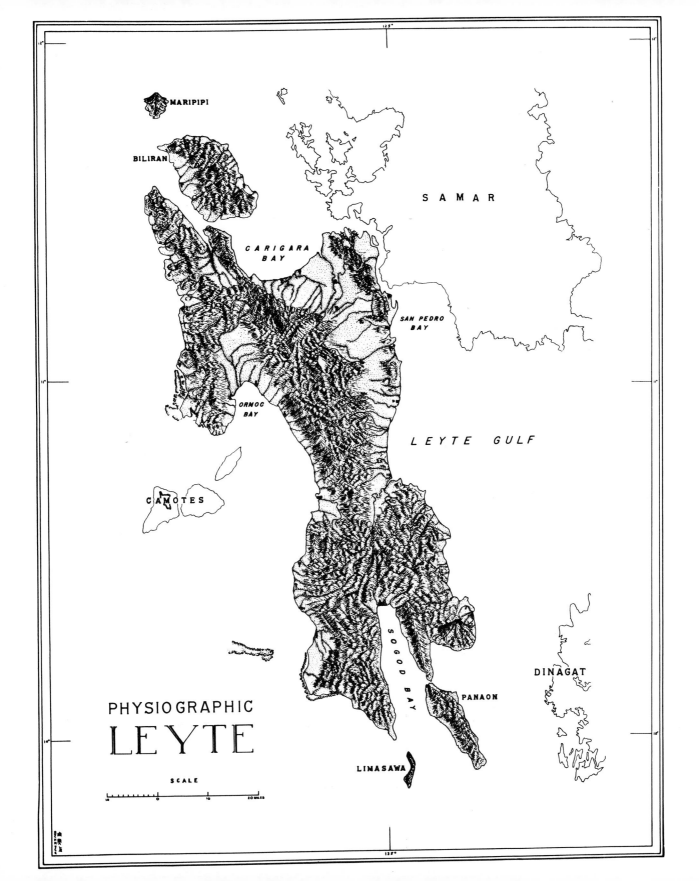

MARIPIPI

BILIRAN

CARIGARA BAY

SAMAR

SAN PEDRO BAY

ORMOC BAY

LEYTE GULF

CAMOTES

DINAGAT

SOGOD BAY

PANAON

PHYSIOGRAPHIC
LEYTE

LIMASAWA

SCALE

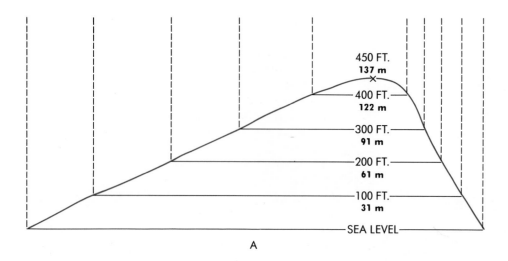

A

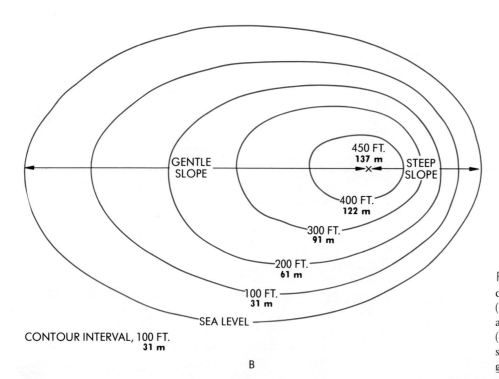

GENTLE
SLOPE

450 FT.
137 m

STEEP
SLOPE

400 FT.
122 m

300 FT.
91 m

200 FT.
61 m

100 FT.
31 m

SEA LEVEL

CONTOUR INTERVAL, 100 FT.
31 m

B

Fig. 2–15   Contours. (A) All
contour lines represent a 100-foot
(30.5-meter) elevation differential
as is obvious from side view.
(B) They are closer together on the
steep slope than on the more
gentle one.

the steep side of the hill the foreshortening of the overhead view would make them appear close together, while the gentle slope would be represented by wider spacing. We know that all these contours have been accurately measured and are at the proper distance above sea level, yet the

plane or map view from above gives us a direct relationship between slope and line spacing (Figs. 2–15 and 2–16).

Let us assume that a road on a map is 1 inch long, that 1 inch equals 1 mile, and that during the course of that mile, the road crosses five 100 foot (31 m) contours.

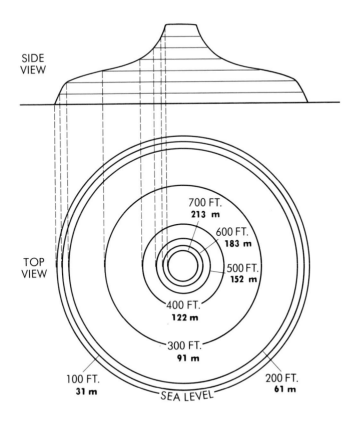

SIDE
VIEW

TOP
VIEW

700 FT.
213 m

600 FT.
183 m

500 FT.
152 m

400 FT.
122 m

300 FT.
91 m

100 FT.
31 m

200 FT.
61 m

SEA LEVEL

CONTOUR INTERVAL, 100 FT. 30.5 m

Fig. 2–16   The map reader should be able to visualize the actual profile of the island from simple contour spacing.

This means a change of elevation of 500 feet (152 m) in 1 mile—quite a steep slope. But elsewhere on the same map, another 1 mile road might cross only two contours and thus climb only 200 feet (60 m)—a much gentler slope. A quick glance at the average contour map can tell us the essence of the terrain immediately. In the areas that appear brown because of great crowding of many contour lines close together, the country must be very rough (or steep), while in the white regions where only a few contours show up, the topography is subdued. Just how rugged the mountains are and what the exact elevations might be will, of course, require somewhat closer scrutiny.

The contours themselves are labeled—often only every fifth line, which is drawn slightly darker—and any place falling directly on a line can be assigned an accurate ele-

vation. Also, prominent peaks, road junctions, lakes, etc., will usually be marked with their exact height. Locations between contour lines must have their elevations interpolated within the limits of the contour interval. The selection of the contour interval is determined by the scale of the map and particularly by the character of the terrain. If a section of the Rocky Mountains is to be mapped, a large contour interval, 100 feet (31 m) more, must be used, since there are huge changes of elevation within very short distances. If in such country a 10 foot (3m) contour interval were used, there would be so many brown lines on the map attempting to show every 10 foot (3 m) change of altitude that they would merge and be indecipherable. On the other hand, a map of almost flat terrain must utilize a very small contour interval in order to bring out the minor undulations. A 100 foot (31 m) interval on such a map would result in no contour lines at all if there were no 100 foot (31 m) change in elevation on the plain.

Theoretically, every contour line must close, that is, eventually come together to form a circle, for all land masses in the world are islands, and since we are measuring from sea level, closure becomes mandatory. In actual practice, however, only limited areas are represented on a single map, and a single closed line or a series of concentric circles means a hill. This is a good general rule to keep in mind as one attempts to interpret contour maps, but there is one exception. A hole in the ground must also be represented by closed contours. In order to distinguish one from the other so that the map can be read intelligently, the depression contours have little *hachures* on the downslope side. For instance, we might encounter a volcano with a substantial crater at the top. The mountain would be represented by a series of closed contours, but the crater, if its depth were more than the contour interval, would also be closed, and the only way we could tell that it was a depression rather than a simple extension of the cone would be by the hachures (Fig. 2–17).

One of the best guides to interpreting terrain slope is an assessment of the stream pattern, for streams flow in valleys and therefore slope will inevitably be up as we move

One of the best guides to interpreting terrain slope is an assessment of the stream pattern.

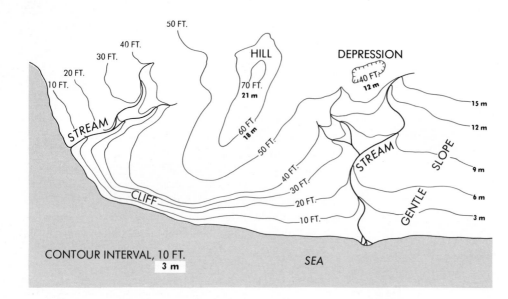

Fig. 2–17　Simplified contour map.

away from the stream. But streams must flow downhill and therefore must cross contours that represent slope. It may appear that contour lines parallel stream valleys, indicating the steepness of the valley sides, but if the map reader carefully follows a given line, it will be discovered that it eventually crosses the stream. Normally, however, each contour line will run far upstream before it crosses, for the elevation of the stream bottom some distance upstream is equal to the height of the bank below.

Remember, then, *contours must cross streams but they will point upstream.* This concept is useful in determining the direction of stream flow on the frequent maps where a stream enters on one side and flows off the other with no other indication of which way it is going.

PART TWO

WEATHER

CHAPTER 3

ATMOSPHERE

The atmosphere is a mix of many gases and it is never clean, but the human contribution to its adulturation has been so varied and ingenious that it will probably never achieve normality again. Copper smelting doesn't do much for the countryside, even with high stacks to disperse the noxious effluvia. Much of the smoke we see is particulate matter (solids and dust) cast aloft, but invisible toxic gases are being introduced into the atmosphere as well— neither is especially beneficial to plants or the human respiratory system.

INTRODUCTION

As the earth wheels through space, meticulously tracing its prescribed paths about sun and axis, it is accompanied by an enveloping atmosphere held closely to it through gravity. A large majority of the planets in our solar system appear to have some sort of atmosphere developed in the early stages of planet formation, and the earth with its "air" is representative of the total system. But as well as we can determine, our atmosphere, reflecting size of planet, distance from the sun, and evolutionary history, is distinctive from all the others in detail of composition. It is a mixture of many gases (and some nongases) that has traditionally been regarded as permanent. However, natural change of composition may be taking place very slowly through such processes as vulcanism and radioactivity at the earth's surface as well as through scarcely perceptible seepage of certain atmospheric elements into space from its ephemeral perimeter. It is also becoming increasingly obvious to those of us who are interested in breathing that the influence of humans is seriously altering the character of the lower atmosphere quite rapidly.

COMPOSITION

Permanent Gases

A rather large array of gases is involved in making up the total air that extends outward from the earth a distance of perhaps 6000 miles (9656 km), depending on definition, but these gases are curious in both their uneven proportions and their tendency to stratification. In sheer volume the atmosphere is absolutely dominated by two gases: nitrogen (78.1%) and oxygen (20.9%). Argon is a bad third at 0.9% and all the many others make up an aggregate of less than 1% (Fig. 3–1).

Variable Gases

Locally these totals change slightly from time to time due to the introduction of variable gases. Two in particular derive from the earth and exhibit concentrations in the lower atmosphere: water vapor and carbon dioxide. Water vapor is the colorless, odorless, gaseous form of water and ice. It may exist in any of these forms in the air, but as a gas it

Percentage of Gases by Volume		Variable Components
Nitrogen	78.1	**Natural**
Oxygen	20.9	
Argon	0.9	
		Water (vapor-ice-liquid)
		Dust (salt, spores, etc.)
		Carbon Dioxide
Neon	Less than	**Generally Contributed**
Helium	0.1%	**by Human**
Krypton	(trace	
Xenon	gases)	
Hydrogen		Dust (soot, etc.)
Ozone		Carbon Dioxide
Methane		Nitrogen Dioxide
etc.		Sulfer Dioxide
		etc.

Fig. 3–1   Composition of the atmosphere.

can achieve up to 4% of the atmospheric mix over tropical oceans (or essentially 0% over Antarctica in winter). A certain long-run balance is achieved since moisture evaporated from the sea as gas is eventually delivered back as water or ice via streams, rain, snow, icebergs, etc. A parallel kind of balance exists for carbon dioxide as well. Its earthly origins comprise such things as combustion, organic decomposition, and respiration, but these are matched by an equivalent plant photosynthesis and oceanic absorption. Here too, a momentary imbalance is normal, causing carbon dioxide to be classified as a variable; but an overall equilibrium must be maintained or we will have altered our atmosphere.

Gaseous Contamination and Smog

This is precisely what seems to be impending. Humans have gotten into the act and have uncaringly or unknowingly become obsessed with combustion. It should be stressed that combustion is normal; forest fires and vulcanism were here long before people. It can even be demonstrated that the long-burning prairie and forest fire of the prehistoric past is under very efficient control with the use of modern firefighting techniques and equipment. Smokey the Bear is not loafing on the job nor are cigarette throwers and little boys with matches always the chief perpetrators. The culprit is the industrial revolution of the past century and a half and the increasing use of inanimate energy to relieve struggling sweating humanity of its grueling burdens—and in the bargain to be immeasurably more productive. But inani-

Fig. 3–2   Overgrazing is a widespread scourge in the semi-arid west. With bare soil exposed to the wind "real" dust can be drifted readily at ground level or swirled to great elevation.

mate energy has been derived very largely from the huge-volume combustion of our fossil fuels, and inevitably there has appeared on the scene, Garys, Birminghams, and Donetsks with their murky palls. The fossil fuels, coal and oil, are accumulated hydrocarbons, stored solar energy from endless time past. When we set out to burn in a century or two the earth's total supply, which required eons to develop, there have to be some side effects.

To compound the difficulty, efficient production and the "good life" of the western industrial nations have spawned the automobile exhaust and ubiquitous garbage to be burned. Human dignity as expressed in the white collar job and "creative" leisure has been bought at a price. The question to be posed is, is it worth it? We just do not know yet, but a part of that price has been the contamination, perhaps permanently, of our lower atmosphere. Too much carbon dioxide for the oceans and plants to absorb and other toxic end-products of combustion, such as carbon monoxide, nitrogen dioxide, and sulphur dioxide, are spewed out into the atmosphere. One result is more rapid absorption of the sun's energy, which could result in a general warming of our climates; another more obvious effect is the increasingly common photochemical[1] smog forming over our cities. Ozone, a naturally occurring gas at high levels, yet toxic and corrosive, develops in appreciable amounts as a component of smog, through the reaction of nitrogen dioxide to solar energy. Precisely what other chemical reactions are occurring in our poisonous urban atmospheres or what can be anticipated for the future has yet to be pinned down. Perhaps the question now should be: can we resolve the difficulties without sacrificing all of the gains of industrial civilization? Or can the technology that has been so successful as a producer be equally efficient in solving the resultant problems?

Dust

Another variable element in the atmosphere is of some significance: the solid particles held in suspension and often transported long distances by the moving air (Fig. 3–2). With the exception of ice, the general term for all solids is dust, but this includes a wide variety of materials. Cer-

---

[1] In photochemical reaction, particle structure changes as a result of radiant energy absorption.

Volcanic Eruptions

Ash fall from El Chichon.

That the cold winter of 1783 in the eastern U.S. could have been related to the large amounts of ash cast into the atmosphere by Iceland's erupting Laki Volcano was a clear possibility, thought Benjamin Franklin. Surely, he reasoned, if solar energy normally reaching the earth is impeded by as much as 5–6%, temperatures inevitably would be affected. In 1815 the Indonesian volcano Tamboro erupted, blowing an estimated 60 cubic miles (252 km^3) of dust far into the stratosphere. That year there was frost every month in New England and frequent crop failures worldwide—it became popularly known as "the year without a summer." Another Indonesian explosion did it again in 1868, Krakatoa, and in the bargain killed 3600 people, mostly from the resulting tidal wave. But around the world vivid sunsets and temperature variations were observed for over three years, indicative of dust circling the earth carried by upper atmosphere winds.

So when Mt. St. Helens went off in 1981 there was immediate scientific speculation as to whether these huge quantities of atmospheric solids might affect temperatures. However, St. Helens, for all the violence of its explosion, blew out the side of the mountain and the greatest part of the massive ejecta was expelled horizontally. The ash cloud that did carry around the world proved to be a relatively minor climatic factor. Not so El Chichon, the little-known 3,000 foot (914 m) volcano in Tabasco, Mexico. When it abruptly came to life in 1982 after a century of quiescence it pumped more dust and debris into the upper atmosphere than any eruption in the last 70 years. The gigantic smog-like cloud encircled the earth and U$_2$ flights brought back samples indicating not only particulate matter but quantities of sulphur and other gasses. Between the equator and 30° N&S as much as 10% of the possible solar radiation was screened out. "This is a significant amount" say NOAA meteorologists, "it may take five years before all of the dust finally disipates." This means that it will be convenient, and perhaps even accurate, to ascribe every weather quirk to El Chichon until 1987.

tainly, tiny particles of soil swept up from the desert or the plowed field come to mind when we talk of dust; but how about plant spores, microorganisms, tiny fragments of salt from sea spray, volcanic smoke and debris, industrial soot, end-products of nuclear explosions, and meteorite minutiae? All dust! It is perceptible when it occurs in great clouds, of course, but the blue sky and red sunsets are also the product of minute dust particles as they scatter light waves.

Most of this material originates at the surface of the earth and a great deal of it is of sufficient size that it cannot be carried very high or very far. But microscopic particles can stay aloft for many years, particularly if they are cast up by massive explosions. So although it may be a great deal "dustier" over the great cities or deserts than out at sea, the lower atmosphere is never clean. For this we are profoundly grateful since there can be no condensation, no rain, no clouds, without dust particles to act as hygroscopic nuclei. Simply said, moisture must condense on a solid surface and a totally gaseous atmosphere contains no such surfaces. In the laboratory this can be demonstrated very easily in an artificially clean air.

Like carbon dioxide, the amount of dust in the air has been considerably increased by human activity and the natural balance is being destroyed. Combustion is one of the degrading processes, but plowing, overgrazing, forestry, and nuclear blasts are also to some degree contribu-

Fig. 3–3   Here is the infamous Baker Day nuclear test blast at Bikini atoll, July 25, 1946. In addition to huge quantities of vaporized salt water, microscopic particles of radioactive soil, used battleships and dead fish were carried high into the atmosphere and wafted downwind to contaminate much of the northern hemisphere.

tions of humans. This last has a particular significance because the vaporized soil is thrown so high that when it condenses into tiny fragments it may be kept aloft for as long as five years and transported around the world by jet streams. But the particular danger is that radioactive isotopes such as strontium 90 have extended half-lives (27.7 years for strontium 90), so that when they are eventually washed out of the air by rain they are capable of contaminating the earth virtually anywhere for long periods of time (Fig. 3–3).

In terms of weather and climate the influence of these three variables—water vapor, carbon dioxide, and dust—

The particular danger is that radioactive isotopes have extended half lives, so that when they are eventually washed out of the air by rain they are capable of contaminating the earth virtually anywhere for long periods of time.

far transcends their limited quantity in the total atmosphere. Nitrogen, making up 78.1%, is relatively inert and simply supplies mass. Oxygen, at 20.9%, is for breathing and we would not be here without it. But for weather and climate, we must have the other three. They all obtain from the earth and hence are concentrated near it, thereby playing a definitive role in absorbing radiant energy. As we shall see shortly, atmospheric temperature at the earth's surface is strongly controlled by these three elements.

STRATIFICATION

One of the characteristics of the atmospheric gaseous mélange is that it is a mechanical (rather then chemical) mixture. This means that the various components can and do arrange themselves into zones of concentration, a series of concentric layers or shells about the earth. Vertical lamination is further served by the ability of air to compress under its own weight, thereby causing the densest air to be

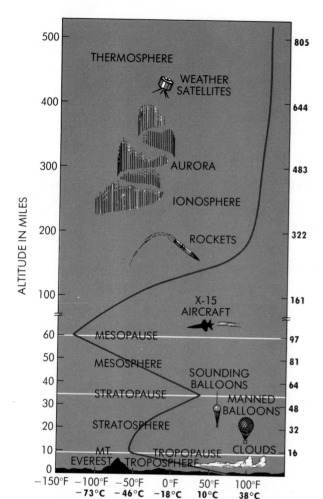

Fig. 3–4   The atmosphere in cross section.

than water—50% of the total atmosphere is in the lower 3.5 miles (5.6 km), 90% in the lower 10 (16 km).

Like all gases, air is mobile as well as compressible. Factors contributing to its animation are chiefly terrestrial, including such things as differential heating of land and water and rough topography. The air nearest the earth is not only the dirtiest but the densest and most turbulent as well (Fig. 3–4).

Troposphere

As we attempt to differentiate and draw at least rough dividing lines between the various atmospheric tiers, a useful tool is found in the vertical temperature distribution curve. Sharp breaks in this curve tend to coincide fairly well with some of the other factors involved. For instance, a sudden bump in the curve appears at an average 8 to 9 mile (13 to 15 km) elevation. This is the *tropopause*, the line separating the *troposphere* from the *stratosphere*.[2] Up to this altitude, temperature decreases with elevation from a ground-level high point at a predictable rate, then suddenly steadies at the tropopause. The troposphere, the bottom of the atmospheric layers, can be identified not only by this characteristic rate of temperature decrease, but by a number of other indicators. It has a fairly uniform mix of nitrogen, oxygen, and argon. It contains the bulk of the water vapor, clouds, carbon dioxide, and dust, as well as the densest and the most turbulent air. As a result, essentially all of the ingredients for the propagation of our day-to-day weather are found in the troposphere.

Stratosphere

The stratosphere begins beyond the tropopause and its upper limit is the *stratopause*, about 35 miles (56 km) above the earth's surface. If we go back to the vertical temperature curve we find that the steady decrease through the troposphere levels off at the tropopause and then maintains a constant reading until, at about 20 miles (32 km) up, it gradually increases to the stratopause. At this point the temperature reaches a level very close to that at the surface of the earth, but abruptly it breaks away and begins to cool again. This great heating at the top, one of the distinctive

found at the bottom of the pile or at the surface of the earth. Some authorities like to speak of "the ocean of air" with humans as demersal inhabitants akin to the halibut. But the magnitude of the compressibility of gas is even greater

"The ocean of air," with humans as demersal inhabitants akin to the halibut.

[2] Since this is a temperature curve it reflects both season and latitude. The tropopause is higher in summer and in the topics [up to 12 miles (19 km)] and lower in winter and high latitudes [as low as 5 miles (8 km)].

features of the stratosphere, is attributed to a narrow band comprising most of the atmosphere's ozone, which intercepts and absorbs the greater part of the ultra-violet spectrum emitted from the sun. Aside from this, however, the typical nitrogen/oxygen/argon mix of the troposphere is generally representative; but significantly lacking are winds, water vapor, carbon dioxide, and dust.

By absorbing significant quantities of ultra-violet waves, the ozone layer serves as one of a series of in-depth upper atmosphere barriers to potentially dangerous solar radiation. But now we face a breakdown of ozone over time, or at least some people think so. The villain of this piece is the commonly used aerosal spray whose gaseous propellent expels deodorant or shaving cream. We're not much worried about the world being engulfed in shaving cream, but when the gas involved is freon (floride or fluorocarbon) then there is a twofold problem: (1) this gas is almost totally inert in the troposphere and therefore can accumulate endlessly, (2) when it eventually finds its way to the upper stratosphere and is exposed to ultra-violet wave lengths it decomposes, freeing chlorine which in turn destroys the ozone.

We're not worried much about the world being engulfed in shaving cream, but when the gas involved [in the aerosal spray can] is freon (floride or fluorocarbon) then there is a problem.

The worry is that increasing ultra-violet radiation escaping the depleted ozone trap may damage vegetation, step-up the incidence of skin cancer, maybe even affect the long-run climate. Why take a chance? Some eminent scientists (not all) suggest that we bite the bullet and rough it as our ancestors did, somehow living happy and fruitful lives sans spray cans.

Mesosphere and Thermosphere

Beyond the stratosphere is the *mesosphere* with temperatures decreasing to their lowest reading in the entire atmosphere at the *mesopause* 60 miles (97 km) high. This zone is dominated by oxygen in the discrete atom form, split apart by short-wave solar energy. Still farther out, to approximately 500 to 600 miles (805 to 96 km), the temperature once again increases, chiefly as a result of ionization[3] giving rise to the term *thermosphere*. The air is so thin at this level and the nitrogen and split oxygen atoms so remote one from the other that there should be no illusion of real heat in what some experts define as the outer limit of the atmosphere. At these heights the gaseous medium becomes so rarified that transiting satellites experience essentially no frictional drag, nor do they accrue increased surface temperature despite the "thermosphere" nomenclature. The distinction to be made regarding the word "heat" is the lay person's *sensible temperature* which is tangible and personal vs. the physicist's *heat equals energy*. Here we are dealing with energy.

But those who select 600 miles (966 km) as this outer limit must admit to gases beyond that point, chiefly helium. To call such rarefied air "atmosphere" now becomes debatable. Above about 1500 miles (2414 km), helium gives way to hydrogen, the lightest of the gases. Perhaps the limit of trace hydrogen is 6000 miles (9656 km). Yet there are those who dislike even the term "interplanetary space," citing tenuous evidence that this greatest natural void of all is not absolutely without gaseous matter.

Elements and Controls

Our point of emphasis in the ensuing chapters is on weather and climate, so that the composition of the atmosphere, its behavior, and the processes involved are merely academic unless they can be brought to this ultimate focus. To clarify the various items at issue, we must organize them into some sort of system. Fundamentally, there are four basics or *elements* of weather and climate:

Temperature.

Pressure (and its resultant, winds).

Moisture (in all its forms—gas, liquid, and solid).

Storms.

Obviously, the orchestration of all four of these elements into endless varieties of concert arrangements is weather and climate. But at this stage, in order to understand them as

[3] At these high altitudes, ultraviolet waves and high-energy particles (cosmic waves) react with oxygen and nitrogen atoms to expel free negative electrons. This process is *ionization*, and an alternate designation for most or all of the thermosphere is *ionosphere*.

The Van Allen Radiation Belts

The earth, as a great magnet, exerts a powerful magnetic field whose influence is felt well beyond the outer margin of the atmosphere. This zone is termed the *magnetosphere* and may extend rather unevenly for more than 50,000 miles (80,500 km) from the earth. Satellites monitering radioactivity in space in 1958 relayed the unexpected information that radiation of uncommon intensity was concentrated in two belts in the magnetosphere; one at about 10,000 miles (16,000 km) above the earth and a second lesser belt at 23,000 miles (37,000 km). They have been designated the *Van Allen radiation belts* after the physicist who first described them.

These radiation belts are interpreted as concentrations of highly charged particles from the sun, trapped by the earth's magnetic field. So here is yet another line of defense in protecting the earth and its inhabitants from dangerous radiation.

When we observe periodic solar flares and solar storms along the sun's periphery there follows faithfully, magnetic disturbances evidenced by short-term difficulties in radio transmission; and it is entirely possible that during each solar storm a larger amount than normal of the sun's highly charged residue finds its way to earth.

Dermatologists have been attempting to correlate outbreaks of skin cancer with observed major solar flares, so far with varied results. But what if the earth's magnetic field should weaken, as it may very well have done each time it reversed itself in the geologic past? Instead of increasing the supply of cosmic particles we postulate a flawed protective screen. Could the dinosaurs have disappeared abruptly because they were bombarded by deadly radiation from outer space?

precisely as possible, we will attempt to isolate and examine each element. In reality they cannot be totally isolated, as we shall continuously discover, but it is worthwhile making the effort.

As we move along in our discussions we will be making the transition from weather (the momentary state of the elements) to climate (essentially a long-time average of the

Yet there are those who dislike even the term "interplanetary space", citing tenuous evidence that this greatest natural void of all is not absolutely without gaseous matter.

daily weather). With this change of emphasis will come an increasing concentration on a second set of weather and climate factors called *controls* How much rainfall does a given location receive on the average and why? What is the reason for extreme temperature variations within a short distance? These are climatic questions and the answers revolve about an understanding not just of the elements but also of controls. For example, latitude is probably the single most important control of temperature; it gets colder if one goes away from the equator. But it also gets colder if one climbs to the top of a snowcapped mountain or, often, if one approaches the sea, so oceans and mountains are temperature controls too. There are moisture controls and wind controls and storm controls—a virtually endless list. But if we correlate an accentuation of elements with weather and of controls with climate, we will not be too far wrong.

CHAPTER 4

TEMPERATURE

The leaf-bare trees proclaim that the season is winter yet under the glass cover is a luxuriant profusion of vegetative growth—the greenhouse provides a neat little demonstration laboratory of solar radiation behavior. Its function is as a transformer of wave length from short incoming waves to longer heat waves. The earth and its atmosphere operate in precisely the same fashion.

INTRODUCTION

The sun is merely one star among the heavenly legions and not a very remarkable one at that, in brilliance or in size. And the earth, over 100 times smaller and 93,000,000 miles (149,669,000 km) distant, intercepts an incredibly tiny amount of the sun's total radiation—actually less than ½ of one billionth. Yet the sun is our exclusive source of energy; the solar furnace, directly or indirectly, drives the earthly engine. What alternate sources could there be? Other stars, some very large and hot, clearly emit enormous quantities of energy but their effect on the earth is neutralized by equally enormous distance. Internal heat released by vulcanism or radioactive fusion/decay can in no way be considered significant except perhaps in an extremely local situation. So we must depend on the sun. How much energy actually arrives at the earth's surface and how it is put to use become the critical factors.

The sun is merely one star among the heavenly legions and not a very remarkable one at that.

THE SOLAR CONSTANT VERSUS INSOLATION

Despite the tiny fraction of solar energy that actually impinges on the outer margin of the earth's atmosphere, *it is our total receipt* and by our earthly measurement standards it is a very large amount. We visualize a plane at this far periphery of the atmosphere (wherever that might be) at right angles to the stream of solar radiation. The receipt of energy at that plane is called the *solar constant* and theoretically amounts to about 2 calories[1] per square centimeter per minute. As the name implies, this figure should be constant. Actually, it has tended to fluctuate 2 to 5% as a

result of varying solar energy output due to sun spots or solar storms, the elliptical orbit of the earth and thus a changing distance from sun to earth, and the difficulties of computation from mountain-top stations 50 years ago relative to sophisticated modern orbiting instruments.

The solar constant nonetheless represents the potential energy available to the earth, but the value of the energy that actually arrives at the ground is very different. This last is called *insolation*. Don't call it sunshine. There is a tremendous difference between the total energy reaching the earth from the sun and that which is visible. Sunshine is merely the visible portion. On an earthly plane parallel to the one at the outer margin of the atmosphere, the difference between the solar constant and insolation would simply be an amalgam of those things that happened to the energy stream as it passed through the atmosphere via the shortest possible route. This difference is on the order of about 50%—in other words, there is a 50% casualty rate of one kind or another to the waves of the sun as they pass through the atmosphere (Fig. 4–1). We must remember, however, that at different seasons and different latitudes both the amount of atmosphere traversed will change and the curved surface of the earth will accept the solar energy at something less than a right angle. So this 50% figure of insolation versus solar constant is simply a working approximation.

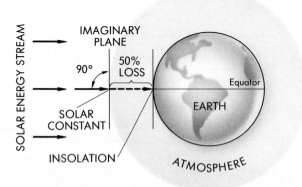

Fig. 4–1   The solar constant versus insolation. There is an approximate 50% loss as the solar energy stream passes through the atmosphere.

---

[1]A calorie is a heat measure equal to that required to raise the temperature of a gram of water 1°C.

Measuring Temperature

As ruler of the Italian state of Tuscany from 1620 to 1670, Grand Duke Ferdinand II was a significant figure in the advancement of scientific knowledge. He founded the Accademia del Cimento (Academy of Experiments), dedicated to the practice of experimental laboratory science and the testing of Aristotalian metaphysical speculation; his resident genius was Galileo, long renowned as a scientist, if not generally respected in ecclesiastic circles.

Into this rich compost of intellectual ferment stepped the young mathematician Evangelista Torricelli, first as Galileo's lieutenant and shortly thereafter as his successor. Within three years he was to preside over the epochal experiment that resulted in the invention of a workable barometer demonstrating that the atmospheric weight (pressure) could be measured and that the weight was not static but varied from time to time. Curiously this critical bit of knowledge opened the gate for the development of an accurate thermometer.

Galileo had shown some 40 years earlier that air expanded when heated and that, in expanding, would draw a column of liquid up a glass column. But this was an open system and the yet unknown effect of air pressure defeated early attempts at accurate calibration. The Grand Duke himself is given credit for realizing now, in the wake of Torricelli's revelation, that a hermetically sealed tube could isolate the influence of temperature from that of pressure. As any good Tuscan would, he used wine as his liquid medium.

In the early 18th century Gabriel Fahrenheit, a German physicist/chemist working in Holland, attempted to improve the then existing thermomenter by 1) creating compactness utilizing heavier mercury as the liquid, and 2) introducing a rational standard scale. He selected the freezing and boiling points of water as his bases but soon discovered that salty water did not freeze at the same temperature as fresh; so after much experimenting he concluded that 0° on his calibration should represent the lowest temperature at which he could supercool water. Selecting a purely arbitrary length for a degree, it turned out that 32° equalled the freezing point of pure water and 212° its boiling point at sea level.

The Swedish astronomer, Anders Celsius, followed up a few years later with what he thought was a somewhat simpler scale and called it *centigrade* (the prefix *cent* meaning 100). This system assigns Fahrenheit's 32° ice point as 0° and his 212° steam point as 100°, one degree thereby almost doubling the size of the F°.

There is yet another scale used widely in the scientific world called the *Kelvin scale*, after Lord Kelvin of Britain. It utilizes the concept of *absolute zero* defined as *that temperature at which molecules cease to move*. The technical definition of heat is related to the speed at which molecules vibrate: the more rapid the movement, the greater the heat. So Lord Kelvin, simply utilizing Celsius' degree spacing and ice/steam reference points, determined that absolute zero would be the equivalent of 273°C. KELVIN SCALE: absolute zero = 0°K, ice point = 273°K, steam point = 373°K.

SOLAR ENERGY PASSAGE THROUGH THE ATMOSPHERE

Planck's law, stated simply, says that the *hotter* a body, the more energy it will radiate and the *shorter* the wave length of that radiation. So although the electromagnetic spectrum (radiated energy) of the *very hot* sun encompasses a variety of wave lengths, all are short (Fig 4–2). This becomes significant when solar energy passes through the atmosphere because of the selective reaction of air molecules, dust, and water droplets relative to wave length.

High in the ionosphere, solar radiation encounters and splits oxygen and nitrogen atoms and heats the atmosphere. *There must be absorption to transform the sun's radiation into heat*, so a fraction is "used up" here to produce heat and will not reach the earth. In this case it is

The hotter a body, the more energy it will radiate and the shorter the wave length of that radiation.

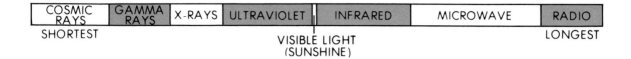

| COSMIC RAYS | GAMMA RAYS | X-RAYS | ULTRAVIOLET | INFRARED | MICROWAVE | RADIO |

SHORTEST   VISIBLE LIGHT   LONGEST
                                            (SUNSHINE)

Fig. 4–2   The solar spectrum of electromagnetic radiation.

advantageous that these high-energy waves are intercepted, for they are deadly. They are picked off along with highly charged particles called cosmic rays. Thus the ionosphere acts as a screen to shield us from harm. At a lower level, specifically the stratopause, the ultraviolet wave range is efficiently absorbed by the top of the atmospheric ozone concentration. Again absorption and resultant heat. But although ozone is receptive to ultraviolet, it is transparent as far as other wave lengths are concerned (*transmission* is the process as opposed to absorption).

Another process occurring in the upper atmosphere should be mentioned—*scattering*. Radiant energy waves encountering infinitesimal gas molecules, and perhaps meteorite dust, may not be absorbed or transmitted but merely dispersed. This change of direction is scattering. The energy is not used or lost but simply redirected. At these higher levels, scattering tends to be selective and only the shortest visible light waves are affected, thereby giving the sky its blue color. Larger particles in the lower atmosphere also scatter selectively the longer red waves, hence a red dusty or smokey atmosphere and red sunsets.

As the somewhat depleted solar energy enters the lower atmosphere (troposphere), the same dust particles and water droplets that cause scattering react in other ways as well. Together with carbon dioxide and water vapor they are all excellent absorbers of certain wave lengths. However, these do not include any except the very longest marginal wave lengths of the sun's spectrum, and hence the amount of absorption and heating is limited. Of considerably more importance in altering the receipt of insolation at the earth is the process of *reflection*—the mirrorlike reversal of the energy stream back into outer space so that it is forever lost to the earth. Cloud tops are the most obvious and efficient reflectors, but dust and some gas molecules can reflect too. When flying above a cloud cover, we are immediately aware that a reflection is occurring. Sunglasses are the order of the day to combat the painful brilliance of the white cloud top (Fig 4–3). Yet it is only the "sunlight" that is observ-

Fig. 4–3   A portion of the sun's short-wave energy stream is reflecting off of these cloud tops, redirected into space and lost forever; some small part is absorbed by the tiny water droplets and transformed into heat; but a very large segment simply flows unimpeded to the earth to be absorbed there.

able; a certain amount of invisible short-wave energy is being reflected simultaneously. There is some minimal absorption here at the cloud, but reflection, like scattering, merely redirects; the difference between the two is largely a matter of acuteness of direction change. It should be made clear, nonetheless, that reflection from clouds, haze, or dust is never 100% efficient. Some energy is transmitted. Again, we have selectivity. Beneath a dense cloud it is not totally dark. A portion of light comes through, some a product of scattering called *diffuse daylight*, as well as much of the invisible shortwave spectrum.

So now at the earth we have attained insolation and in so doing have forfeited an approximate 50% of the solar constant through absorption (added heat to the atmosphere) and reflection (lost forever). But our topic is temperature, and since we inhabit the surface of the earth, our concern is specifically the temperature of the troposphere. The question to ask is, *how much heat does the lower atmosphere gain by passing solar energy through it?* The answer is, not much. A great deal of energy is turned back via reflection and the atmosphere turns out to be largely transparent to the short sun waves that traverse it all the way to the earth. The exceptions to atmospheric transparency as regards absorption and heat are the ionosphere and the upper ozone layer at the stratopause, but these are both outer atmosphere and are not involved in our direct personal experience of weather and climate, largely functions of the troposphere.

Water vapor, water droplets, ice, dust, and carbon dioxide, because of their selective absorbency, combine to contribute *not more than 10 to 15% of the total accrued heat.* Where does the rest come from?

TERRESTRIAL ABSORPTION AND RADIATION

If the sun is our sole source of energy and thus of heat, how does the troposphere acquire 85 to 90% of its heat if passing solar energy through it is so ineffective? Actually, *the sun heats the earth and the earth heats the atmosphere.* The earth functions like a transformer in an electrical circuit, receiving energy in one state and transmitting it in another. An excellent absorber, the surface of the earth is highly receptive to the short waves that were eschewed by the atmosphere above and it translates them into heat. The earth becomes a heated body like the sun but not nearly as hot. It emits long waves called *heat waves* that tend toward the infrared side of the spectrum. As this terrestrial long-wave radiation enters the atmosphere from the bottom, it is now in the general wavelength range of efficient absorption by the water vapor, dust, water droplets, and carbon dioxide concentrated near the earth, and heating is both rapid and massive.

The question is, how much heat does the lower atmosphere gain by passing solar energy through it? The answer is, not much.

The Greenhouse Effect

This concept of indirectly heating the atmosphere from the bottom via a transformer is aptly called the *greenhouse effect* or commonly now the *atmospheric effect*. Growing tomatoes in February can be accomplished by capitalizing on the transparency of the glass roof in a greenhouse to short waves. Substitute the atmosphere for the glass and the simile becomes apparent. Short waves enter unchanged, strike the earth and are absorbed, then are reradiated as long heat waves that find the "glass" a solid barrier forming a trap. Build an instantaneous greenhouse: park your car in the sun for two hours with the windows closed and observe the interior temperatures. You will be experiencing a demonstration of wave-length differential and selective absorbency (Fig 4–4).

Land versus Water

The earth's surface is not composed of a single homogeneous material, and we must make some distinctions relative to the reflective capacities or albedo of strongly differing substances. New snow, for example, will warm very slowly because, like the white cloud top, it reflects rather than absorbs. So will light-colored sand or soil and (to some extent) water, which makes up well over half of the surface of the earth. However, the degree of reflection varies (Fig. 4–5); snow is the most efficient, other light-colored materials less so, and water only when the sun is low. This means that during the middle of a summer day, or in the tropics

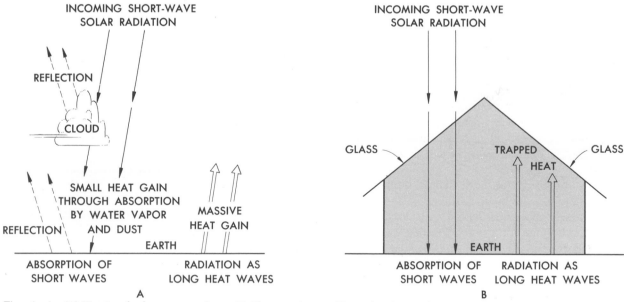

Fig. 4—4   (A) Heating the lower atmosphere. (B) The greenhouse effect. The glass in the greenhouse is the equivalent of the atmosphere in allowing short-wave energy to reach the earth, but trapping the long heat waves as they are reradiated.

where the sun is high in the sky all year, solar energy will penetrate water rather than reflect. So we should probably regard water as not a very good reflector since the low-angle sun's waves that it does reflect are a good deal less effective heaters than direct waves in any case; they not only pass through a greater depth of atmosphere, to their detriment, but disperse their energy over a much wider area (Fig. 4–6). Dark plowed fields, evergreen forests, and macadamized parking lots (which increasingly veneer our world), on the other hand, absorb much more solar energy than

	(Approximate Percentages)
New snow	90%
White sand	40
Light soil	30
Concrete	30
Forest or crops	15
Dark soil	15
Macadam Black Top	10
Water	5–90, depending on angle of receipt

Fig. 4—5   Albedo of selected earth surfaces.

they reflect. But with these endless diversities in mind, we can generalize to the extent of saying that there are only two basic materials making up the surface of the earth: land 29% and water 71%. These two differ fundamentally in their ability to absorb and reradiate solar energy.

Land, on the whole, is a good absorber and in the summer heats up rapidly, while water is a relatively poor absorber and over the period of a 3 month hot season changes temperature only slightly. What causes this great difference in ability to heat between land and water?

The reasons are several. First, land is opaque and the energy arriving on one square foot is strongly concentrated at the surface. A similar square foot (0.09 m^2) of transparent water allows the penetration of energy to a considerable depth, thus dispersing the heat through two dimensions, and the surface heats at only a fraction of that of a comparable area of land. Also, water is moving, both vertically and horizontally, so that no given area of surface remains exposed to the continuing receipt of the same amount of energy as does a comparable land surface. And, of course, the cooling effect of evaporation over the water is more continuous and effective than over land. Add to these the fact that water has a higher *specific heat* than land

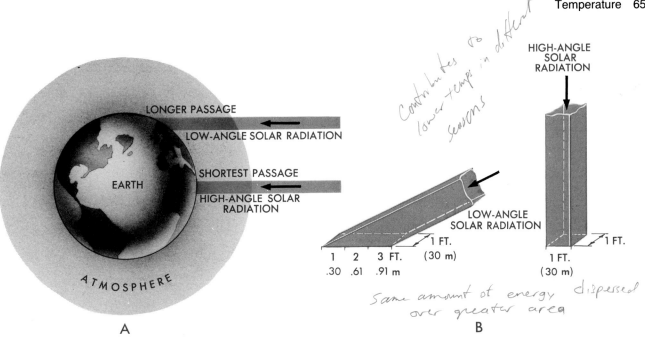

Fig. 4–6 Effectiveness as a heater of low-angle versus high-angle solar radiation. Low-angle radiation loses energy as it passes through an exaggerated depth of atmosphere (A) and then disperses that lesser energy widely (B).

(it is simply a physical law that other things being equal, it requires almost five times as much energy to raise a gram of water one degree as it does a gram of dry earth) and it becomes obvious that a continent heats intensely during the summer relative to the oceans.

A good absorber is a good radiator.

A *good absorber is a good radiator*. This means that during the winter and at night, when the receipt of the sun's energy is reduced or lacking, the earth will rapidly lose its heat, while the water will tend to maintain that which it has managed to pick up, releasing it very slowly. The temperature of the earth's surface in the middle latitudes will vary from summer days to winter nights as much as 150°F (66°C), while that of the oceans may exhibit a change of only 3 to 5°F (2° to 3°C). Since the lower atmosphere takes its temperature characteristics from the earth's surface, we can easily see how oceanic air masses may differ radically from those that are normally resident over land areas. The oceans and their attendant air masses tend to maintain very much the same temperature the year round, while the land

and the air above it will frequently show great seasonal extremes.

Heat Transfer

Conduction and Radiation. There are two related mechanisms involved in the transfer of heat from the earth to the air above it—*conduction* and *radiation*. Conduction involves the movement of heat from a warm body to a cold one in *contact*. Heat will flow from the warmer to the colder until they are of equal temperature. As a simple illustration, place a sterling silver spoon in a cup of hot coffee and note how the handle heats to the temperature of the coffee. Silver is an excellent conductor and will very rapidly assume the temperature of the hot liquid.

This same principle applies to the earth and the atmosphere at their point of contact. The warm earth will heat the adjacent cooler air through conduction. At first this heated layer will be shallow, only a few feet deep, but as the day wears on, it will become both warmer and deeper. At the same time a transfer of heat through space—radiation—will be occurring. This is much like conduction except that contact is not necessary; therefore, a wide zone

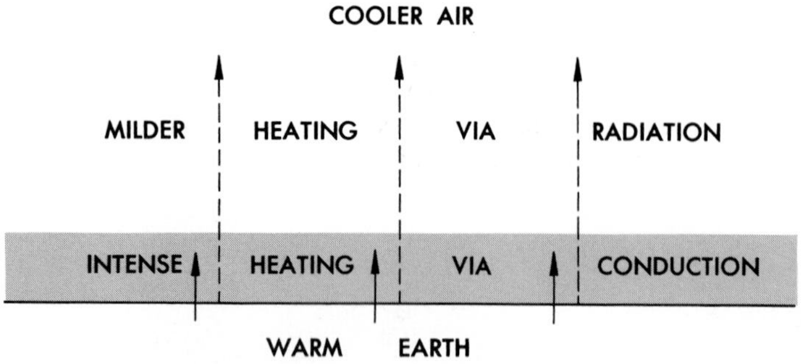

COOLER AIR

| MILDER | HEATING | VIA | RADIATION |

| INTENSE | HEATING | VIA | CONDUCTION |

WARM   EARTH

Fig. 4–7   Heat transfer. Because conduction is much more efficient than radiation, the most intense heating occurs immediately above the surface.

of the lower atmosphere is heated in this manner. But conduction is the more efficient of the two, and the air at the surface will be heated much more intensely. Putting one's hand on the stove will result in blisters—intense heat transfer via conduction. But several inches above the burner it will be pleasantly warm—modest heat transfer via radiation (Fig. 4–7).

In evaluating the effectiveness of these processes in heating the atmosphere, we must again come back to the air's ability to absorb. Water vapor (or humidity) stands out as the most critical element, especially in maintaining the heat in the atmosphere. It may very easily become excessively hot during the early afternoon in a desert climate simply from intense surface conduction and radiation; there are no clouds and the earth receives its full complement of solar radiation. But as soon as the sun goes down, that same clear air and lack of humidity allows the earth's accumulated heat waves to escape at a rapid rate. If, however, water vapor is introduced into the lower atmosphere, the long-wave terrestrial radiation is absorbed by the gas molecules which in turn reradiate some of their heat waves back toward the earth. Water vapor not only closes the escape hatch for the earth's amassed heat of the day, but functions as a baffle to hold it near the surface allowing only very gradual heat loss over a long night.

Compression and Condensation.

In addition to conduction and radiation, there are two other commonly occurring means by which the atmosphere may be heated—*compression* and *condensation*. Compression involves the loss of altitude of an air mass. As it descends, the greater pressure of air from above causes it to compress into a smaller volume that produces heat. This is not an unlikely situation, as sizable masses of air are constantly in motion in the troposphere.

Water vapor not only closes the escape hatch for the earth's amassed heat of the day, but functions as a baffle to hold it near the surface.

Condensation, the change of water vapor into liquid, also releases heat into the atmosphere. Where did the heat come from? It has been there all the time, latent in the water vapor, for in the process of becoming a gas, heat was expended. Let us suppose that a person is perspiring. As that perspiration evaporates from the skin, one feels cooler—heat has been subtracted. Where did it go? It is in the water vapor waiting to be released when the process is reversed and the gas becomes a liquid. This is condensation. It may not seem like a significant amount of heat to be considered, but when literally trillions of raindrops are formed instantaneously in a storm, a great deal of heat is released in a hurry. We can carry this a step further. Heat is required to change ice into water, so each drop of water has an increment of latent energy. Then heat is used up in changing water to gas, and water vapor now has two increments of latent energy. If water vapor is suddenly changed directly to ice, as in the formation of snow (*sublimation*), two increments of energy are released and the air is heated rapidly (Fig. 4–8). These are much more than interesting phenomena; *latent heat of condensation* is credited with being an important element in the formation and perpetuation of most of our storms (Fig. 4–9).

Fig. 4—8 Sublimation. These delicate snow crystals, no two alike, result from the sudden change of gas (water vapor) to ice as the temperature is lowered below the freezing point. They have not gone through the intermediate water stage or they would appear as frozen raindrops.

We can now list four basic means of heating the atmosphere: conduction, radiation, compression, and condensation. Two of these, conduction and radiation, may also function to cool the atmosphere, and *expansion* (the reverse of compression) and *evaporation* (the reverse of condensation) are cooling mechanisms also.

Heating	*Cooling*
Conduction	Conduction
Radiation	Radiation
Compression	Expansion
Condensation	Evaporation

We have found that where the warm earth is in contact with cooler air, there is a transfer of heat from warm to cold and the air is heated through conduction. Now if we turn this about as on a long winter night with the ground snow-covered, the air still, and the day's heat dissipated

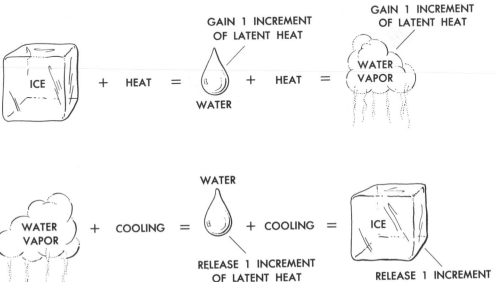

Fig. 4—9 Latent heat.

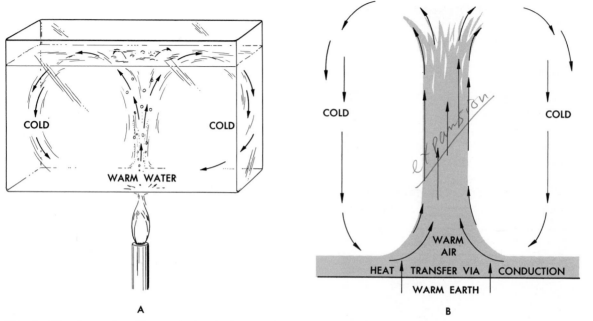

Fig. 4–10   Convection. Heat applied at the bottom of either a liquid (A) or a gas (B) causes a rising central column and sinking side columns.

rapidly via a clear atmosphere, it is entirely possible for the earth to be very much colder than the air above it. When this condition occurs, heat moves via conduction in the opposite direction and the lowest stratum of air is cooled. The air at a higher level can also lose heat through radiation to the colder earth as well as out into space. So conduction and radiation may function at any given time either to warm or cool the lower atmosphere.

Compression warms by causing a mass of air to decrease its volume as it loses altitude. If that same air mass were to go aloft, its volume would increase as expansion took place and its temperature would decrease. And finally, if condensation releases heat, then when evaporation is occurring the air is being cooled, as at the surface of the sea or over a forest of transpiring trees.

Advection and Convection.   Since the air is seldom still, once it has achieved its temperature characteristics in one place, it is likely to be transported elsewhere. Horizontal air movement of this type (advection) resulting from simple winds is normal enough, but vertical mixing or tur-

bulence is also frequent, with a common cause being convection. To understand and visualize convectional circulation, we may set up a laboratory demonstration, utilizing a tank of water with a Bunsen burner heating a point at the bottom. The heated water will rise visibly above this spot, and the colder water from the top and sides will be seen to be moving in to take its place. This rising central column and sinking side columns is convection, triggered by heating at the bottom. The principle works equally well in gas as in liquid, so if the narrow band of atmosphere directly in contact with the warm earth is heated by conduction, it will rise and colder air from above will sink to replace it (Fig. 4–10). Convection is often explosive, as the trapped warm air breaks through the colder air above it suddenly during the hottest time of the day. This mechanism transports intensely heated air to great heights and allows cool air from aloft to be exposed to the heating effects of the earth.

Both compression and expansion of sizable masses of air are involved here too, so the simple movement of warm and cold air vertically cannot be totally divorced from other processes. But generally we can say that neither horizontal winds nor vertical transfers of air in convection adds or subtracts any heat to or from the atmosphere that was not there already. These are turbulence or mixing factors that help to distribute heat throughout the lower atmosphere and cannot be properly called heating or cooling mechanisms by themselves.

THE GLOBAL ENERGY BUDGET

If solar energy continues to arrive at the earth to be absorbed and translated into heat day after day, year after year, eon after eon, there simply has to be a persistent heat buildup unless we can demonstrate elements of at least a roughly equal flow in the opposite direction. All of our longtime weather records reveal, certainly in a general fashion, a strong continuity of temperature characteristics with little to indicate the warming trend that would be mandatory if energy receipt exceeded loss, so such a counterflow must exist. The term "budget" is used because it implies balance—the income column must equal the expenditure column in the household budget or the energy budget if there are not going to be problems.

The term "budget" implies balance in the household or the energy budget.

Vertical Exchange

We need not worry about reflection since this energy is returned unused back into space from the earth and the atmosphere, but the substantial segment of solar radiation that is absorbed by the earth is another matter. On clear days when the humidity is low, only a small amount of the reradiated heat from the earth is intercepted by the atmosphere and energy in this long-wave form is sent directly back toward the sun. Also, the meteorologists talk of "radiation windows" in the atmosphere. This refers to a narrow segment of the long-wave spectrum that is rejected by the atmosphere, especially water vapor, and some heat waves from the earth flow unimpeded into space. It is, of course,

the exception. Normally terrestrial radiation is readily absorbed by the lower atmosphere and then reradiated, some back toward the earth and some into outer space. In the course of this to-and-fro energy exchange from earth to atmosphere and back, a continual seepage of heat into space takes place, and it must ultimately be of sufficient magnitude to balance our budget.

There is no requirement, however, of instantaneous energy transfer; heat accumulation does occur. The afternoon is hot because energy receipt has exceeded loss for several hours, and the maximum temperature reading usually displays a lag of 2 or 3 hours after that of the highest sun. And the minimum temperature is usually experienced about sun-up, at the end of a night of steady radiation loss with no compensating receipt. In most of the world the lengths of day and night vary with the seasons so that energy accumulation versus dissipation exhibits a seasonal as well as a daily cycle. The same kind of maximum/minimum temperature lag that shows up during a day is observable on an annual basis. In the Northern Hemisphere July or August is usually the warmest month, not June at the solstice; and January or February is the coldest, not December.

Even this annual lag of a month or two before a balance is restored is not a long one, for solar energy is called on to accomplish many chores before it can complete its cycle. Solar energy evaporates countless gallons of water each day, melts great masses of ice and snow, succors and sustains the complex food chain, and indirectly drives the great wind and ocean current systems. Even coal and oil are stored energy from the past remaining to be liberated on combustion.

Horizontal Exchange

We must assume also a horizontal balance factor within the earthly sphere, that is, an energy transfer from tropical regions to polar. The tropics obviously receive a surplus of solar radiation from a year-round high sun and lack of seasonality. Observed temperatures are continuously hot—but they are not getting hotter. Conversely, the high latitudes are never warm, reflecting the combined influences of low-angle summer sun and long winter nights—but they are not getting colder because of a long accrued energy deficit. There have to be transfer mechanisms (Fig. 4–11). The

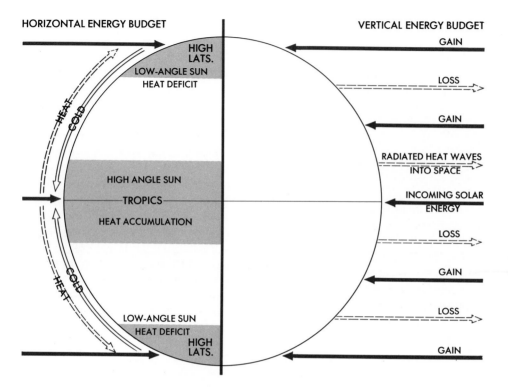

Fig. 4–11 The vertical and horizontal energy budget. All of the solar energy received and·put to work at the earth's surface must ultimately be returned to space (right), and during the period prior to such restitution, the surplus heat of the tropics must in some fashion balance out the polar deficit (left).

ocean currents are one and the planetary wind system another. These can get a little complicated, as we will see later, but essentially they do the job. Imagine a mass of warm air from a tropical ocean being wafted poleward and carrying with it a cargo of water vapor derived from evaporation of the sea. The warm air transports its own heat poleward as it goes, and then eventually the water vapor condenses as it cools releasing the latent heat far from its original tropical source. Such processes are occurring continually as cold polar air interchanges with warm tropical masses. It should not be surprising then that our middle latitudes, the point of contact and sharp contrast, should also be a region of storms and generally unstable weather. Also, reference to the color map of the oceans (Map V) will show that all cold currents flow equatorward and all warm ones poleward. Interestingly enough, these flows of air and water are set in motion basically by the very energy differences that require their existence. The surplus of tropical heat and the deficit of polar heat demand a budget mechanism and the imbalance produces its own salvation.

All cold ocean currents flow equatorward and all warm ones poleward.

Climate Change

All of this is not to say that the earth cannot experience long-term climate changes, for there appears at the moment to be sufficient evidence to more than hint at a number of these in the past. A variety of things are conceivable that might alter our world climates substantially. We know little about sunspots. What if they multiplied sufficiently to cut down on even a minor fraction of solar radiation? In other words, let us postulate a reduction in the solar constant. Or of more obvious significance right now, suppose there were an increase in carbon dioxide and other manufactured atmospheric contaminants to absorb, reflect, and scatter selected energy wave lengths. Or given the immense variation in the absorption and radiation ability

Fusion

Tokamak Fusion Test Reactor, Princeton University.

The provocative elements in a serious consideration of fusion energy are: 1) that the fuel is hydrogen, readily extracted from the ocean and hence virtually unlimited, and 2) that radioactive waste materials are essentially nil. All that has to be done is to heat ionized, electrically charged, hydrogen gases (deuterium and tritium) to temperatures rivalling those of the stars, in excess of 100,000,000C° (180,000,032F°) and then to compress and contain those gases so that the nuclei of the atoms fuse into new and heavier atoms. As this is accomplished the new atoms emit energy to be captured and put to work for useful purposes.

We've already come a step along this route with the man-ufacture of the H bomb, but here the fusion was uncontrolled. The trick is to produce these immensely high temperatures while at the same time achieving containment. *Plasma*, the ionized gases, must be confined for a matter of seconds to allow the atomic nuclei to fuse. Ultimately, to function as a practical energy-producing reactor, fusion becomes self-sustaining and no outside power is required beyond the original start-up charge.

From the very beginning of fusion power research it has been obvious that metal or plastic vessels were far too fragile to hold the superheated plasma, and a significant breakthrough came with the concept of a controlled magnetic field to contain them. The Russians were the first to construct such a device dubbing it *Tokamak* and they are busily working on a much bigger one— so are the Japanese and British. The U.S. version of this larger, new generation, doughnut-shaped Tokamak is a $314,000,000 model at Princeton University, financed by the Department of Energy. Here, in December 1982, American scientists achieved what has been billed as the first controlled fusion in history on their initial try. The burst of plasma lasted just $\frac{1}{20}$th of a second and required far more energy input than was produced. Further, the highest temperature applied to the gases was only in the 100,000C° (180,032F°) range, very cool by fusion stan-dards. When 100,000,000C° (180,000,032F°) temperatures are produced, the energy burst lasts several seconds, and the power production equals the power expenditure, then we will have ar-rived. This is called the "break-even point" and the projections indicate that it is perhaps another five years down the road. From the "break-even point" to large-scale commercial reactors—who knows how long or at what cost?

of land and water, what would happen if we could change their proportions? Major faulting could throw up new con-tinents or depress the ocean basins; the Antarctic ice cap could melt and the sea inundate all low lying coasts. These are only some possibilities.

The entire solar system (and presumably the rest of the universe) is dynamic and ever-changing, but time is essentially endless and change is slow, so given human lifetimes or even historical time we can talk in terms of permanence and not be in any real error. We work with the concept of an energy budget as a useful model in our attempts to relate and understand the major element of the current energy system. We assume it is permanent—at least until we know more about it.

SOLAR POWER

If we can say that coal and oil are simply stored solar en-ergy from the past, then it would seem wholly legitimate to look at our current daily energy receipt from the sun as a source of practical instant power. The potential is there and has long been recognized. Constantly, we are encountering such provocative statements in the press as, "if only 1% of the solar energy arriving in the Sahara Desert each day were converted to electricity, it would supply all of the world's projected electrical energy requirements for the year 2000." Nonetheless, very little has been done in any sort of organized way until fairly recently when the worldwide oil crisis brought home to all the realization that alternative energy sources must be considered seriously.

Of these alternative sources, much fevered argument has centered about nuclear fission with its manifold dangers—if only we could master the technology of the safer nuclear fusion we could forget about fission and would achieve perhaps the quintessential source of power. But from whence did the inspiration for fusion come in the first place? The sun, a mammoth natural fusion reactor, delivers huge amounts of energy on a daily basis to any who would harness it. Solar energy credentials are impressive; it is clean, safe, soundless, does not alter the earth's heat budget, and above all it is free and infinite. Arriving at the surface of the earth in a diffuse form, the problem is to concentrate

and focus it. We have mechanisms today that are capable of achieving this although undoubtedly they could be improved given the imperative. Mirrors and heat collectors can put the solar heat to work directly, while power cells transform electromagnetic radiation into electrical current (Fig. 4–12).

The only essential drawbacks are darkness and cost. Darkness involves not only night but clouds, fog, smog, and winter, each limiting, frequently for long periods of time, a steady predictable energy receipt. So immediately we are faced with a need for short-run storage of accrued energy while conventionally fired thermal back-up units are held at the ready. Batteries for storing electricity, or water, rocks and chemical banks for heat, have all proven to be immensely inefficient. A pile of coal covered with snow during a long winter loses no energy over time, but any modest accumulation of solar energy dissipates at a rapid rate unless used immediately.

In December of 1982 the Department of Energy in collaboration with NASA, published a report indicating that there are no insurmountable technological hurdles to a practical solar power satellite (SPS). The notion is to place giant batteries of solar cells into orbit 22,000 miles (35,406

Fig. 4.12 We are beginning to regard the practical utilization of the sun's energy seriously these days. After all, it's free! This parabolic mirror device at Odeillo in the French Pyrenees, is properly called a solar furnace and is the largest in the world. The reflective surface measures 180 x 130 feet (55 x 40 m) and is a mosaic of 9500 individual facets.

km) above the earth, their movement synchronized with that of the earth so that the cells would face the sun permanently without any obfuscations of night or clouds. On earth, specialized receiving antennae would intercept the SPS energy stream in the form of microwaves, transform it into electricity and funnel it directly to existing regional power grids. Expensive dreaming? Who knows at this stage of the game—at the very least it demonstrates creative imagination.

The closer the sun is approached, the colder it gets.

Today's costs for much less grandiose schemes remain prohibitive, except in unique situations, although no one doubts that they can be lowered with volume production of hardware; while at the same time relative costs of competitive energy sources are inevitably escalating. It appears, then, as we project into the immediate future, that solar energy indeed has some real potential and will very likely become a significant energy source in certain favored high-sun locations. Like mineral resources, there will be "have" and "have not" nations. But wherever practical, homes will be heated, cooled, lighted, and industry powered by the same solar energy that continues to work in the old familiar ways nourishing the world's food chain, driving the wind and ocean current complexes, and tanning tourists.

Fig. 4–13 Normal lapse rate. The seeming anomaly that the closer one approaches the sun the colder it gets, is manifest in this view of 8,260 foot (2518 m) Mt. Egmont in New Zealand. From the subtropical tree fern in the foreground, up past the timberline to the perennial snow, the temperature decreases visibly.

LAPSE RATES

Normal
Under usual circumstances, a thermometer attached to a balloon will indicate a constantly lowering temperature as it rises. The tops of mountains will support snow while the lower levels do not. That is, the closer the sun is approached, the colder it gets—seemingly paradoxical until we remember that the lower atmosphere receives its heat from the earth, and the farther we get from the radiator or heat source, the greater the cold (Fig. 4–13). Visualize a stratified troposphere with the warmest air at the immediate surface and progressively cooler layers atop it. This loss of temperature with increase of altitude (or gain with de-

crease) is called the *lapse rate*, and if we assume the air to be reasonably still and the observer or instrument to rise through it, the rate of temperature loss approximates 3.6°F (2°C) per 1000 feet (305m). This loss is fairly predictable, at least to the tropopause and is called the *normal lapse rate*

Adiabatic
There is another type of lapse rate, however, that should be differentiated from the normal lapse rate: the *adiabatic lapse rate*. Where the normal lapse rate assumed moderately still air, the adiabatic lapse rate involves the lifting of a sizable mass of air. As it rises, its receipt of heat from the earth will become less and less, but since we are dealing with a large mass of air, the temperature loss from simply

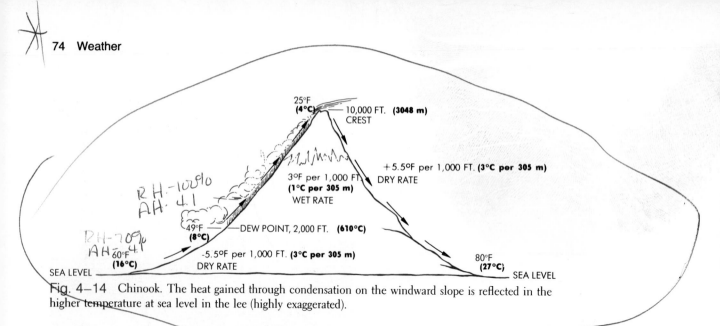

Fig. 4–14 Chinook. The heat gained through condensation on the windward slope is reflected in the higher temperature at sea level in the lee (highly exaggerated).

being farther from the source of heat will be slow and scarcely noticeable for some time. But there is a cause of instantaneous heat loss that will show up immediately; this is expansion, which results in a decrease of about 5.5°F (3°C) per 1000 feet (305 m). It is called the *dry adiabatic lapse rate* and applies to any rising expanding air mass *if no condensation is occurring within it*. If condensation occurs in any of its forms during the ascent of an air mass, then the *wet adiabatic lapse rate* becomes applicable. This lapse rate will vary with situations but is *always less than 5.5°F (3°C) per 1000 feet (305m)*.

Let us consider a *chinook* (United States) or *foehn* (Europe) wind as an example of the practical application of the wet and dry adiabatic lapse rates. These winds are warm and dry air currents blowing down from the mountains. The chinook, experienced along the eastern slope of the Rockies and Cascades in winter and early spring, melts the snow, thus opening up the range for grazing, but also frequently causing serious floods. If we set up a simplified and somewhat exaggerated example, the mechanics will become apparent (Fig. 4–14). We will place a 10,000-foot (305 m) mountain range in the path of prevailing winds, thereby forcing the air mass on the windward side up the slope. If the temperature of this air mass is 60°F (16°C) at the foot of the mountain, as it rises the air mass will lose 5.5°F (3°C) per 1000 feet (305 m) because of expansion.

At the 2000 foot (610 m) level, it will have lost 11°F (6°C) and the original 60°F (16°C) air mass will now have a temperature of 49°F (9°C). If we establish this 2000 foot

(610 m) level as the *dew point*, that is, the point where any further cooling will cause condensation, then from the dew point to the top of the mountain the wet adiabatic lapse rate must apply. Since this rate is always less than 5.5°F (3°C), let us, for the purposes of this example, place it at 3°F (1.6°C) per 1000 feet (305 m). As the air mass continues to rise from 2000 feet (610 m) to the 10,000 foot (3048 m) crest, it loses 3°F (2°C) per 1000 feet (305 m) or 24°F (13°C) for the 8000 foot (2438 m) uplift. Thus the air mass arrives at the top of the mountain at a temperature of 25°F (−4°C). Now if it descends the lee slope, the air mass will be compressed and warmed as it goes. Condensation requires cooling, so condensation will cease at the crest; and as the air mass comes down, it will be warmed at the dry rate of 5.5°F (3°C) per 1000 feet (305 m). The air mass descending 10,000 feet (3048 m) will gain 55°F (31°C), and this added to the 25°F (−4°C) at the crest gives it a temperature of 80°F (27°C) at the foot of the mountain.

By simply taking a 60°F (16°C) air mass over the top of a mountain and down the other side, we have added 20°F (11°C) to its temperature. Where did that heat come from? The only difference in ascent and descent is the zone of precipitation between the 2000 (610 m) and 10,000 foot (3048 m) levels on the windward side, so the heat must have been that released by condensation. There was still a net drop in temperature as the air went up through this zone, but only at the rate of 3°F (1.6°C) per 1000 feet (305 m). In other words, the air continued to expand and lose heat at the usual rate on the way up, but condensation replaced a part of that loss, and the 80°F (27°C) air mass at the foot of the lee slope is not only warmer but drier than when it started out.

Lapse rates, then, are simply the rate at which there is a loss or gain of temperature with a loss or gain of altitude. The *normal lapse rate* assumes a still atmosphere with the observer moving through it, while the *adiabatic lapse rates* apply when an air mass changes altitude and expansion or compression is in operation.

INVERSIONS

On occasion, the normal lapse rate may be reversed; that is, a gain in altitude will result in a gain in temperature. This situation is called a ***temperature inversion***, or simply an inverted normal lapse rate. Inversions are usually of short duration but quite common nonetheless. A long winter night with a clear sky and still air is an ideal situation for the development of the most common type of inversion. The heat of the day is radiated off the earth during the night, and by the early morning hours the surface of the earth is cooler than the air above it. Heat moves via conduction from the lower atmosphere to the earth and there forms a cool stratum beneath warmer air above. This means that if a thermometer were taken aloft, as it passed upward and out of this cool zone, it would indicate warming—just the reverse of the normal situation.

Smoke rising from a chimney on a cool still morning is likely to continue straight upward for some distance; it is warmer than the air surrounding it. But as the smoke rises, it cools and at the same time runs afoul of the warmer air

strata above which it cannot penetrate. It flattens out abruptly here, making visible the sharp line of demarcation between warm and cold air (Fig. 4–15). This is the inversion ceiling and it forms an effective barrier to the normal exhaust of combustion by-products into the upper troposphere. Such a situation could be serious if it were long-lived, but luckily this type of inversion, although common, only lasts a few hours until the sun comes up and begins to warm the earth. Conduction shortly restores the normal lapse rate and all is well.

The worst feature of the subsidence inversion is its persistence.

There are places (the Los Angeles basin is the type example) that occupy a position in the world scheme of pressure systems wherein air is frequently sinking and compressing for lengthy periods of time. There develops what is called a ***subsidence inversion***, compression warming the air to a temperature that exceeds the temperature of the air below it. This means, of course, that no effluvia may escape from underneath the ceiling. If the situation is further complicated by a topographic basin with the inversion lid slapped tightly atop it, then great concentrations of smog can develop (Fig. 4–16). The worst feature of the subsidence inversion is its persistence. High winds or heavy rain can break it up, but usually the unhappy populace must simply endure a lengthy vigil as they wait for the high pressure (descending air) to move on. Without the contaminants there would be no problem, but since the pressure system cannot be changed, we are confronted with the

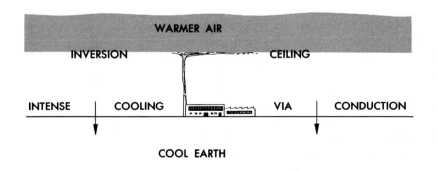

Fig. 4–15   Temperature inversion. The abnormal situation of temperature gain with increased altitude.

Fig. 4–16 Inversion ceiling. From above, the inversion ceiling with its trapped contaminants, is readily apparent. That the ceiling is below the crest of the basin perimeter is critical in the formation of a hermatically secure seal.

choice of continued and accelerated suffering or some considerable improvement of our current pathetic efforts to limit smog constituents.

AIR DRAINAGE

Cold air at the surface, as in an inversion situation, flows under the influence of gravity and the local topography. Being heavy and dense, cold air acts almost like water and moves down the slope to pile up deeply in pockets and val-

ley bottoms. This is called *air drainage* (Fig. 4–17). If any significant altitude is lost, there is adiabatic warming due to compression, but on the relatively small-scale local scene there may be a good many degrees of difference in temperature between valley floors and sloping foothills. It is common practice to plant frost-touchy citrus or peach orchards on slopes so that as cold air develops at the surface, it will flow away. The moving cold air remains shallow on the slopes, affecting directly only the trunks of the trees, but it will often accumulate to a considerable depth in the valley, causing frost damage to the delicate fruit and foliage (Fig. 4–18).

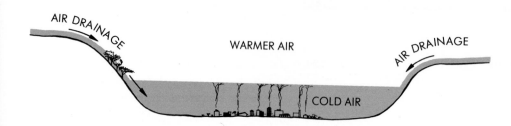

Fig. 4–17 Air drainage. Cold air flows downslope like water, filling up terrain depressions to a considerable depth.

Fig. 4–18   Fighting off frost damage in the orchard. Citrus is highly susceptible to just a few degrees below freezing, especially if the cold lasts for up to six hours. The propeller is designed to produce groundlevel turbulence and discourage development of a fatal inversion. Old-fashioned orchard heaters and smudge pots have largely been replaced with the more efficient and cleaner propellers.

ISOTHERMS

Frequently, it becomes necessary and useful to plot accumulated temperature data in map form to make visual for the map reader the distribution of particular temperatures at a given time. One method would be simply to write the temperature at the locations of the various reporting stations. But imagine the welter of figures on a map of the world or the United States if all reporting stations were represented. It would require long and careful study for any sort of meaningful pattern to emerge from this method. The reading of such a map can be greatly simplified by the use

It is common practice to plant frost-touchy citrus or peach orchards on slopes so that as cold air develops at the surface, it will flow away.

of *isotherms*, lines that connect all points of equal temperature (Figs. 4–19 and 4–20).

Normally, the isotherms are drawn on a map on which monthly temperature averages have been plotted, and in connecting certain selected temperatures, they divulge a pattern of occurrence that is not at all readily apparent from the written figures alone. We see immediately that isotherms exhibit a general east–west trend expressing the influence of latitude as the major control of world temperature. But over the middle latitude continents, they tend to deviate sharply from this trend, especially along west coasts. For instance, the January 32°F (0°C) isotherm in the north Pacific is a fairly straight east–west line through the Aleutians, striking the west coast of North America at about Juneau, Alaska. Obviously, it cannot continue directly eastward into the continent, for we know that the winter temperature averages in northern Canada are well below 32°F (0°C). In order to find as mild a January in the interior of the continent as that of Juneau, we may have to

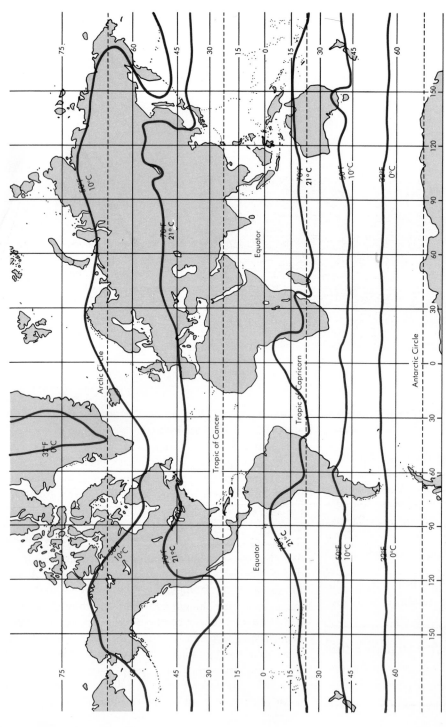

Fig. 4–19   World mean July isotherms (reduced to sea level).

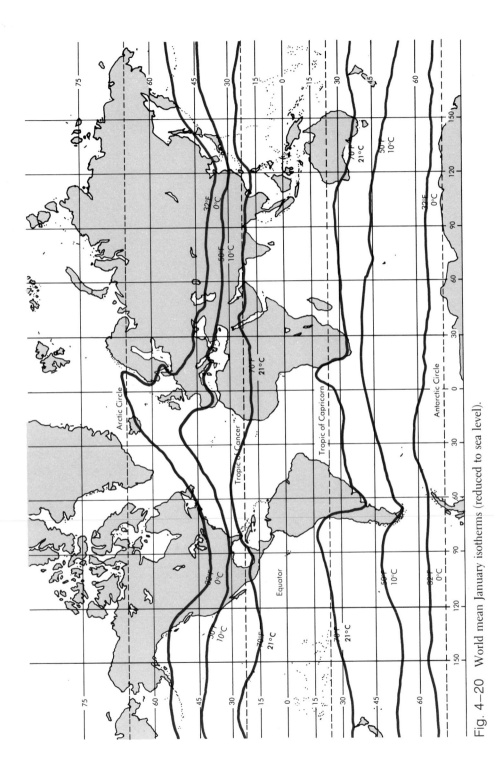

Fig. 4–20   World mean January isotherms (reduced to sea level).

drop as far south as central Kansas. So the isotherm in connecting these similar temperatures will show a sharp break at the west coast and trend almost north/south. In the North Atlantic and western Europe this deviation in direction is even more apparent, reflecting both the larger

In Western Europe latitude as a temperature control is of considerably less importance than coastal vs. inland location.

Eurasian continent and the northward probing warm Gulf Stream. By drawing a single isotherm, we see immediately that in this particular part of the world, latitude as a temperature control is of considerably less importance than coastal vs. inland location. Now check the 32°F (0°C) isotherm in the Southern Hemisphere for July (winter). It is simply a straight line because there are no great continents to introduce the factor of massive variation in seasonal heating between land and water.

Isotherms, like contours, are simply manufactured value symbols to help us interpret at a glance a whole panoply of temperature data plotted on a map.

CHAPTER 5

PRESSURE AND WINDS

Air pressure at the earth's surface is not readily sensed by the human body, but wind which is certainly obvious is the resultant of pressure differences. And wind can be put to work with the proper equipment. Sails to catch air currents and propel ships are "old hat" until we observe plastic, mechanically trimmed sails combined with diesel power on the modern Japanese vessel Shin Aitoku Maru. These are called Sail Equipped Motor Ships and the aim is to minimize the use of expensive fuel oil whenever possible.

INTRODUCTION

Pressure is the total weight of the atmosphere pressing down on the surface of the earth. At sea level in the middle latitudes, a column of atmosphere exerts an average pressure of about 15 pounds (7 kg) per square inch (3cm²) called *one atmosphere*. If something occurs to increase this weight, that is, to cause the air to push down a little harder, the pressure is increased at the surface, and this particular spot becomes a high-pressure center (*anticyclone*). On the other hand, there may be a lifting of the column, thus relieving some of the pressure and causing a low-pressure center (*cyclone*). Temperature, for example, is a common cause of pressure differentiation (although by no means the only one). Air that is cooled at the bottom will sink and increase pressure, whereas warming will cause the air to rise and thus lower the pressure. If we climb a mountain or go aloft in an airplane, we are no longer at the bottom of the atmospheric ocean, and pressures decrease as altitude is gained. A barometer may even be employed as a crude altimeter.

Unlike the other weather elements, pressure is something of an intangible. We cannot hear it, smell it, or feel it; in other words, the human body is a terrible barometer. It is not the best thermometer in the world for that matter, but one can tell general temperature changes, even if humidity variations and the like prevent us from being very accurate. But only those occasional individuals whose rheumatism and war wounds ache or whose noses bleed are able to sense changes in pressure.

THE BAROMETER

Therefore we are strongly dependent on instruments to keep us informed of changing pressure conditions. The critical instrument is the *mercury barometer*, which is not very different from the simple original invented by Torricelli in 1643. This first barometer was merely a glass tube with a vacuum in it, closed at one end. The open end was placed in a dish of mercury. As normal air pressure was exerted on the mercury surface, a column was forced up the tube to a height of roughly 30 inches (76 cm) (sea level and middle latitude). Variations in pressure caused changes in

this height.[1] It seems a little strange at first to measure pressure, a force, in linear units, but the height of the mercury column measured in inches is a direct indication of the pressure of the air. Today's modern barometers have only a few modifications on this original. Inches are subdivided into tenths or less for added accuracy, and a minor correction is made for the effect of temperature on the mercury. Such barometers, because of their awkward dimensions, are usually firmly fixed in a permanent wall mounting.

The human body is a terrible barometer.

A more convenient but somewhat less accurate barometer is the *aneroid*. It is a small metal diaphragm with a partial vacuum inside. As the outside air pressure becomes less, the sides bulge outward. This movement is shown on a dial that indicates the pressure change (Fig. 5–1).

Coming into much more common use today, and officially adopted by the U.S. Weather Bureau in 1940, is another unit of air pressure measurement called the *millibar*. It is a more logical unit than the inch since a millibar is a direct measure of force. A millibar is equal to a force of 1000 dynes per square centimeter, and a dyne is the force that will accelerate one gram of mass one centimeter in one second (one dyne is approximately the weight of one milligram). Since both the inch and the millibar are widely used, the important thing to remember is their relationship. An inch of mercury is equal to 34 mb; that is, *one standard atmosphere is normally said to equal 30 inches or 1,013 mb.*

ISOBARS

Once again the use of isarithms, in this case *isobars*, allows us to transfer data from an instrument to a map so that the actual pattern of pressure differences may be made visible. These isobars, drawn at 0.1 inch (4 mb) pressure

---

[1]Water might be used; it is cheaper than mercury, but it is also lighter, and normal air pressure would push it up a tube approximately 33 feet (11 m).

Fig. 5–1 Aneroid barometer. Compact and portable, the aneroid is widely used although not quite as accurate as the mecurial barometer. The instrument at the left, its face calibrated in inches rather than millibars, also has utility as an altimiter. The inset at the bottom records the regular pressure decrease with altitude gained.

variations, take the form of roughly concentric circles indicating centers of high or low pressure. Remember, pressure is relative—no specific number of inches is always low or high. If, for instance, we have a place with a barometric reading of 30 inches (1013 mb) and the pressure rises in all directions from that spot, then it is a low-pressure center. However, this same 30 inches (1013 mb) might very well be a high-pressure center if the pressures on all sides of it were lower.

WIND

Nature, abhoring inequalities, attempts to compensate for pressure differences by the transfer of air. High pressure represents a surplus of air and low pressure a deficit, and the general lateral movement of air, or wind, is thus from high pressure to low. *Wind* is the resultant of pressure differences, and both its direction and velocity are determined by the relative location and intensity of highs and lows. If we can envision a low center as a lifting of the air above a given place, then "a vacuum of sorts" is formed at the surface that pulls air to it from all directions. The greater the lifting, the more effective the vacuum and the pulling effect of the low. A high, on the other hand, is at the bottom of a column of air being forced down, and the air at the surface is being pushed out away from the high (Fig. 5–2).

Velocity

Intensity of high and low, and thus velocity of winds, is represented on the map by the spacing of the isobars. If

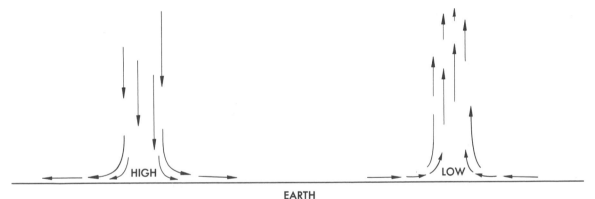

Fig. 5–2 At the surface air flows into lows and out of highs.

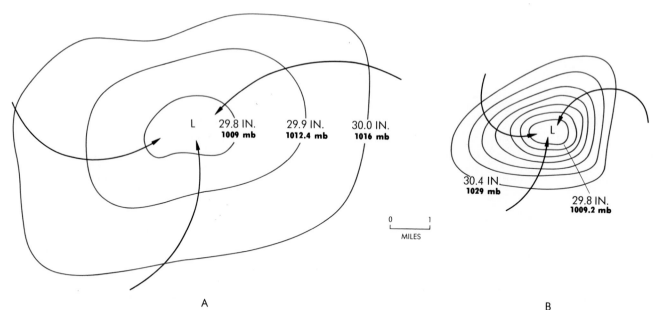

Fig. 5–3 Wind velocity. The spacing of isobars is indicative of the steepness of the pressure gradient. Wind velocities in (A) are very much lower than those in (B).

each isobar represents 0.1 inch (4 mb) pressure difference and the isobars are close together, there is a rapid change of pressure within a short distance, and wind will move across these isobars at a high velocity. This change of pressure as represented by the isobaric spacing is called the *pressure gradient*, and wind will always flow *down* the pressure gradient *from high to low* (Fig. 5–3).

Wind will always flow down the pressure gradient.

It will be recalled that contours or isopleths representing terrain elevation often arrange themselves in concentric circles similar to isobars. Such a pattern indicates either a hill or a depression (with hachures added). If we

Fig. 5–4 The anemometer measures wind velocity. Air currents catching the cups spin them at a rate which will be recorded in miles-per-hour on the dial at its base.

Gaspard Gustave de Coriolis

The Frenchman G. Coriolis, as he habitually signed his name (1792–1843), is often described as a mathematician or phyicist—perhaps he was a bit of both, but his lifelong field of research was in applied mechanics. As a child of the early industrial revolution he was primarily interested in how machines worked and concerned with improving their performance. Educated as a civil servant he shortly turned to science, by 1829 was teaching and publishing in the field of mechanics and had achieved membership in the august Academy of Science.

In an 1831 memo to that body he commented on some new and provocative theories relative to inertial forces and accelerations in composite motions as they applied to hydraulic systems in machines. Refined, these basic natural laws were explicated in his last book published posthumously as "the momentum of relative velocity and rotation of the frame of reference." Although useful and well-regarded these concepts did not take the engineering world by storm. But the delayed and unintended application of Coriolis force to elements in motion on a spinning globe is now regarded as his great and significant contribution.

The earth rotates with greatest velocity at its largest circumference, the equator. Hence any horizontally moving projectile (air current, water current) responding to the pull of gravity and runaway centrifical force will, as it changes latitude, be advancing over a more slowly moving earth. All of these factors at work (Coriolis' momentum of relative velocity) when viewed from the rotating earth (his rotation of frame of reference) will result in a predictable apparent deflection of the moving object from its original straight-line course.

Fluid mass application, borrowed directly from the Coriolis hydraulic systems theory, early aided in our understanding of ocean current behavior. Belated recognition to the man who gave us a rational explanation for observable natural phenomena that had never been wholly understood, was bestowed in 1963 when an official French oceanographic research vessel was named after him. Ironic, perhaps, in honoring Coriolis the scientist, not the engineer.

select a hill or topographic high and pour water on the top it will flow away from the peak in all directions. The moving water will cross the contours and we can say that it is flowing down the topographic gradient (slope), away from the high point toward lower elevations. Similarly, water poured anywhere on the lip of a depression will progress across the contours into the bottom of the hole, or again down the gradient. So, like water, air moves from high to low. The rate at which either will flow is determined by the steepness of the gradient. Contour lines close together describe a steep slope; isobars close together represent a similar steep pressure gradient. Air or water continues to flow in the same downslope direction, but now at an increased velocity across the crowded isopleths (Fig. 5–4).

Wind is the resultant of pressure differences.

Coriolis Force

Once the wind is set in motion by pressure differentiation, it does not flow directly from high to low as might be expected, but follows a somewhat devious course in arriving at its destination. This comes about through the effects of *Coriolis force*. Ferrel's law states the effect of this force: *Any horizontally moving object in the Northern Hemisphere will exhibit an apparent right-hand deflection and in the Southern Hemisphere an apparent left-hand deflection.* At the equator no such force exists, but its effect increases with latitude and reaches a maximum at the poles. This applies to ocean currents, rivers, bullets, baseballs, and, of course, air currents.

The Coriolis force is an apparent rather than an actual effect since its manifestation is dependent on the point of observation. If we can visualize a carousel and a boy throwing a baseball at a fixed target off the whirling platform as an analog, and station ourselves with the boy on the carousel, we will be in the equivalent position of observing the path of a horizontally moving object from the earth. As the boy throws the ball directly at the target, the ball will curve away and miss badly. That this is illusory becomes obvious when we view the identical action from above or, in essence, observe the earth from outer space. We find that the ball's trajectory is perfectly straight and that the boy's aim is faulty, for he fails to compensate for the motion of his platform. The curving trajectory of the ball is only apparent, but if the boy never leaves the carousel it is very real to him. Deflected air currents are equally real as long as we do not intend to leave the earth and its atmosphere.

Once a wind begins to move, two forces affect its flow: the pressure gradient causing it to move from high to low and the Coriolis force causing it to veer off course. Let us consider a Northern Hemisphere high-pressure center with air moving away from it in all directions. *We must face with the wind* to determine which direction is right and then add a right-hand deflection to the general air flow. The end result is air not moving directly out of the high but *spiraling out in a clockwise spiral*. In the low, the winds attempt to move at right angles across the isobars into the vacuum of the low-pressure center but are diverted to the right as they move. They arrive at their destination eventually but follow an indirect course, a *counterclockwise spiral*. In the Southern Hemisphere, these spirals are reversed (Fig. 5–5). The degree to which air currents are forced to

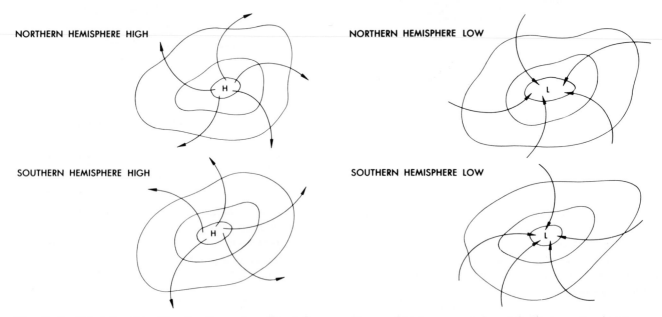

NORTHERN HEMISPHERE HIGH

NORTHERN HEMISPHERE LOW

SOUTHERN HEMISPHERE HIGH

SOUTHERN HEMISPHERE LOW

Fig. 5–5　Wind direction. The attracting and repelling influences of lows and highs, respectively, set the air in motion but its ultimate course is determined by Coriolis force.

angle across the isobars or spiral is determined by latitude (increasing deflection with gain of latitude) and friction of the air with the ground (lesser deflection with increased friction). The rougher the terrain, the greater the friction so that sea winds display greater deflection in response to Coriolis force than those blowing over land, and some upper air currents, which are essentially frictionless, are actually known to turn 90° and parallel the isobars.

WIND POWER

Here is another one of those free offerings similar to solar energy, unique in this world of high and ever-increasing cost, simply there for the taking. Wind is, of course, second generation solar energy—a direct result of differential heating of the earth—and it displays most of the same advantages. Reasonably soundless, pollution free, safe and infinite, there is much to recommend wind as a large-scale power source.

People have been utilizing wind, like the sun, locally far back into antiquity. However, gearing up for a major modern assault presents some interesting challenges. Nonetheless, wind has its proponents who increasingly are people and agencies that speak with some authority. A joint National Science Foundation and National Aeronautics and Space Administration committee suggests that by the year 2000, a major development program in wind power could provide electricity equivalent to the total amount consumed in the United States in 1970.

NASA already has several pilot models in place and private power companies are experimenting as well. Two hundred foot towers with equally long whipping rotor blades may not be far off—but the neighbors of currently existing smaller versions have already been complaining about low-frequency noise pollution, so a few unforeseen engineering refinements are yet in order.

Suitable locations for "wind farms" can be found in many places but probably the west coast, Rocky Mountains, and Great Plains are most universally suitable. How about several thousand giant wind turbines installed in line across the central plains from Canada to the Gulf? Or these same turbines stationed on floating platforms off the coast (Fig. 5–6)?

Fig. 5–6 Old and new near Clayton, N.M. The DOE/NASA 200 kw experimental wind turbine supplies a part of Clayton's electrical requirement, while the old-fashioned windmill continues to draw up groundwater, as it has for generations, to service the local grazing industry.

How about several thousand giant wind turbines stationed on floating platforms off the coast?

Storage of energy becomes a problem on windless days and as yet we have no real solution, but ingenious methods have been suggested to supplement those inefficient ones we already utilize. Flywheels, electrolizing water to produce hydrogen, or pumping water into high reservoirs when the wind blows and releasing it through power generating turbines on calm days may well have application to solar and other power sources as well where present lack of storage inhibits development.

SIMPLIFIED WORLD PRESSURE PATTERN

The entire circulation system of the troposphere is driven by the solar heat engine with its striking energy contrasts between low and high latitudes. Without these contrasts there would be no requirement of horizontal heat transfer to satisfy the energy budget and, ergo, no complex of air and water currents. But, given this pressing need to get the surplus heat out of the tropics in exchange for cooler air, a responsive system has developed. In theory it should be simple: cold air, heavy and dense, would flow along the ground toward the equator from both poles, while warm tropical air would rise and move poleward at higher levels in reciprocity. In other words, there is a giant convectional system. This is indeed the essence of the matter but nothing is ever quite as simple as it first seems, and at least two major complications immediately interject themselves: Coriolis force and differential heating of land and water.

The entire circulation system of the troposphere is driven by the solar heat engine.

Put the second of these aside for the moment and consider only the first. If we can visualize an all-water but rotating globe (to induce Coriolis force), we can come to grips with a predictable pattern of worldwide pressures and winds. There does exist over the major oceans of the world a reasonably predictable series of wind and pressure belts with names and terminology that are often nautical in flavor, reflecting the old-time sailorman's reliance on a first-hand knowledge of the winds. And so if we temporarily erase the continents with their seasonal temperature/pressure/wind changes, we will have extended the oceans and their elementary pattern over the entire globe. There is rationale behind this procedure, for although the resulting system is both theoretical and simplified, merit exists in introducing the generalized outline as a framework on which to hang some of the more complicated variations as they are developed later (Fig. 5–7).

Doldrums (Equatorial Low)

The belt immediately astride the equator and extending for roughly 10° on either side is called the *Doldrums* or *Equatorial Low*. Its outer margins are not precisely at 10° N and 10° S because the entire zone, along with all the others on the earth, shifts north as the overhead sun moves to the Tropic of Cancer and south when the sun moves to the Tropic of Capricorn. Nonetheless, the Doldrums is generally coincidental with the equatorial region and is the result of the constant high temperatures of this part of the world. The overall movement of the continuously heated air is *up*, and thus the pressure is low. And since the air movement is vertical, there are no winds, or at most only variable breezes, so that this zone is characterized by calm. The sailors who first recognized and named these belts dreaded the Doldrums for they were often becalmed for long periods of time in equatorial latitudes. The expression "in the doldrums," meaning to feel blue, has come down to us from these times, for there was nobody as unhappy as a sailor caught in the tropical calms.

Horse Latitudes (Subtropical High)

Some little distance out from the equator both north and south, at about 30° is a second zone of calms. This is a narrow belt called the *Horse Latitudes* or *Subtropical High*. Again, vertical air movement is dominant—in this case sinking rather than rising air. Obviously, temperature is not a factor here in causing the air to descend, as we are at the edge of the tropics and temperatures are high. But the *Antitrades*, upper air winds flowing away from the equator, are deflected increasingly as they move (Coriolis force), and since there is no retarding of this deflection due to friction with the earth, by the time the Antitrades have reached about 30° N and S latitude, they are flowing parallel to the equator. Air piles up and sinks of its own weight. The result is high pressure at the surface.

Trades

Between this permanent high and the permanent low of the Doldrums is a broad zone in which air flows horizontally along the surface toward the equator (i.e., from high to low). This belt in both hemispheres is called the *Trades*. In the Northern Hemisphere as the air moves from the Horse Latitudes, it is canted to the right and takes a

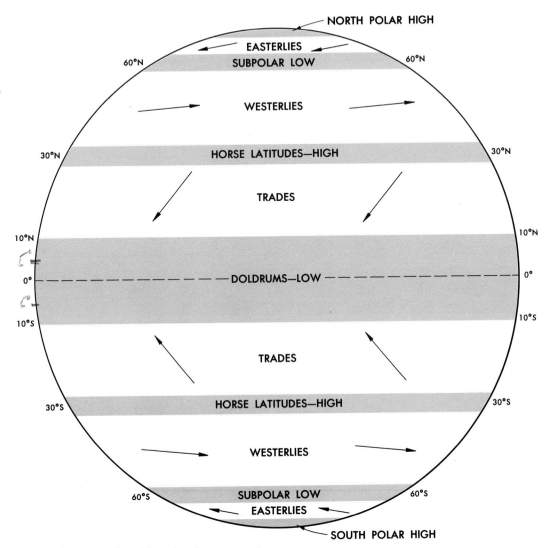

Fig. 5–7   Idealized world wind and pressure pattern.

southwesterly course (or from the northeast). Remember two things: first, winds are named by the direction from whence they come; second, always face *with* the wind when determining right- or left-hand deflection. Applying left-hand deflection to the Southern Hemisphere Trades, we find that they flow from the southeast toward the equator. Characteristic of the trade winds is constancy of flow always from the same direction and moderate velocities—in a word, perfect winds for sailing. Every westward bound mariner took advantage of them if at all possible, and the designation *trade* winds became a truly descriptive term (Fig. 5–8).

Tropical Circulation Cell

Note that in the tropics we have already encountered a breakdown in the easy transfer of warm air toward the poles. There has developed a somewhat self-contained tropical circulation cell: rising air at the equator, carried away at a

Fig. 5–8 Fluttering and bowing before the brisk trades, coconut palms in a favored tropical beach habitat transpire huge quantities of moisture from their wind-blown crowns.

higher level via the Antitrades, only to accumulate and subside at the Horse Latitudes and flow back as surface Trades toward the equator. Coriolis force is at fault as it interrupts the frictionless warm air movement aloft (turns it 90°) and forces it to pile up at roughly 30° north and south (Fig. 5–9).

Westerlies

The tropical circulation system is not wholly closed, however. Some of the sinking air at the Horse Latitudes is turning poleward along the surface in direct opposition to that fraction that sweeps toward the equator as Trades. This current is immediately impelled by the ever stronger Coriolis deflective influence to swing off toward the east, so that finally, the middle latitude surface air flow (roughly encompassing the latitudes 35° to 60° north and south) comes from almost due west. Hence the appellation *Westerlies*.

However, the character of the Westerlies, unlike the Trades, is neither constant nor mild. Frequently gusty and boisterous, even violent on occasion, the winds are capricious and likely at any given moment to blow strongly from directions other than west. This kind of unpredictable behavior is highly typical of middle-latitude circulation in general, for here we have a broad zone of conflict between advancing warm tropical air masses and polar air currents attempting to push their way equatorward.

Subpolar Low

An immediate result of this confrontation is the development of a narrow *Subpolar Low* zone at the mean point of contact. The lighter warm air with its cargo of water vapor is forced to rise above the advancing cold along this line (also called the *Polar Front*) and the released latent energy makes this a major region of storm generation. Although a narrow belt at about 60° North and South is labeled Subpolar Low and long observation has established it as a permanent zone of low pressure, the Polar Front itself will frequently deviate toward the equator. The character of warm/cold air conflict is one of erratic pulsation, surges of cold air pushing the front temporarily far out of line, especially over the large Northern Hemisphere continents. On these occasions any semblance of orderly air flow in the Westerlies is disrupted and the entire middle latitudes are affected by local squalls and storms emanating from this ever restless subpolar arena.

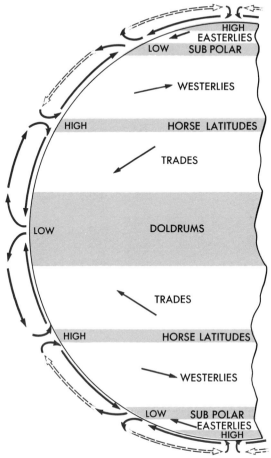

Fig. 5–9   Circulation cells.

Middle-Latitude Circulation Cell

Nonetheless, by generalizing rather broadly, we can visualize a moderately self-contained circulation cell in the middle latitudes comparable to that of the tropics. A part of the sinking air at the Horse Latitudes spills poleward to form the Westerlies, and although in the course of their run across the middle latitudes the Westerlies are deflected sharply and interrupted by frequent storms, they function to eventually transport warm air into the Subpolar Low. At this point the warm air encounters cold and is forced aloft and theoretically flows back at a higher level to help feed the Horse Latitude subsidence. This deduced upper air flow, above and counter to the Westerlies, appears to be as er-ratic and unstable as the Westerlies, but it is probably safe to say merely that there must be an overall drift of air toward the equator to complete the circulation cell.

Polar Highs and Easterlies

At the poles we encounter enduring and well-developed high-pressure zones. Like the Doldrums of the equatorial region where year-round high temperature results in low pressure, the cooling effects of Antarctica and the permanently frozen Arctic seas induce a thermal high. Air, chilled at its base, sinks and flows outward in all directions toward the equator, veering rapidly to the west (from the east) as it goes. These are called *Easterlies* or *Polar Easterlies*, and although a plethora of Arctic/Antarctic weather data is lacking, short-range spot checks seem to prove at least rough easterly currents. These bring cold air to the Subpolar Low to perpetuate the warm/cold confrontation; and again, theoretically, some of the air that lifts at this point flows back poleward to complete a high-latitude circulation cell.

THE EFFECTS OF THE CONTINENTS— LOCAL WINDS

We have been discussing the essential patterns of winds and pressures that *would* occur *if* the world were without continents. It is reasonably accurate over the ocean basins, but once the land masses are included, variations immediately become apparent. However, these effects are not distributed evenly between the hemispheres. Not only does the Southern Hemisphere exhibit a much higher proportion of ocean to land than does the Northern, but its major land masses are in the tropics. The significance of this is in their lack of strong seasonal temperature alternation, a characteristic of the middle latitudes. It is precisely in the middle latitudes that the Northern Hemisphere continents become most massive and the ocean basins increasingly pinched.

The Southern Hemisphere exhibits a much higher proportion of ocean to land than does the Northern.

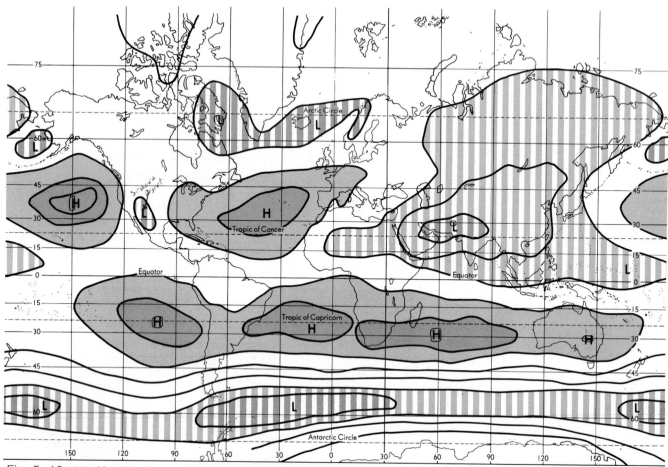

Fig. 5–10 World July isobars.

The excessive cold of winter and contrasting great heat of summer give rise to striking seasonal pressure differences, thereby significantly disrupting the simplified wind and pressure system (Figs. 5–10 and 5–11).

The Doldrum belt of our idealized all-water earth is very little disrupted by the insertion of continents in that, irrespective of land and water, temperatures remain high and pressure low. But in the high-pressure belt of the Horse Latitudes, the heating of the land areas in summer causes continental lows to develop, thus breaking up the earth-encircling high belt into fragments. In the summer the Horse Latitudes become a series of isolated cells over the cooler seas. In the Northern Hemisphere there are two: one in the Pacific called the *Hawaiian High* and one in the Atlantic, the *Bermuda* or *Azores High*. In the Southern Hemisphere, there are three oceans: the Pacific, Atlantic, and Indian, each with its high-pressure cell. During the winter as continents cool and their lows subside, these oceanic highs are connected across the land and the idealized belt reasserts itself. But in the Northern Hemisphere where the continents increase their bulk in the higher latitudes, the winter cold is most severe in Canada and Siberia, causing intense high pressure; the high-pressure belt linking the Atlantic with Pacific becomes badly deformed

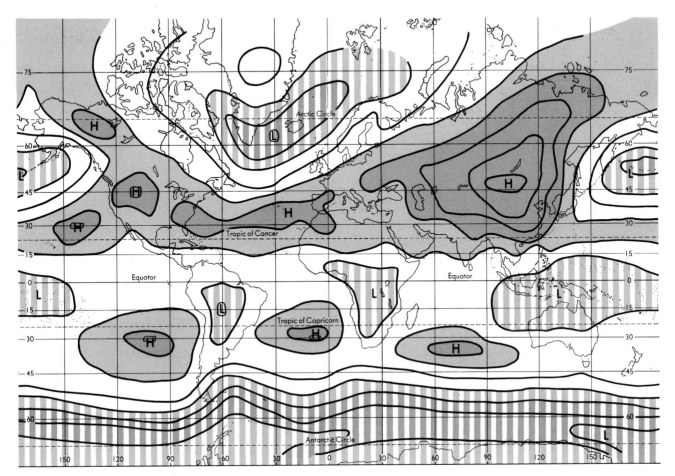

Fig. 5–11   World January isobars.

with great northerly bulges. Winds that are, of course, the resultants of pressure differences exhibit seasonal variations in the vicinity of the continents.

The Subpolar Low Zone is a simple matter in the Southern Hemisphere in that these latitudes completely lack sizable continents, and we have virtually an all-water earth again. But in the Northern Hemisphere the landmasses are huge and the ocean basins severely restricted so that, once again, isolated cells at sea become the pattern. They are called the *Icelandic Low* and the *Aleutian Low* and are permanent in these areas. When the intense winter high asserts itself over adjacent Canada and Siberia, these lows,

despite their small size, are very deep and well developed; but in the summer when the continents are warmer (although not really hot at these latitudes), the lows are quite weak. The polar regions, which continue cold at all seasons, remain as permanent highs.

The Monsoon

It is not surprising that except at sea all sorts of complications to the orderly pattern established earlier will become apparent. An excellent example of the kind of local winds that result are the relatively large-scale monsoon winds of

The Manila Galleons

For 250 years, with scarcely a break from 1565–1815, at least one Spanish galleon made the round trip from Manila, Philippine Islands, to Acapulco, Mexico. Their cargo was a rich mix of Far Eastern luxury goods destined for the grandees of Spain and the New World in return for Mexican silver and Spanish specialized manufactured products. And their route was unvarying—westward before the Trades, a smooth and pleasant journey; and eastward via the Westerlies, the lumbering ungainly galleons pitching and yawing through some of the Pacific's dirtiest weather.

Spain's colonial ventures into the Far East centered on the Philippines, but it became obvious early in the going that although these islands supplied lost souls for the church to reclaim, they were not a productive source of gold for Spanish coffers as had been so many of the South and Central America colonies. It remained for China and to lesser extent India and

Southeast Asia to supply the riches in demand back home. A typical cargo outgoing from Manila might include cigars from the Philippines; silk, porcelain, laquerware, and tea from China; spices from the Indies; gems from India; and ivory, camphor, and teak from Siam. Mexican silver, bullion and coins, paid for much of this—until the mid-1930's the massive silver cartwheel "one dollar Mex" still circulated and was the most valued coin in China.

These great ships, eastward bound in the Gulf of Alaska, were amazingly seaworthy if thoroughly uncomfortable. Through all of those 250 years only a few were lost to the weather. Vaguely, through the drizzle and low clouds the anxious lookout would make his first landfall—it could be Baranof Island in Alaska, Vancouver Island, or the dark and gloomy coast of Northern California; but sailing before the squally Westerlies it behooved the captain to stay well offshore and coast southward now toward better weather and Acapulco. Here is where the English freebooters, Sir Francis Drake and Woodes Rodgers among them, sent their swift vessels in against the battered giants. They didn't always win but the enormous potential booty was sufficient spur.

Acapulco and Manila are 16.5°N and 14°N respectively, both well within the Trade Wind zone the year around. The southern tip of Hawaii is 19°N, also a Trade Wind latitude—did Spanish ships sight these islands? Almost a thousand Manila Galleons plied this route over 250 years, and Pele, the Hawaiian goddess of fire, signals frequently from Mauna Loa. How could they have missed it? Perhaps they didn't. Imperial Spain was never one to publicize her secrets to the world. However, recent research in the archives at Madrid has turned up a map showing some sizable islands in the correct latitude, although the longitude was in considerable error. But Spanish captains were not in possession of practical chronometers until long after the English and accurate longitude measurements suffered.

Asia. Basically, their cause is differential heating of land versus water.

The general features of this concept are as follows. In summer the landmass of Asia heats up much more rapidly than the surrounding oceans, especially in its tropical portion. Local pressures, then, exhibit a seasonal high over the cool sea contrasting sharply with the heat-induced continental low that reaches its maximum development in the desert of northwest India/Pakistan. Reacting to this pressure difference, air flows onshore (from high to low), displaying as it moves a characteristic counterclockwise spiral. The entire procedure is reversed in winter. A cold

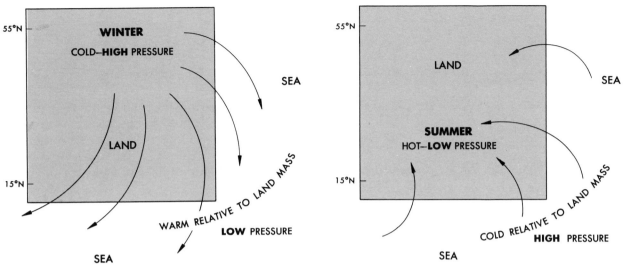

Fig. 5–12   Idealized Northern Hemisphere monsoonal circulation.

continent develops high pressure, especially in north-central Siberia, and the warmer sea, which has changed temperature very little from the summer, becomes a relative low. Now the air flows offshore, spiraling in a clockwise direction (Fig. 5–12).

However, this simple explanation of monsoon circulation, while correct as far as it goes, fails to take into account many of the intricacies of the local scene. For instance, the summer monsoon "breaks" with great suddenness in India, an unlikely situation if it were tied to the gradual progression of the seasons. But, then, India is almost hermetically sealed off from the continent by the great Himalayan barrier and the easy flow of seasonal winds is disrupted, channeled, or even blocked completely. It is entirely possible that India, home of the classic monsoon, has a circulation system virtually independent of the rest of monsoon Asia (Fig 5–13).

Entering into all of this is the role of a well-defined

Fig. 5–13   Monsoon. Summer has officially arrived in Bombay with the "break" of the monsoon.

jet stream, an eastward trending, high-velocity, upper troposphere air current that flows south of the Himalayas in summer and north in winter. More will be said of jet streams in general and their relationship to storms and precipitation, but each year as our knowledge of their behavior is augmented, an increasingly important part has been assigned to the jet in explaining monsoon phenomena.

India is almost hermetically sealed off from the continent by the great Himalayan barrier and the easy flow of seasonal winds is disrupted, channeled, or even blocked completely.

There are also the generally southwest summer winds of tropical Asia, which have been explained as deflected Southern Hemisphere Trades rather than as merely spiraling onshore breezes. The contention is that as the heat equator (as opposed to the geographic equator) follows the summer overhead sun into the Northern Hemisphere, it pulls with it the southern southeast Trades. As they cross the equator, Coriolis force causes them to swerve right, thus becoming the southwest summer monsoon.

So, without question, the monsoon is involved and much is yet to be learned regarding its variations and their causes. Yet the underlying genesis remains, simplified perhaps, but basic—the differential heating of land versus water. By placing a huge land mass in the middle of the ocean, we have developed local winds. Out at sea, the Trades, Westerlies, etc., continue to blow in their predictable paths, but the continental influences dictate the distinctive winds that are so characteristic of Asia.

If the requisites of monsoon development are a land mass of continental size and one with sufficient latitudinal spread to encompass both high latitudes and tropical to subtropical latitudes, then North America appears to be a possibility. Interior North America can get almost as cold as Asia in the winter, and although the true tropics are almost lacking, southern United States does become quite warm in summer. As a result, we have a monsoon. Not as well developed as that of Asia, which is larger, but a monsoon nonetheless. Canadian cold air forces itself clear down

to the Gulf on occasion and sticky tropical air invades the Middle West. We do not have a name for this phenomenon as the Asians do and sometimes its effect is overpowered by other more dominating weather factors, but the general tendency is there and we can see it operate close to home every year.

The rest of the world's continents are large enough, but their mass is concentrated in the tropics. They get hot and pull air onshore in summer, but except for Antarctica there are no cold air masses represented. And yet oddly, Australia, the smallest of the lot, displays a full-blown monsoonal circulation along its north coast. This is explained by its proximity to the huge Asian continent. During the Australian winter there is no really cold air over the continent and only a mild high develops—scarcely sufficient to cause significant offshore air currents. But when it is winter in Australia, it is summer in Asia, and the attractive power of the Asian continental low pulls the air off the Australian north coast. In summer the desert heart of Australia develops the deep low required to complete the annual wind reversal. If Australia were located far from Asia,

---

And yet oddly, Australia, the smallest continent of the lot, displays a full-blown monsoonal circulation along its north coast.

---

it would have no monsoon. Similarly, the central east coast of Africa, from whence the traditional Arab voyages to Asia and back took advantage of the seasonal wind reversal, is like northern Australia. Without propinquity to Asia, there would be no monsoon.

Land and Sea Breeze

On a much smaller scale, this same sort of air circulation occurs along coastlines on a day-and-night basis rather than on a seasonal one. This is called *land and sea breeze* and is characteristic of the tropics during the entire year and the summer season in many other parts of the world (Fig. 5–14). As the land heats up during the day, low pressure gradually develops in contrast to the high over the cooler sea. The mild pressure difference causes gentle cooling breezes to move onshore. At night the reverse occurs and

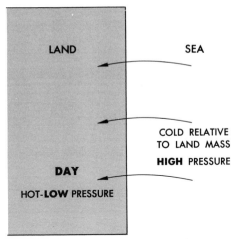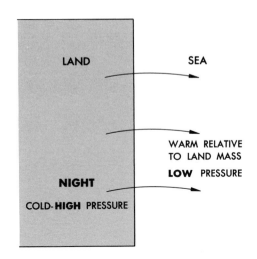

Fig. 5-14   Idealized land and sea breeze.

the breezes blow in the opposite direction. The entire phenomenon is extremely shallow and normally affects only a coastal strip a few miles wide, but often it is the difference between reasonable comfort and oppressive heat.

The monsoon and land and sea breeze are examples of local winds that defy the standard concept of worldwide belts of pressure and winds. Occasionally, they are strong enough to completely obliterate the effects of these belts for a time; more often, they merely modify them.

CHAPTER 6

MOISTURE

Most of the world's water is either on its surface in the liquid or crystalline form, or it is hovering close by in a thin stratum of adjacent atmosphere as water vapor. Condensation is the instantaneous change of form from gas to liquid through cooling. Here dew condenses in jewel-like droplets on the surfaces of a spider web following a cool night. The spider will be pleased when the sun comes out and reverses the process of condensation-evaporation.

INTRODUCTION

In terms of weather, water vapor is undoubtedly the most important of the atmospheric gases. But it is odorless and invisible and the only manner in which the human body can sense it is in conjunction with air temperature—so-called *sensible temperature*. When the air is dry, evaporation of perspiration from the skin cools us; however, when the humidity is high, evaporation is inhibited, perspiration is ineffective as a cooling mechanism, and we are not only hot but uncomfortably sticky. Don't point to a cloud and call it water vapor. It is not vapor at all but liquid droplets in suspension just as is a plume of "steam" issuing from the spout of a teakettle—steam is the proper term for superheated gas and therefore invisible.

Don't point to a cloud and call it water vapor.

Water in its gaseous form is significant in that it is the measure of the precipitation and cloud potential of any given air mass; it contains within it latent heat to be released on condensation; and it is an effective absorber of radiated heat.

EVAPORATION

Water vapor finds its way into the atmosphere via evaporation, primarily from the oceans but also from plants (transpiration), soil, lakes, rivers, and ponds, and even ice. However, the amount of water changing form via evaporation varies widely from place to place. Since temperature is the major control, tropical seas and forests are by far the greatest contributors. The polar ice caps with their permanent low temperature and the tropical deserts, despite their high temperatures, display a consistent lack of substantial precipitation and are largely without evaporation. The middle latitudes fall somewhere between these extremes.

But there is only a specific amount of water in this world, a closed system from which essentially none is lost or gained. Water merely changes form regularly from ice to liquid to vapor. Obviously, we are not talking about most of the earth's water, which stays in the oceans, but only the small part that is skimmed off the top by evaporation and is easily transported by winds to faraway places to form clouds and fall as rain or snow. Water is returned to the sea via streams, springs, and icebergs as part of the *hydrologic cycle* (see Chapter 27) and the rest condenses out directly over the sea.

Inasmuch as we have already looked into the general character of the planetary wind system that transports the moist air, the particular problem here is to identify and measure the water vapor taken into the air by evaporation and then to explore the ways and means of getting it back out again.

HUMIDITY

We call this water vapor *humidity*, and its measurement at a particular time and place may be expressed in any of several ways. The most common type of humidity measurement for the nonprofessional is a percentage or ratio expression called *relative humidity*. Briefly stated, this is *the amount of water vapor in the air relative to that which it is capable of holding at a given temperature. At a given temperature* is a critical part of the definition, for the air's capacity to hold water vapor varies with temperature. Cold air can hold much less water vapor than warm. Thus air at 50% relative humidity is holding half the amount of water vapor that it is equipped to hold (Fig. 6–1).

If we had such a mass of air at sea level and caused it to ascend, the air would cool through expansion. Its volume would increase but its moisture-holding capacity would decrease, and without any gain of moisture, the relative humidity would become greater than 50%. We would be, in essence, decreasing the size of the container, and the same amount of water that filled the larger container half-full now fills the smaller one almost to the top. When further cooling continues to decrease the moisture-holding ability of the air, the air mass will eventually become filled to capacity. This is *saturation*; the air is holding all the water vapor that it is capable of holding, and the relative humidity is 100%. This is also the *dew point* because any

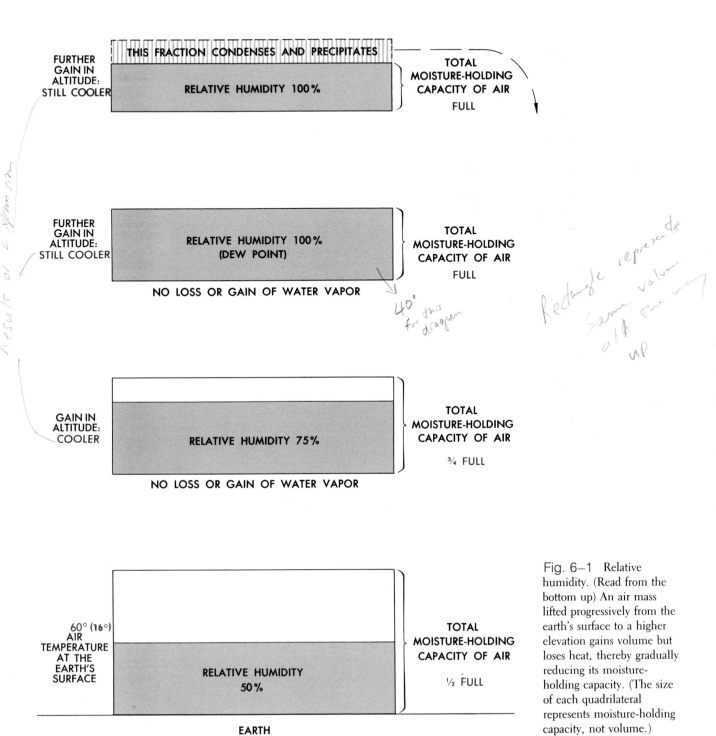

Fig. 6–1 Relative humidity. (Read from the bottom up) An air mass lifted progressively from the earth's surface to a higher elevation gains volume but loses heat, thereby gradually reducing its moisture-holding capacity. (The size of each quadrilateral represents moisture-holding capacity, not volume.)

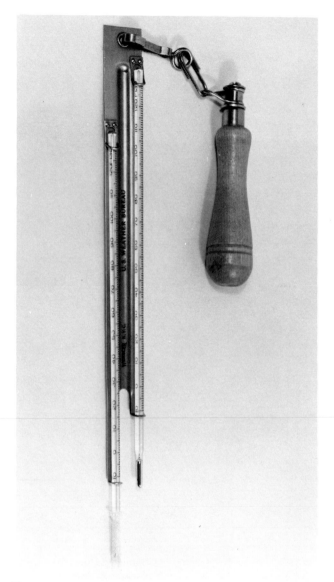

Fig. 6–2 Psychrometer. Nothing more complicated than two mercurial thermometers, but note the muslin covered wet bulb (left). The handle is for the operator to grasp and swing the entire mechanism about his head; this facilitates maximum evaporation and cooling.

will be in the form of rain; if below freezing, snow. Continued cooling will result in continued condensation, and the relative humidity will be maintained endlessly at 100%. If we were to bring this same air mass back to its original temperature, the relative humidity would no longer be 50% but something less, for a portion of the original water vapor has been lost through condensation.

To determine relative humidity, we employ an instrument called a *wet and dry bulb psychrometer* (Fig. 6–2). It is a simple affair, made up of two standard thermometers mounted on a frame, but one has a moistened piece of muslin wrapped about the bulb. We read the difference of temperature between the two thermometers (if any). If the air is saturated, there can be no evaporation of the moisture in the cloth and so no cooling. Under these circumstances, the two thermometers will register the same temperature. However, the drier the air becomes, the greater the evaporation and the greater the cooling of the wet bulb. A significant contrast in the temperature readings will become apparent. This difference of temperature, then, is directly related to the amount of water vapor in the air, and when applied to a table that is published by the Smithsonian Institution, it will give us an accurate dew point and relative humidity measure (Fig. 6–3).

CONDENSATION

There can be no condensation unless a surface is present on which the liquid can condense; thus the significance of the dust in the atmosphere. Not all dust is *hygroscopic*, or capable of uniting with moisture in such a manner as to act as a nucleus for droplet formation, but much of the normal atmospheric dust is hygroscopic, and the usual reaction of cooling below the dew point results in drops of water. In the clouds, where condensation takes place involving tons of water, nobody is too sure exactly what happens. First there is the problem of rain or ice becoming of sufficient size to overcome the updraft and fall as precipitation; it requires roughly a million tiny water droplets to form one raindrop. Collision, attraction of smaller to larger drops or to ice crystals, and electrical discharge are all thought to be common processes. But while this is going on there is often in a cloud great quantities of supercooled

further cooling will make the container too small to hold all the water and the surplus will spill out as condensation.

The dew point is where condensation begins as cooling continues. If this point is above freezing, condensation

Temperature of Dew Point [a]

Air Temp. Dry Bulb (°F)	Depression of Wet Bulb (°F)							
	1°	2°	3°	4°	6°	8°	10°	
20°	16	12	8	2	−21			
30°	27	25	21	18	8	−7		←Read
40°	38	35	33	30	25	18	7	Temperature
50°	48	46	44	42	37	32	26	of
60°	58	57	55	53	49	45	40	Dew
70°	69	67	65	64	61	57	53	Point
80°	79	77	76	74	72	68	65	←Here
90°	89	87	86	85	82	79	76	

Percentage of Relative Humidity [a]

Air Temp. Dry Bulb (°F)	Depression of Wet Bulb (°F)							
	1°	2°	3°	4°	6°	8°	10°	
20°	85	70	55	40	12			
30°	89	78	67	56	36	16		←Read
40°	92	83	75	68	52	37	22	Percentage
50°	93	87	80	74	61	49	38	of
60°	94	89	83	78	68	58	48	Relative
70°	95	90	86	81	72	64	55	Humidity
80°	96	91	87	83	75	68	61	←Here
90°	96	92	89	85	78	71	65	

[a] Excerpted from Smithsonian Meteorological Tables.

Fig. 6–3   Dew point and humidity meteorological tables.

air—air cooled as much as 30 to 50°F (17° to 28°C) below the dew point yet refusing to condense. Airplanes flying through such a supercooled atmosphere encounter immediate heavy icing on the wing surfaces, so apparently, if the wings are such effective surfaces, naturally occurring freezing and condensation nuclei are in short supply.

Cloud Seeding

In order to force rain from such a cloud, pioneer weather modifiers flew above it and dropped dry ice in an attempt to lower the temperature to the point where supercooled air would become ice. The ice crystals in turn would form surfaces for further condensation and freezing, and eventually particles of ice of sufficient size would precipitate.

In the clouds, where condensation takes place involving tons of water, nobody is too sure exactly what happens.

The theory was that nearly all rain left the cloud as ice, melting before it reached the ground. Undoubtedly, this is frequently the case although some tropical and summer situations have proven to be antithetical to such a hypothesis. Nonetheless, the cloud seeders and weather modifiers have met with some success (probably not so much as the commercial firms would claim). They now use silver iodide crystals, which form excellent hygroscopic and freezing nuclei, pumped into the air from ground generators, and although all clouds do not produce rain, the right clouds under the proper circumstances can be made to give up some extra moisture. Control of precisely where it will fall is something else again. A farmer in the next county may not be particularly anxious for the rain or snow the local taxpayers contracted with the cloud seeder to produce.

Dew and Hoar Frost

If cooling will cause condensation, then moist air brought into contact with a cold surface will condense on that surface as the air is suddenly brought to the dew point. This is *dew.* Any surface that is an especially good radiator, such as the metal top of an automobile or broad leaves of plants, will rapidly lose the accumulated heat of the day after the sun goes down and through conduction will then cool the air immediately adjacent until dew is formed. The outside of a glass with a cold liquid in it will "sweat." The glass is not leaking but is removing moisture from the atmosphere by cooling. Similarly, the inside of windows in the kitchen and bathroom are likely to "fog" as the high humidity of these rooms contacts the cold panes. If the surface is below freezing, sublimation will occur and delicate ice crystals form on that surface as frost.

Fog

On calm, clear nights when loss of the earth's heat through radiation is particularly efficient, dew will form rapidly, and then gradually as the night becomes longer, a deeper stratum of air will be cooled below the dew point and tiny droplets of moisture released throughout this zone. These will be kept in suspension by normal minor air turbulence. This is *fog.* It may vary in thickness from a few inches to several hundred feet, depending on the degree of cooling and the moisture content of the air. Most fog is formed in this manner and is properly called *radiation fog* because the lower atmosphere loses its heat via conduction and radiation to the cooler earth. However, if a normal radiation fog is wafted by gentle breezes away from its place of origin and is experienced then in a region where conditions are not suitable for the development of fog, it is called *advection* (or transported) *fog.* The California coast experiences this type of fog habitually. A cold current offshore provides the cooling for saturated oceanic air masses that move onshore at night, and these fog banks forming over the cold current are then brought onto the coast (Fig. 6–4).

Fog is usually a transitory phenomenon, for as soon as the sun comes out, it begins to evaporate. The longer waves in the sun's spectrum are intercepted and absorbed

Fig. 6–4　Looking north along the San Andreas fault, San Andreas Lake in the middle foreground. Every summer afternoon a finger of dank sea fog creeps through the Golden Gate (from left to right at top) to envelop San Francisco and the Bay.

by the top of the fog stratum, while the short waves go through it and heat the earth. So the fog is "burned off" by being heated from both the top and the bottom and is usually dissipated by noon. The last remaining wisps of fog are not at the ground but at some intermediate level.

Fog is "burned off" by being heated from both top and bottom and is usually dissipated by noon.

Precipitation

The formation of dew and fog is all very interesting but somewhat academic from the standpoint of significant amounts of usable moisture. Nowhere in the world is any sizable agricultural venture carried on that is dependent for its moisture on dew or fog. Somehow large masses of air must be forced to give up their moisture in sufficient quantities to cause "real" rainfall (or snowfall). The principle remains the same whether we are dealing with dew, fog, or heavy rain—the air must be cooled below the dew point; but to cause rain, this must be accomplished rapidly and on a large scale. The obvious way is to force a large mass of moist air to go aloft and cool adiabatically. There are three common ways that this may be accomplished, and all of the world's significant precipitation occurs from one or a combination of these methods. They are: *orographic*, *convection*, and *cyclonic* or *frontal* (Fig. 6–5).

Orographic. The word "orographic" means mountains, so in order to have this type of precipitation, there must be a topographic barrier in the path of prevailing winds.

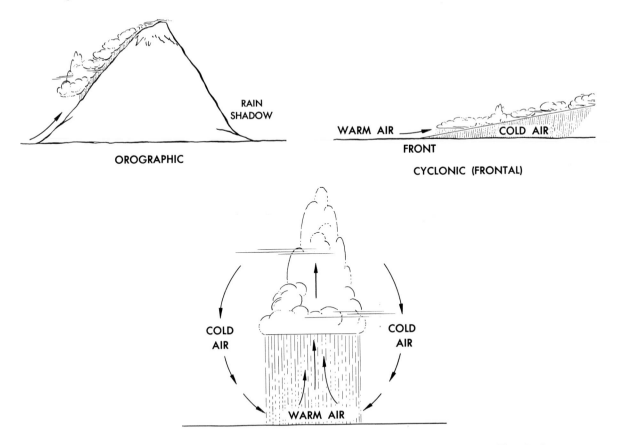

Fig. 6–5   Causes of precipitation.

Thunderstorms

A cumulo-nimbus thunderhead, dark and brooding against the summer sky, is merely a malevolent variation of the common fleecy fair-weather cloud. Both are more often than not the result of hot-season surface heating and the development of a convective cell. The difference between the two is a simple matter of violence of uplift.

If a warm saturated air mass is abruptly lifted several thousand feet in a matter of minutes with great heat as the activator, huge quantities of energy are released through wholesale condensation and a predictable series of spectacular resultants will follow. Or much the same violent updraft can be initiated in an extended line along a cold front—(frontal/cyclonic lifting) or a steep mountain wall (orographic lifting). In every case the top of the cloud reaches great height, usually well above the freezing point even in the tropics and a full-fledged storm is functioning.

The convective concept of a central rising column of air balanced by sinking side columns is correct as far as it goes, but within the energy-rich thunderstorms there are normally several independently rising and sinking elements or cells. Each of these is an explosive entity unto itself with savage wind-shear characterics, so that thunderstorms in reality are complexes of seething internal churning. Caught up in all of this are supercooled water droplets carried alternately above the freezing line and back below to grow in size as each new layer of ice accretes. When they finally become too heavy to be kept aloft (or the force of uplift abates) down they come in a torrent of hailstones. "Hailstones the size of golf balls" may be a frequent exaggeration by the press, but in Coffeeville, Kansas on September 3, 1970 a hailstone was measured at 6 inches (15 cm) in diameter and 1⅔ lbs (.6 kg)—more akin to a softball than a golf ball.

The same internal turbulent fury that produces hail appears to be the trigger mechanism in the onset of thunder and lightening as well. For reasons we know not why precisely, atmospheric violence begets electrical charges and they arrange themselves within a cloud, the positive charges at the top, negative at the bottom. On the earth immediately below, the usual ground negative is nullified and becomes positive temporarily. This has been described as an "electrical shadow", faithfully following the cloud as it moves. Enormous electrical stresses between these opposing poles, as many as 100 million volts generated, eventually overcome the insulating effects of the surrounding air and lightning flashes from cloud to ground or within clouds, functioning as a mechanism to relieve nature's electrical imbalance. And the instantaneous heat release by each bolt of lightening forces air to expand explosively. Sharp staccato reports, booming cannonades, rolling growling thunder off in the distance, all are the indirect sounds of lightning's great heat.

Such situations occur widely, as along the west coasts of North and South America, and this is a common cause of precipitation. As has been discussed previously relative to adiabatic lapse rates, moist air forced against a mountain front ascends, resulting in expansion, cooling, and precipitation. The type and amount of precipitation is controlled by the slope and height of the mountain and the temperature and moisture content of the air mass. In the lee of such a topographic feature, there is a relatively dry zone called the *rain shadow* that is, if the rainfall were sunshine, behind the mountain there would appear a reciprocal shadow. This effect is very striking in many parts of the world: Oregon and Washington, divided north to south by the high Cascades, exhibit thoroughly split personalities in the drear and dripping west versus the sere and sunny east; Trade-wind islands, be they West Indian, Hawaiian or Madagascar, are similar in their sharp rainfall contrasts. Orographic precipitation may occur at any season of the year or in any part of the world where the above conditions prevail. The windward slope of the mountain will be the wet side and the leeward side dry.

Convection.   "Convection" is not a new term but was dealt with previously in conjunction with transfer of heat to high altitudes. It also operates as a major cause of precipitation. Remember that heat is the requirement to set convective currents in motion—buoyant heated air at the surface trapped below cooler air aloft is a very unstable condition. As the hot earth on a summer day transfers its heat to the air above it via conduction, the increasingly warm air may at some point burst explosively through the cooler air above it. In this violent updraft the warm air cools rapidly through expansion as it rises several thousand feet in a matter of minutes, and condensation of a tremendous amount of moisture is virtually instantaneous. The resulting rainfall is violent and usually of short duration. In addition to original heat derived from the ground, the sudden and massive release of latent energy further activates

the air mass until its towering thunderhead may occasionally punch through the colder air above it to the level of the tropopause. Such rainfall (never snow, because it requires heat to set it all in motion), is a tropical or summer phenomenon and is typical of many parts of the world.

Cyclonic.   "Cyclonic" or "frontal" precipitation results when two differing air masses are side by side and one is forced against the other (the details of air masses will be explained later). The lighter air will ride up over the heavier air and thus be forced to cool adiabatically, producing condensation and precipitation. For instance, in the middle latitudes we frequently have warm and cold air in close conjunction. The warm air being lighter than dense cold air will run over the top. This is not unlike the warm air being forced up a mountain slope: it cannot invade the cold air and must move up over its sloping front. Frequently, this slope is low-angle and thus the rate of expansion is slow, with gentle drizzle or snow flurries resulting. This type of precipitation occurs most often in the belt of the Westerlies or near the equator where differing air masses commonly meet (Fig. 6–6).

Fig. 6–6   Snow as well as rain can precipitate at the taction point between warm and cold air. Rural folks stay indoors and await the rescuing snow plow as uncounted delicate little snow crystals thoroughly disrupt normal communication.

Fig. 6–7 Fluffy cumulus clouds in the tropics. These appear to be building up toward an afternoon shower or two.

CLOUDS

Like fog, clouds are tiny drops of liquid water or ice held in suspension by air turbulence; and like fog, clouds are a result of moist air being cooled below the dew point. But a major difference exists in the method of their formation. It is true that clouds are usually encountered at moderately high elevations, while fog is normally at the surface of the earth, but high fogs and low clouds may also be distinguished and properly labeled. However, if the cooling and subsequent condensation results from moist air in *contact* with a cold surface, that is, cooling through conduction and radiation as described previously, then it is called fog. If the condensation results from *adiabatic cooling*, the suspended water droplets make up a cloud.

The usual classification of clouds is on the basis of their appearance. Terminology for the three well-known standard cloud forms or genera has come down to us from the initial effort at cloud classification in 1803.

Cumulus clouds have as their chief characteristic a certain depth. The general appearance is fleecy, fluffy, or cottony. They vary from the towering convectional thun-

derhead, with its flat bottom marking the dew point and its top thousands of feet above, to the tiny fair-weather puffball. The cumulus cloud never covers the entire sky and its bottom at least is seldom at a very high level (Fig. 6–7).

Stratus clouds do cover the entire sky, are usually gray in appearance because they have cut out a large part of the light, and show little of the vertical development of the cumulus cloud. Usually low to medium in elevation, the stratus cloud is commonly formed along fronts and often emits light drizzle or snow flurries.

Cirrus clouds are high-level clouds made up of ice crystals and exhibit a characteristic fragile and feathery appearance. They are invariably fair-weather clouds, and even when covering the entire sky are so insubstantial that they are often described as a cirrus haze (Fig. 6–8).

Once [the] three basic cloud genera have been determined, there is a good deal of common sense involved in identifying variations.

Fig. 6–8   The feathery wind-drifted character of ice crystals in suspension is evident in these cirrus clouds, called "mare's-tails."

Once these three basic cloud genera have been determined, there is a good deal of common sense involved in identifying variations. Simple combinations is one possibility. For instance, if the entire sky is covered and yet the cloud is obviously very high and transparent so as to cause a halo around the moon, then we might call it cirro–stratus since it has characteristics of both. Or if the sky is gray and wholly cloud-covered at a low level, but the cover is made up of a series of coalesced linear cumulus puffs, then the proper designation is strato-cumulus.

Further definition may be desirable to point up peculiarities of shape, height, or structure. Thus we come to a second level of classification, the cloud species. *Alto* means higher than normal; *fractus* or *fracto*, torn or rent as a high

wind might disperse a stratus layer and allow glimpses of blue sky; and *nimbus* or *nimbo*, meaning literally rain and referring to dark "dirty" cumulus or stratus clouds with rain either imminent or present. These terms are simply tacked on to the usual description for clarification. The International Cloud Atlas of 1969 identifies 14 cloud species, many esoteric and seldom used such as *spissatus* (grayish), *uncinus* (comma shaped), *humilis* (flattened cumulus), and *castellanus* (turreted). For the professional cloud observer, this same atlas lists 27 cloud codes. Despite all of these minutiae, cloud classification for most purposes is a fairly simple operation with only a small working nomenclature required.

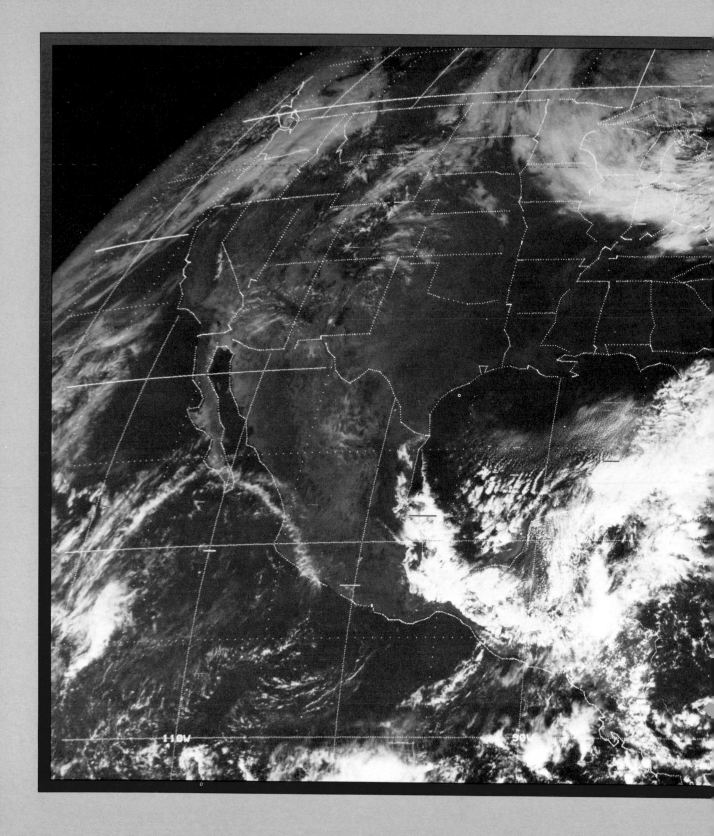

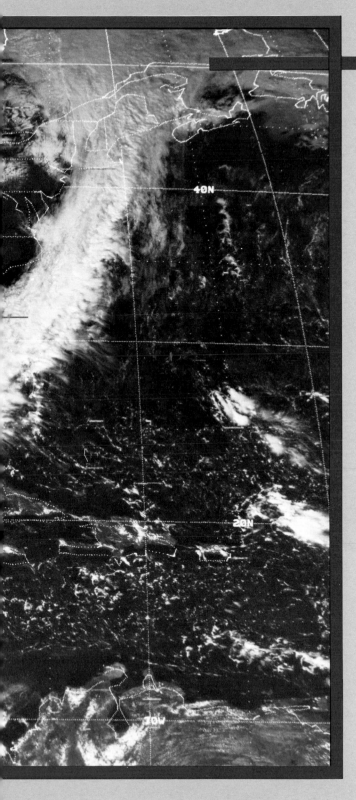

CHAPTER 7

AIR MASSES, FRONTS & STORMS

Air masses in conflict breed storms and it isn't difficult to see on this satellite image where the action is. Dry, bitterly cold air from interior Canada is being driven down across the central plains, midwest and far into the south to encounter warm, moist, tropical air. The line of contact is the sharply defined cloud line called a Cold Front. One may be certain that atmospheric turbulence and severe precipitation of one kind or another is rampant along the front, from New England to the West Indies.

INTRODUCTION

Following World War I a major revolution in meteorology and weather forecasting took place with the introduction of the radical Norwegian school of air mass analysis. It was hailed as an important step forward, and by the 1930s was adopted almost universally. Today, with relatively minor modifications, it remains the most widely accepted basic concept in middle latitude meteorology, although it is being critically tested, especially in connection with new analytic developments such as satellites and computers. There is also some question as to its worth as a research vehicle in the tropics.

AIR MASSES

The key to this system of forecasting/analysis is the recognition and classification of the world's air masses. When we speak of *air masses* we are designating something relatively specific—*a great body of air, continental in size, which displays a singular homogeneity of both temperature and moisture.* And since the earth itself transmits these characteristics to the lower atmosphere from its surface, we are, of course, also talking about a considerable earthly region or *source area* that exhibits more than a little homogeneity of its own. There is one further element to be considered, and that is *time*, for a massive body of air cannot truly reflect the character of the surface it overlies without some reasonable time in residence; that is, the source area should be a quiet haven over which the air mass may linger.

Air masses do not exist just anywhere, and they are not just random bits of atmosphere.

Air masses do not exist just anywhere and they are not just any random bit of atmosphere. In order to develop properly, they have a basic set of requirements that appear to be satisfied most effectively in the permanent high-pressure regions of the world; a tranquil, downward, vertical displacement of air with dispersing surface flow is in strong contrast to the confluent air currents and inherent turbulence of the lows. The *high latitudes* with their generally low temperatures and the permanent high-pressure cells in the *subtropics* would seem to be ideal, and it is here that we isolate the basic sets of air masses for our theorizing (Figs. 7–1 and 7–2).

Because the source area is of such significance in identifying and giving character to the air mass, our classification scheme is as much one of source areas as of air mass, for the two are essentially inseparable. There are two basic air masses—Polar and Tropical—with the strong differentiation of temperature immediately obvious. They are in turn subdivided into continental and maritime, and if it becomes necessary these can be subdivided even further. The symbolism at this point is simple:

cP = Polar continental

mP = Polar maritime

cT = Tropical continental

mT = Tropical maritime

Let us see precisely what these stand for.

Polar Air

Polar Continental.   If we start with *Polar continental* air, the name itself tells us something of location and inferred character. The classic source areas for this air mass are the broad high-latitude stretches of Siberia and Canada. Typically frozen and dusted with snow through a long and dominant winter season, any air mass overlying these regions must become thoroughly chilled. Even the short summer is not actually hot. Thus one fundamental characteristic of cP air is cold—exceptionally so in winter, modified slightly in summer. And since chilled air cannot hold moisture even if the frozen continent could offer any for assimilation, the cP air is dry as well as cold. Again, summer is different. As the snow melts, river ice goes out and marshes and tundra bogs become common features; the somewhat warmer air gains a shallow layer of humidity through increased evaporation. But all of this is relative since we have a second category of Polar air, mP or *Polar maritime*, which is patently moist. If we must characterize cP

The Bergen School of Meteorology

Vilhelm Bjerknes.

There was nothing fancy about the Meteorology School in Bergen's Geophysical Institute. It operated out of a second-story workroom in the home of the director, Vilhelm Bjerknes. But here, during the WWI years 1917–19 in neutral, isolated Norway was spawned a new and innovative concept of ordered air masses, travelling cyclones faithfully following predictable tracks, and fronts (the term borrowed from wartime usage signifying a line of contact and inevitable violent conflict). Bjerknes, his 22 year-old son Jacob and dedicated assistants Tor Bergeron and Halvor Solberg, working at their spare wooden benches with minimal data at hand, brilliantly revolutionized the entire theory of middle latitude atmospheric behavior. It was in retrospect to be termed the genesis of "modern meterology."

Wilhelm Bjerknes (1862–1951) was a physicist by profession who, in 1903 at the age of 41, made the decision to turn his attention to a study of the atmosphere. His aim was to achieve a better understanding of northwest Europe's mercurial weather. It had been recognized for many years that a continuing succession of low pressure centers marched relentlessly out of the grey and blustery North Atlantic to sweep across Scandinavia, and that their arrival brought wind, clouds and rain in varying and sometimes hazardous admixtures. Bjerknes proposed a theoretical study of this observed behavior and the application of physical laws to explain its origin and character—his interest was not in practical forecasting. But a forecaster is what he became, yielding to the vicissitudes of the time. The magic years of 1917–23 produced, for the first time in history, a reliable system for forecasting weather as much as 5 days in advance.

Gratified, but far from exuberant, Bjerknes left these practical matters to others. By 1926 he was back to teaching physics at the University of Oslo, and ever the theoretical scientist, puzzling out new and provocative research conundrums in his laboratory. But he left a legacy to the world, a legacy of understanding the complexities of atmospheric behavior which has been the solid base for all subsequent meteorological research.

air with a minimum of equivocation we say it is *cold* and *dry* (Fig. 7–3).

Polar Maritime. Now we come to mP air masses that could not be called Polar if they were not cold, but whose source areas are the warmer high-latitude oceans (not including those that are habitually frozen): the North Pacific/Bering Sea, the North Atlantic/Norwegian Sea, and the entire sweep of Southern Hemisphere oceans fringing Antarctica. This air, overlying unfrozen seas with frequent probing warmer currents, simply cannot be as cold as the continental masses and as a result is a far more efficient evaporator of readily available moisture. So mP air is *less cold* and *more moist* than cP air.

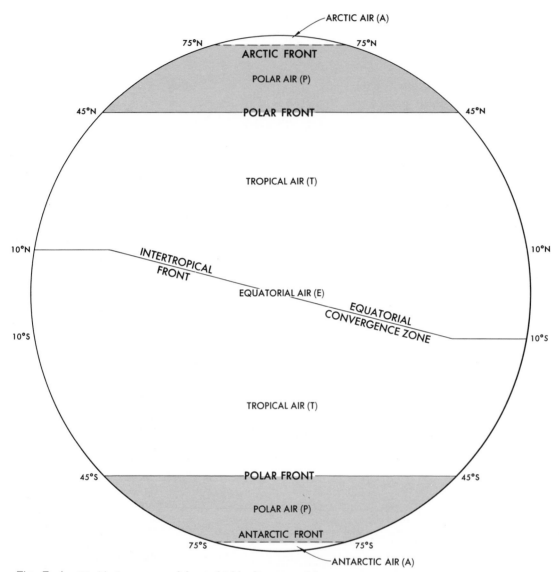

Fig. 7–1   World air masses and fronts (highly diagrammatic).

Arctic Air.   A further distinction is often made by many students of high-latitude meteorology, that is, to identify an *Arctic* (A) air mass. Its source areas are Antarctica and the permanently frozen Arctic seas. Such an air mass would, of course, have much in common with Polar air, but its advocates point out that a thoroughly recognizable difference exists between A air and the warmer and more humid cP summer air mass/mP air mass, and thus an Arctic cat-

egory is viable. If we were to characterize it, Arctic air would be described as *extremely cold* and *extremely dry*, without any semblance of seasonal change.

Regions Affected.   In the United States we commonly experience two of these, cP and mP, both exclusively in the winter when there is a strong tendency for Polar air to push out and invade neighboring territory. In the

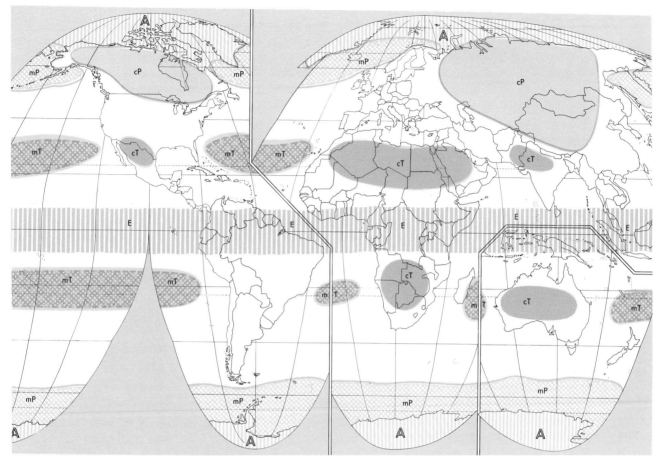

Fig. 7–2   World air masses and source regions.

interior of the country and on into the northeast coastal region, cP air from the north Canadian plains is a regular winter visitor. It overlies the Dakotas and parts of Minnesota, Montana, Wisconsin, and Michigan for long periods of time, and these states might even be regarded as marginal fragments of its source area. Elsewhere in the northern half of the country, cP air is experienced as cold spells as the air mass surges outward in waves or lobes only to retreat again after a short occupancy.

The West Coast is normally protected from invasion by the generally prevailing Westerly circulation and by the Cascade barrier. Cold air moving along the ground in a moderately shallow layer has difficulty overcoming high continuous mountain cordilleras—although it can flow

around compact masses or through the gaps and passes of broken chains. However, the general "grain" of the land in North America, with its north/south trending mountain ranges, is highly conducive to a ready invasion of most of the nation by cold air from the Canadian interior.

cP air brings with it the cold dry character of its source area. It produces sunny crisp days when one can dash from the house to the barn and back in shirtsleeves without great discomfort, only to discover frozen ears.

mP air sweeps in commonly along both coasts of the northern United States. It is a great deal warmer than cP and is also usually close to the point of saturation. A blustery day in Boston or Ketchikan, with the temperatures in the 20's (F) and a cold sleet coming down, chills one to

Fig. 7–3 Winter totally dominates each year in the far Canadian interior. This ice fog is representative of the still and algid cP source region.

the marrow and for most people is distinctly more uncomfortable than frozen ears in eastern Montana.

Europe is exposed to these same two air masses (there is never any Arctic air as far south as Europe or most of the United States) but on a slightly differing scale. Eastern Europe is closest to the Siberian cP source area and experiences the coldest temperatures. But despite the east/west trending character of the European mountain ranges and thus a seeming invitation to cP invasion, this "grain" of the land allows instead easy access to strong Westerlies supplying a continuous mass of mP air from the North Atlantic. mP air along with the Mediterranean and Baltic warming and humidifying influences, generally keeps the cP air at bay.

Western Europe is open, nonetheless, to occasional devastating cold spells of some duration when this balanc-

ing system temporarily breaks down. But it is interior Asia and the Far East that feel the brunt of the greatest of all cP air masses during the winter. Only Japan, Korea, and the immediate coast of eastern Asia experience the slight amelioration brought in by North Pacific mP air.

There is no cP air in the Southern Hemisphere although a great A mass overlies Antarctica; the lightly populated lands of the southern ocean normally experience only mP. South America is closest to Antarctica and in addition thrusts its high Andes athwart the Westerlies, so that mP air masses are common there much of the year. One has only to read Darwin or the accounts of myriad ships' logs as they describe the weather of the Strait of Magellan to appreciate the chill winds and high humidity that typify this part of the world. Even as far north as the Argentine Pampas there are occasional equatorward surges of mP air

Fig. 7–4   By far the world's greatest cT air mass develops over the 3,000,000 square mile (7,769,964 km²) superheated Sahara.

masses. Australia and South Africa are barely within range, but isolated New Zealand experiences mP air as a frequent winter visitor at its southern extremity.

Tropical air has the capacity if not always the opportunity to carry immense quantities of water vapor.

Tropical Air

Tropical air is warm, the precise opposite of the biting polar cold, and by virtue of this, Tropical air has the capacity (if not always the opportunity) to carry immense quantities of water vapor. Its source areas are two: the permanent but somewhat migratory oceanic high-pressure cells of the Horse Latitudes and the great low-latitude deserts. Both are warm but there are some fairly important differences. We point

these up by distinguishing between cT *(Tropical continental)*, the product of the desert source area, and mT *(Tropical maritime)*, overlying the seas.

Tropical Continental.   The cT air mass is basically *hot* and *dry*. Heat may vary a bit with season but the dryness remains immutable. Perhaps the type example is the extensive Northern Hemisphere air mass that develops over the greater Sahara source area, the epitome of heat and aridity (Fig. 7–4). However, since worldwide tropical deserts are frequently limited to narrow coastal strips, only two other source areas can even approach the Sahara in size or climatic extremes: central Australia and southern Africa.

This is a desiccating air mass that when it ventures outside its source area, literally "burns up" the populace and vegetation as it goes. The dreaded *Sirocco* that periodically blights Mediterranean Europe, is cT air drawn out of the Sahara by a passing northerly cyclone. The United

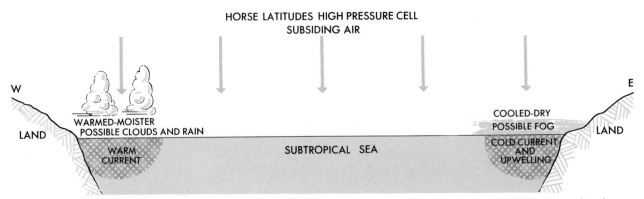

Fig. 7–5   Tropical maritime air. The Tropical oceanic cell displays a drier, more stable eastern extremity as opposed to the greater humidity and cloudiness of the west.

States is seldom afflicted by the invasion of true cT air because of both the elevation and lack of bulk of the northern Mexico source area. Southern Californians nonetheless know and fear the *Santa Ana* wind that occasionally brings small-scale quasi-cT air from the southwest desert.

Tropical Maritime.   mT air is considerably more widespread than cT because of the greater size and prevalence of its source areas. The Pacific Ocean is by far the world's largest single physical feature. In the Southern Hemisphere it is all-encompassing at every latitude and in the Northern Hemisphere, although limited in breadth at the far north, it reaches almost half the earth's circumference in the tropics. The North and South Atlantic as well as the Indian Ocean are also extensive source areas for mT air.

We would anticipate that mT air masses overlying tropical seas would be moist to the point of saturation, but this is not the case universally. One must distinguish between the eastern and western sections of the Horse Latitude high-pressure cell in each ocean basin. At the eastern extreme, decisively subsiding air, warming adiabatically as it descends, becomes slightly chilled as it encounters a sea surface cooled by coastal upwelling and encroaching cold

current. The resulting relatively drier and more stable air contrasts to a degree with the other end of the same mT air mass, where a warm current contributes to warmer and more unstable surface air. This acts to counteract partially the general dehydration of subsidence (Fig. 7–5). However, despite these east–west aberrations, we classify mT air as *warm* and *moist*, and when elements of it push beyond the source area they carry these characteristics with them. One variant, called locally *Tropical Gulf* air (gT), regularly flows each summer up the Mississippi valley to introduce an unwelcome sample of the maritime tropics far into the continental interior. And nominally moderate Shanghai, Sydney, and Buenos Aires, all located at the western extremities of ocean basins, experience the hot stickiness of saturated mT air as oceanic highs shift poleward with the summer sun.

Equatorial Air.   There is also in the tropics a supposed *Equatorial* (E) air mass that some (those who do not recognize an Arctic air mass) will disclaim as too similar to Tropical air to be identified properly. If it exists, it is the child of the Doldrums and is at all times *hot* and *saturated*. Obviously, such an air mass is difficult to separate from western mT, but it does differ, at least slightly, from eastern mT and decisively from cT air.

Modifying Air Masses

Because an air mass is so large we have discovered that a fair amount of time is required for it to achieve a temperature/moisture equilibrium with its source area. It should

The dreaded Sirocco, *that periodically blights Mediterranean Europe, is cT air drawn out of the Sahara by a passing northerly cyclone.*

follow, then, that when a marginal wave probes beyond the source area proper and invades surrounding territory it carries with it the basic characteristics of its home region and tends to resist change. Further, it should be made clear that when we talk of air masses departing the source area we never envision the entire mass, or even a substantial part of it, flowing off and completely abandoning its home. Instead we are dealing only with these peripheral pulsations, advancing and retreating on a relatively short-term basis. But the margin of an air mass is generally shallower than the main body, and if it surges too far or stays away too long it is subjecting itself to ultimate mutation.

In order to point up such an event we have adopted a symbolism for two of the most common types of air mass modification.

A "k" is added to the air mass symbol (as mTk) *when the invading air is colder than the surface over which it is flowing* and the mass is being heated at the bottom (k derives from the German word for cold, *kalt*). A "w" is added (as mTw) *when the invading air is warmer than the surface over which it is flowing*.

This would mean that in the first instance, moist Tropical maritime air perhaps had moved in from a cooler sea over a summer-heated continent. The air mass would be warmed at its base, begin to exhibit instability, and thundershowers might well develop. "k" then means unstable. When the "w" is added we visualize light and buoyant Tropical maritime air encountering a cold surface with chilling at the bottom and the development of an inversion. This would be a stable air mass with fog or smog often resulting.

Therefore although much of the theory of air mass interaction is based on the recognized ability of the mass to retain certain reasonably innate temperature/moisture characteristics as it travels abroad, we must concede that change is possible, even extreme change to the infrequent point of complete assimilation of the alien.

FRONTS

As we have gone through the process of locating and labeling air masses, we have identified high-latitude cold air masses of moderately varying character, and low-latitude warm air masses, but nowhere have we encountered moderate, lukewarm, or tepid air masses. This is understandable because *there simply are no middle-latitude air masses.* In this critical region in both hemispheres we find continual conflict between the incessantly advancing and retreating margins of the masses on either side. This is where intensely differing air masses meet, a line of juncture called a *front*.

Where intensively differing air masses meet, there is established a line of juncture called a front.

One of the features of any front is a strong disinclination of heterogeneous air masses to mix readily—like oil and water, each maintains resolutely its individual identity. To be absolutely accurate we probably should think in terms of frontal zones of 50 to 100 miles (81 to 161 km) in most cases instead of a definitive "line," nonetheless, a front is generally the antithesis of the usual experience in nature where gradual transition is the rule. Here it is essentially a very sharp line between strongly divergent elements and as such is a line of conflict and atmospheric dynamism.

Polar Front

It is significant that the Norwegians originated the concept of air masses and fronts, for their weather, coming to them from off the stormy Norwegian Sea, is usually bad and highly unpredictable. Their concern with comprehending the ingredients of weather and with establishing some sort of scientific means of forecasting is understandable. So when they finally got their whole weather picture together and all of their basic air masses established, it turned out that Norway (and the middle latitudes in general) was directly astride the *Polar Front* much of the year.

The Polar Front is the most decisive of all the world's fronts. It separates Polar and Tropical air, and anyone who establishes permanent residence within its zone of influence is automatically going to encounter "interesting" weather. Warm moist Tropical air is forced to skid up over cold dry Polar air, thereby releasing a great deal of latent energy through condensation. This is the beginning of a whole sequence of events leading to storm formation. In

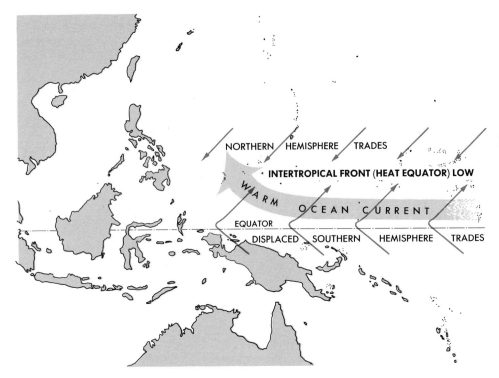

Fig. 7–6 Intertropical Front (equatorial convergence zone). September/October in the western Pacific. The "heat" equator is well north of the geographical equator following four months of a near overhead sun reinforced by a massive northward-probing warm ocean current. The low pressure developed here draws the Southern Hemisphere Trades across the equator and they are turned 90% to the right by Coriolis force to meet the northern Trades head-on.

fancy language, the Polar Front is an active line of cyclogenesis.

Arctic-Antarctic Fronts

There are other fronts too. Remember the Arctic/Antarctic air masses and their somewhat subtle differentiation from the various kinds of Polar air. A couple of problems assert themselves immediately: (1) nobody really knows very much about Arctic/Antarctic meteorology and (2) no matter how one conceives of it, even the relatively warm and moist mP air is really not very different from the A air mass. But if we are willing to concede a mild and sometimes seasonal *Arctic/Antarctic Front*, then it is most clearly defined around the periphery of Antarctica and at the permanent ice line of the restricted Northern Hemisphere ocean basins. In summer it shows up to some degree along the north coasts of Canada and Siberia as well.

Intertropical Front

There is also a theoretical front in the tropics separating E air (if there is such a mass) and T air. It is the rough equivalent of the Doldrums and that is scarcely a front. But if we visualize the general convergence of the Trades from both hemispheres as bringing air masses into conflict, then at least we have a zone something like a front (Fig. 7–6).

The term *Intertropical Front* was introduced into meteorology following the successful application of air mass analysis and frontal theory in the middle latitudes. It was postulated that as the heat equator shifted north and south of the geographic equator with the seasonal migration of the overhead sun, the Trades from the winter hemisphere would be pulled across the equator. Not only would they then have a longer trajectory than the summer hemisphere Trades and thus exhibit greater surface modification of moisture and temperature, but they would also react to the opposite hemisphere's Coriolis force and shift their direction of flow by some 90°. The end product, the Intertropical Front, would appear where the two trade-wind currents met head-on and the lighter was forced up over the denser. Temperature differences are not great, and the Intertropical Front is certainly less well defined than the Polar Front, disappearing entirely over the continents and blurring periodically at sea. And although we are using here the term "Intertropical Front" since it has become fairly well established, there are those who argue persuasively for the substitution of *"Intertropical (or Equatorial) Convergence Zone"* in its place. But whatever the name, all agree on the existence of a tropical trough (low pressure) at the point of theoretical meeting of the opposing Trades; and further, it is agreed that this front marks a line of precipitation, cloudiness, and apparent storm genesis.

STORMS

Just about any low-pressure center can be called a storm, so long as it is localized on something less than a continental-size scale, establishes its own spiraling air circulation, and can be measured on a barometer. It need not blow the roof off the barn before it becomes a legitimate storm; wind velocity is not the measure of nomenclature here. *Cyclone, low, depression,* and *storm* are all good terms and fairly synonymous in our usage. Within the category of storm, however, there are other names describing more specific types. A hurricane or typhoon is certainly capable of taking the roof off the barn, and a tornado will blow the barn away and all the cows as well; but a storm can be a much milder version with modest breezes and a little drizzle as its most violent features.

It need not blow the roof off the barn before it becomes a legitimate storm; wind velocity is not the measure of nomenclature here.

However, storms of whatever persuasion must be generated in some fashion and here we come to the significance of fronts. Most cyclones are developed as a result of unlike air masses in conflict, so we can say that storms display a distinct affinity for the world's frontal zones. By way of systematizing this relationship, we will attempt to correlate the most frequent and important of our storms with the major fronts. But first a word of qualification: all cyclones are not of frontal derivation. The common convectional thunderstorm, which is sometimes very violent, is

an example. Also, there is so little hard data available on certain fronts, like the Arctic and Antarctic Fronts, that any related storm behavior is strictly speculative. Nevertheless, the following storm/front relationships are clearly cogent:

Polar Front

1. Standard middle-latitude cyclones.
2. Tornadoes.

Intertropic Front

1. Weak tropical cyclones.
2. Hurricanes/typhoons.

On or near the *Polar Front* are found what we can call *standard middle-latitude cyclones.* Occasionally, the term *extratropical storms* is applied, simply meaning that their occurrence is "outside" of the tropics. These are large [up to 1000 miles (1609 km) in diameter], slow-moving, relatively mild (as far as wind velocities and general destruction are concerned), and quite common throughout the year. In opposition to these storms in the general area of the Polar Front, there appears a second type of cyclonic disturbance, *tornadoes.* These are different in every way from the standard middle latitude cyclones in that they are the most violent and destructive storms known. They occur in only limited regions at only certain times of the year, are extremely small in size, and move very rapidly.

The *Intertropical Front* also is related to two types of unlike storms: *weak tropical cyclones,* which are large, mild, and common, and *hurricanes or typhoons* (the appellation differing for the same type of storm in various parts of the world), which are small, seasonal, violent, and occur in only certain limited locations.

Jet Streams and Upper Air Waves

Involved in the generation and movements of these various storms to some degree are the so-called jet streams. First discovered during WW II by high-flying aircraft whose pilots were greatly chagrined to find that they were making almost no progress despite full power ahead, the jet stream is a high-velocity upper air-current flowing from west to east roughly above the Polar Front. It is erroneous to speak of a single jet stream, for there are many that are known, but only the original Northern Hemisphere jet has been studied with some care. And even here, much remains to be learned for, obviously, voluminous data on a writhing subject near the top of the troposphere are hard to come by. Nonetheless, enough reliable work has been done to more than hint at a considerable relationship between the various jet streams and the moisture and temperature characteristics of surface climates, and particularly of storm genesis. They are being introduced here because of the almost certain influence of the *upper Westerly jet stream* on the storms of the Polar Front and vicinity, and a very probable relationship between the lesser known *tropical jets* and the cyclonic disturbances on and near the Intertropical Front.

The Northern Hemisphere upper Westerly jet stream varies in both height and velocity with season, being lowest [4 to 5 miles (6 to 8 km)] and strongest [200 to 400 mph (322 to 643 km/hr)] in winter and highest [7 miles (11 km)] and weakest [50 to 75 mph (81 to 121 km/hr)] in summer. It is a narrow twisting band of air meandering wildly at times and often breaking down into several parallel elements rather than a single current. Its main course is much farther north in summer than winter, suggesting an affinity with the similar advance and retreat of the Polar Front below it or the seasonal surface temperature variations that control that front.

During the wild winter of 1976–1977, excessively dry in the west and bitterly cold in the east, the jet stream was credited with directing Pacific storms on a far northerly course into Canada. Once clear of the Rockies, these storms come swooping southeastward into the United States bringing with them Pacific moisture far from home in the Polar Continental Canadian source region. A part of the problem seemed to be a major, winter-long, high-pressure cell persisting off the West Coast, the apparent result of a huge oceanic pool of cold water. At the same time a comparable warm pool was identified farther out to sea. Precisely why these wayward waters ambulate remains unexplained at the moment, but the culmination seems to have been that water vapor accrued from the warm Pacific pool was dumped on Buffalo in crystalline form 12 feet (3.7 m) deep. The jet stream, Polar Front, and storm tracks were all reacting in concert to aberrant surface temperatures.

A similar jet exists in the Southern Hemisphere middle latitudes, presumably acting much as its Northern Hemisphere counterpart, although the lack of large conti-

nents and resultant differing air masses undoubtedly causes variations. Intermittent jets too are known. Thirty miles (48 km) above the poles we find the *polar night jets*, active during six months of bitter winter cold when the sun remains below the horizon, but disappearing with its reappearance. And over the Indian Ocean is a reverse jet stream that changes its direction 180° as surely as do the surface monsoon air currents in response to seasonal heating and cooling of the Asiatic continent.

In addition to jets, there are other kinds of air motions that seem in one way or another to affect storms at the earth's surface. The upper Westerlies, for instance, feature a series of recognizable long waves below the level of the jet. They move about in a quasi-predictable pattern and appear to be an element in the basic atmospheric heat exchange between the polar and equatorial regions. Sufficient research has been accomplished in recent years to point to an unequivocal relationship among these long waves, the jet stream above them, and the formation and movement of low-level cyclones. A weaker but somewhat similar wavelike pattern has also been identified in the Trades, which may play some part in the development of hurricanes.

The fronts themselves are therefore not the sole factor in developing cyclones. Undoubtedly these upper-air perturbations are in some complex manner manifestations of the fronts.

Polar Front Storms

Standard Middle Latitude Cyclones.

Of the four storms previously mentioned, the standard middle latitude cyclone is by far the best understood. Because it has been the major weather maker in the United States and western Europe where the bulk of the world's scientists are concentrated and because of its frequent occurrence and mild slow-moving character, it has been carefully studied and analyzed for many years.

According to the Norwegian frontal wave theory, standard middle latitude cyclones may form along the Polar Front wherever a wave or protuberance of one air mass advances against the other. If cold air advances against warm or vice versa, the warm air is forced aloft and a low-pres-

Soviet Agricultural Planning

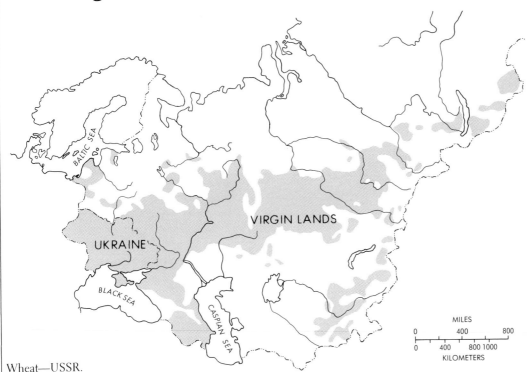

Wheat—USSR.

An attempt at practical application of wave theory has been going on in the USSR for some time now. During the years 1954 to 1960, 100 million acres (40,468,600 ha) of new wheat fields were plowed and planted east of the Volga River and on across the steppes of Kazakhstan—Khrushchev's "Virgin Lands". Traditionally, the semiarid Ukraine, 1500 miles (2414 km) to the west, had been the Russian bread basket. But a combination of increasing population and diversification of Ukrainian farms to crops other than wheat had forced the opening up of new farm land, and despite its marginal rainfall (comparable to the American "Dust Bowl"), the decision was made to go ahead. The planners knew very well that there would be failures, perhaps even 50% over the years, but they reasoned that the 1500 miles (2414 km) between the Ukraine and Kazakhstan was exactly one half the width of an upper troposphere wave length. This meant that, in theory at least, both regions would not be subject to drought at the same time—and, of course, both would not achieve maximum rainfall at the same time either. As it has worked out over the last 30 years, the facts have supported the theory, at least roughly. There has never been a national bumper crop nor has there yet been an across-the-board failure.

sure center comes into being. Around it, the air begins to circulate in a typical Northern Hemisphere counterclockwise spiral and a storm is born.

Certain locations along the front appear to be particularly active in cyclogenesis: the Texas Panhandle, the Gulf of Lions, and the upper Yangtze Valley, to name but a few. But outstanding in this regard, producing more and bigger storms than all other Northern Hemisphere sources combined, are the two permanent Subpolar Low cells, the Aleutian and Icelandic centers. These spew out a steady stream of cyclones, especially in the winter. In the Southern Hemisphere where the Subpolar Low is a belt in the uninterrupted sea, storms typically form almost at random anywhere astride the Polar Front.

Once the cyclone is formed on the Polar Front, its general slow movement is to the east as it is pushed by the Westerlies, its course determined by the front itself and the jet stream above it. As long as the center of the storm "rides" the front, it can maintain itself, for each cyclone depends on the interaction of both cold and warm air to sustain its vigor. If it were to "jump the track," that is, move to either the warm- or cold-air side of the front, the interaction of warm and cold air would cease and the storm would die.

As long as the center of the storm "rides" the front, it can maintain itself, for each cyclone depends on the interaction of cold and warm air to sustain its vigor.

Imagine a cyclone astride the Polar Front in the Northern Hemisphere. The air would spiral about the low-pressure center in a counterclockwise circulation. On the western side of the storm this would cause cold air to move against warm at the front, forcing the warm air aloft, and very likely precipitation would result. This segment of the Polar Front within the storm where cold air is advancing against warm is called the *Cold Front*. On the eastern side of the storm, warm air would attack cold at the front. Again, the warm air would flow over the cold, expand, and cool as it rose, causing precipitation. This portion of the Polar Front where warm air is forced against cold is called the *Warm Front*. Thus the Cold Front, the Warm Front, and the center of the storm where air is rising vertically are the precipitation zones within a cyclone. However, the pattern and character of precipitation differ slightly between the two fronts (Fig. 7–7).

At the Warm Front the advancing light and buoyant warm air experiences a great deal of difficulty in pushing cold air back, but in flowing up and over it, modifies the cold air into a thin, low-angle wedge. Warm air moving up this gentle slope is cooled slowly. Stratus clouds form along the line of contact, and a broad zone of drizzle or snow flurries develops on the cold-air side of the front.

On the western side of the storm where cold air is advancing against warm at the Cold Front, a somewhat different situation prevails. Since cold air can move against warm easily, the depth of the cold air may be considerable, and its leading edge will be close to vertical as opposed to the thin wedge slope at the Warm Front. This means that as the warm air is forced aloft, it follows up this steep leading edge, cooling rapidly and producing high cumulus clouds and heavy precipitation. Frequently, the cold air above the surface moves forward more rapidly than that which is in contact with the ground and slowed a bit by friction, and warm air is trapped below this overhanging bulge. When the trapped warm air bursts aloft suddenly, thunder, lightning, and torrential precipitation result. Because of this, the Cold Front is sometimes referred to as the *Squall Line*. Note that the precipitation zone is rather narrow and in advance of the Cold Front. There may also be precipitation of a much less vigorous character behind the front (or in the cold-air sector) as the warm air swings up over the cold air.

The chief differences between Warm and Cold Front precipitation are these:

1.   Warm Front precipitation is of the light-drizzle variety, while that of the Cold Front is likely to be much heavier.

2.   Warm Front clouds are usually stratus, while Cold Front clouds are cumulus, often thunderheads.

3.   The Warm Front precipitation zone is broad and wholly on the cold-air side, while the Cold Front precipitation zone is quite narrow and is astride the front with the greater part of the precipitation occurring in the warm-air sector.

The Cold Front is crisply defined on the home TV image as a narrow band of cumulus clouds reflecting a great deal of light for the satellite sensor to record.

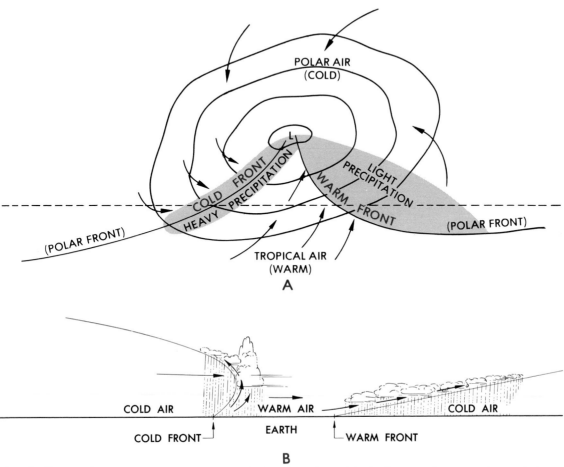

Fig. 7–7 Warm and cold fronts—standard middle latitude cyclone. (A) The fronts as they appear in a typical Northern Hemisphere storm. (B) A cross section of the same storm (along the dashed line).

When the television weatherperson shows us the daily satellite image of winter storms marching in from out of the north Pacific, the protean cloud pattern inevitably reveals two constants: the long trailing southwesterly cold front and a counterclockwise cloud spiral about the low pressure center of the storms. Warm Front clouds, a result of not very decisive lifting of warm moist air, may or may not be there on all occasions. And even when they are apparent, look to be merely an undistinguished and disorganized blob at the cyclone's advancing margin. The Cold Front, on the other hand, is crisply defined, a narrow band of cumulus clouds reflecting a great deal of light for the satellite sensor

to record. Surely there will be precipitation when that element of the storm passes over.

Since the Cold Front advances more rapidly than the Warm Front, it is possible for it to catch up with the Warm Front after a storm has been in existence for several days. When this occurs, it is called an *occlusion* (or *Occluded Front*). As the Cold Front gradually moves in on the Warm Front, the warm-air sector becomes progressively smaller until finally, when the fronts come together, the last of the warm air is forced aloft. At the surface only cold air remains. The rising warm air above the front gives off precipitation, and the vortex at the storm's center may main-

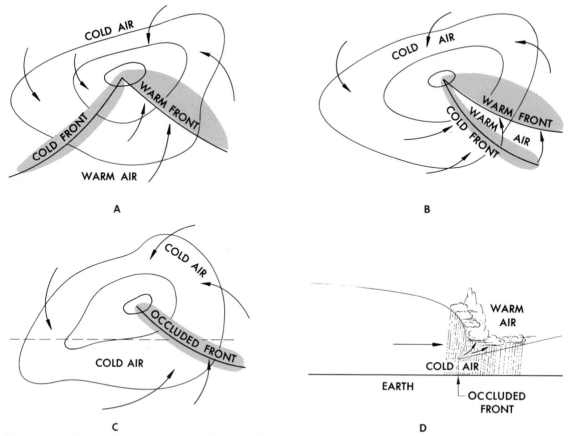

Fig. 7–8 Occluded Front. In (A) and (B) the cold front is advancing more rapidly than the warm, culminating in an occluded front (C); (D) is a cross section along the dashed line in (C).

tain itself weakly, but in essence the storm has "jumped the track"—it is no longer astride the Polar Front (Fig. 7–8). The important interaction of warm and cold air has ceased, and the storm usually dies out gradually. Every cyclone does not develop to the occluded stage, but the possibility is always present.

It is important to recognize that there are two basic air circulations involved in these standard middle-latitude cyclones: (1) the primary Westerlies forcing the general movement of all storms in the middle latitudes from west to east and (2) the local spiral winds about each individual low center. It is this steady parade of slow-moving storms migrating through the middle latitudes of both hemispheres that gives the basic character to our day-to-day weather.

Tornadoes. The second of the storms associated with the Polar Front, the tornado, is quite different from the standard middle-latitude cyclone, and certainly less well understood. Its sudden appearance without much advance warning, violent character, small size, and rapid movement make it extremely difficult to study. What we know of tornadoes is pretty much a matter of merely cataloging the frequency, time of occurrence, and location of all known tornadoes over a period of many years. Why they occur and when and where they do is speculative at best.

Tornadoes are largely all-American storms, the greatest frequency occurring in the central plains states from the Gulf to Canada. They also occur somewhat less commonly east of the plains in the South, Middle West, and even now and then along the eastern seaboard into New

England. West of the Rockies, tornadoes are virtually unknown. The rest of the world is not immune, Australia reporting the largest number, but by far the greatest tornado frequency is in the United States.

Late spring to early summer is the period when most of these storms manifest themselves, although the entire warm half of the year is subject to at least occasional tornadoes. The region of maximum frequency appears to exhibit something of a northerly seasonal migration; that is, the greatest number appear in the south in the spring and gradually move north with the onset of summer. By late summer the number of storms has decreased, but some of these may be experienced as far north as the Dakotas or the Canadian prairie provinces.

As we analyze the ever increasing numbers there is the illusion of more tornadoes than ever before. At least we hope it is an illusion.

It should be noted that with our development of radar as a tracking device, fewer storms are slipping by undetected as they often did in the past. Unless they tore up the countryside in transit we were unaware of them at all, and if they presented an immediate danger, no warning was given. As a frustrated Iowa meteorologist once observed "many of our weather reporting stations are only 9:00 A.M. to 5:00 P.M. operations, but many of our tornadoes are on the night shift." So today as we analyze the ever-increasing numbers, there is the illusion of more tornadoes than ever before. At least we hope it is an illusion. In any case the largest numbers continue to occur in May with June second and April third; and the south central plains states from Texas to Kansas are center stage in this annual drama.

Like the standard middle latitude cyclones, tornadoes move in a general easterly direction, but they are a great deal smaller and advance with greater speed. A typical tornado may be only a few hundred yards across, and tornadoes have been known to wipe out all the houses on one side of the street completely, while scarcely affecting those on the other side. Furthermore, they commonly exercise a curious pattern of skipping as they go. Although the funnel remains active and visible and torrential rains accompany it, the bottom of the funnel lifts off the ground for a while

before coming back down to carve out a path of destruction. Luckily, the storm is normally of short duration, and although in its short life it may run a great distance, it generally disappears abruptly after only a brief period of activity.

The accumulated observations of thousands of tornadoes show a definite pattern relative to their genesis. Almost all (90% of the average 400 per year in the United States) appear to originate somewhat in advance of the Cold Front of a standard middle latitude cyclone. Spring is the turnaround point in the annual North American "monsoonal" circulation. Cold dry Polar Continental masses are still far south of the Canadian border, but warm super-humid Gulf air is just beginning its incipient summer surge northward. At a Cold Front their sharp diversities are magnified. Temperature contrasts on opposite sides of the front seem to be of less importance than the moisture differential. If the warm air temperature is high enough to enable it to hold a large amount of water vapor and if its humidity is near saturation, then the advancing cool dry air, with the upper air moving more rapidly than that at the surface, will cause a violent updraft accompanied by extremely heavy and rapid condensation. Given these conditions in Kansas in May or June, a tornado or even whole clusters are likely. This is the best that the meteorologist can do—forecast tornado likelihood. Just how many will develop, if any, or exactly where is beyond the forecaster's knowledge. Once a tornado has shown itself, then we know that it will quickly outrun the front heading east-northeast, and developments from then on are in the lap of the gods.

Thunderstorms too are formed in this same zone ahead of the Cold Front, and some tornadoes appear to spring from them as they do occasionally from isolated thunderstorms elsewhere. And once again, we must credit the jet stream with an assist in storm formation, although general agreement has not been reached as to its exact role relative to tornadoes (Fig. 7–9).

The destructive power of a tornado is enormous and it is without doubt the most violent storm in nature. Wind velocities certainly are in excess of 200 to 300 mph (322 to 483 km/hr) but nobody knows how much because they have never been measured. The results of such velocities can scarcely be exaggerated, and weird stories are legion—straws driven through trees, babies snatched from mothers' arms, and on ad nauseam. But in conjunction with the wind are

Fig. 7–9   A clear weather dust devil has the look of a tornado about it. This is a particularly big one spiraling tightly about a compact low pressure center. It is the product of surface heating rather than frontal lifting and is probably closely related to the short-lived but spectacular water spout at sea. Both of these are capable of local damage but are not in the same league with a tornado.

Fig. 7–10   The tornado aftermath. Something violent has just gone through here!

other equally powerful forces. The lifting capacity of a tornado as the violent winds spiral upward can, for example, pick up houses off their foundations. The dark appearance of the storm funnel may be at least partially attributed to great volumes of soil, fences, and assorted livestock sucked up into the vortex. Also, since a tornado center is probably the deepest low-pressure center experienced anywhere on earth, as it suddenly envelops a building, the normal air pressure trapped inside is many times that outside and the building explodes. Only a steel reinforced construction with a high percentage of windows is relatively safe from this type of destruction as the glass blows out to relieve the pressure.

Often in the eyes of the general urban public, tornadoes seem remote and slightly rural in character. Dramatic reports come in from the countryside and crossroad hamlets—and damage and loss of life seems generally low—that is, because there is a lot more countryside with small towns than there are cities. However, be assured there is nothing at all to keep a tornado out of your town short of residing in another part of the world and/or sheer luck. In recent decades several sizable cities have witnessed devastation as they found themselves directly in the path of one or more tornadoes: Flint, Michigan, 113 deaths; Waco, Texas, 114 deaths; Worcester, Massachusetts, 94 deaths; Wichita Falls, 54 deaths; Tulsa, Omaha, and Xenia, all damaged severely (Fig. 7–10).

The odds makers tell us that the likelihood of a tornado occurring over any given square 50 miles (130 km) of area is about once in 1000 years, yet the good folk of Gainsville, Georgia, have been visited three times so far in less than 80 years: 1903, 28 deaths; 1936, 203 deaths; and 1944, 44 deaths. This is the nature of odds—many other towns in what we regard as prime tornado country have never been "honored."

Intertropical Front Storms

Weak Tropical Cyclones.   The dramatic hurricane has largely overshadowed any other type of tropical storm, both in the popular mind and in organized research. But there is a much more common type that is receiving belated recognition as not only a significant element in day-to-day tropical weather, but as the seed, so to speak, from which hurricanes flower. The simple name "weak tropical cyclone" lacks real character, but so does the storm that it designates. Scarcely visible on the weather map, a weak tropical cyclone is an amorphous entity, displaying no outstanding wind or temperature features.

*M*any towns in what we regard as prime tornado country have never been "honored."

It appears to be a product of the Intertropical Front and Doldrums, and distinguishes itself chiefly by exhibiting a continuous cloud cover and heavy rain. The mild temperature and moisture differential between the impinging trade-wind currents at the Intertropical Front are seldom sufficient to account for any major updraft of air, but saturated Tropical and Equatorial air masses require very little lifting to produce rain and clouds. These large but mild storms are common features the year round, and once formed, display a general tendency to drift aimlessly. Their progress is slow and halting, and they often stagnate, drip heavily, and after a few days or weeks of existence, unobtrusively fade away.

Hurricanes/Typhoons.   As unspectacular as the weak tropical cyclones appear to be, some of them, at certain places and particular times of year, develop for unknown reasons into very spectacular and dangerous hurricanes or typhoons. These storms are limited to the summer and fall seasons and occur almost exclusively in the western part of major ocean basins in both hemispheres. Specifically, these regions include: the southwest corner of the North Atlantic (Caribbean and southeast U.S. coast), the southwest corner of the North Pacific (Philippines to Japan and southeast Asiatic coast), the northwest corner of the South Pacific (northeast Australian coast and adjacent islands), and the northwest corner of the South Indian Ocean (Madagascar and the east coast of South Africa). The South Atlantic is excepted here, probably because the Intertropical Front does not move south of the equator as it does in the other oceans, for reasons we know not why. Also, less frequent storms may be encountered in the Bay of Bengal, Arabian Sea, off the west coast of Baja California, and the northwest coast of Australia.

Despite the difference in names, both hurricanes and typhoons are the same type of storm, typhoon being the Far Eastern name and hurricane applying in the Caribbean. Other local names also refer to the identical phenomenon, such as *cyclone* (India), *baguio* (Philippines), and *willy willie* (Australia). But whatever it is called, the storm shows up in the general region of the Intertropical Front and seems to develop from what first appears to be another simple weak tropical cyclone. The low center becomes deeper and deeper, the winds spiral about it with higher and higher velocity, and the general dimension of the storm becomes more compact [100 to 200 miles (161 to 322 km) across]. NOAA's Hurricane Warning Center in Florida has decided, rather arbitrarily, that when the winds reach 75 mph (121 km/hr), such a storm shall be classified as a hurricane, although as it continues to grow, winds commonly exceed 100 (161) or even 150 mph (241 km/hr). In spite of these high winds, the storm itself advances quite slowly, on the order of 5 to 10 mph (8 to 16 km/hr).

At one time, violent tropical storm formation was explained quite simply as the result of the interaction of contrasting air masses at the Intertropical Front. In the summer this front migrates a good many degrees beyond the equator in all western ocean basins except the South Atlantic, and nearly all hurricanes/typhoons were observed to originate in this general vicinity during the warm season. However, it has become increasingly obvious with the discovery of tropical jet streams, high-level fronts, and trade-wind upper-air waves that storm development is quite complex, although the details have not been wholly worked out. Like the tornado, the hurricane/typhoon is a bit too robust to study easily, but hurricane-hunter planes have been flying into the centers of these storms for several years in an effort to unravel their mysteries; and the pioneer

TIROS and much more sophisticated Nimbus weather satellites will undoubtedly supply us with a great deal more data than have been available in the past.

We do know that the initial appearance of these storms 8° to 10° off the equator is related to the requirement of sufficient Coriolis force to allow the violent spiraling of the winds about a low-pressure center. The equatorial regions are thus exempt. Heavy concentrations of hurricanes/typhoons on the western sides of oceans may partially be a result of the generally exaggerated poleward shift of the Intertropical Front here, but also because of somewhat higher air and water temperatures. Since the vigor and driving force of this type of storm undoubtedly derive from the latent energy that is released as warm saturated oceanic air is sucked up the vortex to be condensed into torrential showers, the warmer the air and the higher its moisture content, the greater the energy potential.

One of the curious features of the hurricane/typhoon that has long been observed but never explained in adequate fashion is the *eye*—a warm, calm, cloudless region at the very center that has no counterpart in any other cyclone. Any number of first hand accounts are available describing the sudden cessation of the wild wind, warmer temperatures, and the clearing of the sky, with only a heaving sea to hint of storm conditions. This is the eye and its passage is of short duration, with the wind picking up abruptly and this time blowing 180° from its previous flow. Only a column of sinking air could account for all these characteristics of the eye; yet it is unusual, to say the least, to encounter sinking air in the midst of a brisk vortex that is obviously accomplishing mass uplift of huge volumes of moist air (Fig. 7–11).

Once the storm is fully formed, it begins to follow a typical path—a path that is reasonably predictable in its general outline, but wholly unpredictable as to detail. Impelled by the easterly Trades, hurricanes/typhoons, irre-

Fig. 7–11 Filling the Gulf of Mexico from Yucatan to the mouth of the Mississippi, a huge hurricane, with eye clearly visible, sidles slowly toward the northwest. Although its forward progress may be leisurely it spirals about its eye at well over 100 miles per hour (161 km). This one finally came ashore near the U.S. Mexican border, dripped mightily, carried some things off downwind, and finally blew itself out far inland.

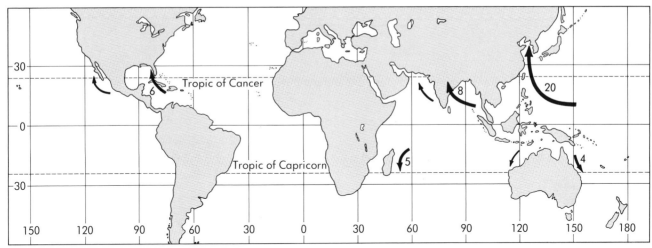

Fig. 7–12   Average hurricane tracks. The number represents annual frequency.

spective of hemisphere, begin a slow movement westward. At this stage they may very well gain in strength and intensity, for they are still over tropical seas and tap a constant supply of warm moist air. But sooner or later, a high percentage of these storms begin to curve away from the equator, generally following an approximation of the western end of the permanent oceanic Subtropical High cell. No two are exactly alike and herein lies the difficulty in forecasting. On rare occasions, one may not curve at all; others will curve at varying rates; and some may even stop or form a loop in their course. The disastrous Cuban hurricane of 1963 crossed the southern end of that island and then turned and recrossed it before resuming its northward curve (Fig. 7–12).

It is unusual, to say the least, to encounter sinking air in the midst of a brisk vortex that is obviously accomplishing mass uplift of huge volumes of moist air.

There are two ways to stop a hurricane; both involve the removal of the moist air that is its fuel. If it invades a large land mass, there is an abrupt decrease in the moisture content of the air sucked in at its base, and the storm will fade out, often very rapidly. Or if it survives at sea to complete its poleward curve, it will finally recurve at higher latitudes and move to the east. Here the ocean is colder and the air above it less capable of holding moisture, so that a tropical storm at these latitudes will gradually become a mere Westerly squall.

The high winds accompanying violent tropical storms are the obvious destructive force that accounts for the awe in which they are held, especially in those regions of the world where such insubstantial construction materials as nipa and bamboo are the rule. But the wind is not all that is to be feared. Heavy seas kicked up by wind and torrential rains are widespread for many miles, far from the center of the storm. The combination of overtaxed surface drainage channels attempting to carry off the rainfall and excessively high tides and storm waves (called the *storm surge*) running up the river mouths from the sea can result in disastrous floods. One of the great natural calamities of all time occurred in this manner on the Ganges Delta in 1737 when over 300,000 persons were killed outright and uncounted others starved later because of flooded fields and subse-

---

Luckily, hurricanes move slowly, and although they cannot be deterred, a careful daily to hourly check on their movements does make a warning system practical.

Fig. 7–13   Typhoon aftermath. Headed grain in the field, ready for harvest, is particularly susceptible to damage by both high winds and pounding rain. These Oriental rice farmers are attempting to salvage a bit of their crop, which represents eight months of labor and next year's food for the family.

quent crop loss (Fig. 7–13). In May of 1965, again in November of 1970, and still again in 1977, almost exact duplicates struck East Pakistan (now Bangladesh), killing many thousands. The newspapers were full of accounts of "tidal waves" engulfing the low delta islands, but the storm surge is what they actually referred to. A tidal wave is something entirely different (see Chapter 20).

Still another type of destruction was displayed when, on October 3, 1964, hurricane Hilda plowed into the Gulf Coast of the United States. A dancing vanguard of tornado clusters preceded it, causing great damage well in advance of the storm proper.

In the United States our hurricane warning system has become increasingly effective through the years in allowing the populace adequate advance time for evacuation and "battening down." Luckily, hurricanes move slowly, and although they cannot be deterred, a careful daily to hourly check on their movements does make a warning system practical. But even in the United States where hurricane warnings are best developed, there was heavy loss of life in Cameron, Louisiana (1957), and near Biloxi, Mississippi (1969), when hurricane veterans refused to believe that they were in danger and attempted to sit it out.

PART THREE

CLIMATE

CHAPTER 8

CLASSIFICATION OF CLIMATE

If we are going to embark on a project to classify the world's climates, we are wholly dependent upon a huge and growing mass of accrued data. But how these data are acquired and their reliability is a significant concern to those who must depend upon them. Fortunately there have been widespread and dramatic improvements in recent years—a rapid evolution from the backyard volunteer to orbiting satellites, batteries of sophisticated collecting devices and the ubiquitous computer. One man seated at a console can command simultaneously the instantaneous happening, the long-time record, every manner of source material, as well as digestion, analysis and prognostication.

INTRODUCTION

We have dealt with the word *climate* in superficial fashion in an earlier section, but it seems to be one of those every-day terms that becomes a little sticky at exact definition time. Fundamentally, it is a blend, amalgam, mix, or average of the ever-changing momentary weather at any given place; and theoretically, then, every region in the world is endowed with its own climatic personality made up of the controlled interaction of the four weather elements. One of the variables that makes it difficult to define climate with great precision is our reliance on observed weather data and the lack of dependability of some of these data.

WEATHER DATA

Time, for instance, is a critical factor in weather observation. Generally speaking, the longer the record the more reliable our conception of a given climate, for a long-term mean tends to iron out the freak year and short-run oscillation. But occasionally we are forced to operate with a badly truncated record, and when this is the case, location of the record station becomes important in any assessment of validity. In most parts of the world only a five- to ten-year transcript of continuous weather observation could easily lead to serious distortion of our attempt to visualize climatic character. New Yorkers are wryly appreciative of the Mark Twain admonition, "If you don't like the weather, wait a minute." In the equatorial and polar regions it appears that the hallmark of climate is deadly monotony and thus the short-run record need not be fatal. We are fortunate that because of the lack of climatic appeal, few people have chosen to live there; hence much of our severely limited weather data are found precisely in those localities where generalization seems to be least risky. For example, although the accumulated observations at the South Pole cover little more than a 25 year period, they show such a day-to-day and year-to-year similarity that long-range projections are being based on them with a reasonable expectation of negligible error.

We should be aware, however, that built into the record that has been amassed over many years is a likelihood of at least minor error. Even in 50 to 75 years, official observations may have been transferred from somebody's backyard, to the city hall lawn, to the airport tower, and finally back downtown to the smogbound roof of the weather bureau. During this same time not only has there been utilized increasingly efficient instruments producing a considerable range of recordings, but changing definitions of such things as "killing" frost, "trace" of rain, rainfall year, and moisture equivalents of snow.

PROBLEMS IN CLIMATIC DEFINITION

There are also weaknesses inherent in the simple average. A climate that regularly features great extremes will not be properly served by an artificial mean "somewhere in the middle," no matter how long the record. Seattle and Chicago both display an annual average temperature of 50°F (10°C), but to cite 50°F as typical of both cities is to imply a totally false picture of the temperature realities of one of them. In Seattle, day after day from summer to winter, the temperature is never very far from 50°F (10°C). In Chicago, most days of each year are far above or below 50°F (10°C). The average simply does not make this clear.

One of the variables that makes it difficult to define climate with great precision is our reliance on observed weather data and the lack of dependability of some of these.

Other kinds of local peculiarities lend character to nearly every climate. Comparing Seattle and Chicago again, both receive approximately 33 inches (84 cm) of precipitation annually. But Seattle's is delivered in the form of a soft drizzle necessitating over 250 days of gloomy, grey cloud cover each year, whereas Chicago's arrives in part as explosive, hour-long, summer thundershowers, or often as wind-driven, deeply drifted winter snow. Another factor could be fog, in most parts of the world a limited and innocuous phenomenon, yet on the occasional west coast a daily occurrence exerting a significant influence. In some regions the frequency of killing frost, above and beyond the mere monthly temperature average, constitutes the absolute determinant of a given agricultural regime.

All of these things are part of the total climatic image and all are the product of climate controls influencing, in one way or another, the fundamental elements. Seattle is affected by the control of a moderating sea while Chicago is continental; recurrent fog is frequently the product of a cold ocean current; and killing frosts, even far into the subtropics, may well result from the control that topography exerts over air drainage from a massive cold source region.

So if we say that climate is essentially the long-term mean of the daily weather conditions, we are on the right track. The climatic controls are at work in our definition of the average condition and the resultant is a *basic* climate picture. But the *basic* climate still requires embroidery with marginalia of special individual detail that make it distinctive from all the others. The climatic definition, then, with the necessity of pointing up and explaining individuality, becomes a complex of many things, not the least of which is a measure of perspicacious interpretation and analysis.

CLASSIFICATION

The initial point of departure here must revolve about the realization that *no two places in this world experience exactly the same climate.* Witness the home garden featuring a wide variety of plants, each with its own specific climatic requirement. The azalea must have shade, the hibiscus full sun; the camellia requires constant moisture, the hollyhock virtually none; the oleander needs great heat, the rhododendron as little as possible; the rose is totally frost hardy, the geranium moderately so, and the poinsettia not at all. And yet every one of these may exist in the same garden, none more than 50 feet (15 m) from the other; change their places and each will expire. Admittedly, the artificialities of mechanical sprinkler and chemical fertilizer give the gardener an edge over nature, but unquestionably there are myriad micro-climates here. Espalier a peach tree on a south-facing white wall in Scandinavia and it will produce a bountiful crop or plant an orchard no more than 3 miles (5 km) from the east shore of Lake Michigan for equally spectacular results. (Fig. 8–1). Again, these are

Fig. 8–1 Cherry orchards near Traverse City, Michigan. The lake acts as a heat reservoir each fall to exend the growing season. Moreover, the same lake, cool in April, impedes the early blossoming that the occasional aberrant warm week encourages. This common false spring is inevitably followed by one last freeze of a dying winter. A few miles from the lake, fruit trees do not do nearly as well in these high latitudes.

All Greeks felt themselves singularly blessed to be rational inhabitants of the moderate Mediterranean world, properly designated as the Temperate Zone.

Fig. 8–2   The temperate world of the Classical Greeks.

microclimates and the fruit trees are reacting to the purely local situation.

If the world's climatic pattern is in reality simply an infinitude of slightly differing climates, how do we even begin to classify? The key is to *generalize*. If we cannot find any extensive regions exhibiting *identical* climates, we will generalize a bit and classify regions with *similar* climates. And setting some relatively broad limits beyond which a given climate cannot trespass, we can arrive at an acceptable if not perfect system of classification, the reasonable alternative to dealing with climatic chaos.

Early Greeks

The early Greeks were the first to give serious attention to the concept of climatic similarities. They had, of course, been long aware of the character of the Sahara desert encountered (no matter what the route to the south) immediately adjacent to the Mediterranean basin. Here was an uninhabitable world of shifting sands and wildly extreme temperatures. They called it the *Torrid Zone* and totally shunned it. Across the mountains to the north was another sharply alien environment of cold winds and snowy forests, the home of primitive tribes. This had to be the *Frigid Zone* and no self-respecting Greek felt any particular inclination to invade it. As a matter of fact, all Greeks, self-respecting or otherwise, felt themselves singularly blessed to be rational habitants of the moderate Mediterranean world properly designated as the *Temperature Zone* (Figs. 8–2 and 8–3). Given the paucity of hard data of the times, the Greeks can hardly be faulted for their climatic naïveté. They were at least recognizing the basic control that latitude exerts on temperature. But when modest variations of the Greek model are perpetuated in our contemporary texts, there is certainly little excuse even in the sacred name of generalization. To generalize the climate of the whole world into three temperature zones is overdoing it to an extreme.

Classification Rationale

There are other elements besides temperature that contribute to climatic character, and even temperature responds to other controls than latitude alone. Any responsible classification must consider, among other things:

I. Moisture
  A. Precipitation
    1. Amount
    2. Distribution
      (seasonality)
    3. Type (rain or
      snow
    4. Cause(s)
    5. Reliability
    6. Extremes
  B. Humidity
  C. Clouds
  D. Fog

II. Temperature
  A. Average
  B. Range
    1. Annual
    2. Diurnal
      (daily)
  C. Extremes
  D. Cause(s)

III. Storms

Very early it dawned on somebody that the natural vegetation of any particular region was a visual synthesis of climate.

How do we pull all of this together? Very early it dawned on somebody that the natural vegetation of any particular region was a visible synthesis of climate. It tends

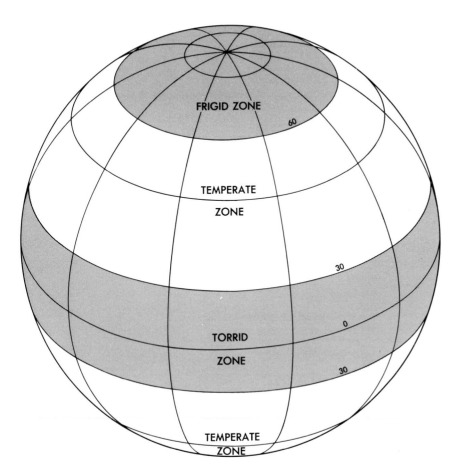

Fig. 8–3 The Greek classification. A slightly modified version of the ancient system that continues to appear in elementary textbooks. It points up the strongly dominant influence of latitude on temperature but is flawed badly by its failure to bring out alternate temperature controls and especially by completely ignoring moisture as a climatic element.

to reflect strongly the complex interrelationship of all of the factors listed above and is a great deal easier to observe and classify than the limited data derived from unevenly distributed weather instruments. So, although it has developed that vegetation regions and climate regions do not coincide precisely, they are sufficiently similar (as we are at great pains to point out in Chapter 14) that many of the standard classification systems are based heavily on this close relationship.[1]

[1]C. Warren Thornthwaite, for instance, in his relatively modern "Approach to a Rational Classification of Climate, 1948," works with a counterpoint of precipitation to evapotranspiration (evaporation from the soil plus transpiration from plants). He is dealing with moisture and temperature (the major control of evapotranspiration) to determine agricultural potential while at the same time classifying the climate.

The Köppen System

The best known and the generally accepted standard of the twentieth century has been the system devised by Wladimir Köppen in 1918 (Appendix A). He too worked with natural vegetation as a perceptive guide and, then, applying quantitative methods, arrived at a climatic classification. Critical boundary lines are given specific values or are arrived at through the use of formulas (boundaries are subject to continuous minor revision as new data become available), and each region is assigned letter symbols. Undoubtedly, there is cogency in such an approach for the advanced student, but at the introductory level pure unmodified Köppen can become a little turgid. When decoded, his symbols are based on German words, which themselves must be translated, and his measurements are

Wladimir Köppen

Wladimir Koppen.

Born in St. Petersburg (Leningrad), Wladimir Köppen (1846–1940) spent his youth on a Crimean estate presented by the Czar to his father, a distinguished historian. The early impress of these formative years in a region of benign climate and subtropical vegetation was to mark his professional life. In moments of retrospective nostalgia Köppen pointed to the Crimean experi-

ence as the spark that ignited his curiosity about the reciprocal relationship between climate and vegetation, and led to his specialization in botany.

At Heidelburg and Leipzig in Germany he studied all of the sciences in addition to botany, and shortly thereafter embarked upon a career as a meteorologist at the German Government Naval Observatory in Hamburg. Forty-four years later he retired; but never during his service as a government technician did he become a bureaucratic drone—always he was an active scholar and publisher. Köppen is renowned in meteorological circles for his pioneer contributions to the mapping of ocean winds, the use of devices for exploring the higher atmosphere, the melding of aerology and synoptic meteorology, and the brilliant anticipation of atmospheric fronts. He even became involved, at the age of 75, with his son-in-law Alfred Wegener in advocating the theory of continental drift.

Geographers, however, remember Wladimir Köppen for his percipient climatic classification based on the vegetation/climate symbiosis. Inspired by Grisebach's map of vegetation regions of the earth (1867) he worked off and on at expressing his empirical theory into map form for over 30 years, finally publishing a sharply botanical "Classification of Climate" map in 1900. It was moderately well-received by his fellow scientists but a major revision of 1918, in his own words "freer from botanical geography and more closely adjusted to pure climatology", was of major significance. The climax of his work was the joint editorship of the great 5 volume *Handbuch der Klimatologie*—this at the age of 84.

in Celsius degrees and metric scales deciphering into awkward fractions. Nonetheless, most commonly used classifications bear a strong family resemblance for in some manner or other they are merely Köppen simplifications.

A Modified Köppen

The classification about to be introduced is one of these. However, we will substitute descriptive titles for Köppen symbolism; we will generalize rather broadly where Köppen is specific; and we will set the limits of most climates somewhat wider that Köppen, thereby incorporating into one a number of his lesser subdivisions. The end result will

be 11 separate climates, each with its own peculiar set of characteristics (see Map I).

If this color map is frightening in its seeming complexity, be assured that there is logic here and a large element of predictability. Note, for instance, that the color representing a particular climate is repeated over and over again but always in the same relative position on each continent. The reddish shade is strongly concentrated near the equator and the bright green always on middle latitude west coasts. This simply means that an identical set of controls is at work in the same place on each continent, so that if we learn the personality of a climate in one location we automatically are familiar with that of every other spot where the same color appears on the map. If we were to establish a huge hypothetical landmass in mid-Pacific, it would be relatively easy to predict, at least in rough fashion, its various climates for they would be mere repetitions of those found elsewhere.

If we were to establish a huge hypothetical landmass in mid-Pacific, it would be relatively easy to predict at least in rough fashion, its various climates for they would be mere repetitions of those found elsewhere.

There are two parts of the world that we will make no attempt to classify—the oceans and the mountains. This does not mean that they do not have climates, but long-time weather records on which to base our system are not available at sea as they are at land stations. And mountains break down into what is essentially a mass of microclimates. Every 1000 feet (305 m) of elevation changes the temperature roughly 3°F (1°C), while shady slopes versus sunny slopes, windward slope versus lee slope, and so on, introduce such myriad variations that it becomes impossible to integrate each individual valley's climate into the world pattern.

THE CLIMOGRAPH

A number of the basic characteristics of each of these climates, as we establish them, can be rendered visual by plotting the data of a representative station in graph form. Such a graph is called a *climograph* and is a condensed version of the main features of each climate region (Fig. 8–4). The graph is made up of 12 vertical columns corresponding to the 12 months of the year. They are always labeled starting at the left with January and following in sequence so that December is the right-hand column. In the Northern Hemisphere the middle columns will be summer and the outside ones winter, whereas in the Southern Hemisphere, the reverse will hold. Along the right margin of the graph, a precipitation scale is marked and the normal monthly precipitation is plotted by blacking in

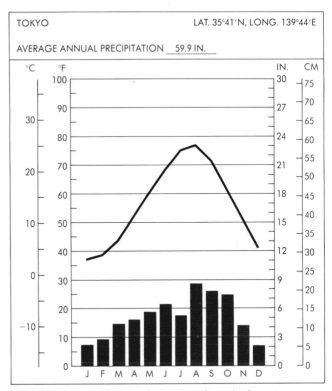

Fig. 8–4   The climograph. In chart form the long-time weather record of an actual station may be rendered in an easily readable fashion. Monthly temperature averages are connected to form a curve and monthly precipitation averages are represented by shaded bars.

each column from the bottom to the proper height as indicated by the scale. When this is done for each month, the total annual precipitation becomes visible at a glance, as does the typical distribution of that precipitation throughout the year.

Along the left margin of the graph is a temperature scale. Related to this scale, the average monthly temperature is plotted by placing a dot in the middle of each column. When all months have been plotted, a line is drawn connecting the dots, and this curve makes apparent the average temperature for each month and the annual temperature range.

Because of their usefulness as immediate visual representations of the more important features of each climate, typical climographs of actual stations will be included for each region discussed in the following treatment.

CHAPTER 9

THE TROPICAL CLIMATES

Those considerable parts of the tropics which receive reliable rainfall exhibit a great potential for agriculture, but that potential has not always been fulfilled; water, great heat, and a year around growing season does not automatically trigger agrarian development. Often the missing ingredients have been organization, capitalization, and a guaranteed market, elements supplied in the 19th century by colonialism. Today these nations are independent, but it is difficult, perhaps even economically inadvisable for them to escape the colonial legacy of commercial mono-cropping. Here is sugar cane in Fiji.

INTRODUCTION

The tropical climates approximate in their location that wide belt astride the equator that the Greeks called the Torrid Zone. As they looked south from the Mediterranean, the northern margin of the Sahara was obviously the outer edge of the excessively hot land and the beginning of a more moderate temperature regime. Others have suggested the Tropic of Cancer and the Tropic of Capricorn as the limits of the true tropics, but actually the Greeks were nearer to the truth when they selected the Sahara margin some 30° to 35° from the equator. Köppen's selection of a cold month averaging 64°F (18°C) comes close to this, as does the poleward limit of coconut palms. Within this zone are three climates, all generally warm as the "tropical" would indi-cate, but differing markedly in their precipitation charac-teristics. They are: *Tropical Rain Forest*, *Tropical Dry*, and *Tropical Wet and Dry* (Fig. 9–1).

TROPICAL RAIN FOREST

The major representatives of the Tropical Rain Forest cli-mate occur within 10° of the equator and therefore are roughly coincident with the Doldrums and at times with the Intertropical Front. Thus the Amazon Basin in South America, the Congo Basin and closely adjacent areas in Africa, and most of the East Indies are the chief land areas in the Doldrum belt and each exhibits a Tropical Rain Forest climate.

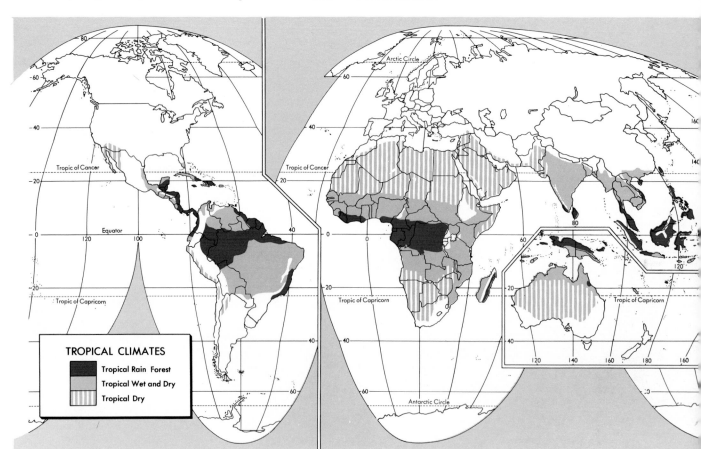

Fig. 9–1 Tropical climates.

The words "Rain Forest" are a key to the precipitation characteristics of this climate. Heavy constant rainfall is required to support a rain forest vegetation. Probably 65 inches (165 cm) per year would be a minimum requirement, while 100 to 150 inches (254 to 381 cm) would not be uncommon. And the distribution of this moisture throughout the year must be even, at least to the extent that no real dry season is experienced. Köppen has suggested that in the tropics with a high rate of evaporation, any month receiving less than 2.4 inches (6 cm) of rainfall must be regarded as dry. So any tropical situation where at least 65 inches (165 cm) of rain falls in a year and no month receives less than 2.4 inches (6 cm) of rain will support rain forest vegetation and must be classified as Tropical Rain Forest. Locally, the populace may speak of the wet season and the dry season. It is entirely possible that one season may experience double the precipitation of another, but if it is true Tropical Rain Forest, there is no dry season and the reference is actually to a wet season and a not so *wet* season (Fig. 9–2, Sandakan).

The typical source of all of this moisture is the almost daily convectional shower.

The typical source of all this moisture is the almost daily convectional shower. Saturated equatorial air heated by a sun never far from overhead gives rise to this phenomenon that occurs in many places, virtually "on schedule" every day during the period of maximum heating—2 to 3 o'clock in the afternoon. In general, mornings are bright and fresh, but as the day wears on, the lower atmosphere, heated by conduction and radiation, becomes increasingly warm and sticky until relief arrives in the afternoon in the form of clouds and torrential showers. The violently ascending convectional column is topped by towering thunderheads and lightning, and thunder frequently accompanies the short but heavy rainstorm. For a while at least, as the rain descends in sheets and the runoff floods the ground, there is a welcome cooling, but the storm is of short duration and within an hour the sky clears, the vegetation steams under a high sun, and the heat returns, accompanied by even higher humidity than before the storm.

Occasionally, a weak tropical cyclone will move through, giving rise to several days of continual cloud cover and drizzle, but these are infrequent and by far the most important moisture source is the convectional shower. Since the heating is constant the year round as a result of equatorial location, and the showers occur regularly, it follows that no dry season will exist and heavy forests will be the landscape feature.

Temperatures are, of course, generally high, but despite being so near the equator, world record high temperatures are not found in the Tropical Rain Forest. The daily clouds during the early afternoon cause the heat curve to level off so that 90° to 95°F (32° to 35°C) is the normal daytime maximum. This is, of course, warm, especially when combined with high humidity, and afternoons are enervating and uncomfortable. But it is a far cry from the over 100°F (38°C) that is typical of many locations much farther away from the equator. The worst thing about these 90°F (32°C) temperatures from a human comfort standpoint is that since there are no seasons, every day is the same. The only real respite is at night when temperatures may drop to 65 to 70° (18 to 21°C). This is certainly not cold, but these are the coldest temperatures ever experienced here. Lack of seasonal temperature change means that there is more than a kernel of truth in the old saying, "Night is the winter in the tropics."

Night is the winter in the tropics.

Notice on the climographs (Fig. 9–2) that the average monthly temperature (day and night both) is never far from 80°F (27°C) and that the temperature curve is virtually a straight line. This is a visual expression of an annual temperature range that never varies more than a degree or two. The daily range, which does not show on the climograph, is considerably larger, averaging as much as 25° (14°C).

Having traced the chief characteristics of the Tropical Rain Forest climate, we will look again at the location map (Fig. 9–1) and notice that although the bulk of this climate occurs in the Doldrum belt near the equator, there are several narrow coastal strips of Tropical Rain Forest some distance from the equator. These are all in the same latitude (20° to 25° N and S) and all on east coasts: the east-central coast of Brazil, the east coast of Madagascar, the

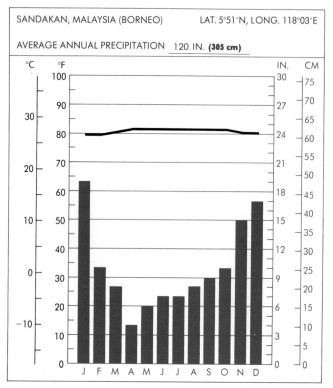

SANDAKAN, MALAYSIA (BORNEO)   LAT. 5°51′N, LONG. 118°03′E

AVERAGE ANNUAL PRECIPITATION   120 IN. **(305 cm)**

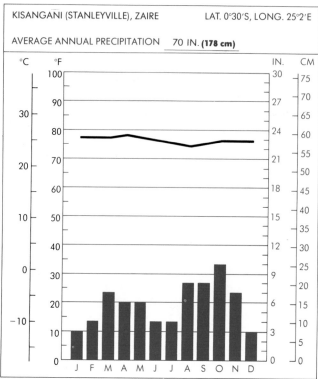

KISANGANI (STANLEYVILLE), ZAIRE   LAT. 0°30′S, LONG. 25°2′E

AVERAGE ANNUAL PRECIPITATION   70 IN. **(178 cm)**

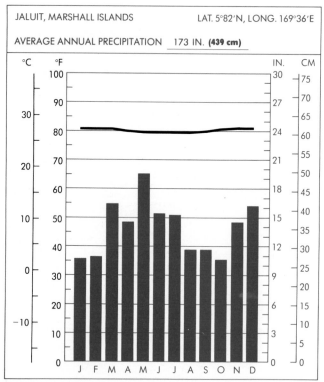

JALUIT, MARSHALL ISLANDS   LAT. 5°82′N, LONG. 169°36′E

AVERAGE ANNUAL PRECIPITATION   173 IN. **(439 cm)**

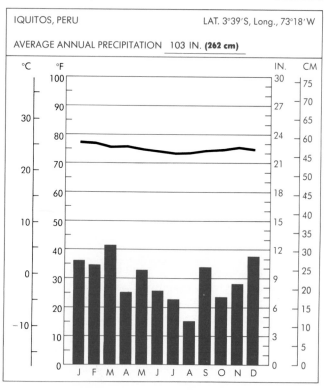

IQUITOS, PERU   LAT. 3°39′S, Long., 73°18′W

AVERAGE ANNUAL PRECIPITATION   103 IN. **(262 cm)**

Fig. 9–2 Tropical Rain Forest stations.

northeast coast of Australia, the east coasts of Central America, Hawaii, most West Indian islands, and the east coasts of Vietnam and the Philippines. (This last region in Southeast Asia is confused by local monsoon circulation, but generally fits the overall world pattern). These areas all have three climatic controls in common that are responsible for their climatic similarities. First, despite their distance from the equator, they are still well within the tropics and thus experience continuously high temperatures. Second, they are all in the Tradewind zone and therefore are constantly exposed to moisture-bearing winds from off warm seas. And third, each of the regions features uplands rising abruptly back of the coast. So as the oceanic air masses impinge on these highlands, they are forced aloft and orographic precipitation bathes the mountain slopes. On the coast proper the heated oceanic air most often gives up its moisture via convectional showers, but the seaward-facing slopes are perpetually cloud-shrouded and drippy. The end result of all this is a climate that is characterized by continuously high temperatures, constant heavy rainfall the year round, and dense rain forest vegetation, all of which is very similar to that of the equatorial regions. There is no alternative, then, but to classify this climate as Tropical Rain Forest despite the noncontiguity of these Trade Wind coasts with the much more extensive Doldrum regions. The causes are different but the result is the same, and in classifying climates, we are concerned primarily with results.

The Tropical Rain Forests, all through history, have shown up as great blanks on the world population map.

If we were to compare a population map of the world with a climate map, it would become immediately obvious that the Tropical Rain Forests have not been sought out by man as a preferred habitat. Along with the deserts and the polar regions, the Tropical Rain Forests all through history have shown up as great blanks on the world population map. This is a difficult region, as many an American or European who has been forced to live there for any length of time will attest. The monotony of moist heat day after day has a distinctly enervating effect on the human body, and diseases are difficult to control, especially those carried by hosts, such as yellow fever, malaria, filariasis, and liver fluke. Mildew and mold flourish, attacking and decomposing textiles and leather in a remarkably short time. Furniture secured with glue falls apart, and termites may honeycomb wooden houses and foundations almost overnight.

Most of these Tropical Rain Forest regions are occupied by small numbers of wandering tribes whose economy revolves about hunting, fishing, and gathering what the forest provides. Some have become sophisticated to the point of clearing small sections of the forest by fire and carrying on a rudimentary agriculture among the stumps and snags. Shortly, however, they are forced to move on and repeat the operation, for soils are typically lacking in soluble minerals near the surface because of heavy rainfall *(leaching)*, and the forest, quick to regenerate itself, crowds out the clearing in a very short time with thick secondary growth that is too much for primitive implements to control (Fig. 9–3).

There are, however, some limited locations within the Tropical Rain Forest that have become the home of large numbers of people and have achieved an important economic position because of certain factors offsetting or complementing the basic difficulties of the climate. Some of these are: superior accessibility, occurrence of valuable minerals, above-average soil fertility, application of capital and technical skills (formerly in colonial holdings), or combinations of these. The island of Java, during the Dutch colonial regime, may be cited as an example. Because Java is in an archipelago, it is easily accessible from the sea, and by virtue of its location off the coast of Southeast Asia it is near important shipping routes. Its mountain backbone of constantly active volcanoes has supplied fertile ash showers that have rejuvenated the soils more rapidly than they can be leached. Economic deposits of minerals, petroleum in particular, have been found. And there has been the application of Dutch capital, direction, and development. All of this resulted in Java becoming a highly productive region capable of supporting a population of many millions. Today the nation of Indonesia has gained a measure of individual freedom at the expense of economic solvency, but the potential remains.

Sri Lanka with her tea and coconut plantations, various West Indian islands with their sugar, Central America with its bananas, Malaysia with its rubber and tin, Surinam with

Fig. 9–3 The tropics have long supported human habitation, but in relatively recent times colonialism and resource development have marked the land with the visible signs of urbanism, industry, and organized production ventures. "Small society" represented by the village and the aborigine may be on its way out, yet here in the tropics it has held its own more successfully than elsewhere.

These are Melanesians in Vanuatu (New Hebrides) whose livelihood is derived from fishing, hunting, gathering and a simple kind of animal husbandry/hoe agriculture based on the pig and the yam.

its bauxite, and Nigeria with its cacao plantations and petroleum are other examples of Tropical Rain Forest areas that have progressed above the average because of one or more special factors. But on the whole, Tropical Rain Forest areas remain sparsely populated.

Despite the backward position of Tropical Rain Forests today in terms of human habitation, they may have considerable potential. Most of these remote areas have scarcely been touched by the geologist in search of economic mineral deposits. It is entirely possible, even likely, that New Guinea or the back country of the Amazon Basin may contain sizable ore bodies. And any place with continuously high temperatures, a year-round growing season, and heavy precipitation certainly has agricultural possibilities. It has been suggested that since the Tropical Rain Forest is one of two of the world's remaining large forest reserves (the other being the subpolar regions), it may become in the not too distant future a major source of timber and wood products. At the moment, the middle latitude forests are adequate and much more accessible to the chief markets, but these sources are being depleted rapidly and at the same time demands are increasing. The basic ad-

vantage of the rain forest is its rate of regeneration—many times that of the middle latitude forests. And although the rain forest is characterized by hundreds of botanical species as opposed to the relatively pure stands occurring elsewhere, there does exist a tremendous volume of basic cellulose capable of reproducing itself at a very rapid rate.

TROPICAL DRY

All of the world's Tropical Dry zones are found in a broad belt at about the same latitude—roughly astride the Tropics of Cancer and Capricorn. Thus they are in the region dominated by the Trade winds and are often called Tradewind deserts. Also, without exception, they have a west coastal frontage. Some, notably in North and South America, are narrow coastal ribbons hemmed in by mountain ranges closely paralleling the coast. Others as in Australia and Africa, lacking this topographical control, extend for long distances inland (Fig. 9–1).

The largest of the Tropical Dry regions by far is that

continuous zone extending from the Atlantic Coast across North Africa, Arabia, and into northwestern Pakistan and adjacent India. Here it has been traditional to apply three different names to the various sections of this extensive desert. In North Africa it is called the *Sahara*; between the Red Sea and the Persian Gulf is the *Arabian* Desert; and in India/Pakistan, the name *Thar* is applied. In virtually identical latitudes, the west coast of North America is also Tropical Dry. Baja California and the coastal strip across the Gulf of California are included here as well as the Imperial Valley and lower Colorado River section of California and Arizona. Frequently, the appellation *Sonoran* Desert is used to refer to the entire area.

The Southern Hemisphere exhibits the Tropical Dry climate too, every continent being represented. In the northern third of Chile and the entire coast of Peru is the *Atacama*, limited to a strip between the Andes and the sea. Also, astride the Tropic of Capricorn and on the west coasts of Africa and Australia are sizable dry regions extending far into the heart of each continent. These are the *Kalahari* and *Australian* deserts, respectively.

Deserts are hot and dry. This is scarcely news.

Deserts are hot and dry. This is scarcely news. But how hot is hot and how dry is dry, and more important, why does this situation exist in these particular locations?

Except for the very high latitudes where evaporation is limited, any place receiving less than 10 inches (25 cm) of rainfall per annum is normally regarded as a true desert. Under these circumstances, there is little or no natural vegetation, only xerophytic shrubs at best, and this cover is widely spaced with considerable areas of bedrock, sand, or gravel interspersed. Actually, in the tropics where evaporation is exceptionally high, even 15 to 20 inches (38 to 51 cm) of rain each year is scarcely sufficient to support more than a skimpy bunch grass cover. And although this may have some small value in good years for nomadic grazing of small flocks, it is so unreliable as to be legitimately classified with the true desert as too dry for more than sporadic human habitation.

Unreliable rainfall is a characteristic of the world's dry regions. As a general rule, *the smaller the total precipitation, the greater the variability.* Lima, Peru, which aver-

ages 1.8 inches (5 cm) of rain per year, recently went over 13 years without any measurable precipitation and then had several storms that brought the average back to normal. Sudden violent storms are typical in the dry regions, resulting in heavy surface runoff on the hard baked earth. *Arroyos* or *wadies* are cut by raging torrents of short duration, adobe buildings melt, and people drown. In some parts of the desert, the rainfall, however unreliable, is concentrated in the winter, as in the northern Sahara or southwest United States. In other places such as the southern Sahara and the Thar, whatever rainfall is received comes in the summer. But the cohesive element that allows us to classify all of this as Tropical Dry is not *when* the rain comes, but the fact that the total is low and even that is usually unreliable. If the precipitation of an area is too little for continuous occupancy by humans, without resorting to irrigation or other artificialities, then we shall consider it as Tropical Dry.

Now why should west coastal exposures at these particular latitudes be so dry? Kalama in the northern Chilean nitrate fields has never recorded even a trace of rain since records were begun 100 years ago, and many other places are almost as dry (Arica, Chile, 0.02 [0.1 cm]; Iquique, Chile, 0.6 [1.5 cm]). One cause appears to be the Trade winds coming in off the sea from the east. Notice again on the maps (Fig. 9–1 and Map I), the Tropical Rain Forest Trade-wind coasts. Here the Trades are forced up over highlands, precipitating most of their moisture at the east coast. In almost every case the Tropical Dry regions are immediately opposite these Trade-wind coasts and thus are in the lee or rain shadow of the highlands and the continental masses. Tropical Dry regions are strongly affected too by the permanent Subtropical High cells offshore where the air is constantly sinking and warming and therefore antipathetic to condensation.

Unreliable rainfall is a characteristic of the world's dry regions. As a general rule, the smaller the total precipitation, the greater the variability.

It may seem a bit strange that the world's driest regions should be in coastal situations. To be sure, the subsiding air masses of the Horse Latitudes discourage precip-

El Niño

Rain can be a disaster along the desert coast of Peru.

El Niño is a warm current flowing southward along the coast of Panama to meet the cold Humboldt in the vicinity of the Galapagos. El Niño, meaning the Christ child, is so named because about every seven years at Christmas time it inserts itself between the Humboldt and the coast and pushes as far south as southern Peru and Chile. On these occasions chaos ensues: torrential rains flood the desert coast—people drown and irrigated crops fail; the rich nutrient mix of the roiling Humboldt is replaced by more sterile tropical waters—plankton eating fish and the seabirds who feed upon them either die or migrate; beaches are strewn with rotting carcasses, the stench magnified as hydrogen sulfide is generated by masses of decomposing microrganisms in the fouled sea.

El Niño is not a northern Pacific phenomenon, but during the excessively wet west coast winter of 1982–83 meteorologists, searching for an explanation, began to bandy about the term "El Niño effect". Waters along the California coast were measurably warmer for unknown reasons, and although fish, birds, and plankton didn't expire in large numbers (the California Current is no Humboldt), still all of this did begin not too long after Christmas.

Now we find extensive masses of unaccounted for warm water far out at sea in both hemispheres. Why invent a new term—let's call it El Niño. So El Niño is now being blamed, in some circles for: drought in Australia and Africa, political ferment in Chad, torrential rains in Japan, and for climatic adversity wherever it may be encountered.

itation, but certainly local land and sea breezes must introduce moist oceanic air onto the heated land from time to time; and so they do, except that an invisible barrier exists to intercept this air and remove its moisture just short of the coast. This is a cold ocean current, and there is one off the coast of every continent in these tropical latitudes. Further, the offshore drift of the general Trade-wind pattern tends to pull surface waters out to sea, allowing colder water from the depths to rise along the coast (*upwelling*) This is discussed in more detail in Chapter 29.

So, since all Tropical Dry regions front on west coasts, cold water is inevitable offshore. And in terms of intercepting moisture, a cold current can be almost as effective as a mountain range. Saturated air masses moving over a broad cold current will be cooled from the bottom, and condensation will occur. Often, it is in the form of dense fog, and as this drifts in over the desert coast, there is the curious anomaly of the air being filled with moisture much of the year in a place that has never recorded more than a fraction of an inch of actual precipitation. Chilling of the air at the surface in conjunction with warming via subsidence at the eastern end of the Horse Latitude anticyclone produces an inversion that makes normal convection virtually impossible. In South America the Humboldt current (the largest of the cold ocean currents) plus the Andes paralleling the coast cause the Tropical Dry zone to extend to just a degree or two south of the equator.

Heat is the other typical feature of the Tropical Dry climate. Here is where the world's record high temperatures are encountered, the current record being a shade reading of 136°F (58°C) at El Azizia, Libya, in the Sahara.[1]

---

[1] Keep in mind whenever dealing with world records of heat, cold, rain, etc., that places with such extremes are very uncomfortable spots in which to live and therefore are lightly inhabited. The cited world records simply mean that for some reason an observer lived there and had the official instruments. It is entirely possible that there are hotter places than El Azizia, but for that very reason they are good places *not* to be and their temperatures go unrecorded.

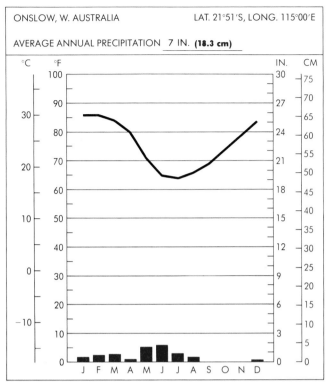

ONSLOW, W. AUSTRALIA LAT. 21°51'S, LONG. 115°00'E

AVERAGE ANNUAL PRECIPITATION 7 IN. **(18.3 cm)**

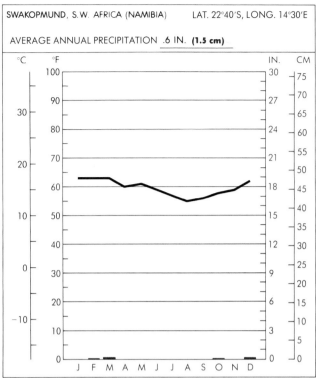

SWAKOPMUND, S.W. AFRICA (NAMIBIA) LAT. 22°40'S, LONG. 14°30'E

AVERAGE ANNUAL PRECIPITATION .6 IN. **(1.5 cm)**

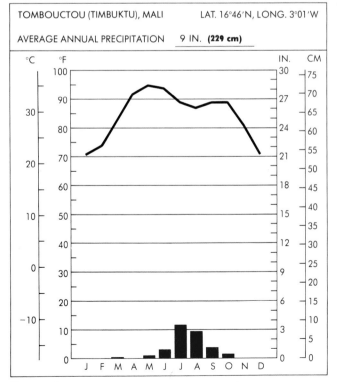

TOMBOUCTOU (TIMBUKTU), MALI LAT. 16°46'N, LONG. 3°01'W

AVERAGE ANNUAL PRECIPITATION 9 IN. **(229 cm)**

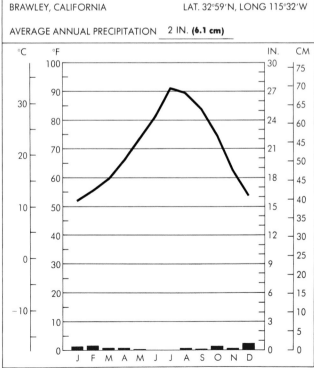

BRAWLEY, CALIFORNIA LAT. 32°59'N, LONG 115°32'W

AVERAGE ANNUAL PRECIPITATION 2 IN. **(6.1 cm)**

Fig. 9–4 Tropical dry stations.

And yet the Tropical Dry region is at the outer edge of the tropics, a considerable distance from the equator. In the Tropical Rain Forest at the equator the temperature almost never exceeds 100°F (38°C) even on the hottest day. The reason, of course, is cloud cover and rain. The sun is directly overhead at both places at one time or another each year, but in the Tropical Dry summer, the overhead sun and the lack of any clouds allow a continued buildup of temperature well into the later afternoon. Only along a narrow band of fog-bound coast is there any amelioration, and this fact has helped to make west coasts preferred locations for permanent habitation in the desert (see Fig. 9–4, Swakopmund).

The popular image of the desert prospector being more common than prospectors elsewhere is erroneous, he is the only person out there and thus commands attention.

The foregoing emphasis on heat extremes refers only to the summer when the sun is high. Daytime maximums are regularly over 100°F (38°C) for two to three months. But in winter the overhead sun is 45° to 50° of latitude away, temperatures are mild, and clear sunny days and crisp nights prevail. Actually, where Tropical Dry locales are easily accessible to populated regions of more severe clime, they have become winter resorts—witness Palm Springs, Phoenix, or Tucson, and major league baseball's spring training program. The annual range of temperature then is sizable, the largest of any tropical climate. But the daily range is even larger. Remember, the control of daily range is moisture in the air. If the atmosphere is heavily charged with water vapor, the radiation of heat from the surface of the earth will be intercepted and absorbed, with the moisture acting as a blanket to hold in the day's heat. Erase this blanket and the daytime accumulation of heat will be radiated off into space in only a few hours after sunset. So the Tropical Dry regions with their predominantly dry air will have a very large daily temperature range. Where summer daytime temperatures may go as high as 110° to 115°F (43° to 46°C), they will drop to 60° to 65°F (16° to

18°C) at night. In winter, something on the order of 80°F (27°C) (daytime) and 45°F (7°C) (at night) is typical.

Consult the climographs (Fig. 9–4) and note that the average for the hottest month is in the neighborhood of 90°F (32°C). This is the mean of both day and night for the entire month. Cool nights act to pull down the 100°F + (38°C) daytime temperatures to this lower average. And similarly, the coldest month temperature is an average of the more extreme day and night maximums.

As has been pointed out previously, the Tropical Dry regions do not possess such climatic charms that people have flocked in to live there, and now, having checked some of the details, it is not difficult to see why. Water is the critical factor. Always where water is available in surface streams, springs, or underground strata, the desert sustains life (Fig. 9–5). A year-round growing season, maximum sunshine, *and water* make the oasis agriculturist a small-scale but extremely productive operator. Traditionally, the oasis has been one of two attractions for people in the desert; the other has been minerals. Given rich mineral deposits, a person will go anywhere and live under the most difficult conditions. On the whole, the deserts are no more productive of minerals than any other climatic region, climate having relatively little to do with the occurrence of ore-bearing formations. To be sure, such water soluble minerals as nitrates and borax could be found at the surface only in dry regions and are usually the result of high evaporation, but the popular image of the desert prospector being more common than prospectors elsewhere is erroneous, however easily understood. The desert prospector is the only person out there and thus commands attention. No large numbers of others have, of their own volition, selected the Tropical Dry regions as their home. But this may not always be the case. Air conditioning, new mineral strikes, or cheap ocean water desalinization could mean new opportunities in the desert.

TROPICAL WET AND DRY

Located roughly between the excessively moist Tropical Rain Forest and the arid Tropical Dry are the regions experiencing the Tropical Wet and Dry climate and, as the name implies, exhibiting some of the characteristics of each of

Fig. 9–5   Brilliant green date palms punctuate a sere and desiccated Arabian landscape. Wherever there is water in the desert, as here in the ancient Saudi oasis of Qatif, people will gather round it in tightly agglomerated little settlements.

their neighbors. The map (Fig. 9–1) shows this intermediate location. Examine Africa first, for this latitudinal lineation is most apparent there, unimpeded by major mountain ranges or coastal indentations. Between the Congo Basin and the Sahara there is a narrow band of Tropical Wet and Dry climate broadly designated as the *Sudan*, while in the Southern Hemisphere a comparable zone (the term *Veldt* is usually applied) shows up between the Congo and the Kalahari. In South America, despite the variations brought about by the Andean cordillera and the Caribbean, the same general pattern is again represented. North of the Amazon Basin, interior Venezuela and Colombia *(Llanos)* and parts

of Central America and the West Indies make up the Northern Hemisphere Tropical Wet and Dry region, while most of southern Brazil *(Campos)*, Paraguay, and sections of Argentina, and Bolivia are representative in the Southern Hemisphere. Much of Southeast Asia from South China to and including India is Tropical Wet and Dry, and the entire north coast of Australia and adjacent Indonesian Islands are of this climatic type.

The "Wet and Dry" in the name places the emphasis in exactly the right place—the annual rainfall seasons. In the middle latitudes, we have a tendency to break the year up on a temperature basis, that is, warm season (summer)

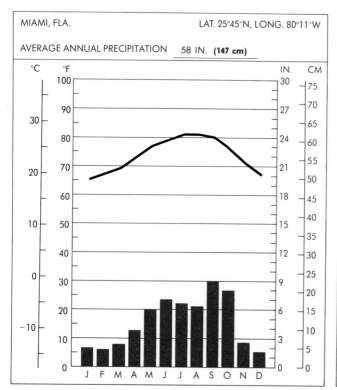

MIAMI, FLA. LAT. 25°45'N, LONG. 80°11'W

AVERAGE ANNUAL PRECIPITATION 58 IN. **(147 cm)**

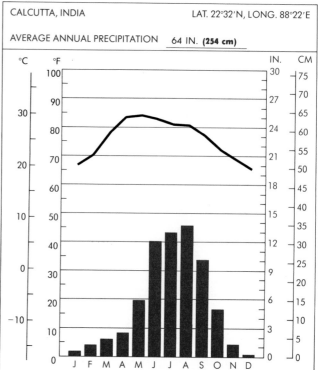

CALCUTTA, INDIA LAT. 22°32'N, LONG. 88°22'E

AVERAGE ANNUAL PRECIPITATION 64 IN. **(254 cm)**

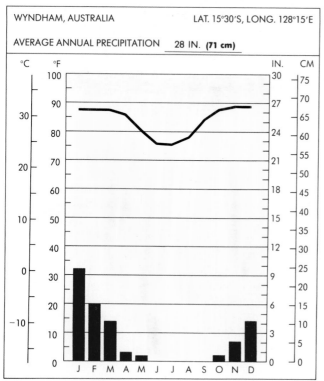

WYNDHAM, AUSTRALIA LAT. 15°30'S, LONG. 128°15'E

AVERAGE ANNUAL PRECIPITATION 28 IN. **(71 cm)**

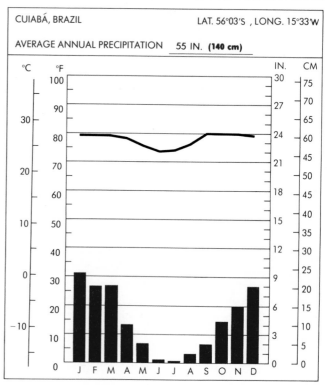

CUIABÁ, BRAZIL LAT. 56°03'S , LONG. 15°33'W

AVERAGE ANNUAL PRECIPITATION 55 IN. **(140 cm)**

Fig. 9–6 Tropical wet and dry stations.

versus cold season (winter) because this is the most prominent seasonal change. However, near the equator where temperature changes are mild, the dramatic difference in seasons is wet versus dry, and the local inhabitants invariably refer to them in this way (Fig. 9–6). How do we account for this phenomenon? The reasons are varied, but generally there are two basic causes—one applying in Southeast Asia and Australia and the other in most of Africa and Latin America.

Near the equator where temperature changes are mild, the dramatic difference in seasons is wet versus dry not hot versus cold.

Let us begin in Southeast Asia. Why the alternating wet and dry seasons in this particular location? This is the region of the monsoon, the all-pervasive seasonal winds whose characteristic circulation is chiefly responsible for the distinctive rainfall regime in this part of the world. Whether it is a result of jet streams, upper-air frontal activity, deflected Trades, differential heating of land versus water, or all of these operating in some sort of concert, the end product is a summer rainy season and a winter dry season. Saturated air drifting onshore from tropical seas each summer readily gives up its moisture on the mountain flanks via convection. The amount varies with distance inland, latitude, exposure, etc., but the general tendency remains for oceanic air to invade the land during the warm months. Concomitantly, colder and drier continental air begins to move offshore with the onset of winter, and although the details vary from place to place, precipitation ceases in large part and the cold season becomes the dry season. In the Southern Hemisphere, but still within the greater Asiatic realm, namely, northern Australia and parts of the African east coast, a reciprocal monsoon also operates as the dominating force. This monsoonal influence, then, produces, throughout a sizable part of the world, a moisture regime featuring rainy summers and dry winters.

In the Tropical Wet and Dry regions of Africa and Latin America the essential cause of rainfall seasons is something else again. Recall that as the overhead sun makes its apparent annual migration from Cancer to Capricorn and back again, all of the world's standard wind and pressure belts tend to follow. Their change in position lags a bit behind the sun and the shift is small (on the order of 5 to 10° of latitude), but to regions along the margins of these belts, even such minor changes can be pivotal in determining climate. The Tropical Rain Forest is almost wholly within the Doldrum zone, and the Tropical Dry is dominated by the Trades and Horse Latitudes, but the Tropical Wet and Dry has no such wind or pressure belt of its own. Instead, it is located astride the boundary separating the Doldrums from the Trades. Thus even slight latitudinal changes of the Trades or Doldrums will cause the Tropical Wet and Dry to be affected.

Consider the Northern Hemisphere summer when the overhead sun is at the Tropic of Cancer. The Doldrums, in shifting slightly northward, has lapped over the Tropical Wet and Dry region, and it is now under the influence of all the factors that give rise to the Tropical Rain Forest climate. At the same time in the Southern Hemisphere the Trades have moved slightly north and are dominating that hemisphere's Tropical Wet and Dry region, bringing Tropical Dry conditions. As the sun moves to the Tropic of Capricorn six months later, this movement is reversed. So, in essence, the Tropical Wet and Dry zones have no real climate of their own but borrow their neighbors' in turn. The Sudan of North Africa, for example, experiences an imported Congo climate in the summer with the attendant daily showers and high humidity, while in the winter arid Sahara conditions move in. The result is a summer rainy season and a winter dry season very much like that of the monsoon. Undoubtedly, this same set of circumstances would prevail in Southeast Asia and Australia except for the local monsoon currents. They are so assertive as to wipe out any traces of the standard wind and pressure system, but climatically (monsoon or shifting belts) the results are the same, and we can logically class all these regions as Tropical Wet and Dry.

This characteristic distribution of precipitation throughout the year is more important than total amount in determining just which areas shall be included as Tropical Wet and Dry. The only limitation on amount is that if a region receives less than 20 inches (51 cm), that region must be classified as Tropical Dry. Any place in the tropics receiving any amount more than 20 inches (51 cm) in one year and experiencing a winter dry season will be Tropical

Rainfall Records

Heavy rainfall on the windward side of Kauai, brilliant sunshine in the rainshadow.

The wettest place in the United States is also billed as the wettest in the entire world. Mt. Waialeale's observatory on the Hawaiian island of Kauai has been recording rainfall data for 50 years and during that time the average annual receipt has been 472 inches (1199 cm.)—that's almost 40 feet (12 m)! We're dealing here now with orographic precipitation in the Trade Winds. Kauai crests at about 5000 feet (1524 m) and the weather station is on the windward slope not far below. All Trade Wind islands and coasts display this same tendency but they must be thrust up to some height—Trades are not likely to give up their moisture over an open sea or a small tropical atoll.

Almost as wet on the average and a good deal more spectacular in its extremes is the 4398 foot (1340 m) Indian hill station, Cherrapunji, where a continuous record has been kept for well over 100 years. The annual average is a modest 451 inches (1145 cm) but in 1861 they measured 905 inches (2299 cm) and during the 12 month period from August 1860–July 1861, 1042 inches (2647 cm)—just short of 90 feet (27 m)! But every year is not a sure thing; only 12 years later a mere 283 inches (719 cm) accumulated.

The remarkable part of the Cherrapunji records is that all of the rain fell in the summer with the midwinter 4 months bone dry. In this part of the world it is the monsoon that is responsible. Again, orographic lifting, as the saturated summer monsoon sweeps in off the Bay of Bengal only to run afoul of a Himalayan outlier, the Khasi Hills.

If it were not for the cooling effect of the approximately 4000 feet (1219 m) elevation, these two locations would be classified respectively as Tropical Rainforest (Trade Wind Coast example) and Tropical Wet and Dry (monsoon example). But the lowland below reflects these same rainfall patterns, if not amounts, and the constant heat is unmodified by elevation, so all of windward Hawaii is Tropical Rainforest and the Ganges delta typical Tropical Wet and Dry.

Cherrapunji is the site of a former British tea plantation and the villagers who reside there do indeed endure these climatic peculiarities. But the Mt. Waialeale station sits in a permanently cloud-bound upland swamp, not an environment that would support any kind of permanent settlement. So it is essentially an automated climate observatory and this brings up a problem—should we regard these as official records? The accepted rule through the years has been that somebody has to live there continuously with the official instruments, otherwise we could blanket the world with automated stations—the top of Mt. Everest, central Greenland, or the middle of the Sahara.

But we have to admit, there is a plethora of moisture at Mt. Waialeale, legal or not.

Wet and Dry—no limit as to total. Some locations exceed 150 inches (381 cm).

The natural vegetation will closely reflect the total rainfall and the length of wet or dry season. If we were to travel from the northern edge of the Congo to the southern margin of the Sahara, a regular progression of vegetation zones would be apparent. Just north of the Tropical Rain Forest, a near rain forest prevails. The trees are more widely

A

B

Fig. 9–7   (A) The monsoon is the pulse of Asia. Its annual rhythm with assured summer rain signals the teeming millions to assemble at the paddy. Each stalk of rice is hand planted and everyone participates because everyone eats. (B) Those who worry less about the serious relationship between hard physical labor and minimal food on the table enjoy the carefree lassitude of Trade wind beach resorts.

spaced and occasional open glades result from the short dry season. Near the desert, where the dry season is dominant, short grass turf and stunted thorn brush are found. In between these two extremes is the savanna, extensive grasslands where the vegetation sometimes reaches a height of 8 feet (2 m). All of these regions are Tropical Wet and Dry because they have in common a winter dry season and a total precipitation of greater than 20 inches (51 cm).

Temperatures, of course, are generally high in any tropical situation, and as might be expected in the Tropical Wet and Dry they are closely related to those of neighboring climates. The summer temperatures are a reasonable equivalent of those of the Tropical Rain Forest. A continual high humidity and afternoon cloudiness combined with 90°F (32°C) days and 70°F (21°C) nights are the average. Heat is greater near the desert margins and the daily range increases proportionately. Winter, being the dry season, displays temperatures that tend to be somewhat like the desert winters, roughly 75° to 80°F (24° to 27°C) days and 55° to 60°F (13° to 16°C) nights.

Paddy rice and the monsoon are highly compatible as long as the rainfall is over 40 inches (102 cm) per year, and the climatic rhythm becomes the pulse of life.

Thus it can be seen that a rather wide range of conditions prevails within the Tropical Wet and Dry regions, and similarly, variations exist in human occupancy and land use. In the Orient this is a highly productive climate from an agricultural point of view. Southern China and India particularly support as many as 3000 people per square mile in some places who are directly dependent on the land for their livelihood. Paddy rice and the monsoon are highly compatible as long as the rainfall is over 40 inches (102 cm) per year, and the climatic rhythm becomes the pulse of life (Fig. 9–7). Sugar cane too, as in the West Indies and India, is adaptable to the alternation of wet and dry seasons and allows these areas to produce a cash crop and sustain sizable numbers of people. Sisal and pineapple are similar in their ability to withstand partial drought. But elsewhere, where the natural grasslands have been put to less intensive use as pasture, much smaller populations are supported. The savannas have proved not to be the ideal grazing land that some have supposed them to be. Isolated, alternately parched and flooded, and plagued by disease, it takes a special breed of livestock to survive here, and more often than not such a rugged animal does not produce particularly tender steak. Only in Australia has any sort of commercial grazing proved successful, and even there, despite persistent efforts, losses are high and profits continue to be small.

But there are always possibilities. Again, mineral strikes such as copper in Zambia, iron in Venezuela, and phosphates, uranium, iron and bauxite in Australia will draw large numbers of people and pay for the construction of modern transport, which will help to defeat isolation. New crops and new agricultural techniques may also help. At Humpty Doo in northern Australia, a joint United States—Australian venture in large-scale mechanized rice culture has been attempted, although with less than complete success.

In these areas and elsewhere, peanuts, cotton, and other tropical crops have shown varied results. With world population continuing to rise at a rapid rate, the relatively empty portions of the Tropical Wet and Dry do have a potential as a productive home for humans.

CHAPTER 10

THE SUBTROPICAL CLIMATES

Some subtropical climates are humid, some are distinctly lacking in moisture receipt, but all have long, sunny, frost-free periods. In Spain the difficulty is sufficient agricultural water at the correct season. When it can be obtained a careful anointing of limited plots flecks the tawny countryside with lush verdure.

INTRODUCTION

Transitional between the tropics and middle latitudes and reflecting some of the characteristics of each, are the subtropical climates. We would expect them to have a little frost, as a limit on man's agricultural endeavors, and, at the opposite swing of the seasons, somewhat milder and more "livable" temperatures than are generally found in the tropics. Large numbers of people, happily occupying the subtropics, proclaim that their climate combines the best features of both the tropics and the middle latitudes. They may be right (Fig. 10–1). In the United States we have popularized the term "Sun Belt" with all of its subliminal positive inferences, to refer to the subtropics and desert. And our demographers and economists point to a steady migration from the more northerly "Snow Belt" with climate admittedly a major influence.

There are two subtropical climates, one occurring in regions facing out on west coasts and the other on east coasts. This opposing coastal orientation introduces some sharp divergencies in climate despite their generally identical latitude.

MEDITERRANEAN

The west coast subtropical climate derives its name from its typical location along the margins of the Mediterranean Basin. There has been some criticism of the use of the term *"Mediterranean"* to denote this particular climate in that it is merely a geographical location and inconsistent with the other climatic names that refer to vegetation or some outstanding feature of the climate. But it does aid the student in location and through long use has become traditional.

Since Mediterranean climatic regions are always west coastal and because they are subtropical and immediately adjacent to the tropics, they must, wherever they occur in the world, share a common boundary with the Tropical Dry climates, for these are the most poleward of the tropical climates and always appear on west coasts too. In the Old World the west coast immediately north of the Sahara Desert is breached by the Mediterranean Sea and the coast extended greatly, so that almost the entire fringe of this inland sea is considered representative of the Mediterranean climate. Normally, however, if the continental west coast is straight, regions exhibiting a Mediterranean climate will

Fig. 10–1   The subtropics—not as hot as the tropics, not as cold as the middle latitudes—are just right for many folks. Everyone can't afford to move there permanently but many do visit and vacation. The French Riviera is representative of world-renowned playgrounds selling sunshine and ambience.

be rather limited, for they must be near the sea and their latitudinal extent is not great. So if there were no Mediterranean Sea, only the coasts of Morocco and Portugal would be so influenced. But with the Mediterranean Sea multiplying the length of the coastline many times, we find here the largest area of Mediterranean climate anywhere in the world—involved are the coasts of Spain, southern France, peninsular Italy, Yugoslavia, Albania, Greece, Turkey, the Levant states, Libya, Tunisia, and Algeria. The Sahara reaches the sea from southern Israel to Libya, but elsewhere the Mediterranean climate completely encircles the Mediterranean Sea and even pushes into the Black Sea littoral in Turkey and the Crimea.

On the west coast of North America, in the same latitudes as the Mediterranean Sea, is another Mediterranean climate region. It includes the California coast from the Mexican border, where it abuts with the Sonoran Desert, northward to just beyond San Francisco Bay, and also the greater part of the Central Valley. In South America, the central one-third of Chile occupies a similar position, and in South Africa, the southern margin of the Kalahari Desert, in the vicinity of Cape Town. Presumably if Africa were longer, the region would be more extensive, but South Africa barely reaches subtropical latitudes and its Mediterranean representative is the world's smallest. An unusual situation exists in Australia in that the Great Australian Bight (bay) indents the south coast so as to present two west coasts in these latitudes, and thus Australia has two Mediterranean climate areas, one in the Perth region of West Australia and one involving Adelaide and environs in South Australia (Fig. 10–2).

Among the outstanding features of the Mediterranean

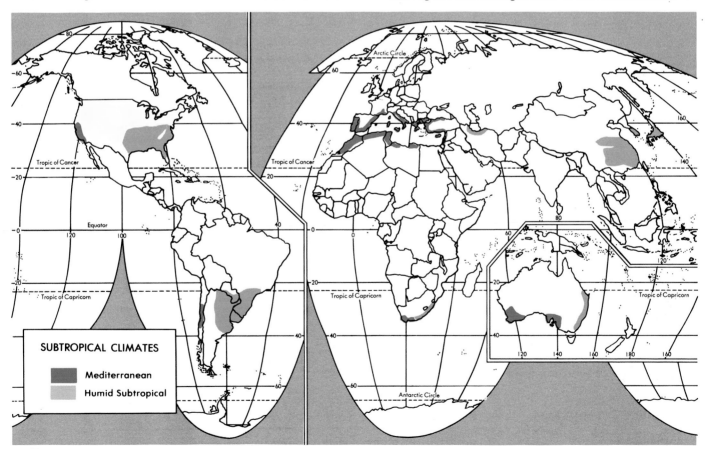

Fig. 10–2   Subtropical climates.

climate is its rainfall peculiarities. Not only is the total annual receipt low, averaging about 15 to 20 inches (38 to 51 cm), but it is highly concentrated in the winter, while the summer is almost absolutely dry. Frequently, the arid season is dominant, reflecting the adjacent situation of the Mediterranean to the tropical deserts. So once again, as in the Tropical Wet and Dry, we find a climate where the year is broken up into two distinct seasons on the basis of rainfall, except that this time winter is the rainy season and summer is dry.

In accounting for such a distinctive rainfall regime, we must analyze the wind and pressure systems of these particular latitudes and recognize again their slight north/south shifting with the seasons. First, dominating the tropical deserts are the Trades, which exert a particularly drying influence on west sides of continents. Then, poleward of the Trades are the Horse Latitudes, regions of sinking air and thus warming drying air. They show up especially as permanent cells at sea, the Hawaiian High off the North American West Coast being representative. And finally, poleward of the Horse Latitudes are the broad Westerlies, characterized by their many eastward-moving cyclones following the Polar Front. These are the wind and pressure systems affecting the Mediterranean regions and, as they shift with the advance and retreat of the summer sun, are responsible for alternating wet and dry seasons.

As an example of a reasonably typical area, let us examine the California Mediterranean region. In the summer when the sun is overhead near the Tropic of Cancer, the northern fringes of the Sonoran Desert move northward to influence California. Offshore the Hawaiian High has shifted to the north also, forcing the storms from the Gulf of Alaska to skirt its poleward edge well north of California. An imported Tropical Dry climate results for the duration of the summer months. But when the overhead sun moves into the Southern Hemisphere, the Hawaiian High tends to follow it south, and the westerly storms from an increasingly active Aleutian Low begin to swing farther south. All of these do not move across the southern half of California but many of them do, and their low stratus clouds and drizzle rain are typical of winter conditions. Occasionally, cyclones will stagnate (blocked by a combination of mountains and a continental high inland) and heavy precipitation will result. At other times, quite violent storms will develop from Polar maritime air detouring far south

over subtropical seas and thus becoming increasingly unstable as it is warmed at the bottom. But between storms bright sunny weather prevails, so that a winter in California is not quite the same as a winter in Oregon, yet frequent cyclones are characteristic and the full year's rainfall is received during the cold half of the year, often in just a few midwinter months. This pattern is essentially the same on all continents in both hemispheres, and its net result is a Mediterranean winter rainfall regime on every continent immediately poleward of the tropical deserts.

Because it is limited to a narrow coastal zone, the high summer temperatures that should prevail at these subtropical latitudes are considerably moderated by oceanic influences. Particularly, the cold current (the same cold current that is found off the coasts of Tropical Dry climates) is an effective moderator of daytime temperatures. Land and sea breezes bring cool air onto the land, and of particular influence are the advection fogs drifting ashore from off the current and shielding the coastal regions from the direct rays of the sun until almost noon each day. Making for even more comfortable living conditions are the cool nights, for summer is the dry season and the accumulated heat of the day is radiated off into space very rapidly during cloudless nights. On the average, summer-month daytime temperatures reach only about 75° to 80°F (24° to 27°C). No wonder the Mediterranean regions have been renowned as resort and tourist areas with long rainless summers and brilliant sunny afternoons, yet mild daytime temperatures and cool nights (Fig. 10–3).

The Central Valley of California, insulated by the coast ranges from the cooling effects of the sea, closely approximates Tropical Dry summer temperatures.

There are exceptions. The Central Valley of California, insulated by the coast ranges from the cooling effects of the sea, closely approximates Tropical Dry summer temperatures. Nights are cool but summer daytime temperatures average near 100°F (38°C) throughout. Latitude has little effect, as Redding in the far north and Fresno in the south display a good deal of similarity—it is distance

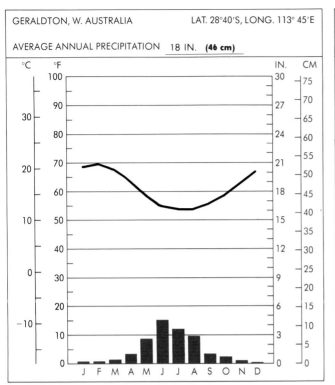

GERALDTON, W. AUSTRALIA   LAT. 28°40'S, LONG. 113° 45'E

AVERAGE ANNUAL PRECIPITATION   18 IN. **(46 cm)**

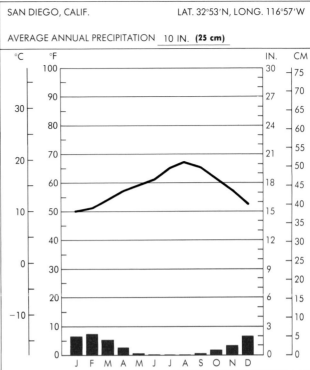

SAN DIEGO, CALIF.   LAT. 32°53'N, LONG. 116°57'W

AVERAGE ANNUAL PRECIPITATION   10 IN. **(25 cm)**

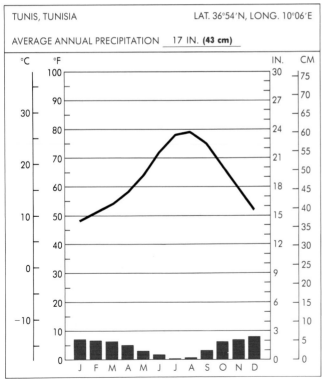

TUNIS, TUNISIA   LAT. 36°54'N, LONG. 10°06'E

AVERAGE ANNUAL PRECIPITATION   17 IN. **(43 cm)**

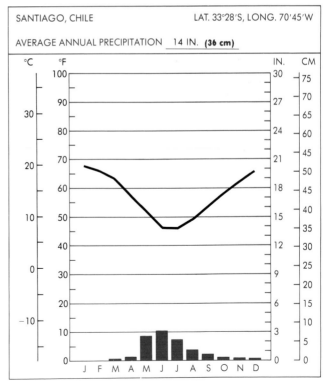

SANTIAGO, CHILE   LAT. 33°28'S, LONG. 70'45'W

AVERAGE ANNUAL PRECIPITATION   14 IN. **(36 cm)**

Fig. 10–3   Mediterranean stations.

from the sea that is important. The question arises, "Should the Central Valley be classified as Mediterranean climate?" In all respects, it is typical except for summer daytime temperatures, and since the only alternative is to establish a separate climatic classification to accommodate it, the Valley is usually included with the Mediterranean. But one should be aware of this single variation.

To a somewhat lesser degree, most of the Old World Mediterranean also deviates from the typical in the same way. The Mediterranean Sea lacks a cold current and, as a landlocked sea, is warmer in any case than the open ocean. Therefore its cooling influence on the adjacent littoral is less effective than elsewhere. Palermo, for instance, in Sicily, is at approximately the same latitude as San Francisco, yet the average temperature for the month of August (both day and night included) is 79°F (26°C), while fog-bound San Francisco reaches only 60°F (16°C) (Fig. 10–3 Tunis).

Short, mild winters and long, warm, sunny summers with virtually a year-round growing season would appear at first to be an ideal agricultural climate, especially for those who are conditioned to regarding California as the great national garden spot. But there is a serious flaw in this agricultural paradise—water. Rainfall is a little short in any case, but more importantly, it comes at the wrong time of the year. The summer growing season is the water-short period, and crops must be limited to those that have the ability to withstand prolonged drought. The natural vegetation reflects this condition in its sparseness and the special adaptation of grasses, brush, and occasional trees to survive aridity. It is revealing that anywhere one travels in the world among the Mediterranean climates olives, grapes, figs, and wheat are common. These are the standard drought-resistant crops and their numbers are not great (Fig. 10–4). To a large degree, the rural poverty of the Old World Mediterranean Basin stems from a numerous agricultural peasant population attempting to wrest a living from the land in the face of this inherent climatic defect.

But given summer moisture, the Mediterranean climate can become highly productive. With irrigation, as in the California Central Valley where exotic streams are fed

Fig. 10–4   The rolling Tuscan rural country near Siena. Grapes and wine are the serious business here, although the agricultural scheme involves olives, grain and vegetables as well. These crops, these methods, and even millions of these people have transplanted themselves very successfully in every Mediterranean locale in the world from South Africa to Chile and California.

Sierra Snow

The Sierra Nevada (Snowy Range) stands as a massive barrier along the entire 350 mile length of California's Central Valley. There are no natural passes across its 8–10,000 foot cerrated crestline as the Donner party and the 49'ers discovered long ago. But if it is an impediment to easy movement of traffic from the east, the Range is equally a major obstruction for Pacific storms and oceanic air currents as they attempt to invade the continent from the west. Orographic precipitation is the result, most of it in the winter during the active storm season, and a large percentage of that in the form of snow at the higher elevations.

These are not dainty, hesitant, tentative little flurries—the Pacific air flow is relatively warm, always saturated and extremely persistent, so that when its only alternative is to rise well above the freezing point snowfall is both inevitable and imposing. 68 inches (152 cm) fell one day in Sequoia National Park,

and the *mere average* for a year in little Alpine County, just south of Lake Tahoe, is 450 inches (1143 cm). If these numbers are impressive, consider the all-time annual record at Tamarack [8000 feet (2438 m), also in Alpine County] of 884 inches (3023 cm). That translates into 74 feet (23 m)—there aren't many places in the world have ever exceeded such a fall.

So here, poised above the hot and arid Central Valley there is a treasure of stored water, glistening in the sunlight and melting at an accelerated rate with the advent of summer. Down the long western flank of the mountains plunge the vibrant rivers, cutting huge canyons as they go to feed the Sacramento in the north and the San Joaquin in the south. This is the dry season irrigation water, tapped to transform the Mediterranean climate from an agricultural liability to a productive miracle.

For continued successful irrigation the key is *water control*. This begins each year with teams of specialists in the mountains sampling snow depth from which is derived an estimate of total runoff for the ensuing season. Dams at critical points on each river hold or release water on demand. All of this works nicely most of the time until the inevitable radically aberrant year arrives—the winter of 1982–3 was one of these.

The early Spanish had found the Central Valley to be alternately an endless marsh and a sere desert; neither appealed to them and they retreated to the more benign coast. In 1982-3 saturation and quasi-swamp conditions had arrived prematurely with a record-breaking rainfall season, now came the melting snow threatening the control dams. Disaster, partially averted by a cool early summer, arrived that year in the same liquid form as the usual salvation. Those old Spanish were more clairvoyant than they knew.

by melting Sierra snows, the long growing season and bright summer sun can be taken advantage of. Now the land, which had been utilized only for sparse grazing or low-yielding dry-field crops, can produce citrus, cotton, sugar beets, and a seemingly endless range of profitable products. However, the normal river in the Mediterranean climate, reflecting the seasonal rainfall, carries water only during the winter; yet even here, given sufficient capital, a dam can be erected to save the winter's water in a reservoir for utilization the following summer. Ideal, of course, is a

nearby mountain range supporting permanent snow, e.g., California, Chile. The melting snows of summer charge exotic streams with abundant flow at exactly the right season. Many would claim that from the standpoint of human comfort, the Mediterranean climate is the world's finest, but without irrigation, it is far from the finest climate for agriculture.

Winters are moderate. Frosts are not unknown but temperatures seldom drop much below 30°F (−1°C), and then only on infrequent winter nights. Such mildness might

To a large degree, the rural poverty of the Old World Mediterranean Basin stems from a numerous agricultural peasant population attempting to wrest a living from the land in the face of this inherent climatic defect.

seem to be expected in the subtropics, but Northern Hemisphere locations in the interior of the continents at equivalent latitudes experience much colder temperatures because they are invaded by outward-moving Polar continental air masses. However, the Sierra Nevada forms a higher barrier against such cold air in North America, and comparable mountain chains ringing the northern Mediterranean Basin are equally effective in blocking out continental air. Occasionally, the cold air will spill over a low spot in the mountains and flow, frequently at high velocity, down the Mediterranean slope. The dreaded *Bora*, at the head of the Adriatic, is a gravity or *katabatic* wind of this type originating in the Danube Basin, but as it loses altitude, it is warmed adiabatically so that when it arrives at the sea, its temperature is a great deal warmer than when it started. It still seems cold relative to the normal temperatures and may be well below freezing, but this type of phenomenon is rare and winter-month averages are usually on the order of 50°F (10°C). In the Southern Hemisphere, continental cold air masses in the middle latitudes are unknown.

HUMID SUBTROPICS

On the east coast of every continent, almost opposite the Mediterranean regions, is the *Humid Subtropic* climate. It too is on the margin of the tropics and shares a common boundary with a tropical climate—in this case, the Tropical Wet and Dry (Fig. 10–2). So if we apply these locational criteria to the United States, we find the Humid Subtropics in the southeastern states. Since this climate, unlike the Mediterranean, is not confined to the seaboard and there are no limiting high mountain ranges, a rather sizable area is encompassed, including all of the so-called

"Old South"—from central Texas to the Atlantic and from the Gulf to a line roughly along the Ohio River to Chesapeake Bay. Comparable to this on the east side of the Eurasian continent is a large part of central China centering on the Yangtze River, and southern Korea and Japan. The same pattern holds in the Southern Hemisphere where southern Brazil, Paraguay, Uruguay, and the Argentine Pampas make up a large contiguous area. But in South Africa and Australia, mountain ranges parallel the coast and limit the Humid Subtropics to a narrow strip. The representative in Africa is very small for this reason and also because Africa scarcely pushes into the subtropics, but in Australia the coastal corridor extends from near Brisbane in the north to almost the southern tip of the continent.

The name "Humid Subtropics" points up the major difference between this climate and the Mediterranean, that is, moisture. Not only is the annual total roughly double that of the Mediterranean, averaging 40 inches (102 cm), but there is no dry season. We are dealing here with the identical wind and pressure systems that we were on the west coast but their effect is different. The Trades, for instance, which in the summer contributed to west coast aridity, are onshore in the east, coming off a warm sea and crossing a warm current. And not merely are the Trades onshore, although they do not migrate far enough poleward to affect the entire subtropics, but the general flow of air into the summer continental low brings moist air off the sea. In the Far East, we use the term "summer monsoon"; however, every continent, even those in the Southern Hemisphere, has a summer monsoonlike tendency and sucks in air from offshore. This air is heated by conduction and the typical warm season precipitation comes in the form of frequent afternoon convectional thundershowers. Heat plus moist air equals convectional precipitation, and the Humid Subtropics achieve this equation each summer.

Heat plus moist air equals convectional precipitation, and the Humid Subtropics achieve this equation each summer.

To add further to late summer/early fall precipitation totals, we should remember that with the exception of South America, these latitudes are affected by hurricane/typhoon influences. The coastal regions especially receive moisture from this last source simply because these storms are in the general vicinity. So while the west coast Mediterranean is experiencing a protracted dry season, the east coast Humid Subtropics are receiving significant rainfall, often, as in the Far East where the monsoon and typhoon are well developed, a definite summer maximum.

Winters are much like those in the Mediterranean. Polar Front cyclonic disturbances swing farther equatorward with the retreat of the overhead sun, and frontal activity gives rise to gray skies and drizzle rain. Only in the Orient is there a variation on this pattern, for North Atlantic spawned storms do not always survive the trip across Asia in the winter. They are compensated for to a degree by new storms generated along the Polar Front in the vicinity of the upper Yangtze Valley, but generally winter precipitation fails to match that of the summer.

In the Northern Hemisphere some of this winter precipitation is in the form of snow as the great continental cold air masses surge out of the interior. They do not dominate the Humid Subtropics by any means, but occasional cold spells are a definite element of the winter climate. Lack of mountain barriers in North America, at right angles to the flow of Canadian air, allows invasions of cold all the way to northern Florida and the Gulf for short periods most winters, and the more northerly locations such as Virginia and Kentucky receive several minor snowstorms every year. Asiatic Humid Subtropic regions, although protected to a greater degree by high mountains, are subject to a more vigorous winter monsoon pushing Siberian air far out of the interior, so that with few exceptions, periodic snow and cold are encountered everywhere. Tokyo, Shanghai, and Seoul all have several cold spells each year with snow a regular winter feature.

Tokyo, Shanghai and Seoul all have several cold spells each year with snow a regular winter feature.

However, the average temperatures for the winter months, as shown on the climate charts (Fig. 10–5), do not always reflect the few days of freezing weather each month, for they are offset by many relatively mild days.

When we compare the averages of Northern Hemisphere stations with those of the Southern Hemisphere, where there are no large middle latitude continents to develop cold air masses, there appears to be no great difference. But the cold spells should be recognized. An excellent example of their effect in limiting agriculture is a comparison of the northernmost limit of frost-touchy citrus on a commercial scale in Florida versus southern California. The lake country of central Florida is the heart of the industry in the Humid Subtropics with the northern limit in the area of Jacksonville. In Mediterranean California sizable orchards are common in the San Joaquin Valley at least as far north as Fresno. There is a discrepancy of 6° to 7° of latitude here, reflecting the frequency of killing frost, yet the coldest month average for both the Mediterranean and Humid Subtropic regions in general is much the same—in the vicinity of 45° to 50°F (7° to 10°C).

Summers are hot—several degrees warmer than the same latitudes on the west coast—for there is no cooling current offshore and no fog. Moreover, the humidity is constantly high because Tropical maritime air is dominant, so that sensible temperatures are uncomfortable, even at night. New Orleans, for instance, features four midsummer months averaging close to 80°F (27°C) and almost daily convectional showers. This is a very near approximation of Tropical Rain Forest conditions and without air conditioning can be a difficult season. Days are often in the 90's (F) (32° to 38°C) and the hottest month average in the Humid Subtropics is seldom below 75°F (24°C). Florida beaches may be very pleasant as the local land and sea breezes keep the air moving, but only a few miles inland, it becomes excessively hot and sticky.

Thus the Humid Subtropics differ in several ways from the Mediterranean climate:

1. Greater total precipitation distributed more evenly throughout the years.

2. Higher summer temperatures.

3. Similar mild winter temperatures, but interspersed with occasional cold spells in the northern hemisphere.

But where this climate suffers a bit relative to the Mediterranean in terms of human comfort, it is basically much more productive agriculturally. A long growing season combined with adequate rainfall in midsummer is the chief

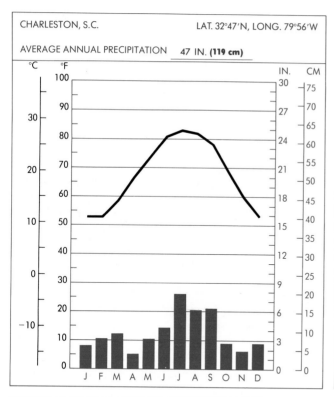

CHARLESTON, S.C.   LAT. 32°47'N, LONG. 79°56'W

AVERAGE ANNUAL PRECIPITATION   47 IN. **(119 cm)**

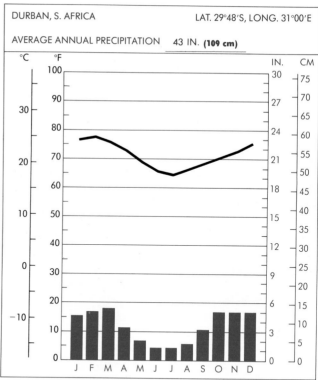

DURBAN, S. AFRICA   LAT. 29°48'S, LONG. 31°00'E

AVERAGE ANNUAL PRECIPITATION   43 IN. **(109 cm)**

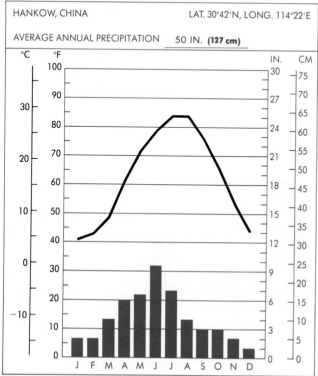

HANKOW, CHINA   LAT. 30°42'N, LONG. 114°22'E

AVERAGE ANNUAL PRECIPITATION   50 IN. **(127 cm)**

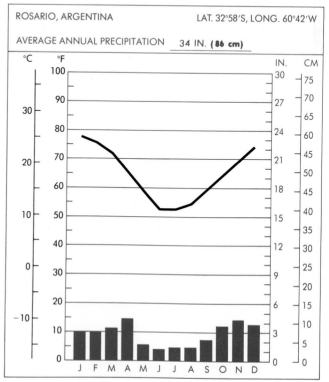

ROSARIO, ARGENTINA   LAT. 32°58'S, LONG. 60°42'W

AVERAGE ANNUAL PRECIPITATION   34 IN. **(86 cm)**

Fig. 10–5   Humid Subtropic stations.

Fig. 10–6   Look what's happening here in the American sunbelt. Dallas used to be a relatively small serene ranch and cotton town with few illusions of greatness. Now it's highrises are visible for miles above the east Texas plain, its southwest accent is tempered by Yankee influences and it is home to "America's Team". Surging urbanism has become a modern subtropical reality everywhere.

advantage, and because the summer precipitation is derived from afternoon showers of short duration, days remain bright and sunny. Everywhere in the world, people are engaged in agriculture in the Humid Subtropics.

Intensive rice culture supports a numerous population throughout the Far East and farming has been a way of life for generations in the U.S. Cotton Belt. Lesser population pressures have led to a more extensive land use in South America and Australia, yet the agricultural potential remains. Where excessive slope or soil inadequacies are a problem, forests often make the land productive and regenerate themselves rapidly, once cut, under the influence

The 13 states of the old Confederacy, taken together, produce more wood products today than any other section of the country.

of a benign climate. Active reforestation of former cropland is widespread in the U.S. The 13 states of the old Confederacy, taken together, produce more wood products today than any other section of the country (Fig. 10–6).

CHAPTER 11

THE MID–LATITUDE CLIMATES

An aerial view of the old walled city of Nordlingen, Bavaria, West Germany reveals an ancient habitation pattern, not too changed today from long ago. Within the walls medieval architecture still prevails although more modern suburbs have spilled beyond. In the outlying farm country multiple cropping is not only feasible but prosperous despite cold middle latitude winters; but tiny fields speak of long occupation and intensive land-use. Farming techniques are a curious blend of traditional hand methods and a few modern machines—huge tractors and combines are simply not compatible with the micro–landscape.

INTRODUCTION

Largely dominated by the Westerlies, the middle latitude climates occupy a broad belt between the subtropics and the high latitudes. Along west coasts they extend in some cases almost to the Arctic Circle, while their dry representative pushes equatorward to abut with a tropical climate. The basic characteristic, as might be expected, is strong seasonality, although there is an important exception; and the middle latitude climates are most widely represented in the Northern Hemisphere where the continents become larger as opposed to their very limited extent at these latitudes in the Southern Hemisphere (Fig. 11–1).

MARINE WEST COAST

The climate of the entire west coast of every continent in the middle latitudes is called *Marine West Coast*. The third of a west coast series of climates stretching from well within the tropics to the poleward limits of the middle latitudes, the Marine West Coast shares a common boundary with the Mediterranean at the margin of the subtropics and it in turn merges into the Tropical Dry. Chile is a neat yardstick illustrating this inevitable series: the northern one-third is Tropical Dry, the central one-third Mediterranean, and the southern one-third Marine West Coast. Only Africa, of all the continents, misses out, for it does not push

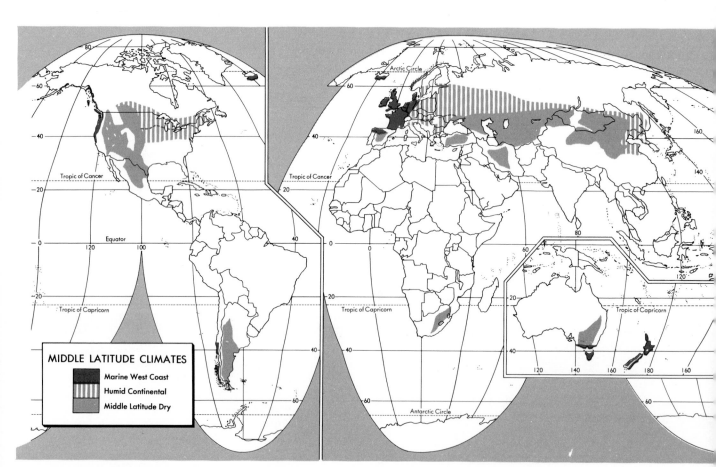

Fig. 11–1   Middle latitude climates.

MIDDLE LATITUDE CLIMATES
- Marine West Coast
- Humid Continental
- Middle Latitude Dry

into the middle latitudes in either hemisphere. Australia barely does, with only Tasmania and the southern tip of Victoria in the vicinity of Melbourne being represented. And New Zealand, in much the same latitude, can be included here with Australia.

The Northern Hemisphere, where the greatest land masses are in the middle and high latitudes, exhibits the most extensive regions of Marine West Coast climate. Along the west coast of North America, the climate is essentially the same from northern California to southern Alaska. The belt is narrow, between the fringing mountains and the sea, but a full 20° of latitude are involved. Western Europe is similar in this respect in that the entire coastal zone, from northern Spain to just short of the Arctic Circle in Norway, is classified as Marine West Coast. Included here are northern Spain, western France, the Benelux countries, the North Sea Coast of Germany, Denmark, southern coastal Norway and Iceland, and, of course, Great Britain and Ireland. However, unlike North America, only in Norway is there a mountain range closely paralleling the seaboard, and the inland climatic boundary is much more difficult to determine. As the word "marine" implies, regions involved in this climate must be closely oriented to the oceans, so that where no mountain barrier exists to demark marine from continental influences sharply, as in Western Germany and France, we must recognize a zone of subtle transition.

Probably the most striking feature of the Marine West Coast climate is the mildness of the winters at these relatively high latitudes. One measure of this is the average temperature of the coldest month, which is typically above 32°F (0°C)—often in the low 40's (F) (4° to 7°C). Another measure is length of growing season (number of days between the last frost in the spring and the first frost in the fall), which averages well over 200 days, and at such locations as Valencia, Ireland, and Cape Flattery, Washington, is over 300 days. This is almost tropical. In the southeastern United States a 200-day growing season is regarded as the minimal limit for growing cotton, so in essence, other things being equal, subtropical crops such as cotton can be grown in, say, British Columbia. Other things are not equal, of course, but it is nonetheless remarkable that such mild temperatures should prevail. Freezing temperatures, if not common, occur every winter and snow is not unknown,

although nearly always it comes in short-lived flurries and the fall melts rapidly. But compared with North Dakota or central Siberia at exactly the same latitude, the Marine West Coast is a virtual hothouse.

Other things being equal, subtropical crops such as cotton can be grown in British Columbia. Other things are not equal, of course.

The oceans are responsible. These great water bodies to the west change their temperature only slightly from season to season, and the air masses above them reflect this moderation. Of equal importance is the fact that the prevailing winds in the middle latitudes are from the west, constantly bathing the west coasts with oceanic air. These keep the excessively cold continental air masses of the interior at bay, and if reinforced by protective mountain chains, as in North America, or if the cold air is lacking, as in the Southern Hemisphere, the west coasts display very moderate temperatures indeed. Western Europe is the least typical in that Eurasia develops the largest and coldest winter air mass in the world (with the possible exception of Antarctica), and there is no continuous mountain barrier to block its occasional invasion of the west coast. But even here, this is relatively rare.

Another definite warming factor, although probably of secondary significance to the basic oceanic influence, is the occurrence of a warm current paralleling these west coasts. This is the same current that was classified as cold off the coasts of the Mediterranean and Tropical Dry climates, but at higher latitudes, flowing through colder waters and retaining somewhat more of its residual tropical temperatures, the current is considered warm (see Chapter 29). Its effect is to exaggerate the already moderating influence of the oceanic air masses, adding perhaps 3° to 5°F (2° to 3°C) to the winter temperature average. In the North Atlantic, where a lesser branch follows the Norwegian coast into the Arctic Ocean, ports are kept ice-free all the way to Murmansk, and the Marine West Coast climate is pushed northward almost to the Arctic Circle (Fig. 11–2).

So in the winter the sea is the source of heat and the continent the source of cold, and the moderation of this

Fig. 11–2 This photo could be in Alaska, Norway or Chile. The latitude is 50° to 60° from the equator, yet the little harbor is ice-free. There is not much physical space here for building a town at the foot of steep mountains, but there is a valuable resource to exploit. The same warm current that discourages icing induces vertical turbulence, as it invades colder water, an abundance of plankton and a numerous fish population.

middle latitude climate is directly related to its nearness to the sea. Cold-season isotherms depart radically from their normal east/west trend on crossing west coasts, especially in the Northern Hemisphere, and align themselves almost north/south; in this part of the world, latitude is of less importance as a temperature control than is coastal orientation.

Summer temperatures are cool, the hottest month usually averaging in the 60's (6° to 21°C). In this season the imported oceanic air masses exert a cooling influence that although not as striking as the winter warming, is readily apparent when compared with summer temperatures in the interior of the continent. The general tendency of the sea to maintain much the same temperature the year round is reflected in the limited annual range of the Marine West Coast climate. Warmed in winter and cooled in summer, the annual temperature range is no more than 15°F (8°C) in many places and seldom greater than 20° to 25°F (11° to 14°C). There are tropical locations with a range greater than that, and it is about the same as the average annual temperature range in the Mediterranean climate. Such a lack of seasonality is one of the chief features of the Marine West Coast climate and is particularly startling when

encountered in the middle latitudes where strong seasonal differences are to be expected (Fig. 11–3).

Warmed in winter and cooled in summer, the annual temperature range is no more than 15°F (8°C) in many places within the Marine West Coast Climate Region.

Despite these very pleasant temperatures, the Marine West Coast regions have not rivaled the Mediterranean in attracting tourists and resorts; they have had a reputation of constant rain and a generally dark and gloomy aspect—a reputation that is not wholly unearned. It is inevitable that a narrow coastal strip with prevailing winds onshore should experience a good deal of precipitation, particularly if it is backed up by high mountains forcing saturated oceanic air masses aloft. And there are many places where the mountain slopes receive in the neighborhood of 200 inches (508 cm) each year. But even where the mountains are lacking, persistent rainfall results from the frontal ac-

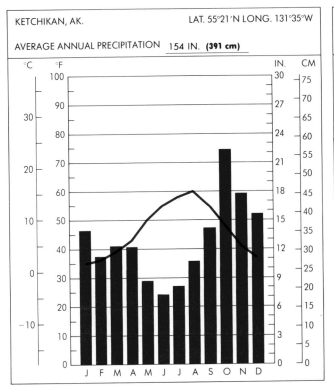

KETCHIKAN, AK.   LAT. 55°21'N LONG. 131°35'W

AVERAGE ANNUAL PRECIPITATION   154 IN. **(391 cm)**

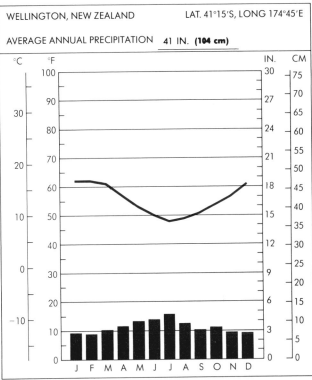

WELLINGTON, NEW ZEALAND   LAT. 41°15'S, LONG 174°45'E

AVERAGE ANNUAL PRECIPITATION   41 IN. **(104 cm)**

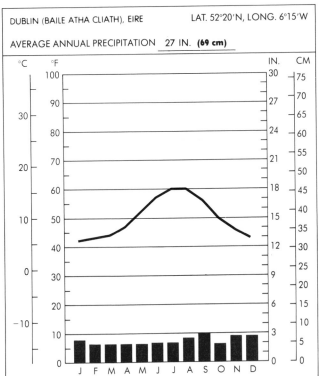

DUBLIN (BAILE ATHA CLIATH), EIRE   LAT. 52°20'N, LONG. 6°15'W

AVERAGE ANNUAL PRECIPITATION   27 IN. **(69 cm)**

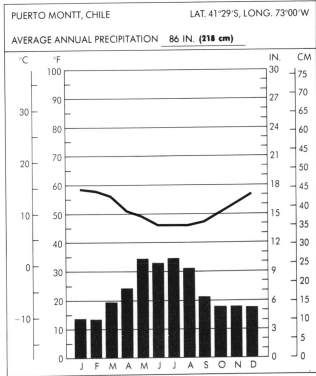

PUERTO MONTT, CHILE   LAT. 41°29'S, LONG. 73°00'W

AVERAGE ANNUAL PRECIPITATION   86 IN. **(218 cm)**

Fig. 11–3   Marine West Coast stations.

tivity of the many cyclones that frequent the Westerlies. This is usually in the form of light but constant drizzle, a good deal of low-hanging cloudiness, and mist and fog interspersed with actual rain. The end product is a great many rainy days each year but a total receipt of less precipitation than might be expected. London, for instance, enjoys a deserved reputation of being a very dank and drippy city. Virtually every day is a rainy day, the ground is always wet, the air saturated, and umbrellas and raincoats are in order for all seasons. Yet London receives only 25 inches (64 cm) of rain each year. In some parts of the world, this would be considered semiarid. But because so much of London's rainfall is in the form of drizzle or mist, it requires about 200 rainy days to produce these 25 inches (64 cm). And because of the saturated air and high rate of cloudiness, evaporation is kept to a minimum and 25 inches (64 cm) is more than adequate for agriculture. Seattle gets only 33 inches (84 cm) of rain per annum, Portland, 44 (112 cm), and Melbourne, 26 (66 cm). These totals are scarcely exorbitant, yet to live in any of these cities is a wet and dreary experience, mainly because of excessive cloudiness. All Marine West Coast regions receive well under 50% of the possible sunshine each year. Scotland, for example, receives less than one hour of sunlight per day in December and only five and a half hours in June despite the very long summer days. It is this unremitting cloudiness that accounts for the general gloom that appalls the casual visitor, but it also is responsible for the lessened evaporation, so that a little moisture goes a long way and maintains the lush greenness of the landscape and the permanent snowcaps on even fairly low mountain peaks.

It is this unremitting cloudiness that accounts for the general gloom that appalls the casual visitor.

Precipitation totals may thus vary widely within the Marine West Coast climate, from a low of 23 to 25 inches (58 to 64 cm) to over 200 (508 cm), but in terms of human occupation, there appears to be a maximum limit of about 60 inches (152 cm). All the large cities and densely populated lowlands are in regions of under 60 inches (152 cm)

of rainfall, while those areas with more are virtually uninhabited. A part of this may be simply that normal drainage cannot cope with over 60 inches (152 cm) and marshes and moors prevail, but in order for a region to receive these high totals, it must have steep mountain slopes immediately adjacent to the sea, and such a coast allows little room for humans to find a toehold. The British Columbian Coast, western Scotland, much of Norway, and southern Chile are all empty, or support only tiny villages of fishermen or lumbermen (Fig. 11–3).

Normally, there is a slight tendency toward a winter precipitation maximum, although the summers are by no means dry, because there is a general lessening of cyclonic activity during the warm months. However, there is one small area that deviates enough from the average situation to merit mention. This is the North American Coast from northern California to southern British Columbia. Here is a distinct, if limited, summer dry season that lasts from about mid-July to mid-September. It is a beautiful time of year with a great deal more sunshine than is usual, and daytime temperatures seldom out of the 70's (F) (21° to 26°C). Lawns and pastures require some irrigation and forest fires are a problem, but generally the dry season is of such short duration that the countryside remains green and the mountain snowcaps glisten without melting away. This unusual situation seems to be the result of a general weakening of the Aleutian Low with a lessened ability to produce vigorous storms, while at the same time the offshore Hawaiian High cell migrates strongly northward, shielding this part of the coast from the few storms that do move eastward. But all of this does not last long. By mid-September, the rains begin and, although somewhat sporadic at first, shortly settle into a 9 to 10-month pattern of clouds and drizzle.

Where flat land exists and the rainfall is below 60 inches (152 cm), the Marine West Coast regions have attracted people. Agriculture is possible, although not ideal, the greatest drawback being lack of sunshine rather than temperature. In many parts of the world, grass has been the best crop.

It is not surprising that most of our domesticated breeds of grazing animals are of western European origin and have been introduced successfully into the Pacific Northwest, New Zealand, and Chile—Hereford, Dorset, Shropshire, Percheron, Holstein, etc. Other crops besides grass are

New Zealand

Marine West Coast New Zealand, over the last century and a half, is perhaps an extreme example of almost total national dependence on the export of animal products to the British market. But Britain is in the European Common Market now and things have gone awry.

Blessed initially with large, temperate, well-watered islands as the fundamental resource, an alien people (Englishmen) came onto the scene to overwhelm the Maori, and set about changing things. They cut the trees, they burned them, and at an ever-accelerating pace accomplished almost total deforestation, exposing the naked face of the land. Quickly, exotic grasses and legumes from Europe were sown to hold the soil, and then alien grazing animals were introduced to transform the velvety green countryside into valuable meat and dairy products. If there is a universal field crop in New Zealand, it is grass. Sown, fertilized, and frequently regenerated and rejuvenated, grass is treated in much the same manner as an agricultural endeavor, but the harvest is accomplished by the sheep and cattle whose intrinsic gastronomic talent is to translate a cheap product into one of great value. Finally, the end-products of this manipulation of the land find their way to foreign markets and earn those economic rewards that support the good life. But that good life is now in jeopardy and New Zealanders are more than slightly concerned.

Leading New Zealand Exports in Order of Value

1.	Wool	10.	Edible offal
2.	Lamb	11.	Casein
3.	Beef	12.	Cattle hides
4.	Dried milk	13.	Sausage casings
5.	Wood products	14.	Tallow
6.	Butter	15.	Apples
7.	Cheese	16.	Veal
8.	Mutton	17.	Crayfish
9.	Sheepskins and pelts	18.	Grass seed

grown, but many of them such as oats and potatoes have the special ability to produce well with lack of sunshine.

It is not surprising that most of our domsticated breeds of grazing animals are of Western European origin: Hereford, Shropshire, Dorset, Percheron, Holstein.

Forests are the natural vegetation in the Marine West Coast climate regions, and their exploitation is an important industry where they are reasonably accessible. (Fig. 11–4). In western Europe most of the trees are gone, having been cut long ago; but in southern Chile and Canada/Alaska, virgin forests remain untouched.

Wood for ships, often rugged terrain, excellent harbors, and a warm current offshore conducive to the growth of sea life, all of these factors have conspired to force men to go to sea as fishermen, and the fishing industry and the maritime heritage as evidenced in extensive merchant fleets and seaborne foreign trade are well represented. Today, trade is tied intimately to manufacturing, and western Europe, with a fortuitous mélange of mineral resources as well, stands out as one of the world's great industrial and commercial leaders.

To a degree, the broken coastlines of typical Marine West Coast climate regions are a result of climatic similarity. All of these regions with the same climate today are characterized by fjorded coasts. Here is evidence that in the past all must have had a similar climate, which was cold enough to support ice at sea level, for fjords are a result of

Fig. 11–4 Forests are an important resource in the uplands and less densely populated parts of the Marine West Coasts. These are massive Douglas fir logs being efficiently handled by modern forestry equipment in western Washington. Reseeding of a new crop is the routine follow-up, either naturally from adjacent forest blocks left standing, or nature can be aided by helicopters.

ice erosion (see Chapter 24). It is a rare situation where in examining a present-day physical map, we can find evidence of climatic similarity by merely looking at the coastal character. Alaska, British Columbia, Puget Sound, Scotland, Norway, southern Chile, Tasmania, and New Zealand all have fiords, evidencing past climatic similarities, and all display a Marine West Coast climate today.

MIDDLE LATITUDE DRY

The *Middle Latitude Dry* climate is very similar to the Tropical Dry, so similar in fact that many classifiers lump all the dry climates together as one. Furthermore, the areas involved in the Tropical Dry climate and those of the Middle Latitude Dry are adjacent and merge almost imperceptibly one into the other so that there are inherent difficulties in deciding exactly where to separate them. But aside from the obvious locational differences between the Tropical Dry and the Middle Latitude Dry, they are the results of entirely different causes and there is one major difference between the climates. We will regard them as separate climates, although we must recognize their many similarities.

It has been traditional to regard the 100th meridian as the 20 inch (51 cm) rainfall line. To the pioneers, here was the beginning of the Great American Desert.

Since the middle latitudes are essentially the Westerly belt, any dry climate occurring here must be divorced from the west coast with its onshore humid air masses (Fig. 11–1). The North American representative comes very close to the west coast but is east of the high and continuous Sierra Nevada/Cascade chain, so that it is in a rain-shadow position and the moisture off the sea is precipitated in heavy amounts on the Marine West Coast seaward slope of the mountains. Here is one of nature's sharpest lines: up to 200 inches (508 cm) of rainfall on the windward slope, while such places as Wapato, Washington, immediately in the lee of Mt. Rainier, and the Owens Valley, California, in the rain shadow of Mt. Whitney, receive 5 inches (13 cm). The entire area between the Sierra Nevada/Cascades and the Rockies averages only 10 to 15 inches (25 to 38 cm). But this is only half the total dry region, for although the Rockies attract some increased moisture through orographic lifting, in their rain shadow on the high plains it is dry once again. It has been traditional to regard the 100th meridian running through the Great Plains as the 20 inch (51 cm) rainfall line in the United States; that is, west of this line, the precipitation is less than 20 inches (51 cm). To the pioneers, here was the beginning of the Great American Desert, and it is so labeled on many old maps. We know now, of course, that the true desert lies much farther west, but in our classification here as in the Tropical Dry, we will include the semiarid regions with the true desert, since water is the critical factor in both. Thus it becomes convenient to designate the 20 inch (51 cm) rainfall line as the eastern margin of the Middle Latitude Dry climate.

A second and even larger Middle Latitude Dry region is encountered in interior Eurasia. It is in the same general latitudes of that in North America but much farther removed from the west coast. In Europe, it will be recalled, no continuous mountain chain parallels the coast to sharply limit all the moisture to a narrow littoral. Consequently, the Marine West Coast in northern Europe merges into somewhat drier climates farther inland, but the decreased receipt of Atlantic moisture with distance eastward is such a gradual process that a 20 inch (51 cm) rainfall line is far back into the interior of the continent. The western margin of the Middle Latitude Dry is in the southern Ukraine, and from there it stretches eastward into the Caspian Basin and Russian Turkestan, interior China north of the Tibetan Plateau, and even into parts of Manchuria and the far north of China proper, almost to the east coast.

The only part of the Southern Hemisphere where a significant landmass is encountered in the middle latitudes is in South America, and even here the area is limited. But Patagonia (the southern half of Argentina) finds itself in the rain shadow of the Andes, which isolate the Marine West Coast region of southern Chile. Here is a close parallel to the situation in North America with the Middle Latitude Dry climate closely approaching the west coast.

Aridity is, of course, the chief cohesive factor in allowing us to classify all these regions together. It does not matter whether the moisture arrives in the summer or the winter, we make no distinction; the important thing is that there is not enough. A good deal of the Middle Latitude Dry is true desert, that is, less than 10 inches (25 cm) per annum, but the fringing semiarid regions receiving up to 20 inches (51 cm) of precipitation are also included. One reason that a district with 15 to 20 inches (38 to 51 cm) of rain can be legitimately classified here is due to the typical erratic nature of its occurrence Fig. 11–5. As in the Tropical Dry zones, *the drier the climate, the less reliable the rainfall*, and this often means great difficulty in carrying on any reasonably sustained and coherent human land use. Yet all of these precipitation characteristics fail to help us find a means of separating the Middle Latitude Dry regions from those of the Tropical Dry, for they are essentially similar. It remains for a certain dichotomy of temperature to aid us in this endeavor.

Winter temperatures in the heart of the Middle Latitude Dry regions, especially in the Northern Hemisphere where continental air masses dominate, are well below freezing, and precipitation is in the form of snow. This is a far cry from the Tropical Dry where cold-month averages are typically in the 50 to 60°F (10° to 16°C) range. In eastern Montana or Chinese Sinkiang, winter days are frequently far below 0°F (−18°C). Here is what makes the middle latitudes a different world from the tropics. On the other hand, summers are hot, often excessively so; but in the Middle Latitude Dry, they are neither as hot nor of as long duration as in the Tropical Dry. Averages run in the neighborhood of 70° to 75°F (24°C) for the hottest month. Nights are cool, reflecting the low humidity; and even in the winter, days may be a great deal warmer than the nights (Fig. 11–5).

To a large degree, the geography of the western United States is oasis geography.

Temperatures, then, particularly in the winter, are the strong divisive element between the Tropical Dry and the Middle Latitude Dry, yet in the zones where these two dry regions merge, there is an extensive area where temperatures take on a median character somewhere between the extremes of the tropics and the middle latitudes. In northern Mexico, southern New Mexico, Iran, Iraq, Syria, northwestern Argentina, and in parts of southern Africa and Australia, we encounter a dry climate, but one whose temperature characteristics are not quite typical of either Tropical Dry or Middle Latitude Dry. Winters, for instance, average just above freezing and summers are in the low 80's F (27° to 29°C). This is further complicated by upland plateaus, as in Chihuahua and Iran; while not high enough to be regarded as mountains they do slightly affect temperatures. Classifiers have always had difficulty drawing a separating line through this territory. One solution is to establish an entire new classification and call it "Subtropical Dry." The problem, then, is that here is an "almost but not quite" region, and such a climate lacks character of its own to strongly distinguish it from its neighbors. The plan

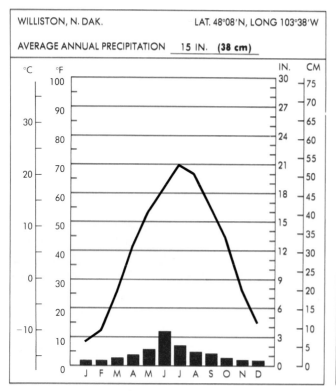

WILLISTON, N. DAK.   LAT. 48°08'N, LONG 103°38'W

AVERAGE ANNUAL PRECIPITATION   15 IN. **(38 cm)**

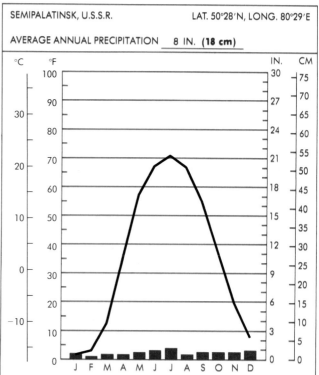

SEMIPALATINSK, U.S.S.R.   LAT. 50°28'N, LONG. 80°29'E

AVERAGE ANNUAL PRECIPITATION   8 IN. **(18 cm)**

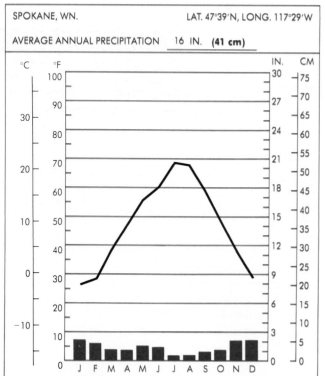

SPOKANE, WN.   LAT. 47°39'N, LONG. 117°29'W

AVERAGE ANNUAL PRECIPITATION   16 IN. **(41 cm)**

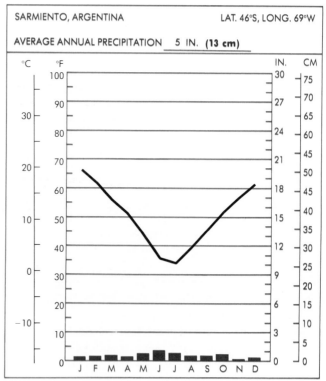

SARMIENTO, ARGENTINA   LAT. 46°S, LONG. 69°W

AVERAGE ANNUAL PRECIPITATION   5 IN. **(13 cm)**

Fig. 11–5   Middle latitude dry stations.

Fig. 11–6   Vivid green alfalfa in this Idaho oasis contrasts strongly with the arid hills. The canal in the foreground supplies irrigation water to only the limited valley bottom.

followed here is simply to draw a line at a midway point through this zone of merging. Recognize that such a line is somewhat arbitrary, and although this border region is often thoroughgoing desert, it is a transitional band between Middle Latitude Dry and Tropical Dry.

As in all dry regions, water is the critical factor in allowing human habitation on any significant scale. To a large degree, the geography of the western United States, for instance, is oasis geography (Fig. 11–6). Yakima, Salt Lake, Reno, Boise, and lower Colorado are all densely populated oases, but in between these limited and widely scattered favored spots are extensive regions essentially unpopulated, although here and there rich mineral deposits support some activity despite a lack of water. But the semiarid desert margins, unlike most comparable regions in the tropics, have been made reasonably productive. Grazing has been traditional, but large-scale grain farming, especially wheat, has taken over extensive areas. The reasonably good accessibility to the middle latitude urban markets is probably largely accountable for the greater development of agriculture, relative to the roughly similar tropical situations, but this is far from ideal agricultural land. Both the high plains of North America and the "virgin lands" in the USSR, as examples, have suffered from recurrent drought, and they have had a continuous history of all too frequent

crop failure. There are those who claim that these regions are not agricultural land at all and should be maintained in permanent grass and that even grazing should be controlled to conform to the inevitable rainfall cycles.

HUMID CONTINENTAL

The third middle latitude climate is the *Humid Continental*, in much the same latitude as the Marine West Coast and the Middle Latitude Dry. It is encountered, however, only in the Northern Hemisphere, for even South America does not exhibit sufficient mass for its full development in the Southern Hemisphere. In North America virtually the entire northeastern quarter of the United States and adjacent Canada is involved. Its southern boundary is the Humid Subtropic, its western boundary is the 20 inch (51 cm) rainfall line or the Middle Latitude Dry, and its northern border is essentially the northern limit of agriculture dictated by a shortened growing season. There is a finger of the Humid Continental that pushes far to the west, north of the Middle Latitude Dry. Decreasing evaporation offsets the slightly lesser precipitation, allowing this region to be classified as humid.

Fig. 11–7   Spring is still over a month away when the de rigueur mid-March snowstorm arrives.

In Europe the Humid Continental is found immediately east of the Marine West Coast and extends well into the interior of the USSR. The area involved is in the shape of a triangle with its broad base to the west. Southern Sweden, all of the Germanys except the North Sea Coast, Czechoslovakia, Hungary, and the Po River Valley of northern Italy form this base, while to the east, southern Finland, Poland, Romania, and much of European USSR are included. The tip of the triangle is north of the Middle Latitude Dry region and pushes as far east as Lake Baikal.

The Far East, even this far north in China or Korea, still reflects the influence of the monsoon.

There is another smaller triangle-shaped area classified as Humid Continental in the Far East; this one has its base along the sea. Most of North China and Manchuria east of the Middle Latitude Dry is included, as are North Korea and the northern half of Japan. The triangle's tip is north of the Middle Latitude Dry region and follows the Amur Valley inland to Lake Baikal (Fig 11–1).

The word "humid" in connection with this climate means an average precipitation of about 30 inches (76 cm), although there are variations from 20 inches (51 cm) at the fringe of the semiarid to over 50 inches (127 cm) near the sea. But nowhere is the receipt of rainfall during the growing season too slight for agriculture. In the winter the bulk of the precipitation is of cyclonic origin from the usual frequent westerly storms, and much of it is snow. Summer rainfall, on the other hand, occurs most commonly as convectional showers. There is a general tendency for summer and winter precipitation to be approximately equal near the coast (except the Far East), but with increased distance inland, a summer concentration becomes apparent. This is explained by the winter dominance of the dry cold continental air masses of the interior from which only limited amounts of moisture are derived. The Far East, even this far north, still reflects the influence of the monsoon, so that strong summer maximums are the rule and winters are almost dry despite occasional winter cyclones. However, Japan, because of its insularity, receives heavy winter snowfall. This pattern of winter cyclonic precipitation and summer convectional showers with no well-defined dry season is reminiscent of the Humid Subtropics, but the annual totals are somewhat less and a good deal more snow is experienced (Fig 11–7).

The word "continental" appearing in the climatic title has definite temperature implications. It is the opposite of "marine," which meant mitigation by the sea of both summer and winter temperatures and a resulting small annual range. Continental has inherent within it the concept of continental control of temperatures—radical cooling in the winter and heating in the summer. And the air masses reflecting this large annual range give their character to the temperatures of the Humid Continental climate. Summers are hot and often sticky, as monsoonal influences bring in tropical oceanic air; nights cool off very little. These are the times when the Iowa farmer must rationalize that his corn is growing nicely even if conditions are far from ideal for human comfort. But it also may be a good deal cooler along the Baltic seafront or in Nova Scotia, and in the far interior the heat is less trying because of drier air despite the high temperatures. Generally, the Humid Continental summers average 75°F (24°C) and are quite warm for their latitudes.

If one wakes up late some May morning it is entirely possible to miss spring altogether.

Winters are even more uncomfortable. Cold-month temperatures must average below freezing in order to qualify as Humid Continental, and although near the coast, they may not be far below freezing; the humidity makes for a penetrating sort of cold. Inland it is drier but temperatures fall to averages of 10° to 12°F (−12° to −11°C) with a good many days below 0°F (−18°C).

Basically, there are only two pleasant months each year—May and October, the transition periods. Indian summer is a glorious time and can be counted on virtually every October, but spring is liable to be a transient affair so that if one wakes up late some May morning it is entirely possible to miss spring altogether.

Long Summer versus Short Summer

Because of its size and the relatively great variation within the region we call Humid Continental, it has become common to subdivide it on the basis of its agricultural potential. This was first done in the United States because of a very obvious "crop line" running east/west across much of the Middle West. The chief crop utilized as a climatic indicator was corn. South of a line drawn through southern South Dakota, southern Minnesota, southern Wisconsin, southern Michigan, the Ontario Peninsula of Canada, northern New York, and southern New England, the growing season is long enough for old fashioned field corn to come to full maturity. The newer hybrid corns allow the farmer a good deal more flexibility, but the climate boundary was drawn well before their advent. Winter wheat, too, adheres reasonably well to the same northern boundary, for its ability to survive in the field (already sprouted) through the winter is closely related to the severity of the winter temperatures (Figs. 11–8 and 11–9). Spring wheat, planted in the spring, must replace it if winters are too cold for survival. The line then largely depends on the length of growing season and, to a degree, the severity of winter temperatures. We have given the names (subdivisions of the Humid Continental) *Humid Continental-Long Summer* to the region south of the line and *Humid Continental-Short Summer* to that part north of the line.

In Europe, although corn is less common, an identical line may be recognized separating the Danube Basin, northern Italy, and the Southern Ukraine from the remainder or more northerly portion of the Humid Continental. And in the Far East, it runs through southern Manchuria and North Korea, and between Honshu and Hokkaido in Japan.

Agriculturally this is a permissive climate, but mere permissiveness is not enough to account for the general large-scale development that has taken place. There is a close relationship to the rise of urban markets and their increasing needs for food (Fig. 11–10). Sea frontage in North America and especially in Japan in the Far East has encouraged the growth of large cities based on foreign trade, and in maritime Europe a similar development has drawn large parts of the adjacent Humid Continental region into its hinterland. But trade is not the only impetus to urbanization; industry too has played a leading role—industry based in large part on coal. It appears that at some period

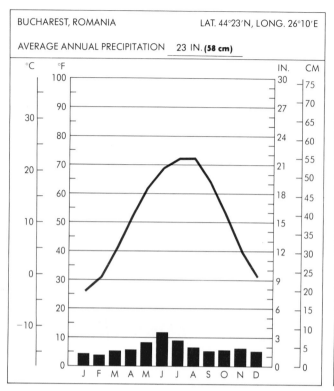

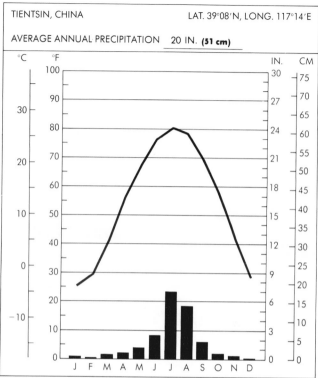

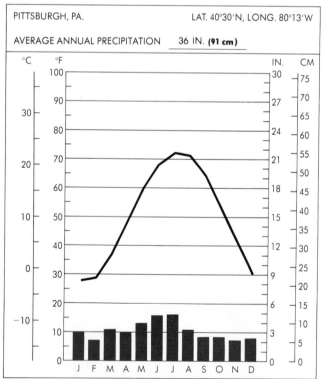

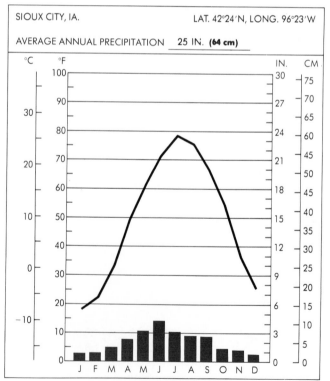

Fig. 11–8   Humid Continental—Long summer stations.

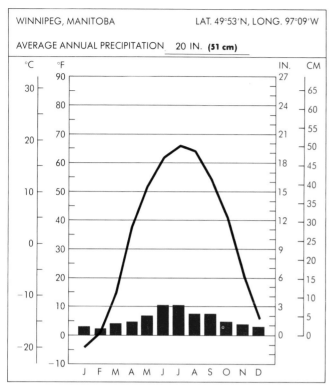

WINNIPEG, MANITOBA   LAT. 49°53'N, LONG. 97°09'W

AVERAGE ANNUAL PRECIPITATION   20 IN. **(51 cm)**

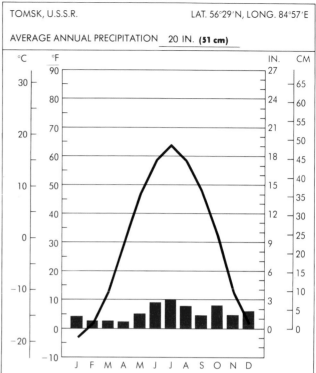

TOMSK, U.S.S.R.   LAT. 56°29'N, LONG. 84°57'E

AVERAGE ANNUAL PRECIPITATION   20 IN. **(51 cm)**

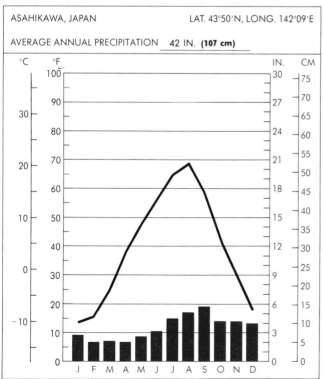

ASAHIKAWA, JAPAN   LAT. 43°50'N, LONG. 142°09'E

AVERAGE ANNUAL PRECIPITATION   42 IN. **(107 cm)**

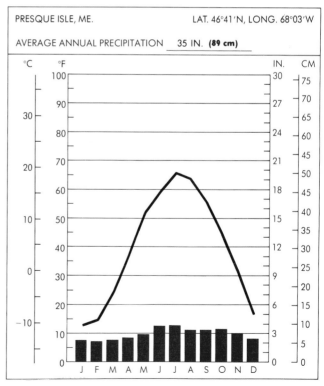

PRESQUE ISLE, ME.   LAT. 46°41'N, LONG. 68°03'W

AVERAGE ANNUAL PRECIPITATION   35 IN. **(89 cm)**

Fig. 11–9   Humid Continental—Short summer stations.

Fig. 11—10 Long days of sunshine offset to some degree the short summer in Saskatchewan. Spring wheat does very nicely.

in the distant past the entire middle latitudes was under the influence of a wet tropical climate, and the accumulation of thick beds of decomposing vegetable material built up in numerous swampy situations. This was embryonic coal, and subsequent pressures eventually formed this into extensive deposits of high-quality coking coal. Not all of today's good coal fields are within the Humid Continental climatic regions but many are: the Appalachians, East Germany, Poland, the Ukraine, Kuznetsk (Siberia), and southern Manchuria. Others are nearby, and this coal availability has led almost inevitably to the establishment of large-scale manufacturing.

Some have said that the fluctuating temperatures and humidity of the Humid Continental climate are challenging and responsible for a particularly vigorous human response, as opposed to the enervating effects of the tropics. Others have held that the undeniable development within this region has been accomplished despite a thoroughly uncomfortable climate. Whatever the cause, some parts of the Humid Continental, especially in North America, must be counted as among the most advanced regions in the world.

CHAPTER 12

THE HIGH LATITUDE CLIMATES

One of the two great virgin forests remaining in the world is the high latitude taiga. In many places it is totally untouched, but in Scandanavia where useful land is both scarce and costly the forest is being invaded. Here are permanent and successful farmsteads set down in forest clearings. The crop formula is hay, barley, roots and garden vegetables—most of it as feed for dairy and other animals. Adjunct winter timber cutting contributes to the family income.

INTRODUCTION

In the high latitude climates we are essentially beyond the limits of agriculture and practically beyond the limits of human occupation as well. These are virgin lands, almost untouched except in a few particularly favored spots. In this aspect the high latitudes are not unlike large parts of the world's dry regions and the Tropical Rain Forest, and the climate is largely responsible. Long, cold winters are the basic restrictive factor. Because of their unattractiveness to humans and their general inaccessibility, the high latitude climates have not been studied with any degree of thoroughness. Recording stations are completely lacking over huge areas, and thus data are limited. Only recently has there been any organized effort to understand these climates, and an immense amount of work remains to be done.

We recognize three high latitude climates: (1) the *Taiga*, (2) the *Tundra*, and (3) the *Polar Ice Cap* (Fig. 12–1). The first two are largely restricted to the Northern Hemisphere.

TAIGA

"Taiga" is a term referring to the distinctive type of vegetation that is characteristic of this climate. It is an open forest made up of a distinctly limited group of conifers, and its occurrence is exactly coincident with the boundaries of

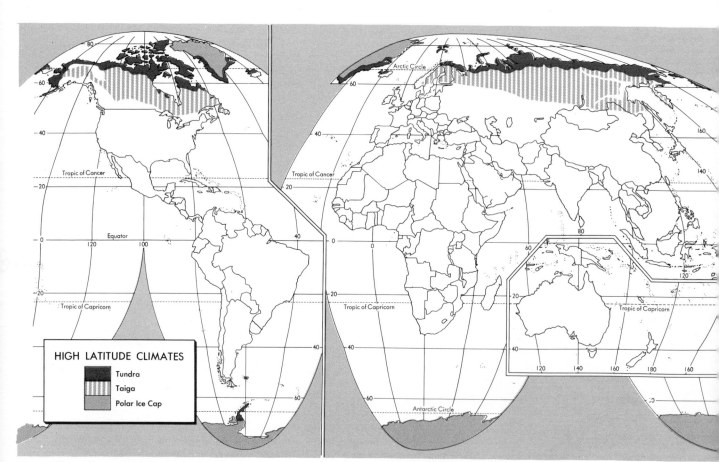

Fig. 12–1　High Latitude climates.

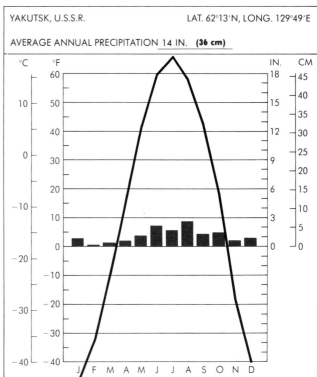

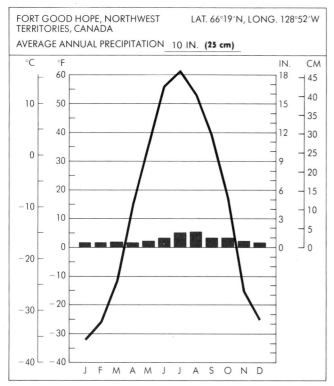

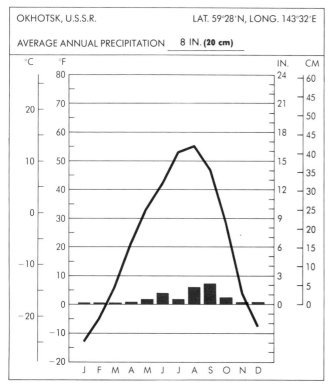

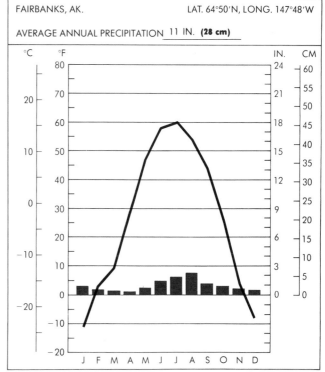

Fig. 12–2 Taiga stations.

the climate. Details of the forest makeup are discussed in Chapter 14, but it is important here to emphasize that only a highly specialized vegetative complex is capable of survival within a region dominated by the rigorous Taiga climate.

The North American representative of the Taiga is a long, relatively narrow band extending clear across the continent between roughly 50° and 60°N latitude. But although it closely approaches the west coast in Alaska, the band does not quite reach it, nor does it touch the north coast anywhere except along the margin of deeply indenting Hudson Bay. Only at the shore of the Atlantic, in Labrador, does the Taiga reach the sea. Thus it is a continental climate and exhibits typical continental tendencies.

At very similar latitudes, a comparable belt of Taiga extends across the much larger Eurasian continent, again touching the sea only at its eastern extremity. From northern Sweden and Finland to Kamchatka on the Pacific Coast, the Taiga region stretches for over 5000 miles (8047 km) and for most of that length shares a common southern boundary with the northern edge of the Humid Continental climate as it does in North America (Fig. 12–1).

Outstanding in the Taiga is the dominance of winter. Not only is it long (eight to nine months), but for many years the Taiga was thought to be the coldest region on earth. Verkhoyansk, a mining community in east-central Siberia, has recorded a minimum temperature of −92°F (−69°C); however, just a few years ago an even colder temperature was recorded at the South Pole, and it may be equally cold at the top of Mount Everest. Notwithstanding, winters are excessively cold in the Taiga, averaging far below 0°F (−18°C). In contrast, summers are surprisingly warm despite their short duration. Fairbanks, Alaska, experiences a few daytime temperatures in the 90's°F (32° to 38°C) nearly every year. Generally, the hottest month average throughout the Taiga is close to 60°F (16°C), making at times a 100°F (56°C) annual temperature range, by far the largest in the world (Fig. 12–2).

Because these far northerly continental interiors are the source region of polar continental air, there is very little moisture available throughout much of the year. Snow falls in limited quantities, but it is dry and powdery. There is an illusion here regarding total amounts; actually, snowfall is very modest. But since no melting occurs for months at a time, the entire winter's receipt is visible all at once, a

Fig. 12–3 Spring comes late in Sweden, but when the ice finally goes out of the rivers a veritable avalanche of saw logs, poles and pulping logs, accumulated during the winter, moves downstream to the mills. Some idea of the size of these logs may be gained by checking against man and dory at the lower right. Compare them with the timber being taken out of the Washington forest (Fig. 11–4).

situation that does not prevail in many areas where snowfall is much heavier. In addition, dry snow is readily picked up and whirled around by high winds, so that blinding blizzards may actually be contributing very little new snow. Drifting, too, against obstructions of any kind in the path of the wind is common. And although snowfall is not excessive, it is difficult to convince an individual who is forced to dig through an 8 foot (2 m) drift to get out of the house that this is the case. Actually, when the total annual fall of this powdery snow is melted down to be measured in inches of moisture, it amounts to very little. By far, the greatest moisture comes in the form of rain from widely scattered summer convectional showers. Precipitation totals average 5 to 15 inches (13 to 38 cm), which would indicate desert areas in most parts of the world, yet this precipitation supports forests here. The greatly decreased evaporation at these high latitudes plus the fact that the bulk of the rain is concentrated during the short growing season allows tree growth (Fig. 12–3).

The entire winter's receipt [of snow] is visible all at once.

Who would choose to live in a climate like this? Apparently not very many, for it is populated only thinly and sporadically. The agriculturalist is conspicuously absent, but the miner is there. Iron in Sweden, gold in Siberia, and uranium in Canada—these attract people. And the forests, although the trees are not large and regenerate themselves very slowly, can be exploited. So far, only in a few places is forestry profitable, and this is chiefly along the populated fringes as in Scandinavia and Quebec. One indication of the remoteness of the Taiga is that it is the last stronghold of the fur trappers; cold winters produce fine pelts. And if animals still abound, there cannot be many people.

TUNDRA

Once again, we have a vegetative title drawing attention to the specialization involved in plants that are able to grow and flourish under the restrictive influences of this high latitude climate. Trees, for instance, are absent, making the boundary between Taiga and Tundra the poleward limit of trees. Only certain sedges, wild flowers, low brush, and primitive mosses and lichens are represented. These are all shallow rooted so that the frozen subsoil is of no hindrance. They can remain dormant during a long, cold season and have the ability to mature and reproduce themselves with great rapidity once the short, cool summer commences.

The Tundra is a coastal climate, never extending very far inland. It involves the entire north coast of North America, including the Canadian Arctic islands as well as the ice-free shores of southern Greenland. In Alaska it is also found facing westward on the Bering Sea, and although the Aleutians and Alaskan Peninsula are not highly representative, they are usually classified as Tundra since they lack trees. The Alaskan south coast, however, in the vicinity of Anchorage and Cordova, is neither Tundra nor Taiga and is probably best included with the Marine West Coast, although much too cold to be typical. A similar merging with the Marine West Coast takes place in northern Norway, and from there the Tundra fringes the entire north coast of Eurasia (Fig. 12–1).

Because of this coastal orientation, the Tundra displays certain marine characteristics. Marine has meant mild temperatures as opposed to continental in our previous usage, but in the Tundra the temperatures are far from benign. However, we must remember the latitudes involved here. Relative to these high latitudes and particularly when compared with the somewhat lower latitude but continental Taiga, the temperatures encountered in the Tundra are really quite moderate. The coldest winter month averages only a little below 0°F (−18°C) despite the fact that offshore waters are frozen much of the time. This may seem cold, yet the Taiga winters are usually a good deal colder. And summers are comparably cool, reaching a warm-month average of about 40°F (4°C), approximately 20°F (11°C) less extreme than interior locations (Fig. 12–4).

We must regard permafrost as fossil ice, never having thawed since the climate cooled to its present extreme from what must have been a far-off warmer time of happily grazing mastodons and romping saber-toothed felines.

Such an annual temperature range, however, results in a condition peculiar to the Tundra, called *permafrost*, a permanently frozen subsoil. The short, cool summers simply do not supply sufficient heat to allow thawing below a surface foot or more. This is not at all a short-run phenomenon—the almost perfectly preserved remains of hairy mastodons have been discovered embedded in the permafrost deep freeze. We must regard it as fossil ice, *never* having thawed since the climate cooled to its present extreme from what must have been a far-off warmer time of

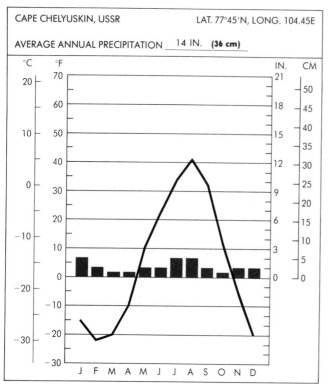

CAPE CHELYUSKIN, USSR   LAT. 77°45'N, LONG. 104.45E

AVERAGE ANNUAL PRECIPITATION   14 IN. (36 cm)

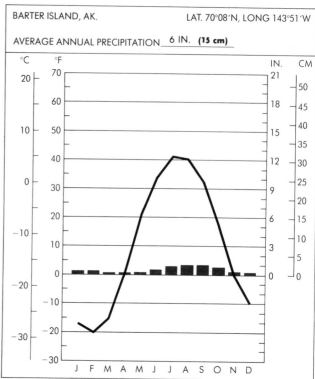

BARTER ISLAND, AK.   LAT. 70°08'N, LONG 143°51'W

AVERAGE ANNUAL PRECIPITATION   6 IN. (15 cm)

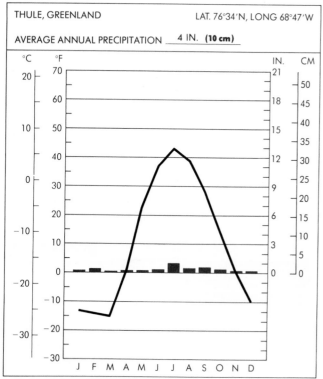

THULE, GREENLAND   LAT. 76°34'N, LONG 68°47'W

AVERAGE ANNUAL PRECIPITATION   4 IN. (10 cm)

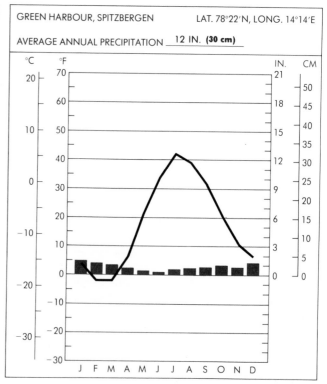

GREEN HARBOUR, SPITZBERGEN   LAT. 78°22'N, LONG. 14°14'E

AVERAGE ANNUAL PRECIPITATION   12 IN. (30 cm)

Fig. 12–4   Tundra stations.

Fig. 12–5   Snaking for hundreds of miles across the tundra the Alaska pipeline rests on an insulated pad. Without the pad as the pipe assembly absorbed the summer solar heat, it would slowly sink into the melting permafrost.

happily grazing mastodons and romping saber-toothed felines (Fig. 12–5).

Permafrost is also, undoubtedly, the determinant of the poleward margin of trees. Farther south in the Taiga, winters are even colder than in the Tundra and assuredly the ground is solid ice. But the forest is dormant over the winter and Taiga summers are of adequate length and warmth that the frozen earth can thaw to a depth of several feet. This permits reasonable drainage of the soil and sufficient root space for trees to manage despite severe winters. Depth of summer ice is critical.

In the Tundra there is usually poorly developed surface drainage on the flat to rolling coastal terrain; the solid ice inhibits any loss of moisture vertically and the soil be-

comes a soupy mud. Alternating shallow ponds and quaking bogs, where tangled roots and matted vegetation "float" on the mud, make the summer tundra landscape a difficult one to traverse. Water is everywhere despite a low receipt of annual moisture amounting to only 5 to 15 inches (13 to 38 cm). As in the Taiga, most of this comes as summer rain, some as convectional showers, but some from Arctic Front storms that are also probably responsible for the common winter blizzards (Fig. 12–4).

Obviously, this is a difficult and monotonous habitat for humans, and traditionally the Lapp reindeer herders of the Scandinavian countries and scattered groups of Eskimos, living chiefly off the sea, have been the only inhabitants. But today, a few scientific, military and big oil company installations are intruding into the Arctic solitude and attempting to solve the problems of modern living under unique and adverse conditions.

POLAR ICE CAPS

The Polar Ice Caps are found in Antarctica and Greenland, the only high latitude land areas permanently covered by masses of ice of sufficient size to induce a distinctive climate (Fig. 12–1). The Arctic Ocean, too, in the immediate polar region, is frozen all year and undoubtedly displays a similar climate, but we have agreed not to attempt a classification of oceanic climates, so this must be ruled out. By the same token, we have not classified highland climates, and much of interior Greenland and Antarctica is well over 5000 feet (1524 m); so in the interest of consistency, only the relatively low parts of these ice-covered lands should be considered. Actually, however, so little specific data are known that it has been usual simply to make a few generalities and to ignore the variations caused by elevation. Gradually, climatic records are being accumulated and eventually we will be able to introduce greater detail.

In 1983, a temperature reading at the Russians' Vostok Station near the South Pole indicated −127°F (−89°C)

The Lapps

The Lapps of far northern Fennoscandia have occupied a high latitude taiga/tundra environment for almost 2000 years, and they have occupied it very successfully. Their numbers are not large, currently about 41,500 and they are spread thinly along the northern periphery of four different countries: 25,000 in Norway, 10,000 in Sweden, 4500 in Finland, and 2000 in the USSR. Slender, dark haired and short-statured among their Scandinavian neighbors, and speaking a Finno/Ugric language (similar to Finnish & Hungarian), it is not at all clear what the Lapps anthropological antecedents may have been. But whatever those origins, they have moved into what for most would have been a very difficult environment and have managed to preserve their race, language, and elements of their culture on the extreme northern fringe of European civilization.

The key to Lappish survival over the centuries has been the reindeer. Although individually owned, reindeer are communally herded in a tightly organized nomadic husbandry system which involves an annual migration of up to 300 miles from the summer mountain pastures to the coast and back. Historically the Lapp culture and the reindeer-herding nomadism have been inseparable.

However, things are changing. Modernism has raised its "ugly head" to those who value the old ways; the "quality of life" has improved immeasurably if one equates modernism with progress. The reindeer is still herded but its range is partially fenced, snowmobiles aid in animal control, national boundary restrictions now limit former unregimented wanderings, and only the herders migrate with their animals, the women and children remaining sedentary in new and permanent dwellings. The old self-sufficiency is threatened by an encroaching European money economy and not all Lapp money requirements to satisfy their current life-style can be met by reindeer herding or selling handicrafts to tourists. Many are being forced out of necessity into wage labor in the mines, construction gangs or as commercial fishermen. For most Lapps though, life is easier than in the *good old days*. Large, modern, totally-planned towns have brought mid-20th century services and facilities to the very heart of Lapland.

Although outnumbered in their own land, like the American Indians and Canadian Eskimos, the Lapps are not a dying race. Their numbers have actually increased in the last few centuries. It is their traditional way of life that faces change, if not possible extinction.

Primarily, we know that the Polar Ice Caps are very cold. The old Little America station established by Admiral Byrd on shelf ice at the Antarctic continental margin has recorded a coldest month average of $-34\text{F}°$ ($-38°\text{C}$). But in 1983, a temperature reading at the Russians' Vostok Station near the South Pole indicated $-127°\text{F}$ ($-89°\text{C}$). Since this is a minimum figure rather than an average and since the South Pole is near 10,000 feet (3048 m) in elevation, such a reading is scarcely typical. Probably, a winter average somewhere between the Little America and the South Pole figures would be about right—colder than the Tundra and possibly even colder than the Taiga. But of greater significance than the winter cold is the fact that the warmest month averages do not reach as high as $32°\text{F}$ ($0°\text{C}$). Even at relatively mild Little America, the summer average is only $23°\text{F}$ ($-5°\text{C}$). This means, of course, minimal thawing and a maintenance of the ice mass despite low precipitation and only slight possibilities of vegetation even if soil were exposed (Fig 12–6).[1]

---

[1] Certain lichens have been encountered on rocky outcroppings, so even here vegetation is not absolutely lacking.

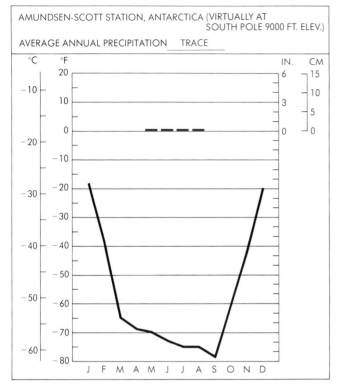

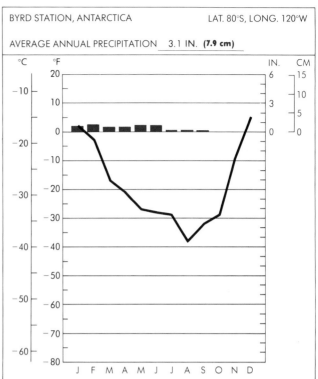

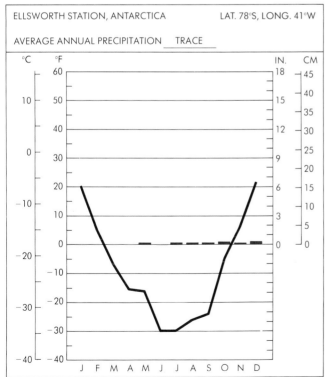

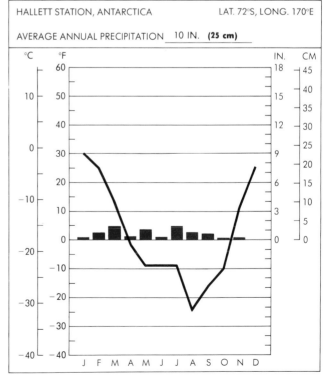

Fig. 12–6 Polar Ice Cap stations.

Fig. 12–7   There are permanent settlements these days in several parts of Antarctica, one is this U.S. Navy base at McMurdo Sound. These little communities are drawn here through scientific curiosity and a felt need to establish a political presence rather than Antarctica's climatic aesthetics.

Precipitation has not been accurately measured, but it is unlikely that the cold air masses are capable of giving off more than 10 inches (25 cm) per annum, and this is in the form of dry snow and minute ice crystals. But since there is a constant loss of ice as these great continental glaciers move out toward the sea, discharging bergs at their periph-ery, there must be sufficient replacement through precipitation to sustain them. Extreme wind and blizzard conditions are more frequent here than in the Tundra, as accelerated air drainage causes high-velocity gravity winds to blow down from the elevated interiors (Fig. 12–7).

PART FOUR

NATURAL VEGETATION AND SOILS

CHAPTER 13

INTRODUCTION TO NATURAL VEGETATION

Vegetation on the land is almost universal. Even as forbidding an environment as a middle latitude desert, where both aridity and bitter winter temperatures militate against casual propogation, exhibits permanent plant communities. Bailey's yucca appears healthy and confident against a background of sage.

INTRODUCTION

Virtually the entire land surface of the earth supports some sort of vegetative cover. Ice fields, barren rocky outcroppings, sterile sand dunes, and alkaline salt flats are occasionally completely devoid of any growing plants, but these situations are relative rarities and excite some comment, for vegetation is the norm. And even here a close examination will often reveal that many rocks support mosses and lichens; certain salt-tolerant plants will grow, albeit thinly, in highly saline soils; and the seemingly driest desert will, after an infrequent shower or even heavy fog, display a rapid flowering of long dormant but still viable spores. It is this natural vegetation that gives a certain distinction to our landscape so that we think not only in terms of terrain or climatic variation from place to place but of vegetational variation as well. Grassy plains, forested hills, or impenetrable jungle are part of our standard vocabulary in describing and transmitting to others the character of a given region.

The seemingly driest desert will, after an infrequent shower or even heavy fog, display a rapid flowering of long dormant but still viable spores.

Fig. 13–1   European antecedents are revealed in the architecture of central Chile.

HUMANS AND NATURAL VEGETATION

The earth as a human habitat is the prime concern of geography, and natural vegetation as an element in humans' physical environment has long played an important role. For example, when the Spanish first attempted the occupation of central Chile, they found familiar Mediterranean open grassy and low brush country that reminded them of home. They understood a grazing economy based on this type of vegetation and discovered that the vine, grains, and olive introduced from Iberia flourished here. But as they pushed south into the increasingly forested littoral of south-central Chile, the Spanish were repelled by the unfamiliar character of the country (and the Indians occupying it), and without the spur of gold to urge them on, they stopped abruptly at the forest edge. Many years later, German immigrants, familiar with a forest environment and the skills to cope with it, moved readily into the accessible portions of southern Chile and developed its latent resources. Their potato and dairy economy and their knowledge of forestry combined to make them feel very much at home in a region that appeared to the Spanish extremely inhospitable. Today, the high-gabled frame homes in forest clearings contrast strongly with the tile-roofed stucco and adobe dwellings in the villages of central Chile (Figs. 13–1 and 13–2).

This is not to say that other factors are not as important as vegetation in controlling where and how humans shall live. The very fact that vegetation can be removed or altered makes it a less compelling determinant than climate or terrain, but it is a factor everywhere and a highly visible one, and certainly a part of the total physical environment that cannot be ignored.

Everywhere humans have laid their hands the original vegetation has been changed.

In attempting to analyze and classify the vegetation of the world under the title "Natural Vegetation," we are, of course, immediately synthesizing the situation, for there is very little *natural* vegetation remaining. Everywhere humans have laid their hands the original vegetation has been

Fig. 13–2 Population and Forest—south central Chile.

changed. They have cut the forests, plowed the prairies, drained the swamps, and introduced extensively new and alien vegetative varieties. There are sizable regions in China, India, and Europe where dense human occupation has been of such long standing that no general agreement has been reached today as to precisely what the original vegetation must have been. Even in lightly populated or wholly unpopulated areas, the balance of nature has been upset by the destruction of animals that were a part of the total ecological environment or by the introduction of exotics (Australia's rabbits are a well-known example). Fire too, frequently of human origin, has caused noticeable variations. Nonetheless, we will make an attempt at classifying natural vegetation, partially because to catalog the pattern of farm crops around the world takes us out of the realm of strictly physical geography and into the myriad ramifications of economic geography and, more importantly, because of the concept of *climax vegetation.*

CLIMAX VEGETATION

The idea behind the term "climax" is that, given a particular plant environment (especially climate and soil), the vegetation association that develops will display a certain character that represents an equilibrium with that environment and will continue unchanged indefinitely. However, if the original vegetation is destroyed or altered, a rapidly growing secondary association of an entirely different type will spring up to replace it. But this is not permanent. These plants will establish a local environment that will encourage their replacement by a second series and these in turn by a third, until finally this evolution of succeeding associations will culminate in the original climax vegetation, which if the basic environment continues static, will maintain itself.

An example might be the destruction of a climax Douglas fir forest by fire. Almost immediately among the blackened snags and ashes, there will appear fireweed, rank grass, and blackberry, to be in turn replaced, as their very growth fosters conditions favoring them, by alder, willow, and bracken fern. These too will establish a local environment that will allow still others to take their place, and so on through several stages until the Douglas fir complex is

reestablished. This may take many years, but if the overall environment remains unchanged, the climax vegetation must reassert itself. Thus if humans and their influences were to disappear from the face of the earth, theoretically, the original climax vegetation would take over everywhere. This, of course, is not absolute. There are many places where humans' work could not be undone entirely—and then too, no natural environment is wholly static. Nature is, in the long run, dynamic and changes are constant. However, it is the pattern of the initial natural vegetation, in some places substantially unchanged and elsewhere largely erased, that we will classify, describe, and analyze.

FIRE

Just a word here regarding fire, the mortal enemy of vegetation—or is it? The common Jack pine, for instance, is one of several pines that actually requires fire to survive. Its cone is so hard and needs such high temperatures to crack it open that no natural phenomenon short of fire can effect the release of viable seeds. There does not have to be a fire every year; the tree is long-lived and continues to produce cones. But without an occasional flash groundfire to rupture the cones and to prepare a scorched earth that is barren of competitors and receptive to the young seedlings, the specie will expire. It is entirely possible that a crown fire will destroy the entire mature forest, but even as it extirpates the parent, it releases the seed of the next generation (Fig. 13–3).

The notion that all fire is of human derivation and totally destructive is simply not true.

Fire is a normal and functional part of nature. It must be, or entire plant species would not have evolved dependent wholly on fire for their survival. The notion that all fire is of human derivation and totally destructive is simply not true. And anyway, humans are God's little critters and if they decide to set a fire to flush game or encourage new grass shoots for their flocks, who is to say this is abnormal?

Fig. 13–3   The swift passage of a flash fire surely was fatal to this mature forest, but in the process its great heat must have activated some seed materials on the ground. There seems to be no question that the young conifers are outstripping all competitors and a new forest is well on its way.

In an earlier chapter the suggestion is made that perhaps modern civilized folks, with their passion for *putting out* fires, are seriously tampering with the natural environment. Recent studies in Sequoia National Park have tended to show that although millions of trampling tourists' feet around the boles of the big trees may not be particularly beneficial, of far more significance to their general well-being is the lack of sporadic fire. Indications are that the virtually fireproof *Sequoia gigantis* needs fire to control its pests and enemies. And in parts of the prairie southwest the natural vegetation has changed radically—not because of overgrazing or plowing and abandonment but because of artificial control of the common, old-fashioned prairie fire.

The U.S. Forest Service has carefully weighed and analyzed the thrust of these accumulating studies and in the last few years decided to experiment on a limited basis with controlled forest fires. Part of the thinking is to clear out accumulated forest trash so that if an accidental fire should occur, it would prove easier to contain. Summer visitors to some western National Parks and forests have been startled to observe official Forest Service personnel busily *setting* fires. One's natural inclination, after a lifetime of

conditioning, is to rush out and empty the thermos jug of lemonade on the flames. But these are *constructive* fires—we are now being asked to recondition our old, carefully nurtured reflexes.

Probably we can say that grasslands are expanding at the expense of forests, for grass, a very simple plant with much of its functional system protected underground, reasserts itself rapidly after the fire has passed. Especially if the fire is set deliberately again and again at frequent intervals, trees have little chance. The prairie peninsula of Illinois/Indiana and the Pampa of Argentina may well be a result of repeated firings in prehistory, and many areas of tropical savannas and scrub are fired regularly today.

If fire is normal, the resulting vegetation has to be regarded as natural vegetation. The long southern California summer arid season, for example, literally invites fire—resinous conifers, thickets of chaparral, and parched grassy slopes. Given the correct low humidity/high temperature amalgam, simple spontaneous combustion will ignite the critical spark. Over the centuries, under these circumstances, it is difficult to conceive of the natural vegetation being anything other than fire residual. It must be remembered too that most of the world outside of North America

Fig. 13–4   Summer fire in a coniferous forest. This is the scene a fire lookout dreads. If the fire fighters can't get to the spot fire in a hurry it can easily develop into a major burn—and despite the case for fire as normal, a useful element in forest ecology, a huge long-burning forest fire probably should not be allowed to burn endlessly. Not many years ago half the pandhandle of Idaho burned while authorities debated the merits of fire as a therapeutic.

and western Europe does not have effective means of controlling fire and so it continues to burn unabated, sometimes for many months at a time (Fig. 13–4).

VEGETATIVE ASSOCIATION

All natural vegetation occurs in plant communities or *associations*, each species mutually dependent on the others. In some associations, one specie is strongly dominant and only a few others are represented, while elsewhere literally hundreds of species will occur. It may appear that a grassy savanna or a high latitude pine forest represents only a single variety, but close examination will reveal that in the savanna, despite the dominance of rank grasses, many other plants grow among and below them and other trees and undergrowth are represented among the pines. This is universal in natural vegetation. Only in the case of planted field crops, where humans expend great effort to eradicate weeds and competing plants, do large-scale stands of a single specie prevail.

TIERING

One of the peculiarities of vegetative communities is a strong tendency for the various elements to arrange themselves into *tiers* or stories (Fig. 13–5). Light (or solar radiation) is the chief stimulant here, for all green plants require light as the basic energizer in photosynthesis:

$$\text{Light} + \text{water} + \text{carbon dioxide} = \text{sugar} + \text{free oxygen}$$

This merely says that a plant breathes in carbon dioxide from the atmosphere, sucks up water through its roots, and combines these with light on its leaf surfaces to produce stored sugar—the surplus carbon dioxide is "exhaled" back into the atmosphere. This simple process, functioning in conjunction with the green-pigmented catalytic compound *chlorophyll*, is *photosynthesis* (*photos*, Greek for light; and *synthesis*, putting together) and its end-product is sugar, the primary food of all living organisms. Intense light builds sugar to a maximum degree; the fruit of the date palm growing in the desert oasis is so high in sugar content that

Unwelcome Neighbors

Strangler fig.

One reason the plant association is almost a universality is that plants in a community are of considerable help to each other—they prosper in concert more successfully than alone. But just as in human communities, there are always the hangers-on and the leeches, those who contribute very little and some take advantage of their law-abiding neighbors. The worst of these are the *parasites*.

True parasites live on their host, drawing all of their nourishment from it by tapping into its vital systems. The common mistletoe possesses no true roots of its own, yet flourishes high amongst the branches of the host tree by inserting sharp-tipped, rootlike appendages through the bark. Equally damaging, oftentimes to the point of fatality, are such vigorous climbers as the tropical strangler fig. Wrapping its host in an obsessive embrace, the fig with roots of its own but lacking a rigid trunk, uses the tree as a means to climb toward the light above. In the process it eventually strangles its benefactor.

Epiphytes (epi-Greek meaning *upon*; phyte—from the Greek phyton meaning *plant*) are the so-called air plants. These "perched" individuals are simply utilizing the host tree as a means of escaping the deep shadows of the forest floor and to attain a place in the sun. They have abandoned all connection with the ground, but do not draw sustenance from their host. Mosses and lichens are epiphytes—so are orchids and Spanish moss (neither Spanish nor moss). Usually their seeds or spores are light and wind carried, and their nutrition derives from minerals supplied by rain from the branches above, or from shallow bits of humus accumulated in a tree crotch.

it is almost nonperishable. In contrast, the strawberries of the Matanuska Valley in Alaska develop to tremendous size as a result of the almost continuous receipt of light during the 20 hour summer days, but they tend to be tasteless and low in sugar because of the weak low-angle sun even in midsummer at these high latitudes.

Plants reach for light in a frantic competition to sustain themselves. Those that can grow most rapidly and withstand the greatest heat and light intensity win out and form the highest strata of the vegetative laminate. Below, cut off to some degree from the direct sunlight, have evolved other plants with various degrees of shade tolerance. Those that require great light intensity and cannot grow rapidly enough have simply expired, leaving the arena to plants with the ability to accomplish photosynthesis with lesser amounts of light, even to the extremes of ferns and mosses that occupy a position of perpetual gloom. But there must always be some light for green foliage to maintain itself.

Fig. 13–5   Both the association of several species and the common tiered character of vegetative communities are visible in this interior photo of a western Oregon forest. The massive boles and tallest trees are Douglas fir, with white pine and hemlock forming a second story. Younger members of all of these make up a third level while several varieties of woody brush, Oregon grape, huckleberry and fern adorn the floor.

Only the *fungi* without chlorophyll or photosynthesis can survive in darkness (Fig. 13–6).

THE SIMPLE PLANTS

Among the many vegetative associations that occur throughout the world certain types of plants seem to be omnipresent—*grass, fern, lichen,* and *moss* are common examples. They appear to be able to adapt to virtually any

Fig. 13–6   Fungi, encountered in the dark cool recesses of every forest, are a part of the decay process that rebates borrowed plant nutrients to the soil. They are living and growing without chlorophyl and photosynthesis.

environment and, if not dominating the plant community, they at least show some representation. Lichens are found at the equator and in Antarctica; grass completely dominates perhaps 40% of the earth and shows up as cane, sedge, or reed nearly everywhere else; while ferns and mosses almost materialize from out of the air if water is present. The particular character of each of these seems to be that it is of primitive origin and simple in structure. It would appear at first glance that normal evolution would phase out the

Grass completely dominates perhaps 40% of the earth.

Fig. 13–7   Equiseta, an ancient seedless plant, flourishes amongst its modern neighbors. Popularly called "scouring rush" by the early settlers it was utilized as an abrasive (because of its high silica content) to shine up metal pots.

All green plants require light as the basic energizer in photosynthesis.

to produce specialized individuals specifically fitted to such an exclusive environmental niche that they are totally inflexible. This leaves a place in the vegetative ecosystem for the general plant, one that can withstand disasters and still come back to fill any and all available voids, sometimes in competition with the specialists. A select few of the simple mechanisms and ancient phyla have not only survived but thrive in the modern plant community as they have proved their adaptability.

Giant tree ferns and *equiseta* (horsetails) dominated the floral landscape during the age of the dinosaurs (Fig. 13–7). Today, smaller varieties with seedless reproduction frequently hold an advantage as they compete for space with seed plants. Mosses and lichens seem to need only light and moisture to do well. Deprived of these essentials, they do not die as readily as more delicate plants but merely lie dormant for long periods of time—reproduction is by simple fragmentation or disjointure. The many kinds of grasses display simple and effective seed dispersal as well as an ability to sprout and grow rapidly after long drought, searing fire, or seasonal freeze.

A select few of the simple mechanisms and ancient phyla have not only survived but thrived in the modern plant community.

simpler plants in favor of newer, more complex and efficient models, and on the whole there is plenty of evidence to support such a point of view. But evolution also tends

CHAPTER 14

CLASSIFICATION AND NATURAL VEGETATION REGIONS

This is an exotic planted forest—Monterey pine (pinus radiata) from California flourishing in New Zealand. Powhataroa butte and the surrounding countryside had never supported more than a sparse scrub vegetation because of coarse, infertile, volcanic ash soil; but when the Monterey pine was introduced it developed faster, larger, straighter than it ever had in its California home. This is afforestation. Extensive regions of cutover land around the world could benefit from reforestation as well.

BASES FOR CLASSIFICATION

There is an obvious very close relationship between the distribution of natural vegetation over the world and the distribution of climatic regions. It will be recalled that the climatologist Köppen utilized vegetation as perhaps his most important criterion in establishing climatic boundaries. Soil fertility, too, as we will shortly discover, is largely a product of climate and whether we regard vegetation as strongly reflecting the character of the soil in which it grows or the soil as being effectively modified by the vegetation growing in it (both being true), the fact remains that in terms of world distribution, a comparison of the climatic, natural vegetation, and soils maps will reveal striking similarities (Maps I, II, and III). Once again, we will make no attempt to classify the vegetation of the world's major mountains for the same reasons that we ignored them in the climatic classification—not because they are without vegetation but because of the multitudinous variation. There is a general tendency for high mountains to display a vertical zonation of vegetation somewhat analogous to the latitudinal zonation that is typical of the rest of the world. But the complications introduced by windward slope versus leeward slope, sunny slope versus shady slope, etc., become so detailed that it is beyond the scope of this general treatment to do more than group them all together as undifferentiated mountain vegetation.

In terms of world distribution, a comparison of the climatic, natural vegetation, and soils maps will reveal striking similarities.

The definitive lines on the map marking the boundaries of one vegetative type from another may be misleading, for there are few sharp lines in nature. Rather, as in climate, there are zones of transition or merging wherein one vegetative association changes gradually into another. Also, it should be kept in mind that in the interest of rational classification, although the actual botanical genera and species may vary widely from one continent to another, they can nonetheless be classified within the same basic vegetative family; that is, desert vegetation, wherever it is found, is sparse and has the special adaptations of root and leaf, allowing it to conserve a limited water supply. So whether it is the sage brush of North America or the blue bush of Australia, we will include it within one grouping.

Basically, there are three types of vegetation associations: (1) *forest*, (2) *grasslands*, and (3) *desert plants*. It should not be startling news to anyone to be informed that forests are made up of trees. They may be *evergreen* or *deciduous* (dropping their leaves during one season of the year), broadleaf or needleleaf. Generally speaking, *coniferous* (cone-bearing) trees are needleleaf and evergreen, while *broadleaf* trees may be either evergreen or deciduous. But as long as the vegetative type has a woody bole and grows to some height, it is a tree, and if trees are dominant in occurrence over other members of the association, it is called a forest.

Grasslands feature grass as dominant. Bunch grass, short grass, long grass, rank grass (bamboos and canes), or sedges—they are all grass, and probably as a group are as widely represented in plant communities as any vegetative type in existence.

The third category, desert plants, is made up of a wide variety of vegetative types, but to whatever genus they may belong, they must have special adaptations of moisture conservation and use to allow them to survive under conditions of limited precipitation.

We probably can add still a fourth grouping here called *tundra*. As will be described later in this chapter, these many species of plants, somewhat like those of the desert, must display certain special adaptations to deal with the particularly stringent environment that they occupy.

VEGETATION REGIONS

Tropical Rain Forest (Selva)

As the name implies, the Tropical Rain Forest climate and the vegetative complex called *Tropical Rain Forest* are virtually identical in location. The Amazon Basin, Congo Basin, Guinea Coast, Ceylon, Malaya, most of the East Indies, and the large islands of Melanesia constitute one of the largest zones of nearly unmodified climax vegetation anywhere in the world, and certainly the greatest area of dense forest vegetation. In addition, the east coasts of con-

Fig. 14–1   Observing a Tropical Rain Forest from above is not very revealing of its structural character. All that is visible is the endless plane of its coalescing crowns. The opening in the forest here has been cleared for access to the Mayan ruins at Tikal, Guatemala.

tinents and islands in the trade-wind zone and minor lesser regions are also clothed with this type of forest. It should be explained that in these minor lesser regions, the typical Tropical Rain Forest climate is not always absolutely necessary for the growth of a true Tropical Rain Forest. The galeria forest is of this type and is discussed a little later in this chapter. And in parts of the Guinea coast of west Africa, southeast coastal India, and the west coast of Burma and adjacent Bangladesh, there is a dry season, and climatically these regions are classfied as Tropical Wet and Dry. However, the rainy season is so very wet [200 or more inches (508 cm) of rain each year] that the soil never dries out, thus allowing the growth of Selva despite a definite rainless period.

Stimulated by constantly high temperatures and heavy precipitation, vegetative growth is unceasing, and the resulting forest is of a particular type duplicated nowhere else on earth. Tall [100 to 150 feet (31 to 46 m)] straight trees, growing close together and reaching for the light, form a solid canopy far above the ground as their foliage coalesces

(Fig. 14–1). In shadowy half-light below, branches fail to develop so that typically the trunk is a featureless shaft flowering and producing leaves only at its upper extremity. The base is often buttressed or fluted to support the tree in the absence of a deep root system that does not have to search for water. Lacking climatic seasons of any kind and supplied abundant moisture, these trees are broadleaf evergreens that continually regenerate their foilage and drop dead leaves throughout the year. But the forest floor is remarkably clean because the hot moist climate fosters microbiotic decomposition at a rapid rate. Even fallen timbers are disposed of with dispatch by decay and a large array of termites.

Stimulated by constantly high temperatures and heavy precipitation, vegetative growth is unceasing.

Well beneath the dominant canopy, there is normally a sparse second story of shade-tolerant trees averaging only 25 to 50 feet (8 to 15 m) in height. This is their preferred habitat and they cannot survive if exposed to direct light. The common cacao from whose beans chocolate is made is one of these, and when it is planted in commercial plantations, great care must be exercised to assure it continual shade by retaining many of the taller trees as the forest site is prepared for plantation use. At the ground, where the shade is deepest, few plants can survive. There is no underbrush as we know it in the middle latitude forests, and there are considerable expanses of bare soil. Rattan, a woody vine that runs along the ground, and a few low fernlike plants are the typical light vegetation. Lianas, or climbing vines, which have the ability to cling to trees, can make their way upward to the light of the upper canopy, and one of the features of the forest interior is the many hanging lianas (and pythons) draped like great ropes among the trunks of the tall trees.

The Tropical Rain Forest is a distinctive world unto itself, with perpetual gloom, a dark and moldy odor, and still moist air.

But notwithstanding this frenzy of continued vegetational growth, the interior of the Tropical Rain Forest is relatively open country and a distinctive world unto itself, with perpetual gloom, a dank and moldy odor, and still moist air. The active life zone is remote in the high canopy where arboreal animals and birds spend their entire life cycles and the trees flower and fruit. At the ground, where deep shade inhibits growth, there is no great difficulty in moving about from place to place, and the great myth of thousands of square miles of impenetrable jungle is just that, a myth. There is such a thing as a jungle and it is highly impenetrable, but it is merely a phase of rain forest, limited to certain favored localities, and by no means dominant.

Jungles come about when light is allowed to reach the forest floor. If a storm or fire removes some trees causing a break in the canopy or, as is more frequent, if humans open a clearing to plant a garden, the ensuing light brings countless dormant seeds to life. Grasses, saplings, bamboo, and low brush take over very rapidly, and within just a few years the region is a dense tangle of interlocking vegetation. Often, elements of the original planted garden, abandoned by migratory agriculturalists as the secondary growth becomes too rapacious to cope with, are found competing with the other species for survival. Left to itself, such a jungle will evolve eventually back to the original climax selva. But if fire is utilized repeatedly to clear a given plot, it will usually develop into a semipermanent rank grass *savanna*. Many parts of Southeast Asia and Africa are visually distinctive because of the great frequency of these grassy openings in the forest. The dense grass strongly suppresses the establishment of tree seedlings, but presumably, given time, even this will finally revert to the climax vegetation as the rain forest slowly encroaches from all sides.

This sort of thing is, of course, merely a temporary jungle but, given a situation such as a fairly steep slope where the trees at various levels do not have interlocking crowns, more light than normal will find its way to the forest floor and a permanent jungle will result. More common than this are riverine jungles that develop along the banks of streams wide enough to cause a rift in the overall canopy. Such dense vegetation, merely fringing the stream course for many miles, has given rise in the past to reports of dense jungle over extensive areas. Early explorers utilizing the waterways as the easiest means of penetrating the interior observed nothing but the densest of jungle from their craft and quite naturally assumed that the back country was of the same character. Even greater misconceptions came about when these travelers, still clinging to the streams, journeyed well beyond the margins of the rain forest proper. Here a dry season each year results in a much lighter, more open vegetation, but fringing the river where water is available to tree roots even during the annual rainless period is a continuation of the riverine jungle called a *galeria forest*. A true picture of the situation can be seen from the air during the dry season. The brilliant green of the galeria, only a narrow band faithfully following each turn of the river, pushes long fingers into the more open, brown-colored vegetation of the surrounding countryside. But voyagers on the river have no means of knowing that they have left the Tropical Rain Forest far behind.

Also permanent in nature are the coastal jungles where a dense wall of trees and underbrush faces out to sea at the inland margin of the beach sands. Many of the usual forest species are represented here, supplemented by vegetational

Kipukas

The Hawaiian term *kipuka* refers to an island of life in a sea of barren new lava. As the relatively fluid molten rock flows from its volcanic vent, there is a strong tendency for each glowing river to follow existing valleys and declivities—it is deflected by obstructions and runs in braided channels as it works its way rapidly downslope. Inevitably, high points and ridges are bypassed and left behind as vegetated islets.

Each kipuka supports its own assemblage of life, isolated now from every other. Easy interchange of seeds, spores, or pollen is no longer possible and the stage is set for incestuous evolution. Some kipukas are very large—Kipuka Puaulu, or Bird Park, in the Big Island's National Park, encompasses more than 100 acres (40 ha). Here in the deep and fertile basaltic soil is a forest of mammoth koa and ohia trees with many unique and rare plants as a part of the association. It is a lush little world unto itself surrounded by a black rock desert.

But some birds, even though the kipuka is home, manage to range far enough to establish contact with other vegetated regions, as do occasional insects and wind-carried spores, so absolute isolation can never be achieved. And the rich and varied life of the Kipuka also acts as a reservoir or source region for pioneer plants slowly establishing themselves in the nooks and crannies of adjacent weathered igneous rock.

varieties whose seeds have the ability to remain viable for long periods of time when immersed in seawater as they are transported longshore, or whose root systems can cope with some mild soil salinity, such as a number of the palms and casuarina. A variant on the coastal jungle is the frequently encountered amphibious *mangrove*. It actually requires brackish water, preferably a few inches to 3 feet (1 m) deep, and thus flourishes offshore, making it difficult to determine the exact point where land meets water (Fig. 14–2).

There is probably no other widespread plant association with as many different species represented.

Aside from its amphibious habits, the mangrove is atypical of the Tropical Rain Forest in another respect—the entire jungle is made up of a single specie. This may not seem particularly odd to those of us familiar with middle latitude forests, where a single type of tree is often strongly dominant, until we recall that the plant community is almost universal. In the Selva the number of individuals making up a community are likely to number in the thousands. There is probably no other widespread plant association with as many different species represented as the Tropical Rain Forest. One of the many difficulties in attempting to exploit the great economic potential of the Selva, be it rubber, Brazil nuts, or mahogany, has been the fact that none of these valuable trees grow in even small groves, much less forests.

The greatest area of untouched virgin Tropical Rain Forest is to be found today in the still lightly populated and inaccessible upper Amazon Basin. Elsewhere in the world, its occurrence is somewhat more spotty. Southeast Asia and much of Africa display the effects of human's hands to an increasing extent. Generally, where population pressures are heavy or along the great river valleys and accessible coastlines, secondary growth or permanent plantation development has changed the original vegetation. But in the backcountry of interior New Guinea, upper Amazonia, or Borneo, the virgin forest remains intact.

Tropical Deciduous Forest and Scrub

The climatic limit of the Tropical Rain Forest is, in nearly all cases, the beginning of a winter dry season. When its severity becomes sufficient to interrupt plant growth for even a short period each year, it is reflected in the character of the natural vegetation. Widespread tropical grasslands are the usual resultant of an extended annual dry period or erratic and unreliable rainfall from year to year. But where the moisture receipt is too great for continuous grass yet insufficient to support a true Selva, a variety of intermediate associations develop as a transitional phase, grouped rather loosely here under the general heading of *Tropical*

A

B

C

Fig. 14–2 Funny roots in the Tropical Rain Forest. (A) A Virgin Island kapok tree is representative of the flanged butt so common where deep water-searching roots are unnecessary. (B) The pandanas palm is a member of a strand association and utilizes the stilted roots as a mechanism to withstand frequent brackish water inundation. (C) And if a land plant intends to literally march out to sea, as does the mangrove, its roots must be equipped with "knees" for breathing.

Deciduous Forest and Scrub. The outstanding features of this broad classification are the continuing dominance of woody trees, a tendency toward more open country than in the Tropical Rain Forest, and a loss of foliage during the dry season indicating dormancy.

In large parts of Southeast Asia and to a lesser extent in South America, Africa, and Australia, where rainfall is heavy and reliable, considerable forests of large trees occur. They are more widely spaced than in the Tropical Rain Forest and since their crowns do not coalesce into a continuous canopy, there is a fairly heavy development of bushy undergrowth. Brown and barren during the winter and admitting light to the forest floor, these deciduous woodlands have an altogether different appearance from the nearby Selva. Also, despite having represented among them some of the same species as in the Selva, there is a much greater tendency for the Tropical Deciduous Forest to display large stands of a single specie. Teak has been outstanding in this respect in Burma, Thailand, and adjacent regions, and because it is a valuable cabinet and construction timber, this

has led to widespread forestry and some alteration of the climax plant community. Fire too is a hazard during the dry season and is believed to have been a factor in limiting the occurrence of this type of forest over much larger areas of Africa and South America.

In regions of lower precipitation or where the dry period is more extended, true forest gives way to a great variety of scrub associations. To dignify such vegetation by the term "forest", despite the dominance of trees, as is sometimes done, is to evoke an image of something much more lordly than actually exists, for the trees are stunted and seldom reach heights of more than 25 feet (8 m). Most display the deciduous character in response to the dryness of their environment, but others develop *xerophytic* (moisture-preserving) tendencies, such as thick bark, small pulpy leaves, and protective thorns, and maintain their leaves the year round as evergreens. Some trees, such as the flat-topped acacias of east Africa, spread out rather thinly with grass intervening and no bushy undergrowth (Fig. 14–3). Some group themselves in groves with sizable grassy openings

Fig. 14–3 Dominating the ground-hugging thorny scrub is a grotesque baobab. These distinctive trees are slow growers but drought resistant and usually appear as individuals spotted amongst the deciduous thickets.

separating one grove from another, while others, although fairly widespread as individuals, develop a dense thicket of thorny undergrowth. In northeastern Brazil, such an association, called *caatinga*, extends for hundreds of miles and is virtually impenetrable. The remarkable evergreen eucalyptus family of Australia (of which there are several hundred varieties adapted to nearly every climate on the continent) is found, along with acacia, in a virgin scrubby woodland that encompasses the whole of the north coastal region.

To dignify such vegetation by the term "forest", despite the dominance of trees is to evoke an image of something much more lordly than actually exists.

Altogether, the entire category of Tropical Deciduous Forest and Scrub represents a gradual change from the dense Selva of the Tropical Rain Forest climate region to the Savanna grasslands that occupy the heart of Tropical Wet and Dry climate region. Based on this concept of decreasing size of tree and denseness of undergrowth with increase of length of dry season, it would appear that the various types of deciduous tree and scrub vegetation should be arranged in bands parallel to the outer margin of the Selva. To a certain extent, this is the case, but less so than might be expected. Causative variables include: total amount of annual precipitation (not merely length of dry season), reliability of rainfall from year to year and season to season, terrain, soils, and, of course, the activities of humans, particularly with respect to fire. The distribution map (Map II) illustrates these variations and emphasizes the difficulties involved in assessing with any real integrity what the climax vegetation of large areas might actually be once humans have occupied that region.

Savanna

The *Savannas* are tropical grasslands representing the drier phase of the Tropical Wet and Dry climate. Where moisture is insufficient to support woody trees and bushes as the dominant plants, the grasses take over. Only in a few places, however, are trees of one sort or another entirely absent; the Savanna normally features scattered acacia, palm, brush, or even giant cactuslike plants along the desert margin (Fig. 14–4). In this respect, and several others, the tropical grasslands are somewhat different from the more familiar middle-latitude prairies. The grass is taller and coarser, sometimes reaching the heights of 8 to 10 feet (2 to 3 m), and during the long dry season, these sharp-edged blades become parched and harsh. Neither does the Savanna form a turf, but more typically is arranged in tufts or bunches with patches of soil visible in between. The sharply divisive wet and dry seasons control a vegetative rhythm, alternating between the tall rank brownish mature

Fig. 14–4   The savanna grassland in Africa is seldom an endless sea of grass—skinny little flat-topped acacias seem always to be about either singly or in open groves. They are less affected by fire than most other woody varieties.

grass and the rapid growth of new green shoots with the onset of the summer rains. If adequate surface drainage is lacking, this same rhythm is also reflected in an annual change from a dry drab landscape to one of flood and seasonal marsh. In late winter and spring comes the fire season, usually deliberately set by the graziers and thereby encouraging a full and early sprouting of the new grasses. The common use of fire has made it difficult to determine precisely the boundary between the Savannas and the forest and scrub country—no two distribution maps are exactly alike. But authorities are in rather general agreement that the grasslands are expanding at the expense of the trees.

The sharply divisive wet and dry seasons control a vegetative rhythm.

Much the greatest extent of unbroken Savanna occurs in a wide latitudinal band in northern Africa between the Congo Basin (where in places it abuts with the Selva) and the southern margin of the Sahara where short bunch grass begins to grade into Desert Shrub. Comparable zones, although somewhat smaller, are found in much of the Veldt country of southern Africa, the Campos of interior Brazil, the Llanos of central Venezuela and adjacent Colombia, and a broad belt just south of the north coast of Australia. Lesser widely scattered pockets of tropical grasses, including even the Florida Everglades, are difficult to show at the scale of most maps.

In comparing the map of world vegetation distribution with the map of world climate regions (Maps I and II), note that although only one type of vegetation (Selva) coincides with the Tropical Rain Forest climate, two distinctive vegetation complexes (Tropical Deciduous Forest and Scrub and Savanna) occupy the region of Tropical Wet and Dry climate. Once again, this serves to point up strongly the transitional character of this region between the very wet and the very dry within the tropics.

Desert Shrub

All desert plants must be equipped to withstand the vicissitudes of a harsh and difficult environment, and as a result, they do not grow in any great profusion, individuals being widely scattered and nowhere forming a continuous

Fig. 14–5   What could possibly grow in the poisoned ground of Utah's Great Salt Desert? Obviously not much, but these brave little strugglers are alive and well.

mantle. It is amazing that a great variety exists at all throughout the world's desert regions. But nature has provided these plants with a number of special mechanisms to achieve survival in an almost waterless habitat, and there are only very limited regions that fail to support some kind of vegetative cover (Fig. 14–5).

The true desert begins where grass ceases to be dominant, but in the slightly better watered regions and where precipitation is reliable, bunch grass persists, intermingled with shrubby vegetation. Many of the bushes and shrubs exhibit deciduous characteristics as a means of survival when rainfall is distinctly seasonal, and a large number of flow-

Coping in Death Valley

Death Valley is a microcosm of plant adaptation to the most radical of environmental extremes. Since weather records have been kept at 280 feet (85 m.) below sealevel Furnace Creek, there has been one year with no rainfall at all, one year when the shade temperature reached 134°F (57°C), one year when there were 134 successive days over 100°F (38°C), and one year when the minimum temperature reached 15°F (−9°C). The difficulties are immense for any vegetative life to maintain itself under these kinds of adverse conditions.

A salt marsh now covers most of the valley floor below sealevel, but there are evidences of a 116 mile (187 km) long, 600 foot (183 m) deep lake filling the entire depression as recently as 20,000 years ago. A cooler, moister period fostered a lush and varied vegetation, but this was to change as the current desert conditions began to assert themselves. Some plants simply died and were replaced by invading Mexican arid-land species, some adapted in a variety of ways to cope, and others

found sanctuary in shady canyons with dripping springs. Today we find little enclaves of temperate vegetation, responding to the cooling effects of altitude, stranded atop mountainous islands in a desert sea, e.g., the bristlecone pines on Telescope Peak. Within Death Valley proper are 21 endemic plants, most are endangered and all rare. Included among them are:

Death Valley Sandpaper Plant   Hollyleaf Four Pod Spurge
Death Valley Monkey Flower   Rockalady
Pamamint Locoweed   Rattleweed
Napkin Ring Buckwheat

The chief adaptation for survival in the hot and arid valley is for a plant to develop a xerophytic character, ways and means of seeking out the moisture that does exist and/or conserving that which it finds. Tiny leathery or hairy leaves and thick bark inhibit transpiration; deep roots to tap ground water or a mass of widespread surface roots utilize runoff. The *succulents*, of which cactus is representative, can store up to a year's water supply in their tissue.

Halophytes are yet another type of xerophytic adjustment, in this case the ability to utilize water with a high salt content. Often about a Valley salt pan will be a concentric vegetative pattern reacting to salinity tolerance. Nearest the center the succulent pickleweed will concentrate for it can withstand up to 6% salt; then a zone of arroweed where the groundwater is only about 1% salt; and finally mesquite or salt bush whose tolerence is still less.

Annuals may exhibit a life span of only a few months as they sprout, flower, and broadcast their seed during the cooler late winter and early spring. The new seed is varnished with a growth-inhibiting substance which washes off in the first rain of the new year, and a temperature activated growth enzyme ensures that frost will not damage the delicate new seedling.

ering annuals display the ability to mature and produce seeds within a very short time following periodic showers. But the largest numbers of desert plants are xerophytic, with special adaptations to conserve moisture or to get along on very little. Since we are including here the natural vegetation of both the tropical deserts and the middle latitude

deserts, still another ability must be possessed by many of the plants—that of surviving severe winter frost.

Most of the desert supports some kind of plant cover, and often in much richer variety than is generally appreciated, ranging from the spectacular saguaro cactus and Joshua trees that may grow up to 50 feet (15 m) tall and

Fig. 14–6   The beautiful palm *livistonia mariae* is a sturdy survivor from a wetter, more benign past. It is encountered today only in Palm Valley in the Macdonnell Ranges of central Australia.

appear from a distance as forests, to the simple lichens giving color to rock faces (Fig. 14–6).

The Sahara and contiguous deserts extending through Arabia and the Middle East into Central Asia form what is by far the largest dry zone in the world supporting Desert Shrub vegetation. Other, but lesser, regions include all continental west coasts in the Tradewind belt, Patagonia (southern Argentina), and the Great Basin of North America.

Mediterranean Scrub Forest

The rainfall of the Mediterranean climatic regime is adequate to support a fairly dense vegetation, even sizable trees, when it is remembered that it is heavily concentrated during the winter season when evaporation is low. But the summer dry season is long and often quite warm so that plants and trees taking advantage of the winter moisture must adapt themselves to withstand extended drought to assure survival. The deciduous habit and complete dormancy is a means of combatting seasonal aridity in the Tropical Scrub Forest, but the *Mediterranean Forest* is broadleafed and evergreen and cultivates a variety of xerophytic adjustments to resist drought. The cork oak with its thick bark, the eucalyptus with its deep tap root, and the California oak with its tiny leathery leaves are examples. The entire plant community has a gray and dusty appearance during much of the year as it continues to grow, but only very slowly, during the long summer. However, with the coming of the winter rains, new shoots appear and the landscape takes on a somewhat greener aspect as the rate of plant growth accelerates. Winter frosts are not unknown, but they are usually mild and of short duration and seldom cause damage to the tender new buds.

Trees, although common in the Mediterranean Scrub

Fig. 14–7   Maquis (matorral) along the south coast of Spain. Although a widespread element in the Mediterranean Scrub Forest maquis is emphatically scrub rather than forest.

Forest, are not dominant. They are usually encountered widely spaced or, in certain specially favored areas such as draws or seasonal water courses where deep roots can reach a higher water table, in limited groves. Moderate height and gnarled trunk and branches are typical. Oaks of many kinds are native to both North America and Europe and have been introduced successfully into the Southern Hemisphere. Conversely, the Chilean pepper tree with its grotesque bole and weeping character and several of the Australian eucalypti have been very popular in California and many parts of the Mediterranean Basin. These last trees grow taller, straighter, and more rapidly than most Mediterranean trees and are an important source of timber in west-

ern Australia. A few of the scrub forest trees, notably the olive, fig, and chestnut, have been removed from their native context and cultivated as important sources of food. Still another representative of the trees, found in small numbers in the scrub community, is the needleleaf evergreen. These trees, such as the digger pine, giant redwood, and Lebanon and Aleppo cedars, are usually encountered along the slightly wetter foothill margins. Others, like the Monterey pine and cypress (pre-ice relicts) and the coast redwood, require the special environment of the low temperature and high humidity of the foggy coasts to survive in the Mediterranean climate.

But despite the wide occurrence of trees in the scrub

forest, the dominant member of the plant community is a low woody brush understory, generally called *chaparral* (United States) or *maquis* (Europe and South Africa) (Fig. 14–7). The many shrub varieties that make up this complex are beautifully adapted to their environment, with both deep roots to tap the summer water table and widespreading shallow roots to take immediate advantage of winter rains. They also spread themselves through sucker propagation from the roots and thus expand their domain in all directions. In the *mallee* (scrub eucalyptus) country of southern Australia, they have proven almost impossible to eradicate short of the monumental task of digging out every involved root system. This scrub, often a tangled intertwined mass, is, however, very susceptible to summer fires, and if it is burned repeatedly may be succeeded eventually by grass. Large parts of California, especially the drier hills, are grass covered—brown in summer and green in winter with the colorful California poppy and blue lupine intermixed.

A *few of the Mediterranean scrub forest trees, notably the olive, fig, and chestnut, have been removed from their native context and cultivated as important sources of food.*

Inasmuch as the Mediterranean climate is limited in its distribution to mere coastal regions of the Mediterranean Basin and tiny west coast exposures poleward of the tropical deserts, the Mediterranean Scrub Forest is not a widespread association. And even within the Mediterranean climate regions, humans have been in residence for such an extended period of time that a typical scrub forest in its virgin state is difficult to reconstruct or encounter today.

Subtropical Coniferous Forest

There is only one locality in the world where coniferous trees are the dominant species within the subtropics—in the southeastern coastal United States. An extensive region forested almost exclusively by several varieties of pine is found here. Stretching from Chesapeake Bay in the north, along the coast in an ever-widening band into eastern Texas, the coniferous forest appears to be strongly coincidental with

Fig. 14–8   Loblolly pine in coastal North Carolina—probably 75 years old. But despite their age these are not large trees and the forest is open. The chief virtue of the southern pine as a commerical timber tree is its ability to occupy land that is unproductive for agriculture and to reproduce itself rapidly.

the sandy soils of the low-lying coastal region (Fig. 14–8). The deep alluvia of the Mississippi bottoms and the heavier soils of the interior support a mixed or broadleaf forest, and since these develop under what is essentially a climate identical to that of the coniferous forest (Humid Subtropic), the soil factor is almost certainly the basic determinant; that is, the vegetation is edaphically rather than climatically controlled. Furthermore, although the Humid Subtropic climate is found in every continent, only in a couple of these areas is there any hint of a true coniferous forest. In southern Brazil there are some limited stands of Araucaria pine that may have been more extensive at one time. Today they are found only at slight elevations that partially moderate the subtropical climate. South of this region the grasslands of Uruguay and much of the Argentine Pampas are sufficiently wet to support trees, although they have not within historic time. Very probably, they did at one time before repeated fires wiped them out, but there is no way now of knowing whether these imagined forests were coniferous.

The [subtropical coniferous] vegetation is edaphically rather than climatically controlled.

Within the pine forest area of the southeastern United States, loblolly, yellow, and shortleaf pine are all represented, often mixed one with another. They grow fairly widely spaced and mature into a medium height tree of less than 100 feet (30 m) tall. Grass and occasional evergreen shade-loving broadleaf shrubs such as rhododendron occupy a remarkably clean forest floor where heat and moisture promote the rapid decay of fallen needles and litter. One of the features of the subtropical evergreen forest has been its reproductive ability after the trees have been removed. With the advent of technologies allowing the utilization of southern pine forest for a wide range of wood products, the regrowth of mature timber trees within 30 years after cutting (less for pulpwood) has given the southern lumberman a real advantage over competitors. Much of this area has been in crops at one time, but impoverished soils and increasing demand for timber has led to extensive plantings of farm woodlots and reforestation of eroded slopes. Today, probably more of this region is in timber than at any time in the last 100 years.

Middle Latitude Coniferous Forest

Like the Subtropical Coniferous Forest, the *Middle Latitude Coniferous Forest* occurs only in North America. It appears to be a product of the Marine West Coast climate and is limited to a narrow coastal strip from northern California to southern Alaska. But there are other areas in every continent, except Africa, exhibiting a Marine West Coast climate, and although they are forested, conifers nowhere constitute the dominant species. Probably the short summer dry season with its attendant forest fires, which shows

Fig. 14–9   Many of these virgin redwoods in northwestern California are over 200 feet (61 m) high. One holds the world record at 368 feet (112 m). But these are *sequoia sempervirons*. The fabled "big" trees of the western Sierra slopes are *sequoia gigantis* and although not quite as tall, are much longer lived and of huge girth.

up only in North America, is responsible. Whatever the cause, our Pacific, Northwest and adjacent Canada and Alaska support some of the world's heaviest forests. Not only are the trees huge, coast redwood and Douglas fir reaching over 200 feet (61 m) in height and 30 feet (9 m) in girth, but they grow very close together (Fig. 14–9). On the west side of the Olympic Peninsula and Vancouver Island, what might well be described as a Middle Latitude Rain Forest exists, with as heavy a vegetation growth as any Tropical Rain Forest. Under conditions of 150 to 200 inches (381 to 508 cm) of rainfall and mild year-round temperatures, the giant Douglas fir grows mightily. Below, an understory of western hemlock, dwarfed by the Douglas fir but a sizable tree in its own right, fills in the narrow intertrunk spaces. In some places the ground cover is bracken fern that on occasion reaches 6 to 8 feet (2 m) in height, or scattered fern is often interspersed with huckleberry, rhododendron, and other low herbacous growth. In the less well-drained areas, western red cedar supplants the Douglas fir; and beneath the cedar/hemlock association, devil's

Fig. 14–10 A snow-capped monarch looms over an extensive stand of old-growth Douglas fir. Logging in the foreground will pay some salaries and local taxes but it certainly detracts from the pristine character of a classic natural landscape.

club and skunk cabbage are common. This tangle of growing vegetation is rendered even more impenetrable by the accumulation of moss-covered downed trees that fail to decompose as rapidly as in the tropics.

Beneath the cedar/hemlock association, devil's club and skunk cabbage are common.

By no means does all of the North American northwest coast support this heavy a growth or even all of these specific plants and trees, but only minor variations on this are the rule. For instance, northern California features the redwood, while the Douglas fir fades out north of Oregon and Washington and spruce and hemlock take over. Also,

since the mountains very closely approach the coast throughout most of the region, the forest changes its character in a regular sequence with gain of altitude. But remember, we are not classifying mountain vegetation and must draw an upper limit to the typical coastal forest, probably well below 1000 feet (305 m). Certain portions of the Middle Latitude Coniferous Forest have been removed permanently as in the Willamette Valley and Puget Sound country, but in less accessible regions and on the steeper hillsides, millions of board feet of timber remain despite commercial cutting and frequent summer fires (Fig. 14–10).

Middle Latitude Broadleaf and Mixed Forest
This type of forest, displaying many variations, occupies regions exhibiting three different climates—Humid Sub-

tropic, Humid Continental, and Marine West Coast. All of these have in common relatively heavy precipitation, thus allowing the development of trees as the dominant vegetation.

A fair amount of light can reach the forest floor, especially since this is a deciduous forest.

In the United States, much of the South, inland from the sandy coastal plains, supports a deciduous broadleaf forest featuring an oak/chestnut association. In the Middle Atlantic States and the Middle West this merges into a quite similar oak/hickory or walnut/poplar woodland. These are trees of considerable girth but only moderate height, and although they grow quite close together, they do not form a canopy after the fashion of the broadleaf Selva. A fair amount of light can reach the forest floor, especially since this is a deciduous forest, and the growth of saplings and young trees is thus encouraged (Fig. 14–11). Brushy undergrowth is not dense, but windfalls and the accumulation of several inches of leaves, twigs, and general forest debris are typical. In poorly drained areas, swamp associations such as the cypress/red gum of the southern Mississippi River floodplain replace dryland forest and may cover many square miles with their jungle-like tangle of supporting brush and vines.

North of this broadleaf deciduous forest, in the region of the Great Lakes and New England and roughly coincident with the Humid Continental Short Summer climate, there is a gradual change to a mixed forest; that is, conifers begin to appear as individuals or in sizable groves or clumps among the broadleaves. Birch, beech, and maple take over from the oak and hickory, and various pines, spruce, and fir show darkly among them. Sandy or rocky soils often are responsible for extensive stands of conifers, as in northern Michigan, New Jersey, and Maine, while elsewhere they are much more generally intermingled.

Europe is somewhat a counterpart to North America in terms of correlating climatic regions to vegetative type except that the Marine West Coast climate here supports a broadleaf deciduous forest and the Humid Subtropic climate is lacking. Unlike the Marine West Coast climate of North America where coniferous forests predominate,

Fig. 14–11   Deciduous trees allow a good deal of light to reach the forest floor albeit generally at the wrong season to encourage heavy growth. But in the spring there is life everywhere as the delicate new leaves impart a limpid quality to the gently filtered sunshine.

western coastal Europe from northern Spain to southern Norway had an original, deciduous forest on the order of that of the eastern United States. Frequent conifers seen today are either at some elevation, on rocky or sandy soil, or have been planted to replace the slower-growing hardwoods. But the Humid Continental-Long Summer climate of northern Italy and much of the Danube Basin corresponds to that of the United States in producing a broadleaf deciduous forest, and the Humid Continental-Short Summer regions of northern Europe are mantled by a mixed forest.

China, Korea, and Japan demonstrate quite clearly the

general progression from broadleaf deciduous and semi-deciduous in the south to mixed forests in the north with increased severity of winter and a shorter growing season.

Broadleaf and mixed forests are represented in the Southern Hemisphere although to a more limited degree. The narrow coastal strip of southern Chile, with its Marine West Coast climate, displays a dense broadleaf deciduous forest where beech is the dominant species, while along the Humid Subtropic coast of southeastern Australia, the ubiquitous eucalyptus, broadleaf but evergreen, is the most common tree.

The character of the original climax vegetation is mere conjecture. In northern China not a vestige remains.

Without a doubt, humans have destroyed the Middle Latitude Broadleaf and Mixed forest more thoroughly than any other vegetative type in the world—in places so successfully that the character of the original climax vegetation is mere conjecture. In northern China, where for centuries not only have the forests been cut again and again but the roots systematically grubbed out for fuel, not a vestige remains. Soil erosion is, of course, inevitable unless some sort of crop or substitute vegetation is introduced to intercept the surface runoff. Reforestation is being pushed with considerable success in many parts of the world, not only to minimize soil erosion, but to make productive those steep slopes and rocky or sandy areas that are not used for other purposes. Northeastern United States, Europe, and China are heavily populated and timber-short and have long had to import wood products from other parts of the world or to go without. Since the original forest was slow growing, replacement is usually with more rapidly maturing conifers.

Middle Latitude Steppe and Prairie

Once again, in the middle latitudes as in the tropics, grasslands show up as transitional vegetation between the climates that are moist enough to support vigorous tree growth and the moisture-deficient deserts. The tall grass prairie is adjacent to the forest, and it frequently interdigitates with fingers of tree-lined streams pushing well out into the prai-

Fig. 14–12   Short grass prairie in Wyoming.

rie, while grassy openings become common along the forest margin. The prairies themselves feature grass averaging 2 feet (.6 m) high, although in places reaching well over that, with the root systems merging into a solid turf. Many lesser herbaceous plants are represented in the prairie also, some of them with showy annual flowers, but woody growth is distinctly lacking, and a prairie landscape is one of endless vistas of waving grass (Fig. 14–12).

As the margin of the desert is approached, the height of the grasses decreases to 6 or 8 inches (15 or 20 cm) and the species change, but the turf continues. This short grass prairie is commonly called *steppe*.[1] Generally, when bunch grass and woody scrub with bare soil between make their appearance and the continuous turf no longer is in evidence, the dry margin of the steppe has been reached. Such grasses as the distinctive American buffalo grass are typical of the steppe, although if overgrazing is allowed, they often are gradually replaced by a less nutritious secondary growth

[1] Obviously, in the tropics too, the height of the grass is lower as precipitation lessens and a kind of steppe develops, but since the term was first applied in Russia to describe a middle latitude short grass prairie, its use has been confined to that particular type of vegetation.

that does not change the appearance of the grassland but is no longer the climax vegetation.

In North America the prairie coincides with the drier western part of the Humid Continental climate region, while the adjoining steppe is found in the moister eastern fringe (chiefly east of the Rockies) of the Middle Latitude Dry climate. This gives us the world's largest continuous middle latitude grassland from Alberta and Saskatchewan south almost to the Gulf. Minor representatives occur in the Washington Palouse county and in eastern Texas. Almost as large a region is found stretching from Hungary through the Ukraine and vicinity and north of the Caspian Sea in a long narrow band far into central Siberia. Sizable outliers occur in Asia Minor, Iberia, and Manchuria. The situation in Argentina, Uruguay, and southern Brazil has already been discussed, but this has been prairie as long as western people have known it and must be so classified. Australia exhibits a large region of middle latitude intermixed scrub and grassland in the interior southeast (chiefly New South Wales and Queensland) and small areas of prairie occur in the Orange Free State of South Africa and in New Zealand.

The long grass prairie has proved to be a highly productive agricultural region since farmers have acquired the implements to break the heavy turf. As might be expected, the food grasses, chiefly corn and wheat, predominate. The steppe, on the other hand, with its lesser total moisture and cyclic precipitation pattern has been traditionally a grazing area, although with improved techniques, small-grain agriculture is at least partially successful. This means that in terms of existing unmodified steppe and prairie today, there are few places in the world where either the plow or overgrazing has not wrought considerable change.

Taiga (Boreal Forest)

The *Taiga* is the world's greatest forest, at least in terms of area involved, for it stretches in a wide band the entire breadth of North America from Alaska to Labrador and across the much greater breadth of Eurasia from Scandinavia to Kamchatka. And it is virtually a virgin forest, partially because of its remoteness from major population centers and partially because of the character of the trees as potential saw-timber. Only along its periphery have commercial lumbering (and Christmas tree plantations) modified the climax forest to any significant degree.

The Taiga is the world's greatest forest, at least in terms of area involved.

The Taiga is basically a coniferous evergreen forest made up of a relatively small number of species. Extensive stands of spruce, fir, and pine are common in North America, while pine dominates the Eurasian forest except for a large area in northeastern Siberia where larch, a rare deciduous conifer, prevails. The extremely long cold winters and limited growing season are not an ideal habitat for trees, and those few species that do exist are simply the ones that have demonstrated their ability to survive. As a result, the trees are small, even stunted, seldom reaching over 50 feet (15 m) in height and 8 to 10 inches (20 to 25 cm) in diameter. Widely spaced, the pointed conifers do not form a high canopy, but the receipt of light at lower levels allows the development of branches and foliage well down the bole, and these skirts tend to coalesce and form what amounts to a low canopy effectively cutting out light at the forest floor. Thus limited by lack of light and cold temperatures, undergrowth is very sparse. A thick carpet of needles and many dead branches and downed trees demonstrate the very slow rate of decay of organic material in the absence of high temperatures.

This accumulation of trash plus the resinous character of the trees make the Taiga highly susceptible to fire, and each summer forest fires burn over extensive areas. Secondary growth of aspen, alder, and birch appears out of place among the dark conifers and, if left undisturbed, will eventually be succeeded by them. But growth is slow, relegated to just a few summer months each year, and the rate of replacement in the high latitude forests contrasts strongly with that in the tropics.

Growth is slow, relegated to just a few summer months each year.

In some places, notably the Ob River Valley of western Siberia, northward flowing streams that are frozen during the winter maintain the ice at their mouths long after the upper courses have thawed. The blocked mouth thus

Fig. 14–13   Much of Labrador has been heavily glaciated, the exposed rocky surfaces covered here with lichen. Thirty foot (9.14 m) spruces strive mightily to form even a patchy forest.

causes the river to overflow each spring and seasonal swamps are formed over large areas, which prevents the establishment of normal forest. Coarse grasses, sedges, and spotty low brush are the result.

Also, in the regions that have suffered severe glaciation, as in much of Quebec, bedrock has been exposed at the surface. Although the trees display remarkable tenacity in establishing their roots in cracks and crannies where even a little soil may be available, true forest occurs only in scattered pockets between the expanses of barren rock (Fig. 14–13). This is a patchy forest, if a forest at all, and locally breaks up the general Taiga landscape of endless trees extending out beyond the horizon.

Within the Taiga, the largest trees and the better developed forest are found along the southern fringe, gradually deteriorating until in the far north trees fade out or mature into mere saplings. This is not because the winters are colder, for the trees are dormant in any case and whether the temperature is −30°F (−35°C) or −50°F (−46°C) makes little difference. Rather, the controlling factors are length of growing season and summer heat. Both of these decrease with distance north, and as roots are forced to compete with permafrost, it is reflected in the size of the vegetation.

Tundra

Now we have gone beyond the tree line, roughly analogous to climbing a high mountain, through the forest to its upper margin. Between this tree line and the permanent snows of the mountain peak is a vegetation zone that is very similar to the *Tundra* and has developed under much the same conditions. The two-month summer season with its low temperatures and even occasional frost is simply inadequate for trees. Botanists will say that there are trees in the Tundra and produce examples of alder or juniper, but they are 6 inches (15 cm) tall and likely to be recumbent in character; to most of us these are not trees. Woody plants of all kinds are rare, and Tundra vegetation is a highly specialized association of species that have the ability to withstand (or produce seeds that can withstand) long bitter win-

Fig. 14–14   Summer tundra. Barren of snow now and frequently poorly drained, the dogs can no longer pull a sled and everyone becomes a pedestrian. Permanent ice is not far below that hummocky surface—vegetation size and variety is strictly limited by cold damp soils and restricted root space.

ters, high winds, constantly cold moist soils, low-nitrogen soils, and shallow permafrost, and to come rapidly to maturity and to reproduce themselves during a short, cool summer (Fig. 14–14). Such conditions rule out most plants. The ever-present grasses are represented, chiefly in the form of rank sedges, and some mosses and the simple primitive lichen are everywhere. These, with sundry other low herbaceous plants, some flowering briefly, many reproducing through buds or shoots from their perennial root systems, make up the Tundra vegetation.

It is a low, ground-hugging vegetation on a flat to gently rolling terrain. Surface drainage is inadequate at best, and vertical soil drainage is lacking entirely as a result of permafrost, often less than a foot underground in midsummer. So despite only an average of 5 to 10 inches (13 to 25 cm) of precipitation per year, the surface is constantly wet and boggy during the growing season. Corridors of low trees may push out from the Taiga along stream courses in response to better drainage and lower permafrost levels, for paradoxically, flowing streams mean drier surface conditions.

During the summer, from seemingly out of nowhere, come great clouds of gnats, no-seeums, and mosquitoes that somehow have survived the killing winter cold.

During the summer, from seemingly out of nowhere, come great clouds of gnats, no-seeums, and mosquitoes that somehow have survived the killing winter cold and can literally eat alive anyone who attempts to inhabit the Tundra. Winter is the time to travel here, despite the bitter cold, for the insects are gone and the surface has frozen solid and is no longer one great bog.

The Tundra is a north coastal vegetation and is found in a virtually unbroken series along the north coast of North America and Eurasia. It is nonexistent in the Southern Hemisphere except in a few tiny high latitude islands. The Falkland Islands off Patagonia and the Aleutians in the North Pacific are commonly included in the Tundra grouping despite their relatively low latitudes, for they do not support trees, although grass is more extensive here than in true Tundra. High winds and cool summers with occasional frost are probably the deterrents to trees.

CHAPTER 15

INTRODUCTION TO SOILS

This was once a productive agricultural field. Today we understate the case when we simply classify it as severely eroded. Soil erosion is insidious at first, nobody really notices that the topsoil is being carried away bit by bit, so that by the time moderate gullying begins much of the damage has been done. Erosion is a cancer on the land—if caught in the very early stages sometimes it can be controlled; but more often than not it is much too far along to heal when the danger is recognized. Obviously this farm in Bolivia has reached the terminal stage. If the inhabitants of the little farmhouse step out the front door they are in danger of falling into a 50 foot (15 m) deep badland chasm.

THE NATURE OF SOIL

Never shrug off soil as "just dirt." It is an important element in the human physical environment. Certainly, its ability (or inability) to produce living growth impinges upon their effective utilization of the earth as a habitat. So we should not only be interested, but at least rudimentarily informed as to its character. But what is soil? Merely rock broken into tiny fragments? If we were to take a rock and crush it mechanically, we would not have achieved soil, no matter how fine the particles. We would have the raw material from which soil is formed, for this breakdown of rock by natural means is the first step in the development of soil, but other steps must follow, and other elements, beyond simple rock fragments, must be present before true soil evolves.

It is traditional to regard soil (*solum*) as the result of the interaction of four elements: (1) mineral (rock fragments—*regolith*), (2) organic (both plant and animal), (3) gas, and (4) water. The mineral constituent not only supplies the great bulk of the soil but is one of the basic determinants of fertility and contributes importantly to *soil texture* (size of particle from fine clays to coarse sands) and *structure* (the arrangement of particles from granular to blocky).

Organic influences involve the breakdown of dead plant and animal remains and their integration into the soil, as well as the probing of plant roots and burrowing animals, the changes involved in the passage of soil materials through the digestive tracts of earthworms, and microbiotic activity in general.

*N*ever shrug off soil as "just dirt".

Growing plants require a certain root aeration and most soils, depending on density and drainage, have some air in them. But gases are also produced by decomposition and chemical change, and many of these, including air, further facilitate and encourage the activity of microorganisms and chemical reactions.

However, of all these soil-forming elements, water is singularly important. Without water there would be no vegetation, virtually no chemical reactions, and a paucity of microorganisms. And since water in soil is a direct result of climate, it can be said without equivocation that, given the initial broken rock as raw material, climate is the chief determinant of soil character. Compare the map of world soil distribution with that of climate (Map I and III) and note the basic similarity.

This intimate relationship among soils, their fertility and formation, and climate is a closer relationship than was generally recognized for many years. It appeared in the early days of soil science (*pedology*) that a knowledge of the geology of a region would lead to an understanding of the soils, for after all, soil appeared to be fundamentally nothing more than decomposed rock. To a degree this sort of logic proved out, but it shortly became apparent that there were major flaws in the reasoning. Extensive areas of quite similar soils were found to overlie and have been derived from many dissimilar parent rocks and, conversely, in comparing several different regions, soils differing markedly were demonstrated to have originally derived from the same type of parent material.

Eventually, a Russian researcher put forth the suggestion that possibly climate had an important effect on the development of soil characteristics, and his subsequent soil distribution maps of parts of Russia were found to coincide almost perfectly with the climate maps of the same area. Modern soil science now recognizes this close relationship between soils and climate. Slope, drainage, vegetation, and parent material are among the many other factors contributing to the final end product that we call soil, but climate, especially precipitation, is of prime importance in determining the fertility of most soils. And in geography where the emphasis lies in soil's utility by humans, fertility is of paramount significance.

MINERAL COMPOSITION

Since nearly all known elements occur in the rocks of the earth's crust, it follows that these elements will be represented in the soils derived from this crust. The chemical composition of a typical soil should reflect initially something of the average proportions of the elements in the crust.

There are about eight rather common ones: oxygen, silicon, aluminum, iron, calcium, sodium, potassium, and magnesium. All others may be regarded as trace elements, although several, notably phosphorus and nitrogen, are of great importance in the ability of a soil to support plant growth. The critical quality here is that the minerals constituting the vegetative nutrients in the soil must be soluble in water, for a *plant receives its nourishment through its roots in solution.* With a few unusual exceptions, such as plants that catch flies or feed directly from the air, this is universal. If we check the common elements in most soils with regard to their solubility, we find that silicon, aluminum, magnesium, and iron (and their various compounds) are not readily soluble or soluble only under certain conditions. On the other hand, calcium, sodium, and potassium along with nitrogen and phosphorus in certain forms are highly soluble in clear water. These then must be the chief plant foods, and if one were to read the list of chemicals on the label of any common garden fertilizer, it would be discovered that these elements make up the bulk of the product that is used for feeding plants. This is not to say that iron, aluminum, or many of the minute trace minerals are not required for vegetative growth; most plants need small amounts of even such minerals as copper and cobalt. But the nutrients that are required in large quantities must be readily soluble.[1]

LEACHING VERSUS EVAPORATION

Let us assume a region of heavy rainfall. As the moisture falls to the ground, some of it evaporates and some runs off as surface drainage, but always a certain amount of it enters the soil as groundwater and percolates downward in response to gravity. The constantly downward moving groundwater takes into solution the plant foods of the topsoil and carries them beyond the reach of normal root systems, thus, over a period of time, washing out the bulk of the nutrients and leaving only the nonsolubles. This process is called *leaching,* and when carried to an extreme results in generally infertile soils. In relatively dry areas, leaching is cut to a minimum and, assuming a normal complement of minerals present, the soluble mineral content and fertility of the topsoil are correspondingly higher.

A plant receives its nourishment through its roots in solution.

There is a factor, however, that tends to offset leaching by causing groundwater and its accompanying minerals in solution to move upward in the soil in opposition to gravity. This is evaporation from the surface. As the top layer dries from evaporation, it sets in motion *capillary action,* a blotterlike effect that pulls the groundwater upward. If evaporation is roughly equal to precipitation, there is little or no loss of fertility from leaching. And if evaporation somewhat exceeds precipitation, not only are the surface solubles replaced as rapidly as they are leached, but in addition the salts from deep in the soil are brought up to within reach of the roots, thus increasing the natural fertility (Fig. 15–1). In the United States a north/south zone in the Central Plains states receiving about 15 to 20 inches (38 to 51 cm) of precipitation per annum and experiencing high summer temperatures has the ideal balance of precipitation and evaporation. Farther west in the desert and its margins, soil fertility is frequently excellent, but there is a danger that the evaporation potential may be so high that following infrequent storms (or inadequate irrigation), capillary action will operate so efficiently as to concentrate all the soluble salts on or near the surface in sterile alkali flats (Fig. 15–2).

It would seem that, generally speaking, *the humid regions exhibit infertile soils because of leaching and dry areas possess fertile soils*—the ideal being just short of the desert where evaporation somewhat exceeds precipitation.

---

[1] The actual mechanics of a soil's moisture-holding capacity and the ability of soil water in turn to take minerals into solution and transfer them to plant roots, is closely tied to the size of the individual soil particle. Both organic and inorganic particles eventually become reduced by normal soil-forming processes to submicroscopic dimensions. Chemical changes at this point lead to the formation of *colloids,* tiny particles that when combined with water, become gelatinous in consistency. These colloids display an ability to absorb ions from the soil solution and then release them in exchange for other ions (termed "base exchange"). They act as a type of fertility bank, releasing stored nutritive ions as the soil solution balance is destroyed by plant withdrawal.

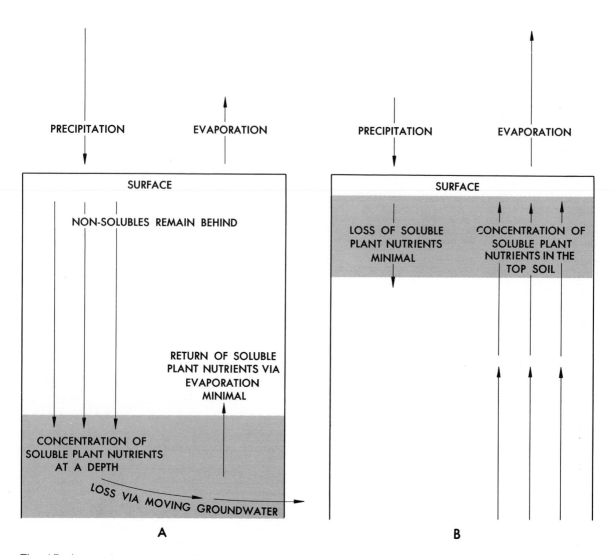

Fig. 15–1   Leaching versus the effect of capillary action. (A) Depicts a situation where precipitation exceeds evaporation. In (B) the reverse is shown. Capillary action offsets leaching when evaporation is greater than precipitation and the solubles are concentrated near the surface.

ACIDITY VERSUS ALKALINITY

Soil water in humid regions is likely to be acid. Rainwater may absorb small quantities of carbon dioxide from the atmosphere, more contaminants will be acquired as it filters through a rotting mat of vegetative debris, and finally it absorbs acidic end products of rock decomposition in the soil itself. Attempting to neutralize these acid tendencies are the alkalies, but unless the parent material is particularly rich in calcium, sodium, or the like, alkalies are seldom in adequate supply. Not only are many alkalies readily soluble and therefore susceptible to reduction by leaching but once a soil becomes acid, acid-tolerant vegetation such

Fig. 15–2   During infrequent showers this desert basin fills with a shallow sheet of water. As it evaporates its salts are deposited on the surface. In addition, for several weeks thereafter further soluable minerals are pulled from the damp soil by capillary action, and the result is a thoroughly toxic soil from the standpoint of growing crops.

as conifers, begins to flourish; the conifers drop their acidic needles to further sour the soil and complete an acid perpetuating cycle.

Once a soil becomes acid, acid tolerant vegetation such as conifers, begins to flourish; the conifers drop their acidic needles to further sour the soil and complete an acid perpetuating cycle.

In drier areas where there is less vegetation, slower decomposition of both organic and inorganic materials, reduced leaching, and an increasing dominance of evaporation, soils exhibit more alkaline tendencies. Grasses rather than trees respond to the sweeter soil conditions and in turn contribute their alkaline remains to be integrated back into the soil.

A good general statement that recognizes the prominent influence of leaching on soil acidity and the easy solubility of most alkaline minerals is that *humid regions generally display acid soils and dry ones alkaline*. Both of these tendencies can be overdone since most normal field crops prefer a reasonably neutral soil, if anything a little on the

alkaline side. But overalkalized soils, so typical of many desert regions, can be managed. The salts are soluble and can usually be washed out by heavy applications of irrigation water—always assuming, of course, adequate soil drainage. And the excessively acid soils of the humid areas can be made productive too by the addition of fertilizers containing proportionately large amounts of lime.

SOIL COLOR

The color of soil is one of its most obvious characteristics, and there is considerable temptation to attempt to relate certain vivid or striking colors to fertility. Locally, this may be possible, but there are few generalizations that can be made establishing such a relationship. The farmer of the Deccan district in India recognizes that his black soils are much more fertile than those of the surrounding areas because they have been derived from dark alkaline volcanic parent material, but Japanese farmers have learned to shun a dark volcanic ash soil that is highly acid in character despite its color and the popular conception that volcanic soils must be fertile. The wheat farmer of the Ukraine finds his black prairie soil to be somewhat alkaline; the bog farmer of the Sacramento Delta finds his to be black but acid. And desert alkali wastes may be either black or white. In Brazil and Vietnam certain bright red soils (terra roxa and terra rouges, respectively) have a reputation for great productivity, while elsewhere in the tropics a reddish hue often means excessive leaching, and the darker alluvials of the river bottoms are preferred. And so it goes; color can be a clue to fertility, but only if the peculiarities of a certain local situation are known.

There is a considerable temptation to attempt to relate certain vivid or striking soil colors to fertility.

One very common source of dark color in soil is fixed carbon, an end-product of organic decomposition. In the prairie, for instance, the grassy turf with its myriad hair roots dies each year, and this vegetable matter, immediately attacked by microorganisms, becomes a part of the soil. In a partially decomposed state, this material is called *humus*, and as it breaks down it releases carbon to color the soil. It would appear that a plant of any kind, receiving its nutrition from the soil, would return exactly that which it had removed when it died and once again became a part of the soil. This is not exactly the situation, however, for humus is in large part colloidal and thus becomes an important seat of topsoil base exchange. Further, nitrogen, an important nutritive element for most plants, is generally not available to the soil directly from the air, but nitrogen-producing microorganisms thrive in the presence of decaying vegetation, and thus humus aids in supplying this vital element. Light-colored desert soils, being unleached, are often potentially very fertile, but their inability to produce a significant vegetation because of lack of rainfall denies them humus and therefore nitrogen. To be made useful for agriculture, desert soils usually require nitrogenous fertilizers. And finally, humus improves the structure and texture of the soil (makes it crumbly or friable and easy to work), aids aeration in dense soils, and being spongelike, adds greatly to the water-holding capacity (Fig. 15–3). So the darker the color, the greater is the amount of humus. But remember, dark coloration may come from sources

Fig. 15–3 Black and friable in texture, this field is rich in humus from repeatedly plowed in organic debris. Bits of last year's waste stems and straw are visible in the foreground, yet to decompose into humus.

other than humus, perhaps parent material, and black soil and fertility need not always go hand in hand.

Red color means iron in soils. It could be that the parent material was high in iron content and passed its coloration on to the soil from which it formed; but, more commonly, much lesser amounts of iron in the soil oxidizing under conditions of high temperature and heavy rainfall is responsible. This means, of course, that reddish to yellowish soils will be most characteristic of the humid tropics and subtropics and will display at least mildly acidic tendencies as a result of heavy leaching.

Nitrogen, an important nutritive element for most plants, is generally not available to the soil directly from the air.

Probably, all things considered, the finest agricultural soil in the world is a series first identified in the eastern Ukraine and adjacent central Asia called *chernozem* (from the Russian, meaning black soils). In the United States these soils are found in a band from Canada south through the central Dakotas into Kansas, and they also occur in similar climatic situations in Argentina, Australia, and South Africa. Here is the previously described region of ideal evaporation/precipitation ratio and, in addition, since the natural vegetation was prairie, a high humus content. Yet only low-yielding grains are grown in these places despite their agricultural promise, because the very moisture factors that are required for this soil to develop are insufficient for most agriculture. Thus is seen the unhappy reality of natural soil fertility versus agriculture—*where enough rainfall exists for successful general farming, the soils suffer through leaching, and where it is dry enough for evaporation to balance leaching, not enough rain is present to allow any but limited agriculture.*

THE SOIL PROFILE

It has been said that a soil is not deserving of the name "soil" until it has developed some sort of profile (Fig. 15–4). Only then has it evolved into a proper life-supporting material, and the profile is the measure of its maturity. Such a theory is subject to question, but undoubtedly all soils are living things and do exhibit an evolutionary progression that usually culminates, sooner or later, in a climax stage and profile development. So one reasonably reliable clue to a soil's character is its profile.

A vertical cut through most soils, from the surface to bedrock, will display a cross section with prominent horizontal bands called horizons. At the top is the A horizon or zone of *eluviation* (leaching). This is the topsoil that experiences the greatest loss of both soluble minerals and finer soil particles through washing and thus tends to have a somewhat coarser texture and lighter color than the adjacent horizon. However, often at the immediate surface, there is a narrow band, black to brown in color, designated as the A_1 horizon, which is made up of the partially decomposed leaves, stems, and general litter of the covering vegetation.

A vertical cut through most soils will display a cross-section with prominent horizontal bands called horizons.

The B horizon, below the A, is the recipient of the minerals and clay particles leached out of the topsoil and is called the zone of *illuviation* (accumulation). It tends to be basically more fertile because of this and is also more darkly colored and denser in structure, even to the point of developing into an impermeable hardpan.

Still farther down is the C horizon, which is only one step removed from solid rock and is thus not yet true soil. It is made up of weathered and broken rock particles and will, as time goes on, become at its upper margin a part of the B horizon. Below this is the parent material itself, sometimes called the D horizon.

Obviously, no two soils will have identical profiles, and it is by the recognition of these differences that soil types can commonly be sorted out and boundaries drawn between them. The A_1 horizon, for instance, is frequently lacking where vegetation is minimal, as in the desert, or where decomposition is so rapid as to prelucde its development, as in the humid tropics and subtropics. But the chernozem soils are so dominated by the A_1 horizon that the remainder of the A horizon is not recognizable. In some soils the margin of each horizon is sharply defined, while in others it is blurred and amorphous. Yet most soils have some kind of profile, and it is a recognizable feature of soil development. Only in newly formed or transported soils, or where some influence such as a steep slope or poor drainage interrupts their proper formation, do soils entirely lack profiles.

THE INFLUENCE OF TEMPERATURE

So far, emphasis has been on the precipitation factor in climate, since soil moisture is undoubtedly the major control of soil fertility. But temperature too has some effect, as in the determination of the rate of evaporation. Temperature is also a partial control of vegetative decomposition in humid lands and thus affects the soil. For example,

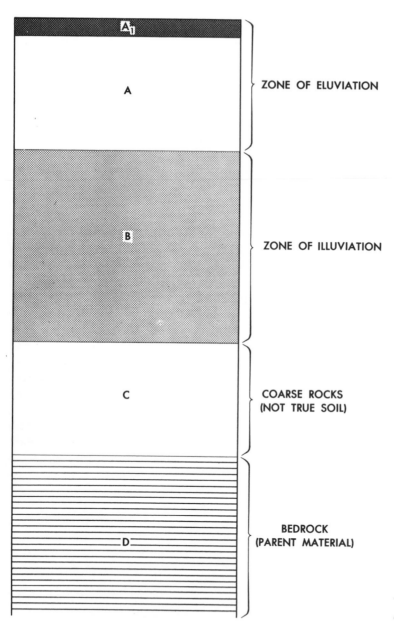

ZONE OF ELUVIATION

ZONE OF ILLUVIATION

COARSE ROCKS
(NOT TRUE SOIL)

BEDROCK
(PARENT MATERIAL)

Fig. 15–4   Idealized and simplified soil profile.
These four horizons are encountered, at least to a
degree, in nearly every mature soil.

let us compare the forest soils of northeastern United States
and southeastern United States. In both areas, heavy precipitation results in forest vegetation and leached soils, but
in the north the winters are much longer and colder than
in the south and the rate of microbiotic activity in the vegetative litter on the forest floor is quite different. In the
northeastern United States there is a permanent mantle of
needles and leaves that accumulates more rapidly than it
decomposes (Fig. 15–5). This means that precipitation entering the soil as groundwater must filter through this rotting litter and becomes strongly acid. The water dissolves
the readily soluble plant food and removes it, but also at-

Fig. 15–5   Any surface water that enters the underlying soil becomes a modest acid, for it must percolate through a spongy, decomposing accumulation of acidic needles on the coniferous forest floor. Conifers are acid-tolerant and will respond favorably to a sour soil thus setting up an acid perpetuating cycle.

tacks and removes at a slower rate the iron and aluminum, leaving in the topsoil a concentration of silica. Such a soil, fully developed, displays a light-colored and relatively thin A horizon, reflecting the absence of soil-coloring elements, although a brownish A_1 horizon, tinged by the carbon of incompletely decomposed ground litter, is always present. It is called a *podzol* or *podzolic* (from the Russian, meaning ash colored) and the process is called *podzolization*. These soils are typical of the middle and high latitude forested lands of North America and Europe.

In the humid southeast, where temperatures are not only higher but winters are shorter, decomposition of forest litter is more rapid and continuous. Consequently, the forest floor is clean and precipitation can reach the soil as clear water with a high oxygen content. Such groundwater removes the normal solubles and also slowly attacks the silica, so that after an extended period the topsoil becomes high in iron and aluminum and the oxidizing iron frequently gives it a reddish color. Carried to its extreme, these soils are called *laterites* or *lateritics* and the process is called *laterization*.

Another extensive area where temperature is an important soil-forming element is northern Canada and Siberia where the Tundra climate is dominant. During the long cold winter, vegetative decomposition is held to a minimum and spongy masses of muskeg develop to a depth of a foot or more. Beneath this, the subsoil is permanently frozen, impeding normal drainage and giving these Tundra soils a peculiar character duplicated in lower latitudes only above the timberline in high mountains.

IMMATURE SOILS

The foregoing discussion has revolved about *mature* or *zonal* soils, soils that have been in one place long enough to react to and reflect local soil-forming influences, especially climate—in other words, a climax soil just as we had climax vegetation. But there is also a general category of *immature* or *azonal* and *intrazonal* soils that, for one reason or another, have not yet reached an equilibrium with these soil-forming influences of the local environment, and here the previously developed generalizations do not always hold true. Newly formed soils, that is, rocks that have only recently decomposed through weathering and erosion, are likely to have a moderate alkaline content for a while despite heavy rainfall, simply because they have not existed long enough to be thoroughly leached. Or swampy conditions with their lack of groundwater movement will impede the normal development of the soil.

Probably the largest group of immature soils owe their characteristics to the fact that they have been transported recently out of the region in which they originally developed.

Probably the largest group of immature soils owe their characteristics to the fact that they have been transported recently out of the region in which they originally developed. They may have been mature soils there but now, in their new environment, no longer reflect the local conditions. Running water is an effective agent of transportation, carrying quantities of silt from an exotic region and depositing it in deltas and floodplains. For instance, the Missouri River, called "The Old Muddy," too thick to navigate and too thin to cultivate, transports an immense load of silt from the dry regions of the west into more humid areas. Its lower valley and the floodplain of the Mississippi immediately south of St. Louis have high fertility ratings because of these imported soils. Carrying dry-land soils into adequate rainfall regions (or the reverse, taking irrigation water into the desert) accomplishes agricultural production miracles by combining sufficient moisture for crops with fertile unleached soils.

Similarly, wind-transported soils called *loess* are immature and renowned for their fertility since they are frequently derived from the desert and deposited in better watered areas (Fig. 15–6). In north China the upper Hwang Ho or Yellow River traverses a region containing the world's largest deposits of loess. Here on the leeward side of the deserts of central Asia, prevailing monsoon winds have piled up a great plateau of fine silt several hundred feet thick and completely masking the underlying rougher terrain. The river, cutting easily through these unconsolidated materials, becomes heavily laden with yellow sediment and carries it off to the sea. Through the centuries the continuous deposition of this silt at the continental margin has built up the extensive North China Plain. The fertile soil is immature, having been transported twice from its place of origin, first by wind and then by stream.

Loess also is encountered beyond the ultimate margin of the great North American and European ice sheets. In the process of advancing and retreating, an immense quantity of erosional debris built up in front of the glaciers, much of it finely ground rock powder highly susceptible to wind removal. As the air above the ice mass cooled through contact (conduction), it flowed off the glacial front as a constant, moderate velocity, katabatic wind that swept up great clouds of fine dust. In northern Germany and Poland this deposited loess occupies a narrow belt along the northern foothills of low mountains, piled up against this barrier like so much drifted snow. Here again, there is a reputation for great fertility because, although the North

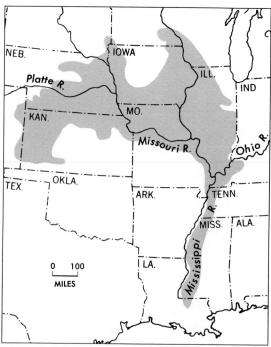

Fig. 15–6 Loess accumulation in the United States. Deriving chiefly from the dry regions to the west, wind-carried soils pile up along the Mississippi. Lesser amounts were distributed by gravity winds off ancient continental ice sheets that swept up fine rock powder from glacial deposits.

European loess does not come from the desert, it is nonetheless relatively unleached as a result of its recent formation as a rock flour ground in the glacial mill. This agricultural region has long been the area's finest, readily delineated by identifying fields of wheat and sugar beets as opposed to the poor-soil crops of rye, barley and potatoes.

Caves in the cliff wall house the populace while they farm the roof of their home.

A further peculiarity of loess is its ability to maintain a steep cliff face-while, paradoxically, being at the same time extremely easy to cut and erode (Fig. 15–7). Its fine powderlike consistency is responsible for both of these qualities. Its own weight compacts it into an almost solid

Fig. 15–7 The flour-like consistency of loess causes it to compact under its own weight to the point that it can maintain a vertical wall as here in Missouri. Cliff swallows have discovered how simple it is to carve caves in loess.

mass, yet it is easily worn away. Unpaved roads traversing a loess base will quickly incise themselves into a steep-sided canyon as the friction of many wheels easily reduces the roadbed to its original dust. In China the Hwang Ho and its major tributaries have rapidly cut through the loess to bed rock, yet the valley sides tend to keep their sheer character. Easily carved caves in the wall often house the populace while they farm the plateau top, essentially the roof of their home. But a major problem develops when the forests or grasses are removed in anything resembling a humid region, for this ease of cutting means that any tiny stream or rivulet can quickly form a deep gully and before long the entire region erodes into a fantastic badland.

THE INFLUENCE OF PARENT ROCK

Another variant affecting soil fertility is the chemical composition of the parent rock. The previous generalizations have been based on the fairly valid assumption that most rocks have a normal complement of minerals, but occasionally a soil will be formed from parent material that is definitely lacking in several important minerals. Such a rock is quartzite, almost pure silica that becomes a sterile sand when it breaks down. Climate, whether dry or humid, will have little effect on a soil of this type. The so-called *rendzina* soils are also examples of the dominating effects of certain parent rocks—Kentucky's Blue Grass region may serve as an example. Here in a humid area where one would expect to find leached acid soils and a forest cover, was an original island of grass amid the trees. The soil is derived from strata of soluble limestone, and despite heavy precipitation, it maintains its alkalinity. The grasses, with their affinity for alkaline soils, were visual evidence of this inherent fertility. No wonder the Blue Grass was a magnet for early trans-Appalachian farmers, notwithstanding its reputation as the "Dark and Bloody Ground" (Fig. 15–8).

It becomes apparent that the relationship between climate and mature soils is intimate. Exceptions are found, but the exceptions are most easily understood and explained in the light of the climatic relationship. Once again,

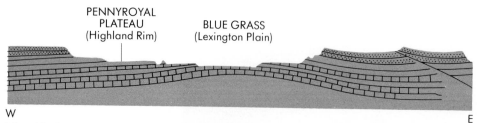

Fig. 15–8 Erosion at the apex of this modest structural arch has exposed a fertile limestone stratum below. The impressive rural prosperity of Kentucky's Blue Grass, based upon rich rendzina soils, ceases abruptly at the Highland Rim.

the thumbnail rule to keep in mind is *dry climate equals fertile alkaline soils* and *wet climate equals infertile acid soils.* The concept illustrated on the map (Fig. 15–9), although originating in 1929 and a bit old-fashioned today, nonetheless expresses this still valid generalization. Dry-land soils are called *pedocals* (*ped* from pedology, *cal* from calcium).

Humid-region soils are called *pedalfers* (*ped* from pedology, *al* from aluminum, and *fer* from iron). Pedocals then are high in alkalies, especially the most common one, calcium; pedalfers are leached of their soluble plant foods, leaving behind concentrations of aluminum and iron (Fig. 15–9).

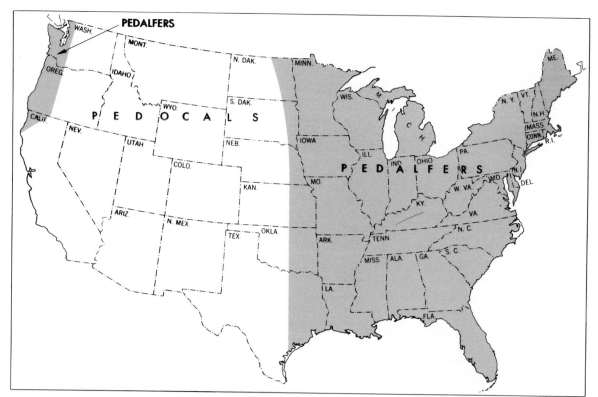

Fig. 15–9 Pedalfers versus pedocals in the United States. The coterminous United States breaks down rather neatly at the 20 inch (51 cm) rainfall line into a dry half and a wet half. Only in the mountains and the Pacific Northwest are acid pedalfers encountered in the West.

Legumes

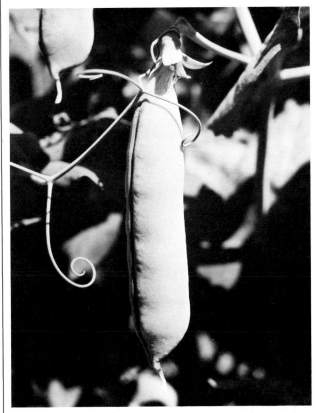

The great family of *leguminosae* develops on its roots nodules (or tubercles) which contain a nitrogen-fixing bacteria, so that by merely growing in a field the plant increases the fertility of that soil. All decaying plants are capable of a tiny amount of this, but the *legume* can remove large quantities of nitrogen from the atmosphere and fertilize the field, while at the same time producing a useful crop. No wonder that legumes are frequently utilized as "nursury" plants—red clover planted with other pasture seeds, or lupines placed beside young pine seedlings on migrating sand dunes. And they are an invaluable part of farm rotation schemes. All field crops remove a panoply of nutritional elements from the soil, but only one returns any of these if the crop is harvested. This doesn't mean that fertilizers are not necessary, but if a legume is rotated to occupy any given field every few years a good deal of fertilizer money is saved in the farm operation.

The most common of the legumes are the peas and beans (or pulses). They all have a pod, a legume, which opens along two sutures when the seeds are ripe. As a significant basic food in many underdeveloped nations, it is important to recognize that the seed contains more protein than any other vegetable product and of the type that is closest to meat protein. Usually beans and peas dry and store well, often have a useful oil by-product, and the pods and vines are frequently edible for humans and animals. Consider the soy bean: a rapid grower, the seeds all mature at the same time and can be threshed; seeds are edible in many ways (milk, curd, flour, sauce, etc.); an important commercial drying oil can be expressed; both the beans and the vine are excellent cattle fodder; and of course, it enriches the soil while it grows.

Pulses

Field pea	Broad bean	Soy bean
Garden pea	Velvet bean	Peanut
Chick pea	Horse bean	Lentil
Cow pea	Kudzu bean	Cajan bean

Forage legumes (no pods with beans)

Alfalfa	Vetch	Lupine
Clover	Lespedeza	Medic

Tree Legumes (Pods with Beans)

Mesquite	Honey locust	Nitta
Carob	Rain tree	

THE HUMAN ELEMENT

Once human culture evolves beyond simple hunting and gathering, nature's delicately balanced soil ecology is disrupted. Perhaps there are domesticated grazing animals forming herds and running in large numbers on the natural grasslands. If the price of wives is measured in cows, there will be an incentive to overgraze the carrying power of a given range and to burn the rank growth encouraging new green shoots and temporary better feed. But now the natural vegetation has been altered, humus burned away and erosion encouraged: the soil-forming processes, closely linked to their covering vegetation, will change also. Or maybe the need for more substantial housing will lead to a lumbering operation and the removal of forest trees. Certainly, agricultural endeavors of even the smallest magnitude require a clearing of the land as the first preliminary step, and immediate corollary soil changes are inevitable.

Even if the agricultural product is run through a cow and returned as farm manure to the soil, all of the initial fertility has not been put back.

With the introduction of the plow, the B horizon is turned up, forming ready-made channels for surface runoff with its furrows, and the loose soil will be carried away. The original vegetation, although removing nutrition from the soil as it grew, returned those same nutrients when it died and decomposed. But humans replace this vegetation with their own in the form of crops, and then, each year, take those crops off the land to be consumed elsewhere. Even if the product is run through a cow and returned as farm manure to the soil, all of the initial fertility that was removed has not been put back, for there stands a healthy animal—evidence that some of the nutrition has been used up. In most parts of the world, agricultural soil fertility is declining, often at an accelerated rate, simply because the farmer cannot afford the chemical fertilizers that are necessary to maintain it after he has removed the crop.

It is possible, of course, to improve soils from the standpoint of fertility as well as to exploit them. Crushed lime to counteract acidity, cover crops plowed back in regularly to offset lack of humus, and tiled fields to facilitate drainage are all methods for making the soil more productive (Fig. 15–10). And chemical fertilizers are being applied today to not only maintain fertility balance, and even increase the fertility each year, but to improve texture and structure as well. In every case, however, humans become an element in soil formation and the original soil is changed.

Soil erosion, the actual physical removal of soil, is probably the most striking and best publicized of the deleterious effects of humans on soil. But it should be remembered that wind and water are constantly moving sed-

Fig. 15–10   Aerial topdressing (fertilizing) of pasture land in New Zealand.

Fig. 15–11   Strip mining simply destroys the land. Even levelling and reforesting does not return it to its original state (although strongly recommended), for the soil horizons have been scrambled and the drainage rearranged.

Fig. 15–12   Strip cropping and contour plowing can both defeat the efforts of surface drainage rivulets to run at high velocity directly downslope (across contours). These are the little streams that carry off the topsoil and eventually initiate gullies in the field.

iments from one place to another, and it is only when humans implement and encourage this activity so that large quantities of soil are involved that they can be called to account. Fundamentally, it is the removal of the natural vegetation that is at fault. It may be logging, mining, or agriculture that requires this, but once the bare soil is free of the anchoring effects of plant roots, it can be very effectively attacked by the elements. Great gullies form rapidly on steep slopes—miniature canyons biting down to bed rock. More insidious, and therefore in some ways more dangerous, is sheet erosion that occurs on clean-tilled gentle slopes following a heavy shower; and the surface runoff, carrying topsoil with it, is in a sheet rather than following a well-defined channel. Wind too can carry large quantities of fine soil following overgrazing or the unwise plowing of the turf in semiarid regions.

But erosion can be countered, to a degree at least, once it is recognized and considered to be urgent and if the re-

Erosion can be countered once it is recognized and considered to be urgent.

sources and techniques are known and available. Strip miners are now required to fill their cuts and plant trees in the more enlightened societies, and hydraulic mining has been declared illegal (Fig. 15–11). Lumbermen replant as they cut, and reforestation of barren slopes, whose timber was removed even centuries ago, is more widespread than ever. On the farm, contour plowing, terracing, strip cropping, and cover crops are all helping to cut down erosion (Fig. 15–12). Yet even here in the United States, where we regard ourselves as efficient technicians, one can still stand beside the tawny Mississippi and watch somebody's farm go by every few minutes.

CHAPTER 16

CLASSIFICATION AND SOIL REGIONS

Plowing the field and garnering the crop is accomplished with thousands of variations around the world; the irrigated oasis in Algeria is profoundly different from an Iowa mechanized farm. With the onset of the summer monsoon, an Indian farmer turns up the soil to begin yet another speculative cropping cycle. That the methods are ancient and the scale of the enterprise small does not mean that this farmer's aims are so very different from every other—a primary concern that the soil maintain the ability not only to produce a current crop, but endless ones into the future.

BASES FOR CLASSIFICATION

The first modern soil classification was devised in its main outlines by Russian scientists about the turn of the century. Translated into English in 1927 by C.F. Marbut (then chief of the U.S. Soil Survey), it marked a major breakthrough from the former domination of geologists in American pedology and their almost total reliance on parent rock in interpreting soil development. Marbut went on to disseminate widely and to refine the Russian approach and his ensuing classification became the world standard. In his 30 or so *Great Soil Groups* (subdivided into *families*, *series*, and *types*) is recognition that (1) soil is a dynamic and singular part of the landscape, (2) soil is the product of a symbiotic relationship among climate, vegetation, parent material, topography, and time, and (3) the soil profile largely synthesizes all of this visually.

However, by the early 1950s it was becoming increasingly evident that this now aging classification, for all of its pioneering merit, could no longer accommodate the unforeseen masses of new soil findings worldwide and that a more up-to-date system was required. Begun in the United States as a cooperative venture by both government and university pedologists, the effort proved to be tedious and contentious. At birth in 1960, after a long gestation period in which it had been revised at least seven times, the new classification was christened "the *7th Approximation*" (minor revisions have appeared as supplements in 1964, 1967, and 1968) and officially adopted by the U.S. government in 1965.

The 7th Approximation classification is as objective and as quantified as possible, and if not philosophically, certainly in its mechanics and terminology, it is disassociated in every way from the earlier model. For instance, there is no distinction made between zonal and azonal soils, and soils disturbed by current agriculture are treated equally along with all others. The nomenclature, although realistically derived, is voluminous and requires careful study particularly by those steeped in the older terminology.

Without any doubt, the 7th Approximation is destined to become the definitive classification of the future, but by the same token, undoubtedly, it is much too complicated and all-inclusive for the beginning general student. The approach to soil classification presented here is based on Marbut's Great Soil Groups replete with old-fashioned terminology, which at this date continues to be more frequently encountered, both in the literature and in the field, than that of the 7th Approximation. Students can sample the flavor of the new system by turning to the tables in Appendix C.

To simplify for this brief treatment, we will deal with only zonal or mature soils, thereby pruning Marbut's Great Soil Groups to 15. These in turn are modified and combined into a very basic 10.

SOIL REGIONS

Laterites and Lateritics

These are soils in which the laterization process is dominant, and they are characterized by great depth, efficient leaching of the soluble minerals including silica, a minimum of organic material, indistinct horizons in the profile, and a reddish to yellowish color. Warm humid climatic conditions are necessary for this type of soil formation, so it naturally follows that the laterites reach their greatest development in the Tropical Rain Forests. Here the heavy daily rain tends to wash the solubles completely out of the soil, which are removed by groundwater drainage so that no real B horizon exists. The chemical disintegration of the parent rock is so accelerated that the C horizon is very deep and difficult to recognize. Neither is there an A_1 horizon, for despite the continuous falling of large amounts of forest debris to the ground, its decomposition is much too rapid to allow significant accumulation.

It seems anomalous that the Tropical Rainforest, one of the most luxuriant naturally occurring vegetative associations anywhere in the world, should flourish on soil so lacking in basic fertility.

The removal of virtually all the minerals from the soil except iron, aluminum, and magnesium sometimes causes these nonsolubles to be concentrated into workable ore bodies called residual ores. But this tremendous leaching

Fig. 16–1   The very climatic conditions that foster this most luxuriant of all vegetative associations, the Tropical Rain Forest, rob the soil of its basic fertility for agriculture.

also means that Tropical Rain Forest soils are largely deficient in plant nutrients and have proved to be infertile in the extreme for agricultural purposes (Fig. 16–1). It seems anomalous that the Tropical Rain Forest, one of the most luxuriant naturally occurring vegetative associations anywhere in the world, should flourish on soils that are so lacking in basic fertility. But trees have different requirements than cultivated crops, and once established, they appear capable of maintaining a fertility cycle that is broken only when the land is cleared. Rapid and continuous weathering of rock materials partially offsets the loss of solubles through leaching. And although not as deeply rooted as most forests, the tree roots of the tropics do probe greater depths than the root systems of cultivated plants, thus tapping more remote strata for nutrients. As these are drawn up through the roots, they nourish heavy foliage, which supplies a constant rain of leaves and forest debris to decompose with great rapidity and further feed the forest. As a result, the soil water is only moderately acid despite torrential rains and heavy leaching. But once the forest is removed for cultivation, the cycle is broken and it can no longer maintain this minimal topsoil fertility. Large amounts of fertilizer are required for any sort of successful agriculture. Even these added plant foods leach away at a rapid rate and must be applied frequently if fertility is to be maintained.

Beyond the margins of the Tropical Rain Forest, a dry season breaks the daily rainfall pattern and evaporation begins to offset leaching to some degree. The soil profile reflects this in exhibiting a sharper definition of the B horizon. But as long as the rainfall remains heavy for a large part of the year and organic material fails to accumulate at the surface, the lateritic trend continues. These soils develop under tropical scrub forest and even tall savanna grass, becoming gradually more alkaline as the tropical deserts are approached.

Curiously, the term "laterite" was first applied to soils occurring in the monsoon region of Southeast Asia where there is a distinct dry season each year. In certain widely scattered localities, an iron-rich hardpan is found at the top of the groundwater table. When dug up, it is soft and malleable, but as it becomes exposed to the air, it quickly hardens into bricks that can be used for construction. Hence *laterite* derived from the Latin, means brick. In today's usage, laterite has come to mean the heavily leached soils of the Tropical Rain Forest, and the old bricklike laterite soil is merely a variant of this and not even especially typical.[1]

---

[1]Because of continuing arguments as to the proper use of the word "laterite," some authorities have introduced such terms as "latosol" and "ferrallite," reserving laterite for the infrequently occurring construction material.

These lateritic soils developing under climatic conditions featuring a dry season are somewhat more fertile than the true laterite. They still require careful management and fertilization to allow cropping, but in many places, particularly the Orient, intensive agriculture has supported large populations for centuries on such soils.

Within the lateritic soil grouping should be included the Humid Subtropic climatic regions with the exception of Uruguay and the Argentine Pampas. Here again, are warm humid conditions and a forest vegetation. There are those who claim that the short cool winter slows microbiotic activity sufficiently to allow some minor accumulation of litter on the forest floor and thus the beginnings of podzolization. Admittedly, this is a transitional region between the lateritics and the podzolics, but the red-to-yellow soil coloring strongly points to laterization being dominant. The yellow of the sandy coastal plain in southeastern United States indicates greater leaching and smaller amounts of iron in the parent material than elsewhere. Among the laterites and lateritics, this soil of the Humid Subtropics is probably the most fertile, displaying the same shortcomings of its tropical relatives but to a lesser degree.

Podzols and Podzolics

Cool to cold winters, fairly heavy precipitation the year round, and forest vegetation are the requisites for podzolization, and these conditions are met over broad areas in the Northern Hemisphere middle and high latitudes. The soils of the Taiga appear to represent the ultimate in podzol development. Here the relatively small numbers of decay-producing microorganisms, active only during the short summer, fail to remove the thick carpet of needles on the forest floor. Rainfall, percolating through this soil cover, inevitably becomes an oxygen-poor acid and, as has been described previously, attacks the mineral elements in quite a different fashion from the clear water of the tropics. Leaching of the A horizon is complete, the only mineral remaining relatively insoluble being silica, and its concentration here just below the surface gives podzol soils their typical white to gray color. At the top, the A_1 horizon may be quite dark, but at most is only a thin veneer. In the zone of illuviation, which receives the material washed down from above, fine clays and iron complexes produce a considerably darker and heavier soil, often a dense hardpan (Fig 16–2).

Its topsoil leached of alkaline plant nutrients, the unimproved podzol appears to be acceptable only to acid-loving conifers, which in turn drop their highly acid needles to the ground to perpetuate the cycle. Obviously, this soil does not have a great agricultural potential, yet it can be utilized if properly managed. Deep plowing to turn up the more fertile B horizon, alkaline fertilizers, and the addition of organic matter have made podzol soils fairly productive in parts of northern Europe. One reason for their poor reputation is shortness of growing season at these latitudes, which rules out the maturation of many crops. But concentration on rapidly maturing truck crops and grains such as rye and buckwheat, combined with reasonable accessibility to market, can lead to a greater use of podzols for agriculture.

Outside the Taiga the middle latitude forest soils are called podzolic rather than podzol, indicating that the podzolization process is dominant, but not to the degree that it is in the Taiga. The lower latitudes of the forested Humid Continental and Marine West Coast climate regions ameliorate the severe temperatures of the Taiga—longer hotter summers in the former, milder winters in the latter. In both cases this means more rapid decay of forest litter and somewhat less acid groundwater. The type of forest reflects the degree of podzolization. Broadleafs prefer a more alkaline condition and in turn drop less acid debris than do conifers, so that in the broadleaf forest regions the A horizon, although high in silica content, displays a brownish to dark gray color. In the mixed forest districts, stands of conifers usually occupy areas where the parent material is more acid than normal. An example is the pine barrens of New Jersey, which have developed on sandy soils.

Basic fertility is greater, obviously, in the podzolics than in the podzols, especially those of the broadleaf forests, and the combination of an adequate growing season plus reasonable fertility has resulted in the rapid removal of the forest and in large-scale cultivation of crops.

Chernozems

The chernozem soils could rather easily be grouped along with their near relatives, the prairie soils and the brown chestnut soils, into a single category, because all three normally develop under a middle latitude grass cover and are, generally speaking, mere variants of one another as far as the soil-forming processes are concerned. However, the

Fig. 16–2   Podzol profile. The most striking feature of the podzol is the ash-colored A horizon. If an A_1 exists it is extremely thin, but the B horizon is easily identified by its dark brownish shade. This example of a podzol soil developed on fine-textured glacial drift in Michigan and displays a very gradual merging of B into C. (See idealized soil profile Fig. 15–4.)

chemozems figure so prominently in the pedological literature because of pioneer Russian soil research that they are regarded as a classic example, and every physical ge-

ography student should be able to identify them explicitly. Therefore they are treated separately here.

These are the black soils (introduced previously) where

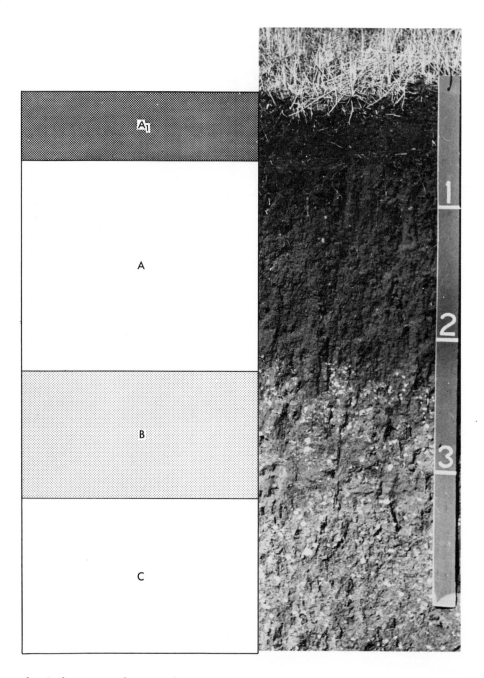

Fig. 16–3 Chernozem profile. Developed on a South Dakota glacial till, this Chernozem profile is typical in its total dominance by a deep A horizon. Close inspection will reveal a slightly darker A_1, but these two elements of the A horizon are together well over two feet thick. (See idealized soil profile Fig. 15–4.)

the A_1 horizon makes up a large part of the deep A horizon (Fig. 16–3). The B horizon is brownish and breaks off sharply to a light-colored C. Characteristic of the chernozem is the zone of lime concentration well within reach of the plow, and this tendency to accumulation rather than leaching, combined with a high humus content, makes this soil as fine an agricultural soil as can be found anywhere. The humus imparts to the topsoil a friable crumblike texture that will not clod and promotes soil aeration and water availability to plant roots. The only drawback to the chernozem as a cultivatable soil is the somewhat low precipitation necessary for its formation and the tendency toward extended periods of drought.

Humus imparts to the topsoil a friable crumblike texture that will not clod and promotes soil aeration and water availability to plant roots.

On the warm and humid margins of the chernozem zones, soils of a reddish-chestnut color are found developing under partial forest cover. They are very similar to the typical chernozem but lack some of the lime concentration near the surface, display less humus, and have greater amounts of oxidized iron. Sometimes referred to as degraded chernozems, these soils are very close to the chernozem in location and character, and it is both convenient and logical to classify them as one.

Black Tropical Soils

There has long been a temptation to call these soils chernozem on the basis of color alone. They are certainly black like the chernozem, but the color does not arise from their humus content, which is quite low. They appear to be a result of the breakdown of black basaltic rock under the influence of tropical weathering, possibly in combination with inadequate drainage. Northern peninsular India and east-central Africa where a Tropical Wet and Dry climate supports a scrub forest are the chief regions of occurrence of this rather specialized soil. If the parent material is the dominating factor in soil formation, then these black tropical soils are intrazonal; but we are not sure and they are noted here because, although their fertility is not equal to that of the chernozem, among tropical soils they stand out as extrmely fertile and have long been utilized for agriculture with minimal fertilization.

Prairie Soils

Very similar to the chernozem, the prairie soil marks the transition between the humid forest podzolics and lateritics of the middle latitudes and the semiarid grasslands of the chernozem. These are the soils of the tall grass prairie, and

Fig. 16–4 Despite its inherent fertility from lack of leaching, the chestnut and brown soils, more often than not support this kind of land use. Too little and erratic rainfall make farming a chancy proposition.

although they have a high content of humus in the A horizon, they are dark brown rather than black. Somewhat greater leaching takes place with an average of 30 inches (76 cm) of precipitation, so that there is less lime accumulation and it is distributed more widely through the B horizon. These soils are almost as fertile as the chernozem, and the greater amount and increased reliability of rainfall make them tremendously productive for crops.

Chestnuts and Browns

On the dry side of the chernozems and merging into the middle latitude desert are the short grass prairies or steppes under which develop highly alkaline brownish soils. Here again, there are strong similarities to the chernozem except that more limited vegetation supplies less humus, and the salt accumulation is more exaggerated because of high evaporation.[2] Fertility is generally quite high, but low and

---

[2] Except for accelerated decay of organic matter resulting in less humus and also a reddish cast to the soil, the short grass tropical desert margins are also usually included in this same category.

sporadic rainfall makes agriculture precarious, although special dry-farming techniques have met with some success despite a strong tendency to wind erosion (Fig. 16–4).

One other region, the Mediterranean climate zone, exhibits a similar soil that might well be classified here also. Frequently, these soils, which are formed under conditions of subtropical temperatures and a scrub woodland vegetation, are called lateritic. But despite a tendency to reddish color and only moderate humus, the long dry season is not conducive to laterization. Perhaps the question is academic in that the Mediterranean climate regions are limited in area and hilly in character so that zonal soils are scarce, but we will include them here with the chestnut and brown soils on the basis of low precipitation resulting in alkaline accumulation.

Here too is a fertile soil basically requiring chiefly nitrogen fertilizers to offset its shortage of organic matter. From an agricultural viewpoint, the chief drawback is the Mediterranean climate with its rainfall at the wrong time of the year. Given sufficient moisture during the growing season through irrigation, these soils are very productive.

Desert Soils

Theoretically, the desert soils should have a great deal of inherent fertility in that leaching is nonexistent, but there are some considerable prohibitions to agriculture even as-

Fig. 16–5 If the implication seems to be that deserts are endowed with endless latent fertility, it just plain isn't so. Water on the desert soils can show spectacular results, but there are difficulties too—sand, salt, coarseness, clays, lack of both nitrogen and humus. Here are two not very promising dry land milieus.

Irrigation

Sharply demarked irrigated plots in the Coachella Valley.

The contention has been made that irrigation is at best a temporary expedient, and that no matter how carefully the operation is controlled, the seeds of its own destruction are inherent in simply putting water on the land. Twin scourges are waterlogging and salinity buildup, and there is no project old or new that has not suffered from both to some degree.

If any kind of shallow hardpan exists in an irrigation site it immediately impedes the vertical drainage of water. This, of course, leads to a saturated soil ultimately "drowning" the field crops. But when adequate drainage is achieved the moving ground water rapidly takes salts into solution. It must be discharged efficiently and the water kept moving or these salts will cumulate and poison the soil.[1] In the Imperial Valley the Salton Sea acts as an outlet reservoir, in Utah the Great Salt Lake, in the Jordan Valley the Dead Sea. Waste water may also be channelled out to sea or into a river flowing to the sea. This last is hard on the folks downstream—witness the trickle that is left of the Colorado River when it finally enters Mexico, much too saline for irrigation. By international treaty the U.S. is now obliged to rectify this.

In the San Joaquin Valley of California, aerial photos show white sploches in many places, stark against the darker cropland, where the fields are "salted out." An estimated 400,000 acres (161,874 ha) have already been lost. Suggested solutions have been: (1) a 316-mile (509 km) ditch to the upper reaches of San Francisco Bay; (2) sacrificing a huge area of agricultural land as a permanent dump site; (3) recycle the waste water through a purifying plant so that it can be used again. All are expensive.

It may well be that the ancient civilizations of the desert Middle East and Pakistan, based on involved irrigation systems, perished as much from dwindling food supply as the bellicosity of their neighbors. One of the oldest of the lot, Egypt, is still functioning, but its salvation has always been the natural flood bringing rich new volcanic soils from Ethiopia to rejuvenate the fields. They are now being deposited behind the Aswan dam.

---

[1] Even drip irrigation, which utilizes much less expensive irrigation water, sets in motion an immense latent evaporation and capillary action that carries salts to the surface.

suming that irrigation water can be obtained (Fig. 16–5). One of these is soil texture, which tends to coarseness. Since mechanical weathering is dominant, its end products tend to be rough and angular, and these are strongly represented through the desert soil profile. Also, wind erosion at the surface removes much of the finer material, leaving behind, to make up the topsoil, the pebbles and larger rocks. These may even be cemented together by the saline remains of evaporating moisture into a *desert pavement*, or these same salts often accumulate through the capillary movement of groundwater into a rocklike hardpan at a shallow depth.

Theoretically, the desert soils should have a great deal of inherent fertility, in that leaching is non-existent.

The general grayish (middle latitude deserts) or reddish (tropical deserts) color of desert soils indicates the deficiency of humus where sparse vegetation predominates, and this in turn means that desert soils lack nitrogen. Without humus or even rudimentary leaching, soil profiles fail to develop. All of this does not mean that desert

Fig. 16–6   Silt originally destined for deposition in the Gulf of California, now accumulates behind the Imperial Dam on the Colorado River. Before this amber-hued, stored water can be introduced into the local irrigation system (see canal at the top of the photo), it must be processed through these desilting devices to forestall rapid clogging of the water distribution channels.

soils do not have agricultural potential, but often in the past, irrigation has been unsuccessful because the character of desert soil was not understood. Finely textured soils that can be easily worked are not entirely lacking in the desert, and if sufficient water and adequate drainage are supplied, surplus salts can be flushed out. There are many examples in all parts of the world where the deserts "blossom like the rose," an indication that their soils are fertile (Fig. 16–6).

Tundra Soils

Because it is so widespread, tundra soil is always classified as a zonal soil, yet it is definitely immature in most of its characteristics. In many ways it closely resembles a typical bog soil of lower latitudes. Normal drainage is impeded by generally flat terrain and particularly by permafrost, so that no real profile develops. It must be remembered that the entire soil is frozen during 10 months of each year and that the soil-forming processes are operative only during the short

Fig. 16–7 Polygons or patterned earth. This view of the winter tundra shows the peculiar phenomenon of patterned earth very clearly. It is probably the result of expansion of summer water to winter ice in the thin soil layer above the permafrost. The center of each polygon is a hummock, pushed upward to accommodate expansion.

Tundra soil is always classified as a zonol soil, yet it is definitely immature in most of its characteristics.

summer season and only in a shallow layer above the permafrost. Leaching is entirely lacking and decomposition of the thick vegetative mat at the surface, although giving the soil a dark grayish color, is extremely slow. There is some vertical movement of soil elements in the late summer when the surface first freezes, and hydraulic pressures build up between this ice layer and the permafrost. Surface heaving, often resulting in the ordered hexagons called *patterned earth*, will culminate on occasion in surface rupturing and outpourings of mud (Fig. 16–7).

An Organization of Soil Regions

The following table is a rational ordering of the several soil regions, to illustrate their place in the spectrum of precipitation and temperature and to indicate their relationship with pedocals and pedalfers.

Zonal Soils

Pedocals (In order of decreasing precipitation)		Pedalfers (In order of decreasing temperature)
1. Chernozems (and related)	Prairie soils	1. Laterites and lateritics
2. Chestnut and brown		2. Podzols and podzolics
3. Desert soils		3. Tundra

PART FIVE

LANDFORMS

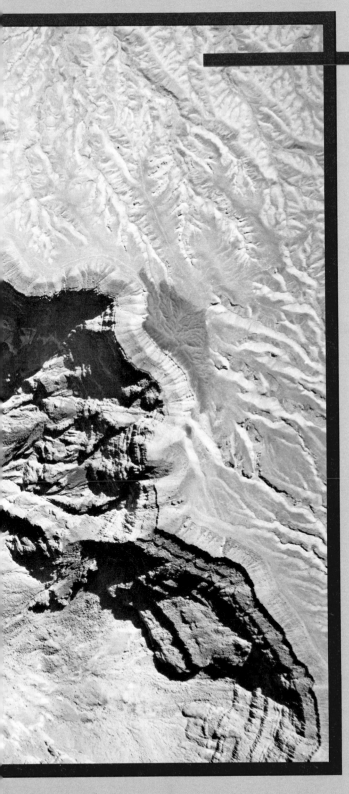

CHAPTER 17

INTRODUCTION TO LANDFORMS

Many kinds of landforms are visible in this aerial view of an arid countryside—all of that vegetative greenery encountered in more humid climes tends to obfuscate and soften the stark edges of landform detail. This is south Saharan Chad—Trou au Natron.

INTRODUCTION

The word "landforms" scarcely requires definition—it is not an arcane scientific term derived from the Latin or Greek and properly difficult to pronounce. When we are dealing with landforms we are simply taking a long hard look at the myriad wrinkles on the countenance of the earth and assaying them singly and in aggregate, as to character. In the lexicon of the geographer, however, "landforms" does have a specific implied meaning: the study of landforms, whether the gross features of continent versus ocean basin or the endless detail of local valley, mountain ridge, or coastal embayment, is aimed ultimately at the earth's surface as the human habitat.

LANDFORMS AND HUMANS

It is not difficult to visualize some very direct and immediate influence exerted by landforms on human affairs—the historic role of the Appalachians in restricting American colonials to the immediate eastern seaboard; the development of advanced Indian cultures on the highland plateaus of Latin America and the striking concentration of the contemporary populace in the identical locations; the almost perfect correlation between the rice-paddy people and the riverine lowlands of the Orient; the fascination of the Willamette Valley for Oregonians as a chosen abode, to the virtual exclusion of the rest of the state. Landforms are there. Our task is to adjust to them, perhaps to change them if it is physically possible, or simply to abandon those situations that are beyond our technology.

The role of national borders in separating states, protecting the state within while at the same time allowing reasonable congress with neighboring countries, has always presented a problem. More often than not, an attempt has been made to utilize some natural feature of the terrain, but seldom with complete success. Here is a universal human problem that almost inevitably leads to human/landform interplay. With France as an example, let us examine several variations on this simple theme of the effectiveness of landforms as *political* boundaries.

The massive Pyrenees wall extends across the narrow neck of the Iberian peninsula, almost perfectly sealing France from Spain. It is pierced only here and there by tunnels and costly winding roadways, and has, all through history, been a mutually abhorrent region from the standpoint of human habitation. As a border, the Pyrenees effectively separates diverse people and strongly discourages invasion of either one by the other. But in so efficiently accomplishing this function, easy and peaceful traffic with good neighbors is automatically precluded. A less formidable border might be to the advantage of both nations.

Diametrically opposed to the obstructive high mountain range as a national boundary has been the sodden, almost dead-flat plain of Flanders in the far northwest (Fig. 17–1). For centuries the borders between France and adjacent countries have fluctuated wildly, with people and cultures moving every direction at will and intermingling freely. With them came the frequent invading armies consciously choosing this easy route of movement. Borders were drawn almost at random across the featureless low country, and inevitably the French were excluded from France and reluctant outsiders were brought in. Today, Belgium, which occupies the heart of this plain, has a populace half French-speaking and half Flemish (Dutch)-speaking, half doctrinaire Catholic and half casual Catholic, and as a fragile national entity engages in continuous border disputes with all of its neighbors. It seems that absolutely free movement across a national border may be worse than none at all.

The fascination of the Willamette Valley for Oregonians as a chosen abode, to the virtual exclusion of the rest of the state.

The problem of the border where France faces Germany in the east, along the Rhine, is still different from the contrast of mountain versus plain. The line of demarcation faithfully follows the center of the river channel, which itself occupies a deep flat-bottomed trench between uplands on either side. This is a rich and productive valley with an excellent navigable river running down the middle of it, but one side is German and the other side is French. The people, cut off to a considerable degree from the rest of their respective nationals by mountainous topography at their backs, are drawn together by mutual isolation and their

Fig. 17–1   Flanders.

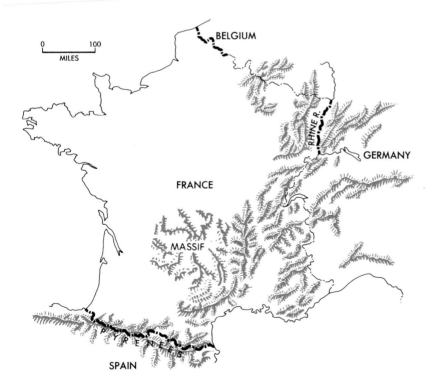

Fig. 17–2   Three critical French borders.

strong dependence on the Rhine as an artery of commerce. They speak a border patois of French/German and regard themselves as a people apart.

So here in France we find in the one small instance of national borders, attempts as adapting existing landforms to human need—or vice versa. France has been the classic cohesive nation of Europe and one contributing factor has been the occupation by the French people of a natural land-unit with, by and large, excellent functional borders. Yet we have seen some considerable variations in the efficacy of three different kinds of landforms as borders (Fig. 17–2).

CLASSIFICATION

Any attempt to inject a sense of the rational into the virtual infinity of features that make up our total landscape must eventually come to some sort of classification. This is good scientific procedure, a recognition of similarity among the various elements, and thence an arrangement into categories where each can be pigeonholed. The point of origin is classification at a large scale—tossing into big buckets everything that demonstrates even superficial resemblance. But this is useful only as the preliminary step to refinement. Next we employ batteries of ever smaller buckets where the lesser variations can be increasingly isolated.

In the case of landforms we usually involve ourselves in a three-stage hierarchy beginning with the simple distinction between continent and ocean basin. The second step is a more sophisticated subdivision of the continental land surfaces into plains, plateaus, hills, and mountains; these require definition. Everyone knows that plains are flat and so are plateaus, while hills and mountains are not. Plains are low and plateaus are high; hills are not as high as mountains. Maybe we should put some empirical limits to our categories, like, plains are under 1000 feet (305 m) in height, but we begin to run into trouble immediately. Only some plains meet this requirement. The high plains that cover much of central North America reach altitudes of 5000 feet (1524 m) over extensive areas. The so-called Appalachian plateau is only 1000 to 2000 feet (305 to 610 m) high and has been so highly dissected by streams that

it looks like broken hill country to most people. Plains are not always flat either. One struggling classifier has established a category for "plains which aren't really plains at all because they have bumps on them that might be called hills or mountains." This begins to get silly. Should we classify the Appalachian mountains with the Andes or Himalayas, or should we talk of little mountains and big mountains? The European Alps crest at over 15,000 feet (4572 m), the Australian Alps at 7000 feet (2136 m). Mt. Lofty near Adelaide, Australia, sounds impressive, but at 1500 feet (457 m) elevation it would be a hill in most other settings. So it seems that one person's hill is another's mountain, and the farther we go into this, the worse it gets. Let us just come back to the original description: plains are generally low and flattish, plateaus are generally high and flattish, mountains are generally high and steepish, and hills are little mountains.

The third and last of the levels of landform classification brings us face to face with the infinitude of detail that we observe every day in our immediate landscape—the cove, the bar, the creek, the delta, the ridge, the dale. How do we arrange these in any kind of ordered system? The approach at this stage must go beyond mere description and identification; we come to an organization according to *mode of formation*—a genetic system of classification.

The European Alps crest at over 15,000 feet (4572 m), the Australian Alps at 7,000 feet (2136 m).

To do this, we must recognize two basic forces constantly at work on the surface and within the earth's crust. They are called the *tectonic forces (diastrophism)* and the *gradational forces*, and they work in opposition, each attempting to cancel out the other. The tectonic forces are the great forces that distort, buckle, and break the surface of the earth, causing mountains and depressions. All about us are evidences of diastrophism at work—earthquakes, volcanic explosions, and high rugged mountain ranges that must have been formed just a short while ago. The gradational forces attack these features and attempt to destroy

A

B

Fig. 17–3 The opposing influences of tectonics versus gradation. (A) Busy Mt. Sakurasima in Japan is still building, the tectonic forces at work. (B) The Spanish meseta stretches to the horizon, an endless near-flat plain. This is a peneplane, the worn down roots of older mountains, evidence that gradational forces have been in ascendancy here for millenia.

them. They wear down the high places and fill in the low ones. Left to itself, without the offsetting tectonics, gradation would smooth the surface of the earth until, theoretically, it would be completely covered with water. In certain parts of the earth, such as in the Amazon lowland or western Australia, tectonic forces have not been active for an extended period, and the land is low and featureless. In other areas where the tectonic forces have, temporarily at least, achieved major ascendancy over the gradational, high mountains or great ocean deeps are the present-day features (Fig. 17–3).

The two oppugnant influences are as follows:

Tectonic Forces	Gradational Forces
1. Folding	1. Running water
2. Faulting	2. Moving ice (glaciation)
3. Vulcanism	3. Wind
	4. Waves

Since these forces, both tectonic and gradational, ultimately produce landforms, our classification system will recognize the genesis as well as the morphology of each feature. All of this is the science of geomorphology (geo-earth, morph-formation).

THE CAUSES OF DIASTROPHISM

In several ensuing chapters we will examine the end-products of tectonics, but before we attack that it might be well to inquire into the origins of the tectonic forces or diastrophism. "Whence cometh these mighty energies that tear asunder the earth?" The answer has to be, "Only the Lord knoweth," for mortals must qualify their statements and speak in terms of hypothesis and theory. Not long ago the idea of a cooling, shrinking earth with a rigid crust attempting to adjust was as acceptable as any. Today, with drifting continents a virtual certainty, thermal convection and massive plastic rock flow or skidding plates triggered by internal heat seem to be a bit more likely. Or a variation called "*phase change*" wherein it is postulated that certain

Whence cometh these mighty energies that tear asunder the earth?

dense rocks increase their volume when heated (probably by pressure or radioactivity) thus causing vertical rupture of the surrounding rocks, may be a factor. Less dramatic in nature is the measurable uplift of extensive regions attributed to '*unloading*." Melting of continental glaciers or the removal of a great volume of overburden will relax the pressure of gravity to allow uplift. Conversely, the accumulation of sediments in an adjacent sea as erosion unloads the continent may equally force the ocean bottom down. This is essentially the old theory of *isostasy* in action—uplift in one place must result in a reciprocal sinking elsewhere. However, at our current point of comprehension we must simply recognize that diastrophism is a reality, to be explained in the most plausible fashion by perhaps all or some combination of the above. An approach to improving our understanding is to continue to study each earth feature resulting from the tectonic forces as the most probable source of clues.

GRADATION

The great rock masses thrown up by faulting, folding, and vulcanism are a formidable challenge to the erosive capacity of running water, moving ice, wind, and waves, but they can and do wear away the world's mountains as has been repeatedly demonstrated, given the implements with which to work and especially given unlimited time. Scientific observers have constantly searched for some kind of repetitive pattern, some way of systematizing the whole function of gradation, not satisfied that it is merely random or accidental in its application or result. Such a method of organizing and classifying the various stages of an erosional cycle was worked out ultimately by the renowned Harvard geographer William Morris Davis early in the century. He called it the *geomorphic cycle*. As a tool for the student in the field in recognizing and describing landforms, it has been highly effective over the years. Far from perfect and

William Morris Davis

Central to a lifetime of contribution to the science of geography was William Morris Davis' (1850–1934) concept of the "Cycle of Erosion" which took the form of a theoretical model. It reduced in a single stroke the great mass of unrelated and often bewildering detail of landscape description to a systematized and predictable pattern. Accused of rigidity by his detractors, Davis

vigorously defended his ideas, claiming again and again that his model was a perceptive tool and that infinite departure from it was the norm. Moreover, he readily admitted his debt to Powell, Gilbert and others in the original formulation, although the terminology was his: *youth, maturity, old age, peneplain* and *monadnock*. In Europe, where modern geography had its beginnings, as well as in the United States, Davis was regarded admiringly as Mr. American Geographer.

William Morris Davis graduated from Harvard (1870) with a degree in engineering, spent his early years in Argentina as a meteorologist, and returned to Harvard as a geologist/physical geographer (1876)—academic labels in those days were seldom restrictive on a man's academic endeavor. Increasingly, however, he thought of himself as a geographer and was frustrated that so few others involved in related research did not apply that label to themselves. In 1904 he issued a call to all American scholars who were geographers at heart, no matter what their formal titles, to join him in organizing a professional society. The Association of American Geographers came into being later that year with Davis as its first president. Forty-eight geographers joined him, many newly out of the closet—geologists were the most numerous among them but represented were oceanographers, botanists, agronomists, meteorologists and economists. Davis was elected president twice more, in 1905 and 1909, a measure of his esteem among his colleagues.

probably more theoretical than real, nonetheless, the geomorphic cycle is introduced here as a widely applied basic concept.

THE GEOMORPHIC CYCLE

We must first assume a new landform, perhaps a flat-topped plateau just raised out of the sea by faulting. But no sooner is it exposed to the atmosphere than the gradational forces begin to attack it and to alter its outline as they attempt to lower it back to sea level. Newly estab-

lished streams flow across its face, cutting as they go and carrying away their cuttings to the sea. But for a while at least, depending on the hardness of the rock to be worn away, the basic tabular outline of the plateau remains intact, scarred to be sure, but easily recognizable as a flat-topped block. Between the narrow stream valleys, there still remain extensive areas of the original surface. At this stage the erosion cycle is called *youthful*, that is, relatively little progress has been made in altering the landform (Fig. 17–4).

But erosion is inexorable, and eventually, as the streams cut deeper and multiply their tributaries, the flat top of the plateau will become so mutilated that it loses completely

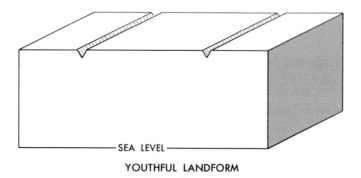

YOUTHFUL LANDFORM

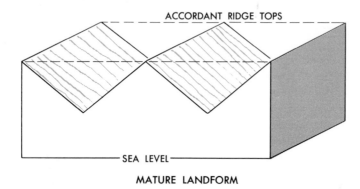

MATURE LANDFORM

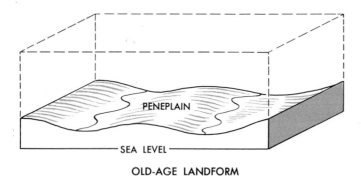

OLD-AGE LANDFORM

Fig. 17–4 The geomorphic cycle (highly diagrammatic). A youthful landform is only slightly modified from the original, a mature landform carved dramatically, and an old-age landform appears as an erosional plain reduced to near sea level.

its earlier appearance. Interstream divides have now been reduced to sharp ridges or peaks, and the only indication that the surface was once at a constant level is the *accordance* (same height) of these ridges. Now the entire landform exhibits a maximum of relief—slope is everywhere and no flat land exists. Such a highly dissected landform has reached a *mature* erosional stage, and a significant part of

it has been cut away and removed, leaving its form greatly altered (Figs. 17–4 and 17–6).

Finally, as the rivers approach the completion of their work, they have cut their valleys down to a point not far above sea level, and in so doing have lost much of their velocity and vigorous downward cutting ability. Now they flow languidly across broad plains, and the rolling inter-

Fig. 17–5   The Caineville badlands at the north end of Capitol Reef National Park in Utah. There is *no* flat land here, hence an emphatically mature landform.

stream divides are far below the level of the original plateau. This is an *old-age* stage of erosion and the landform has been planed away. A nearly flat surface once again appears, but this time at a level very near that of the sea. It is called a *peneplain* or an erosional plain—not quite at sea level and not quite flat because the streams require gravity to erode and at this stage have lost that ability. Often, standing above this rolling surface, there will be a hard-rock remnant that did not erode away as rapidly as the rest. This is called a *monadnock*, but it too will disappear given time (Fig. 17–6).

A peneplain, not quite at sea level and not quite flat.

Fig. 17–6 The top of every butte is at the same height and each is capped with the identical hard-rock stratum. This is all that is left of a higher surface, its hard rock breached and the softer material underneath quickly cut away. The buttes are mere erosional residuals called *monadnocks*, standing temporarily above the general level of the plain. In Europe they are labeled *inselbergs*, island mountains, perhaps a more descriptive term.

Now all of this is theory—assuming an abrupt new landform, assuming that it stands still and does not move, assuming that sufficient time elapses for the entire process to run through its sequences. In nature this seldom happens. But that does not negate the classification as a tool. And do not forget that although running water and a plateau were used here as examples, a volcano can be sculpted by ice, or even groundwater can etch caverns and their subsequent collapse give rise to youthful, mature, and old-age landscape. It is a useful concept and widely applicable in sytematizing the gradation process.

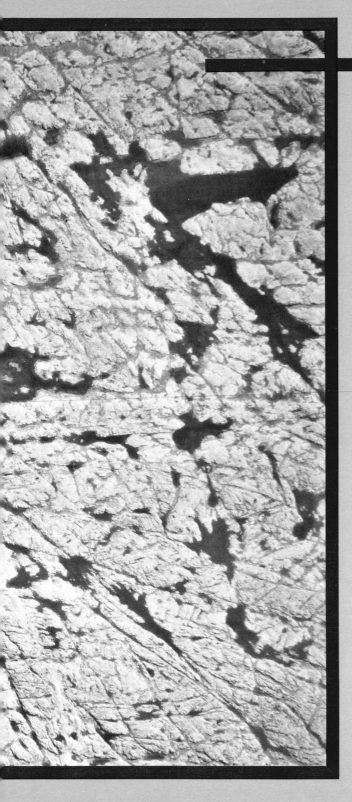

CHAPTER 18

THE EARTH'S CRUST

The oldest rocks in all the world are found in crystalline shields. When erosion has laid them bare we have been able to get at them and apply a panoply of dating procedures, and as of this moment have worked our way back over 4 billion years. Here, the Canadian (or Laurentian) shield has been scoured by continental glaciers, scratching, gouging and polishing as they went. It is a unique and barren landscape.

ORIGIN OF THE EARTH

Since nobody has ever been able to locate a first-hand witness, all of our speculation as to the manner in which the earth came into being is just that—speculation. This does not mean wild guessing. We are well beyond the point of fire deities and water gods to account for phenomena beyond our comprehension. We are, however, still hypothesizing, deducing as best we can from earthly clues. The problem is not yet resolved, but as our knowledge increases with the refinement of research tools and technology, it would be satisfying to imagine that modern theories are closer to the truth than the hoary ones of antiquity.

All are agreed that the sun is the mother of the planets.

All are agreed that the sun is the mother of the planets and that any doctrine explaining the earth must also explicate the solar system. Within these limits there are at present two families of hypotheses: one says that the earth is in the process of cooling from an original body of incandescent gases, and the other, that the earth has solidified from a swarm or dust cloud of tiny solid particles. This latter hypothesis probably squares better with the increasingly accepted theory of the "Big Bang" e.g., the entire solar system is residual from the debris of a tremendous intergalactic explosion. But there are serious objections to each of these approaches. We would like to think that we are on the right track and that modest additions (moon-rock analysis and the Pioneer probes of Jupiter, Venus, and Saturn for instance) and adjustments will resolve the disagreements, but it is not impossible that further fundamental theorizing is still in order.

THE EARTH IN CROSS SECTION

The single most frustrating problem in speculating about earthly origins is that we have so little absolute knowledge of the earth itself. Occupying only the surface, we have plumbed depths in mines and drill bores of less than five miles. But the radius of the earth is almost 4000 miles (6437 km)—a long way to go before the materials of the core can be sampled. Actually, we have reason to believe, through the use of the seismograph and gravity studies, that the crust, which is the only portion to which we have first-hand access, is very thin and highly atypical. The educated guesses of the experts lead us to the conclusion that the earth is a lamination of concentric layers with the hottest/densest substances at the center under greater pressure and the lightest/coolest materials at the periphery (Fig. 18–1).

In the center the core is very hot in the neighborhood of 5000°F (2750°C) and extremely dense, under a pressure of 3.5 megabars or 3.5 million times atmospheric pressure (by comparison, the pressure of water in the deepest part of the ocean is 4 kilobars or 57 thousand times greater than the atmosphere at sea level). It has been traditional to regard the heavy compressed core as high in nickel/iron content from our experience with meteorite substances; probably a solid inner core despite the high temperatures and a molten outer stratum where pressures are slightly reduced. But this assumed character is being questioned today in the light of recent experiments with hydrogen under immense pressure in the laboratory. Physicists have achieved solid hydrogen, a transparent icelike substance, and anticipate metallic hydrogen with the use of their diamond faceted presses. We know that oxygen too may become metallic given sufficient pressure. Could the earth's core be metallic oxygen or an alloy of iron and oxygen?

Physicists have achieved solid hydrogen, a transparent icelike substance, and anticipate metallic hydrogen.

Beyond the core of the earth is the *mantle*, appearing to be a reasonably homogeneous mass made up of some dense basic mineral like olivine, and beyond the mantle, the crust. Between these last two is a very acute line of separation called the *Moho* (a shortened version of Mohorovičić, the Yugoslavian geologist who first identified the crust/mantle boundary in 1909), easily discernible, as seismic waves, either natural or artificial, are analyzed. As these waves travel through the various crustal substances,

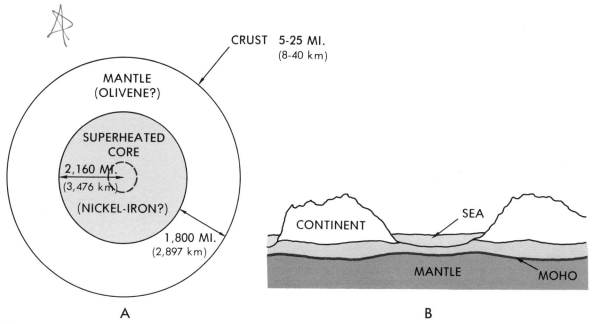

Fig. 18—1 Composition of the earth and its crust, showing a generalized cross section (A) and the basic materials that make up the crust proper (B).

they average close to 4 miles (7 km) per second, but as they cross the Moho discontinuity, they abruptly elevate to 5 miles (8 km) per second and maintain that constant through the entire mantle. Obviously, the crust and mantle are sharply differing layers. But although the crust is thin, with the Moho tantalizingly close, no one has yet tapped the mantle.

The crust cradles the oceans in its surface cavities and thrusts snow-capped mountain peaks as far from the superheated core as possible.

Discussion of the subject has not been lacking. The much ballyhooed Mohole project was to have been an official United States government effort to drill through over 3 miles (5 km) of sea bottom in ocean depths of 14,000 feet (4267 m) west of Hawaii, but the entire program was aborted in 1966, at least temporarily, in the face of escalating costs. This particular sea-bottom location was chosen (despite the obvious hazards of an unstable platform and seasick crew) because the crust is thinnest on the ocean

floor, averaging from 3 to 7 miles (5 to 11 km). On land a comparable project could involve a bore of up to 25 miles (40 km) in depth.

To visualize the crust, one must conjure up an image of a thin, two-ply veneer comprising a concentration of the lightest and most brittle substances in the entire earth; this outermost layer, floating, as it were, on the denser mantle below, cradles the oceans in its surface cavities and thrusts snow-capped mountains peaks as far from the superheated core as possible. The lower ply, made up of dark basic basalt, encircles the entire earth, flooring the oceans and underlying the continents. The more massive, lighter, chiefly granitic continents constitute the broken outermost crustal layer.

CONTINENTAL DRIFT

Map readers for many centuries have been intrigued by the nicety with which the great easterly bulge of Brazil and the West African Gulf of Guinea seem to fit one into the other. There they are with only the Atlantic between, but conti-

Alfred Wegener

Anthony Hallam, *A Revolution in the Earth Sciences* (New York: Oxford University Press, 1973), p. 113. Quoted by permission, Oxford University Press.

It is easy to cast back today and reflect on the brilliance of Alfred Wegener's drifting continents concept as expostulated so concisely in his several editions of *The Origin of Continents and Oceans* (1915, 1920, 1922, and 1929); but during his lifetime he was subject to not only the normal scientific criticism to be expected in new and radical theorizing, but his hypothesis was savagely ridiculed from every quarter and the man himself attacked as not quite bright, a pamphleteer rather than a scientist. Oxford Geologist, Anthony Hallam, writing four decades later, looked back too and attempted to account for the particularly brutal assaults on Wegener and his theory. "The trouble must partly have been that he was not an accredited member of the professional geologists' club; We of course now see it as a positive advantage that Wegener had not been brainwashed by the conventional geological wisdom as a student. The key role that an outsider can play in transforming the world view in a particular subject is now well recognized."

Alfred Wegener (1880–1930), born in Berlin, was educated in astronomy, and attained a Ph.D. in that field in 1904. But he was soon captivated by the infant science of meteorology and did advanced work at the Prussian Aeronautical Observatory in Tegel. This led to an assignment as meteorologist with a 1906 Danish expedition to Greenland where he spent 2 years— it was an experience that proved pivotal to his career for he developed a fascination for Greenland that never diminished. Twenty four years later he was back again as the leader of a German party, gathering scientific data on the icecap itself. There he perished in a fierce storm on his 50th birthday, unaware that one day his notions of drifting continents would receive worldwide acclaim.

The idea itself was born of a flash of insight by a trained and curious scientist, but one who was not a geologist. Wegener simply stumbled across an article advancing a theory of land bridges between continents to explain certain inalienable similarities. The rest was a career-long tour-de-fource of logic, researched and reinforced as the years went by. To his credit Wegener never faltered in his beliefs—his supporters were few and usually lukewarm at best. Only his elderly father-in-law Wladimir Köppen was enthusiastic and loyal from the very beginning. He wasn't a geologist either.

nents cannot just float apart—or can they? The first serious suggestion that exactly this kind of thing did happen was put forth by the German Alfred Wegener in 1915. He had evolved an intriguing theory to support his contention, some of which was based on an impressive body of facts, some on surmise, but he was a meteorologist not a geologist and obviously before his time. For 40 years Wegener was held to be, by even the most respected scientists, at best a dreamer—more frequently, the rankest kind of irresponsible charlatan. Today, not only is the idea of floating landmasses far from scientific heresy, but an increasing mass of reputable research findings has all but convinced the skeptical critics.

First, come back to the mantle. If the crust has skidded here and there, the underlying mantle must be capable of a certain amount of plastic flowage. The concept of a solid becoming plastic is not new or revolutionary, especially under the influence of pressure. Brittle ice, for instance, will flow under fairly minimal pressure, but the commonest example and the one offered most frequently as a simile to mantle behavior is that of a block of roofing tar. When struck with a hammer it will shatter, but simply left on the ground without any addition of pressure or heat, it will slump or flow downslope. If heat or pressure that produces heat is added, the tar will flow even more readily. So it is, supposedly, with the mantle. Given some sort of force to spread the continents, the character of the mantle is permissive.

Next we must look for an applied force to wedge these great landmasses apart. Studies of sea-bottom sediments conducted at Princeton during the 1950s, led to a clue. Why were the sedimentary layers so thin and of such recent age? If the rivers of the world had been delivering endless detritus from land to ocean for all of these millennia, as was demonstrably the case, then surely piles of sediment must floor the sea to tremendous depths. From this question came a hypothesis of mantle/crust thermal convection currents. Rising heat in mid-ocean, often expressing itself in sea-bottom vulcanism, spreads outward toward the continents on either side, to finally sink at land's edge. There would be a sort of endless belt to transport sediments outward and into the abyss of the oceanic trench that so often flanks each marginal landmass. In addition to explaining the lack of sediment accumulation, this theory also offers answers for medial mountain ranges, earthquakes, and lateral trenches.

As further checks were run, it was found that deep-sea water temperatures were higher in midocean and that bottom rocks progressed from younger to older with distance from the center. This last bit of data was obtained through the use of the relatively new technique of *paleomagnetism*. For unexplained reasons the magnetic field of the earth reverses itself from time to time—some 171 times

in the last 76 million years. How do we know this? The record is frozen into the rocks. At the time of rock formation, especially those cooling down from the molten form, ferrous minerals align themselves with the magnetic field so that all rocks of the same age become permanent compasses of their moment of solidification. In theory, then, rock samples from the sea floor should display bands of changing magnetic fields as the continents are approached from midocean. This is precisely the kind of pattern discovered in 1968 by the U.S. oceanographic ship *Glomar Challenger* as it drilled a controlled series of bottom cores.

Whether the convection system functions exactly as hypothesized, whether the requisite heat obtains from deep seated or shallow radioactivity, or simply from pressure—these specifics are minor. The significant fact is that the scientific community has finally acceded to the thought that continents may indeed drift and now has turned its attention from scoffing at Wegener to searching for proof. Researchers look to more and higher quality ocean-bottom research of the kind outlined in the foregoing discussion, and to attempts at matching the rocks of South Africa/Brazil and Australia/Antarctica on the bases of fossils, glacial evidences, and paleomagnetism.

Increasingly dovetailing evidences of every kind have come to light, each new find further solidifying the claim that surely, through the ages, landmasses have changed their lateral positions relative to each other.

SHIELDS (CRATONS)

The continents as we know them today were blocked out roughly very early in geologic time. Each continent has a stable hard rock core made up of ancient granitic and metamorphosed rocks that although subjected to eons of stresses and pressures, have maintained themselves as relatively stable masses. Erosion has worn them down to low featureless plains or pleateaus, where they can be observed on the surface, but their margins are often masked by more recent sedimentary mantles. Along the edges of these *shields*, as they are called, a great deal of frequent activity has occurred. Extensive regions have been alternately above and below the sea, and it is here that most of the world's

Marsupials

The recent discovery of fossilized bones of a small *marsupial* in Antarctica confirmed the oft-speculated ancient connection of South America, Antarctica, and Australia. It had long been known that marsupials originated and flourished in South America and further conjectured that there had been much unfettered and leisurely strolling across the Antarctic land bridge.

But Australia broke off and sailed majestically into the Indian Ocean with its cargo of marsupials. Then Antarctica skidded away, about 50 million years ago, into the high latitudes freezing its passengers out of existence, leaving South America as a single unit. The last act in this drama was the linking of North and South America via the Isthmus of Panama.

Stephen Jay Gould, the Harvard paeleobiologist, paints a vivid word picture of the moment of joining of the two great land masses. Poised at the north, slavering hordes of *placental* (mammals whose young develop in the womb through an extended gestation period) predators pawed the ground and gnashed their teeth; and to the south peaceful little communities of docile, innocent, marsupials went about their daily duties, blissfully unaware that they were shortly to be the evening meal. A bit overdrawn perhaps, but we know that prior to the continental coupling, 29 families of marsupials inhabited South America while 27 wholly different placentals occupied North America. Following the union 22 mammalian families survived, nearly all North American.

One marsupial asylum remains today. Australia, in splendid isolation from the moment it parted company with Antarctica, retains its pristine pouched mammal population, a fascinating and disparate society of wallabies, kangaroos, opossums, Tasmanian devils, wombats, phalangers, koalas and bandicoots. The world's only *monotremes* (egg-laying mammals), the platapus and echidna, share Australia with their pouched neighbors. In the evolutionary scheme of reproductive efficiency, the monotreme precedes even the marsupial.

---

mountain systems are found; the shields themselves remain largely unaltered (Fig. 18–2).

In North America the hard rock core is called the *Canadian* or *Laurentian Shield*, and its surface exposures are evident over a large part of eastern Canada, coastal Greenland, and the Great Lakes region. Undoubtedly, it extends beyond these limits below the surface. The *Fennoscandian Shield*, exposed in Scandinavia and northwestern Russia, is the European equivalent. Asia exhibits a shield (or shields) in central Siberia, southern China, India, and Arabia; South America, in the southern two-thirds of Brazil; Australia, in the western two-thirds of the continent. Antarctica may have

a shield under all the ice, and virtually the entire continent of Africa is a single shield.

Today's existing shields are essentially the oldest rocks on earth.

Since today's existing shields are essentially the oldest rocks on earth, it would seem reasonable that any attempt at reconstruction of what might have been, prior to continental drift, should begin with these blocks. If we postulate

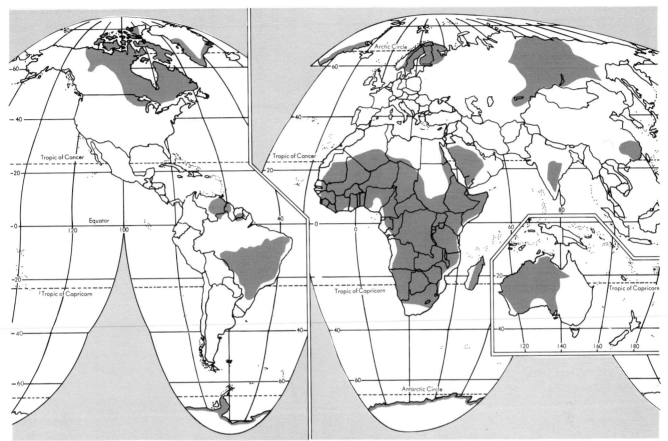

Fig. 18–2 Generalized outline of the surface exposures of the world's shields. Every continent has its stable hard-rock core or shield.

two massive continents, one in each hemisphere, we can regard our contemporary landmasses as fragments of the original, broken off and rafted in all directions at a rate of as much as 4 inches (10 cm) a year. How did they fit together? The shapes of the continents themselves give some help in the reassembly puzzle, as do the proper alignment of ancient poles of magnetism and correlated rock structures. One problem is the general lack of fossils in rocks of this age, but although this constrains certain lines of investigation, others yield interesting results.

The reassembly of the present parts into an ancient entity is still conjectural at best, but Fig. 18–3 shows two examples of the kind of work that is being done. The Southern Hemisphere restructure has been, on the whole,

a slightly simpler puzzle than that of the Northern. These ancient speculative continents have been named *Gondwanaland* and *Laurasia*, respectively. And both of these were coalesced at one time as a single landmass, *Pangaea*.

PLATE TECTONICS

Now that the early apostasy of continental drift has evolved into true gospel, the geologic fraternity has moved on a step further in its theorizing. Not only are continents capable of drifting, but they are parts of larger systems. Visualize the brittle, rigid crust of the earth broken into a series of

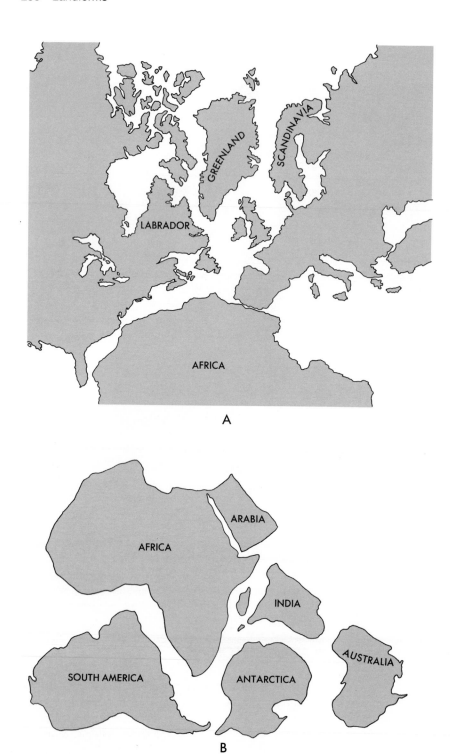

Fig. 18–3   Reassembly of ancient continents. Two attempts at solving the puzzle (A) Laurasia. (B) Gondwanaland. (*Source:* After Hurley and after King, respectively.)

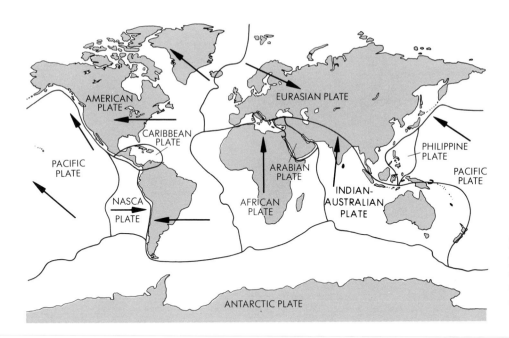

Fig. 18—4 Massive crustal plates in motion. This is merely one version of what is transpiring, but it varies from others only in detail. The theory is generally accepted.

freely flowing plates, most larger than a continent and carrying the continents and shields on their backs (Fig 18–4). The sea bottoms are either included as portions of the greater continental structures or, as in the Pacific, may be plates in their own right. As they jostle and jockey for position there are inevitable collisions and "here is where the action is." The cardinal theorem in plate tectonics is that diastrophism is the product of plate interplay and hence a phenomenon of the margins: its corollary is that the rigid plate mass generally does not get itself involved in radical mountain-building processes.

As the crustal plates jostle and jockey for position there are inevitable collisions and 'here is where the action is'.

Several variants are possible as plates impinge. If both are advancing one against the other, as in the South Pacific where the *Nasca* plate moving east runs head-on into westward trending South America, something has to give. It is the Nasca that is overwhelmed and wedged in underneath the American plate. At sea a deep linear trench along the coast of Chile and Peru is evidence of the steep angle of subsidence of the oceanic plate's leading edge, while the parallel high Andes, replete with smoking cones and seismic spasms, are reflective of violent mechanical disruption, friction and internal heat release.

A somewhat different product of direct plate collision is apparent in the Mediterranean region where Africa encroaches on Europe. Here the long accumulated sediments from the Mediterranean and its much larger predecessor, have been rumpled into a series of impressive fold mountains—the Alps, Pyrenees, and Atlas (Fig. 18–5). The process continues and today's tiny depleted Mediterranean Sea is about to be phased out totally.

But plates do not have to charge into conflict head-on—lateral motion, one to the other is not uncommon. California's west coast, that part west of the San Andreas fault, is shifting to the north while at the same time the American plate is either still or slowly sidling southward. As they grind along, cheek to jowl, tearing and fracturing produces a spectacularly active zone of earthquakes and fault block mountains.

With multiple plates in motion in multiple directions there have to be some places where plates are moving apart. The midocean ridges are located at these lines of tension.

Fig. 18–5 The magnificent Himalayas are the crumpled consequence of an Indian plate pushing northward into rigid Eurasia.

Now, whether thermal convection forces the plates apart or moving plates release pressure thereby bringing molten rock to the surface is not of import here; the point is that when plates pull apart, active vulcanism renews the trailing edge with virgin rock. At the opposite plate margin this new material is exactly countered as the plate dives dramatically into an oceanic trench and is consumed by the mantle. This last is called *subduction* (Fig. 18–6).

When plates pull apart, active vulcanism renews the trailing edge with virgin rock.

So the basic theory tells us that most mountains are formed by plate collision, even those encountered today far inland. The Urals, a modest range supposedly marking the boundary between Europe and Asia, is a series of folds resulting from an Asiatic plate maltreating the sedimentary fringe of Europe during the formation of Pangaea. Currently, Pangaea has disintegrated, leaving the ancient independent European and Asian elements welded into one. If all of this could really happen, then certainly perambulating plates on a globe must have had the opportunity to merge into diverse Pangaeas in the past only to break up and reform unknown times.

There is still some argument as to exactly how many crustal plates we are working with. Some authorities delineate as few as 6, others, up to 20. Undoubtedly, plates are at this moment undergoing deformation, renewal, and fragmentation, so that a precise count becomes difficult. Nonetheless generally speaking, we have both the geometry and the kinematics of plate motion fairly well in hand—it is the mechanisms that drive the plates and particularly the nature of the driving forces that are yet to be perceived clearly.

Why not isolated plumes like a blow torch from below burning their way through to create volcanoes in midplate.

Thermal convection deriving its heat from radioactive decay in the mantle has already been suggested; some think the moon, its tidal forces combined with gravitational forces, may be of consequence. Or if released mantle heat can well up in midocean ridges, why not in isolated plumes with plates moving over them—then like a blow torch from below burning their way through to create volcanoes in midplate instead of at the margin? Hawaii is probably of this origin. So is the Yellowstone thermal district.

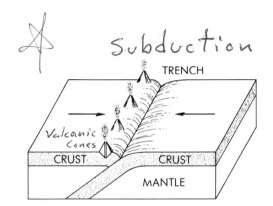

Subduction

TRENCH

Valcanic Cones

CRUST CRUST

MANTLE

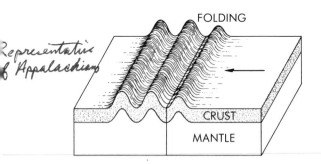

FOLDING

Representative of Appalachians

CRUST

MANTLE

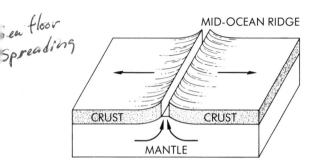

Transcurrent Fault

TRANSFORM FAULT

SanAndreas Fault

CRUST

CRUST

MANTLE

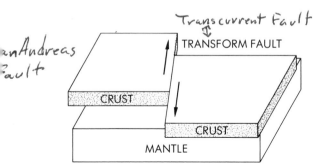

Sea floor Spreading

MID-OCEAN RIDGE

CRUST CRUST

MANTLE

Fig. 18–6 Variations on the resultants of crustal interplay.

GEOLOGIC TIME

In the context of geologic time we are usually dealing with processes that appear to the average person to be operating with painful slowness if at all. Literature is replete with references to "timeless mountains" and "rocklike qualities," implying that in our dynamic world of instantaneous social change, there remains that epitome of permanence, the rock. We instinctively use as the measuring stick of time the four score and ten of human lifetime or the few thousands of years of recorded historical time. In this frame of reference, rocks and mountains may very well be virtually immutable. But geologic time, the aggregate measurable age of the earth, is an immense span almost beyond comprehension and we must constantly remind ourselves that the wearing away of a mountain range or the drifting of a continent is to be calibrated in millions of years—a tiny fraction of the earth's total lifetime (see Appendix D).

In almost unbelievable retrospect, at the turn of the 20th century geologists were in fairly general agreement that the earth's true age could be no more than 100 million years.

The Age of the Earth

It has always been a tricky bit of business, this determination of just how long the earth has existed. We still do not know how the earth originated; if we did, that would help. And only during the last 150 years or so has the problem been approached in any kind of scientific manner. For a long time, decisions on these matters were exclusively ecclesiastical; others were discouraged from putting forth their views. Even when the church/science deadlock eased a bit, the technology of dating remained both primitive and tentative. In almost unbelievable retrospect, at the turn of the 20th century geologists were in fairly general agreement that the earth's true age could be no more than 100 million years. Today, the educated guess is something like 4.5 to 5.5 billion years (Fig. 18–7).

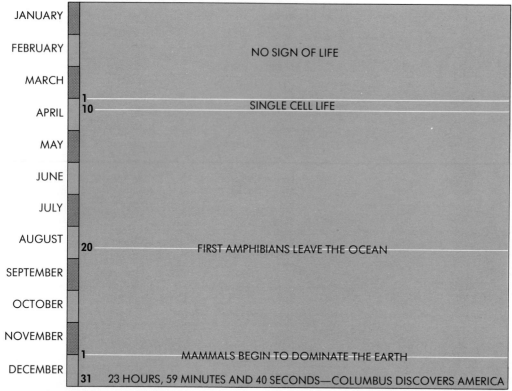

Fig. 18–7   Equating one calendar year with the 4.5 to 5.5 billion years of the earth's existence.

DATING TECHNIQUES

Generally, four families of evidence are cited to support specific claims of age: (1) fossils, (2) extension into the past of contemporary observed processes, (3) radioactive dating, (4) paleomagnetism.

Fossils

Although fossils had been known and recognized for what they were for many years, their systematized study and interpretation is more modern and has given rise to a wholly new branch of geology, *paleontology*. A practitioner in this occult art has to be a modern botanist/zoologist, and perhaps even a bit of an archeologist, and be further blessed with the inquiring mind of a detective. His or her job is to establish a system of evolving life orders (many completely extinct today) to enable careful dating of the rocks in which

each fossil was found. It has been a big chore but one that has now been accomplished in its main particulars, and once put together as a working system, it has proven to be amazingly accurate. Fossil evidence takes us back rather specifically for about 600 million years (Fig. 18–8). Beyond this point the lesser life on the young planet simply has not been fossilized in a significant amount in the old contorted and crystallized rocks such as those encountered in the continental shields; here the fossil as the traditional dating tool breaks down. We know that there are rocks in existence much older than life on the earth, but how much older is difficult to say. We must be willing to get into surmise and accepted tolerances of a billion years or more.

Extending Today's Processes into the Past

The second method of age determination, extending backward into time the processes we observe today, has been used in one way or another to ascertain relatively short time

New Discoveries

In 1980 an American/Australian scientific team identified the fossilized remains of tiny bacteria-like organisms in Western Australia that lived more than 3.5 billion years ago. They had been preserved miraculously in ancient, contorted shield rock because their tube-shaped cells were filled with hard silica. Upon close examination with powerful microscopes, 5 different kinds of organisms were revealed, which suggests strongly that even at this early date there was a relatively complex ecosystem on earth—the origins of life must be more remote than anyone has guessed.

In 1983, in these same Western Australian mountains, geologists discovered grains of rock that are 4.1 to 4.2 billion years old, by far the most ancient ever found on earth. The rock bits are tiny *zircons*, a compound of zirconium, silicon and oxygen. Zircons are found in virtually all granites and many sedimentaries, but because of their hardness and durability, survive long after their parent rocks have been altered beyond recognition. Geologists have sought them out as "tracers". These particular zircons were accurately dated by the new *ion microprobe* technique wherein a tightly focused beam of high energy ions is directed at each individual crystal. Atoms "sputtered off" the surface are, after several intervening steps, susceptible to dating analysis.

periods. Often cited as an example is the behavior of the Nile, of which we have rather carefully kept records for over 3000 years: its rate of cutting at a particular location, rate of delta growth and sedimentation, etc. These are useful data for comparison with other rivers and for assumed Nile character of several thousand more years into the past. Another common technique is the analysis of *varved clays* in modern or ancient lake bottoms. Each varve, or two-tone layer, represents one year of sedimentation and there are instances where a careful study of accurately dated ancient lake deposits has revealed a 25,000-year history.

The basic revolution in this whole approach to measuring time has come with the introduction of the computer. If we know, for example, the average receipt of salts in solution by the seas from all the rivers in the world we have a clue to the age of oceans. If we can measure the total sediments carried by rivers draining the Andes we can apply this rate of denudation to comparable streams cutting away ancient mountains. Wherever huge numbers are involved and tedious calculations, the computer is invaluable. This kind of analysis is hardly precise, but it provides insight into great unanswered problems and can frequently be used as a check or test against older accepted fossil dating.

Radioactivity

The third of these geochronological processes, radioactive dating, is without doubt the most accurate and is particularly useful in its potential for clarifying the ages of the ancient nonfossilized pre-Cambrian rocks. By this method, rocks at several locations in Souh Africa and Greenland have been assigned ages of almost four billion years. There is a flaw in the system, however, in that most igneous rocks are not susceptible to radioactive dating, and this is particularly unfortunate, for these same rocks are totally lacking in fossils too.

The method is based on the known rate of decay of certain radioactive minerals into stable end-products, and since neither heat nor pressure appears to have any effect on radioactivity, we have discovered what is essentially an

Fig. 18–8   A highly detailed fossil fish who somehow swam into a muddy spot and became a part of the shale. This specie no longer exists but was common in tertiary seas. He has been identified and dated by his presence in a known rock stratum from Wyoming—and vice versa.

unbreakable clock. Uranium and thorium are the chief radio elements (there are others) because they are relatively common in occurrence and possess a long half-life compatible with the age of the earth.[1] They are called parent elements and decay through several unstable generations into stable daughter elements; in the case of uranium and thorium, the resulting elements are helium and isotopes of lead. Their abundance in a rock can be measured against the rate of decay of the parent.

Carbon 14 should also be mentioned here although it has the major drawback of a utility of a mere 40,000 years. Its value lies in the fact that it is drawn from the atmosphere by living plants and their remains can then be dated accurately. For the archeologist concerned with the ages of seeds and spores or wooden implements it is invaluable.

Knowledge of radioactivity, or emission of rays by certain elements, goes back only to 1896, when Marie and Pierre Curie pioneered great advances in radium research. But the practical and widespread application of radioactivity as a dating device had lagged until well into the post-World War II period. It is now routinely used and is recognized as a valuable tool.

Paleomagnetism

Finally, we come to the very new method of utilizing paleomagnetism as a means of assaying the ages of rocks. The basics are outlined in the previous treatment of continental drift; and it is merely worth repeating here that such a system of measuring time has already proved itself in a number of applications, especially in closing the igneous rock loophole.

CRUSTAL COMPOSITION

Throughout the foregoing discussions of the earth's crust and its age there have been casual references to rocks. Everyone has thrown rocks, broken rocks, or dug into rock materials, and we tend to give the same label to the many different varieties of the "hard stuff" that seems to be the earth itself. But there are better and more precise defini-

---

[1] A half-life is the length of time it takes for one half of the atoms of the parent to decay and change.

Igneous rocks are not susceptible to radioactive dating and they are totally lacking in fossils too.

---

tions, as well as a system for sorting out and classifying all the many variations of rocks.

Elements

The place to begin is with the elements. Since all known matter is made up of elements (atoms), the earth, the water on it, and the atmosphere have an elemental or atomic base. Ninety-two of these are naturally occurring (10 more have been produced in the laboratory) and each is a distinctive combination of protons, neutrons, and electrons with an assigned name and letter abbreviation, for example, iron (Fe), oxygen (O).

Of these, by far the most abundant in the earth's crust are oxygen 46.59% and silicon 27.72% (by weight). Several others are represented in amounts of over 1%, but if we choose to classify less than 1% as trace elements, then most elements must be so classified, for only 8 elements make up over 98% of the crustal total. Not included among these critical eight are some of our highly utilitarian nonferrous metals such as gold, copper, lead, and zinc. Obviously, when we talk in terms of human exploitation of a resource, we must consider value and physical concentration into economic deposits as well as sheer abundance (Fig. 18–9).

Crustal Elements		Percentage by Weight
Oxygen	(O)	46.59%
Silicon	(Si)	27.72
Aluminum	(Al)	8.13
Iron	(Fe)	5.01
Calcium	(Ca)	3.63
Sodium	(Na)	2.85
Potassium	(K)	2.60
Magnesium	(Mg)	2.09
Total—eight minerals		98.62%
All others		1.38%

Fig. 18–9 Crustal elements.

Fig. 18–10   Quartz crystals are invariably vitreous, very hard and hexagonal.

Only 8 elements make up over 98% of the crustal total.

Minerals

Minerals are ordinarily naturally occurring combinations of elements in the solid state, and their composition may be expressed in a chemical formula, for example, NaCl (sodium chloride or common table salt) or SiO2 (silicon dioxide or quartz). In addition, most minerals exhibit a distinctive crystalline form and degree of hardness that aid in their identification (Fig. 18–10). Occasionally, a mineral may be a single element such as graphite and diamonds, which are pure carbon (C) in two different forms, or so-called native gold (Au) or native copper (Cu) may occur in nature in the pure form rather than as ores. But these are relatively rare and most minerals are combinations of elements.

Rocks

Rocks are combinations of minerals. But if 92 elements may be merged in a virtually endless number of combinations

to produce minerals, then combining these in turn to form rocks gives us an infinity of possibilities as far as the chemical composition of rocks is concerned. It is true that because of the natural proportion of elements in the earth's crust, a very large number of rocks are compounds of oxygen, silicon, and one or more of the 6 minerals represented in excess of one percent. Nonetheless, the varieties of existent rocks are immense.

To inject some order into this multiplicity, we make no effort to classify rocks according to their chemical composition, instead we adopt a *genetic classification.* In other words, we break down the rocks in the earth's crust on the basis of *how they were formed.* All rocks then fall rather easily into one of the three following general categories: *igneous, sedimentary,* or *metamorphic.*

Igneous.   *Any rock that has cooled and solidified from a molten state is an igneous rock.* Therefore if the earth's history is that of a superheated sphere in space, all the rocks making up its crust may well have originally been igneous and thus the ancestors of all other rocks. Even today, approximately 95% of the entire crust is igneous (Fig. 18–11).

Fig. 18–11 Here's a jumbled broken mass of recent lava that covers extensive areas of Idaho's Snake River plain.

As molten material periodically wells out of the earth's interior to invade the surface layers or to flow onto the surface itself, it cools into a wide variety of igneous rocks. In the molten state, this liquid is called *magma* as it pushes out into the crust, but the name *lava* is applied when it ejects itself onto the surface.

Although all magma is made up basically of a variety of silicate minerals, the chemical composition of any given flow may differ radically from that of any other, and thus the resulting rocks will reflect these differences. In addition to sharp chemical differences, igneous rocks also exhibit considerable variation in texture. *Granite*, for instance, is a course-grained igneous rock whose individual mineral crystals have formed to a size easily visible to the naked eye. A slow rate of cooling has allowed the growth of the crystals to this size. Normally, retarded cooling results from the invasion of the crust by a magmatic body that remains buried well below the surface. We may encounter granite as a surface rock in the contemporary landscape, but we know from its coarse texture that it must have formed through a slow coagulation at a considerable depth and then been laid bare by subsequent erosion. All igneous rocks of this coarse-grained character have been given the name *plutonics*.

> *We may encounter granite as a surface rock in the contemporary landscape, but we know from its coarse texture that it must have formed at a considerable depth.*

On the other hand, if this identical magma flows onto the surface and is subject to the rapid cooling effects of the atmosphere, the resulting rock will be fine-grained and quite different in appearance from granite and is called *rhyolite*. These two rocks have identical chemical compositions, as do *diorite* and *andesite*, *gabbro* and *basalt*, and *pyroxenite* and *augite*. There are other couplets of this type differing only in texture. Certainly, the extreme in fine-grained rocks is volcanic glass or *obsidian* where no crystals have had a chance to form because of virtually instantaneous cooling. The well-known black obsidian cliffs of Yellowstone Park are the result of an ignescent surface flow of basalt running head-on into a glacier. Some of the glacier melted at the point of contact we may be sure, but suddenly there appeared also a huge black mass of vitreous stone.

Once again, irrespective of color, texture, or chemical composition, if a rock has formed from the solidification of magma or lava, it falls into the igneous category.

Sedimentary.

All sedimentary rocks have in common a history of once having been simply an accumulation of loose unconsolidated rock materials that have become compressed or cemented into solid rocks. It does not matter whether the individual particles are of organic or inorganic origin, what the circumstances of the accumulation were, or what the chemical composition might be; if they have gone through this transformation, they are classified as sedimentary. The word "sedimentary" implies the settling of the basic rock materials to the bottom of a body of water. Loose sands and silts may be accumulated by wind, stream, or ice ac-

Fig. 18–12 Cappadocia, Turkey, where the local residents have taken advantage of a soft sedimentary called *tuff* and carved out cave homes. Tuff is something like loess in its powdery consistency and distribution by the wind, except that its almost microscopic particle size is a product of violent volcanic explosion. The resultant ash and cinder are carried downwind and deposited selectively according to size. Buried beneath sufficient weight of overburden, the fine ash has become compacted into a soft rock. In some parts of the world blocks of tuff are used as building bricks because of its ease of quarrying.

tion on the land and eventually become sedimentary rocks; evaporating seas may form massive salt deposits; or organic matter may be solidified into rock (such as coal); or even solid particles of volcanic ejecta may be cast up into the wind by violent explosion and deposited as loose debris (Fig. 18–12). But by far the largest part of the earth's sedimentary rocks were formed (and are being formed today) by sedimentation in the sea, so the choice of *sedimentary* as a descriptive term is an apt one. This constant deposition of material on the relatively flat floors of continental shelves has resulted in the *layered* or *stratified* appearance of the rocks developed from them, which is so characteristic of sedimentary rocks (Fig. 18–13). Incidentally, virtually all of the world's fossils are found embedded in sedimentary layers. Without them the interpretation of the history of the earth would be severely hampered.

Fig. 18–13 Sedimentary rocks nearly always identify themselves by this characteristic parallelism. Deposited layer upon layer in some ancient seabed, these imposing ramparts have been uplifted en masse and then exposed through erosion.

A typical example will serve to illustrate four basic types of sedimentary rocks. Let us assume that a stream is discharging into the sea, whose bottom is of a gentle shelving character. As the stream strikes the sea, its velocity is checked rather abruptly, and since a stream's velocity is the chief control of its ability to carry rock material in suspension, it is forced to drop the heaviest portion of its load, pebbles and larger rocks, near the shore. Beyond this, as the stream's impetus continues to slow, coarse sands, fine sands, and finally minute mud particles will be deposited progressively. This sorting by size is typical of stream deposition. But beyond the influence of the stream, further sedimentation is occurring. Tiny microscopic plants and animals with a life span of a few days or weeks (called *plankton*) are constantly raining down on the ocean floor to mingle with the organic remains of bottom dwellers. Commonly, the shells and skeletons of this sea life are made up of calcium carbonate or lime, so that the resulting product is a black ooze high in content of both carbon, from the decomposing organic matter, and lime. Lime may also be precipitated directly from the sea by changes in water temperature (Fig. 18–14).

We have, then, on the ocean floor a succession of loose unconsolidated sediments ranging from coarse pebbles to fine muds and lime-rich organic remains. This is not yet rock, but if compression and cementation follow, these sedimentary raw materials will become sedimentary rock.

Shale, when struck with a hammer, tends to break into thin planes or foliations.

Pressure alone is normally sufficient to transform the fine-grained and somewhat colloidal, limy ooze and muds into rocks. A handful of mud, if squeezed firmly to wring out the moisture, can become a compact mud ball. If this simple pressure is multiplied many times, by the piling of layers of sediments on top of the original or by the folding and flexing of the ocean bottom, the muds and lime will be compacted into solid rock. In the limy ooze the liquid portion (embryonic petroleum) is squeezed out, and the resultant solid calcium carbonate rock is called *limestone* (Fig. 18-15). The muds lose their water content and become a rock called *shale*, a chief characteristic of which is

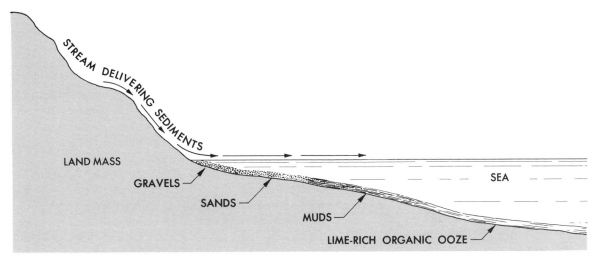

Fig. 18–14 Sorted, unconsolidated sediments on the seafloor—sedimentary rocks in the making. If the proper sequence of events follows this simple deposition, each of these categories will develop into a distinctive rock.

to develop *foliation* at right angles to the pressure that formed it; that is, shale, when struck with a hammer, tends to break into thin planes or foliations parallel to each other.

Simple pressure, however, unless it is of unusual force, normally is not adequate to consolidate sands and pebbles into rock. Sand is not only larger in size than the mud

particles but is harder and more angular. The great bulk of the world's sand is silica, one of nature's hardest widely occurring minerals. Most minerals when worn down to this size through weathering and abrasion disintegrate into silt, but silica can maintain its identity at sand size and even resist smoothing and rounding. Thus through natural se-

Fig. 18–15 The distinctive English chalk cliffs being nibbled back by stormy Channel waves working at their base. Chalk is a soft limestone varient, the compacted remains of the microscopic sea creature forminifera.

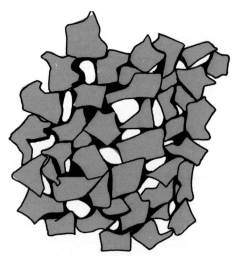

Fig. 18–16 Sandstone. The cement (black) fails to completely occupy the large intergrain spaces formed by angular sand particles. The hardness of the cement largely determines the resistance of the sandstone, and its degree of filling the interstices determines the ability of the rock to hold and transfer liquids.

lection, most particles of this size are light-colored, angular silica. The individual grains of sand do not fit tightly one to the other, and the many pores and interstices make it highly porous and allow ready movement of water through it. Gradually, various minerals in solution or materials in suspension in the water build up in the intergrain openings and set as a cement. The resulting *sandstone* takes a great deal of its character from the cementing agent. For instance, reddish sandstones probably have oxides of iron as a cement or dark sandstones may have mud. These are relatively weak stones because the cement is less resistant than the sand grains. On the other hand, if silica is deposited as the cement, the sandstone will be light colored and very hard. Seldom are all the interstices filled completely with cementing material, and sandstone very frequently is porous and capable of holding liquid; ancient sea brines, groundwater, or oil (Fig. 18-16).

Pebbles and larger stones also commonly are cemented together as are sand grains. Each pebble is a sizable piece of rock with its own characteristics of color and hardness, and it is usually worn down to a rounded form through stream abrasion before deposition. The large number of possibilities of rock character resulting from the many different cementing agents and variety of pebbles makes it difficult to generalize, but any coarse-grained rock with this history of pebbles cemented together is called a *conglomerate*.

These then are the four basic types of sedimentary rocks: (1) limestone, (2) shale, (3) sandstone, and (4) conglomerate. There are also numerous subdivisions. Remember that sedimentary rocks do not have to be formed from sedimentation on ocean floors nor do they absolutely have to show stratification, but most do and the foregoing explanation of their formation is highly typical.

In the surface rocks of the continents, the sedimentaries are strongly represented. Although they make up less than 5% of the total earth's crust, they form a widespread thin covering over much of each continent. In a few places where deposition has been continuing for long periods of time, as in the Ganges Valley, sedimentaries are estimated to be at least 50,000 feet (15,240 m) thick.

Metamorphic. The word "metamorphism" means change of form. *Thus metamorphic rocks are already existing rocks that exhibit a change of form.* Sedimentary, igneous, and even previously formed metamorphic rocks may be changed in appearance, mineral distribution, size, and complete molecular structure if they have been affected by certain forces in sufficient quantity. These forces are pressure and heat, operating in unison. The end results are new rocks that did not exist before.

There are many ways in nature in which intense heat and pressure may be generated, thereby causing changes in the rocks involved. Sediments may pile up in a sinking trough to great depths, thereby applying tremendous pressures on the bottom layers—and pressure produces heat. Or deep low-angle fractures (or faults) may occur in rock layers, allowing skidding along the fault plane. Heat from friction combined with the pressure of overlying strata may alter the character of the rocks in the immediate vicinity. Igneous intrusions into existing rock masses will frequently furnish adequate heat and pressure at the zone of contact (and often some chemical activity as well), and the folding and deformation of original flat-lying beds into involved uplands will be equally effective. Most major mountain systems have experienced all of these effects in the general process of mountain building, so metamorphism is not rare and metamorphic rocks of many kinds abound in the earth's crust.

Given sufficient heat and pressure, fine-grained limestone will be changed into coarse-grained *marble*, a completely different rock. The metamorphic rock will be both harder and denser than the rock from which it was formed, as in coal where *anthracite* is the metamorphosed variety. Shale will become *slate*, harder and more emphatically foliated, and the metamorphism of sandstone fuses the individual sand grains into an extremely hard silica rock, *quartzite*. Distinctive coloring often results from the various cementing materials and impurities, such as carbon or iron particles in limestone causing the typical dark or reddish streaks in marble, or iron-rich cements in sandstone giving quartzite a pinkish hue.

Igneous rocks too may be metamorphosed, for despite their having been under considerable pressure in their original molten state, subsequent pressure and heat applied to the solid igneous rocks will cause them to change their form. A wide variety of rocks called *schists* and *gneisses*[2] are commonly a result of this metamorphism. The schist is characterized by the formation of mica or similar crystals that are aligned at right angles to the pressure. This lineation is visible to the naked eye and gives the rock a rough foliation, as well as a shiny, almost metallic appearance as the light reflects off the many cleavage surfaces (Fig. 18-17). Gneisses too exhibit a lineation of minerals, but typically are banded; that is, they show wide sharply defined bands of alternating dark and light materials. Streaking and banding, as in marble or gneisses, is also visual evidence of non-molten mass plastic flowage.

Theoretically, remetamorphism of metamorphic rocks is also possible but is often difficult to recognize without radioactive dating in the laboratory; any rock is susceptible to metamorphism given sufficient heat and pressure. Recently, we have come to consider seriously a process called *granitization*, which is essentially the final metamorphism of a wide range of rock into granite. If this occurs on any large scale at all, and many think it does, it helps to account for the great continental granitic masses that heretofore have been attributed to vulcanism alone.

The crust of the earth as we know it today then is a

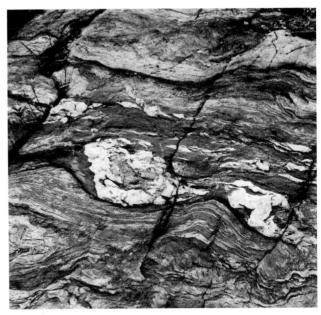

Fig. 18–17 Typical banding in schist; visible plastic flow lines speak of its reconstituted origin.

scrambled combination of these three types of rocks. The original crust may have been igneous as the heated earth cooled at the surface (if we are willing to accept such a theory). But immediately, weathering and erosion began to break this crust up into small particles and individual minerals that were carried off, sorted, and deposited as unconsolidated sediments, while at the same time, organic remains were accumulating in certain favored locations. As these became cemented and compressed into rock, they were frequently uplifted into mountains and continental masses. The forces exerted by such constant crustal deformation caused many of these igneous and sedimentary rocks to be metamorphosed. This cycle, repeated over and over, supplemented by fresh magmatic intrusions and continuing at this moment, accounts for the great variety of rocks that are encountered in the earth's crust, and especially in the continental regions. But notwithstanding this great variety, each of these rocks can be placed in one of the three basic rock categories: (1) igneous, (2) sedimentary, or (3) metamorphic.

---

[2] Schists and gneisses may also be formed on occasion by the metamorphism of sedimentary rocks.

CHAPTER 19

FOLDING

Who says hard brittle rock can't be folded? No one has ever seen it happen—it takes time and our attention span simply isn't long enough. But there has been some tremendously involved folding here in these Idaho mountains. It sometimes requires more than a little faith to believe that those parallel sedimentaries were laid down in horizontal beds below sea level when we encounter them now at 7000 feet (2134 m) and in every attitude but horizontal.

INTRODUCTION

Folding is a slow and deliberate process. No matter how carefully an observer watches the rock it cannot be seen to fold, and yet we know it does because we see folded rock all around us if we are but sensitized to the obvious clues. Nonetheless, the concept takes some thinking about. It scarcely seems possible that strong brittle rocks may be bent into sharp folds without first reducing them to a molten mass. Certainly, if we placed a slab of rock in a powerful vise, it would become cracked and broken. But there are examples everywhere of what must have been originally flat-lying beds with, for instance, marine fossils as evidence of their formation on the sea bottom, which we can observe now folded into a variety of flexures. *It appears that if strong lateral pressures are exerted over a very long period of time, while simultaneously the stratum involved is buried at some depth beneath the confining pressures of overlying layers, folding may be accomplished in all of its many forms.* In other words, both vertical and horizontal pressures are required—and a great deal of time in the formation of even modest folds.

Folding is a slow and deliberate process.

Folds, as we find them existent today, may vary from mild arches and domes scarcely visible to the eye to involuted and recumbent folds where the rock strata have been squeezed into tight corrugations that have overturned and even curled about themselves. They do not have to be symmetrical; in fact they seldom are. Folds are simply the product of compressional forces exerted laterally against the edge of flat-lying strata. The situation is much like pushing against the edge of a rug. If it is stiff and heavy, a broad warp may result, but if it is light and pliable, a whole series of parallel plications will build up, eventually collapsing over each other as the pressure continues (Fig. 19-1). The important element missing in the rug simile, however, is erosion that begins to bite away at the top of each flexure as soon as it is exposed at the surface.

THE FOLDING OF HARD ROCK

Although all types of rock can be folded, given the proper set of circumstances, it is the sedimentaries that are the most striking and readily observed, for they are by definition subject to particle-by-particle gravity accumulation into horizontal beds. Once this flat-lying character is distorted by even a few degrees, it becomes immediately obvious. Furthermore, since most sedimentaries ae formed in the marine environment, they almost surely contain sea-bottom fossils. Any change of sea level, any sort of crustal disruption (folding among others), will expose the fossils at higher elevations. This could, of course, mean that Noah's flood had covered these regions high on the flanks of Mt. Everest, Mt. Ararat, or Mt. Blanc, but it could also mean, as Leonardo da Vinci (1452–1519) pointed out in a demurrer to contemporary philosophy, that the fossilized shell

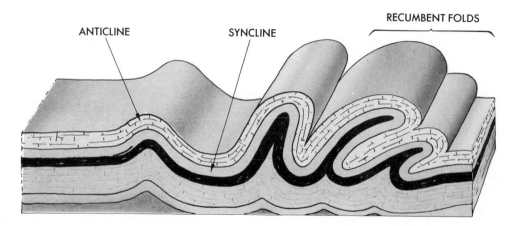

Fig. 19–1 Folding.

Fig. 19–2   Anticlines and synclines.

in the Tuscan hills was evidence of uplift, especially since the sedimentary stratum bearing it was contorted at a steep angle.

Anticline Versus Syncline

The general terminology applied to folds is *anticline* versus *syncline*. An upbowed stratum on a relatively small scale is called an *anticline*, while one that is bent downward is a *syncline*. If this same sort of folding occurs on a larger scale involving perhaps hundreds of miles, the strata are called *geanticlines* and *geosynclines* (the prefix from the Greek word *geos* = earth, emphasizing the size of such structures). Geosyncline carries with it the implication too of a sea-filled basin, great thicknesses of sediment accumulating, and their weight further depressing the bottom (Fig. 19-2).

It does not necessarily follow that anticlines are evidenced as hills on the earth's surface, for their formation is so slow that erosion may easily wear away the top of the fold as rapidly as it is uplifted. Nor is it uncommon to find a syncline at the top of a hill in present-day topography and an eroded anticline in a valley. For instance, the Ap-

palachian ridges and valleys of eastern Pennsylvania are some of the most striking features in the United States. Oriented northeast to southwest, these symmetrical linear ridges, all of the same height, are a seemingly endless roller coaster to the traveler driving cross-grain. Somebody shoved the edge of the rug, as it were, and multiple parallel flexures resulted. Actually, the thick sediments that had piled up in an ancient sea were relentlessly squeezed by an advancing crustal plate into major convolutions. But those immense folds were probably whittled away almost as rapidly as they rose by the agents of erosion, and we now encounter their mere roots. What seems to be a very straightforward series of dramatic but modest folds is in fact the residual remnants of the original—second generation corrugations formed as the steeply tilted strata was etched by running water. Today we see the harder rock layers standing up as ridges while the weaker soft rock strata are the valleys (Fig. 19-3).

Domes

Another variation on the concept of simple folding is the dome, especially easy to recognize when it has been

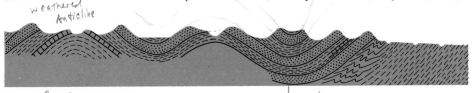

Fig. 19–3   A cross section of the Pennsylvania ridges and valleys.

Fig. 19–4 Breached dome. With the top of the dome cut away, the exposed butts of angled sedimentaries form a series of concentric circles—steep face toward the center and gentle slope away.

breached; that is, as its top erodes away, the remains of the sloping limbs are left exposed in the form of a circle (Fig. 19-4). An excellent example is the Black Hills of South Dakota where breaching has laid bare a hard core, which stands as the highest part of the feature. The scenic sheer cliffs, always facing the center of the hills, are the eroded butts of the strata originally overlying the dome.

Through the centuries 'erosional windows' have widened exposing in the Nashville and Bluegrass basins, the underlying supremely fertile limestone.

Even the gentle warp of the Cincinnati arch has encouraged a moderate wearing away at the crest. Through the centuries "erosional windows" have widened exposing in the Nashville and Bluegrass basins, the underlying, supremely fertile limestone. Surrounding each basin is a highland rim of unproductive shale sharply defining what is in essence an oasis of rendzina soil.

The extensive upland involving parts of Utah, Colorado, New Mexico and Arizona is by any perceptual measure a plateau; we call it the Colorado Plateau. Certainly it is high, averaging roughly 7000 feet (2134 m) and the rock structures, dominant in any panoramic vista for lack of masking vegetation in this desert environment, display a strongly developed horizontal stratification—flat topped buttes and mesas on every hand, canyon walls with parallel and varicolored laminations. And yet, if one could stand back far enough and view the "plateau" as a single construct, it would become evident that it is in fact a modest, if extremely large-scale, flexure. Most of those horizontally appearing rocks are angled ever so slightly.

The history of the Colorado Plateau discloses episode after episode of shallow seas accumulating layers of sediment with occasional periods of huge dry-land sand dunes. As these, now sedimentary rocks, were uplifted slowly enmasse, there appears to have been a minimum of contortion or sudden rending of the structure.

The Colorado River, with its sister drainage arteries, is part of the modern erosion cycle, but is probably less than 10,000,000 years old. Displayed along its convoluted

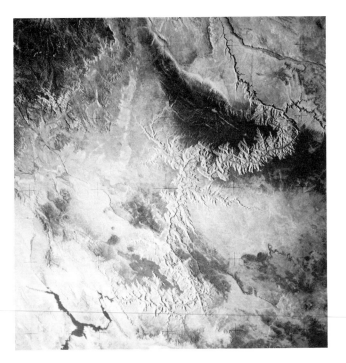

canyon walls is a revealing segment of the earth's history. At the bottom of the mile-deep Grand Canyon, the active river is cutting into dark Archeozoic rocks that may be as much as 3 billion years old (Fig. 19–5). The progressive layers to the top show development of fossil forms, each more sophisticated than its antecedents, and these allow for accurate dating. The surprise in the dating comes when it is discovered that hard Kaibab limestone at the canyon lip was formed 300,000,000 years ago. Certainly much more modern than down where the river is cutting, but shouldn't there be even newer strata on top of this?

Farther north in Zion National Park, where the Virgin River etches out another deep gorge, the familiar Kaibab limestone from the *top* of the Grand Canyon shows up here at the *bottom*. Rising above it along the canyon walls are another series of distinctive colorful, but younger,

Fig. 19–5 Viewed from 270 miles (435 km) high the Grand Canyon is still impressive. That's Lake Mead in the lower left-hand corner. At the upper left are elements of the Grand Stairway.

Fig. 19–6 Bryce Canyon is not really a canyon at all—it is badland erosion along the edge of an escarpment. The soft clays of the Wasatch formation were deposited in an old lake bed and it is their ready erosion that leads to the pinnacled panorama; but the added attraction that mesmerizes the viewer is a fairyland of delicate pink, rose and white coloration.

sedimentaries which culminate in a hard capping layer. This is the Carmel formation and it can be dated at only 170 million years old. Moreover, the Carmel or top of the Virgin valley, is lower by several hundred feet than the south rim of the Grand Canyon.

---

*I*n *Zion National Park, the familiar Kaibab limestone from the top of the Grand Canyon shows up here at the bottom.*

Another 50 miles north are the distinctive rocky chromatics of Bryce Canyon National Park (Fig. 19-6). The strata exposed at the *lowest* point in Bryce is the Carmel formation, and ranging above that ever newer rocks. The Wasatch formation at the rim of Bryce Canyon is about 55 million years old.

So there *are* more recent rocks than those exposed in the Grand Canyon and the supposition is that they were at one time in place overlying it. But where the Colorado River now flows was the top of an uplifted dome and erosion was most active at the top. Stripped away is everything above the Kaibab limestone. Along the flanks of the dome are the exposed, gently sloping residual butts of each rock strata that was removed from above. As one progresses northward from the Grand Canyon a series of distinctive cliffs are en-

countered—here are those angled stubs and each is younger than its predecessor. They have been called the *Grand Stairway* and designated according to dominant color: from south to north, the Belted or Chocolate Cliffs, Vermillion Cliffs, White Cliffs, Grey Cliffs, and Pink Cliffs. The Pink Cliffs are the Wasatch formation which dominates Bryce; the White Cliffs are the Navajo sandstones forming the pinkish stained white walls of Zion Canyon. (Fig. 19-7)

Recent Folding

Lining up in an east/west direction across southern Europe and Asia are a series of high mountains exhibiting some of the most spectacular and severe folding to be found anywhere in the world. The Pyrenees, Alps, Caucasus, and Himalayas are all basically folded mountains (Map IV). At one time, just south of these mountains in the general area of today's Mediterranean and Indo-Gangetic Plain, there existed a great sea-filled geosyncline called *Tethys Basin*. For millions of years, sediments from the adjacent continents filled this basin and formed many thousands of feet of sedimentary rocks.

---

*I*n *the general area of today's Mediterranean and Indo-Gangetic Plain, there existed a great sea-filled geosyncline called Tethys Basin.*

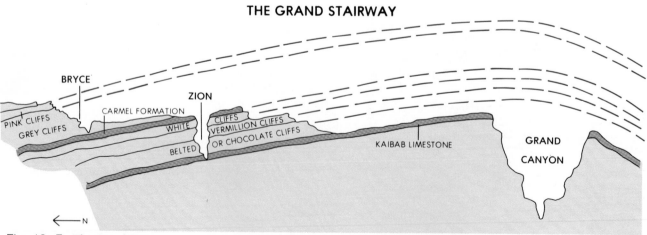

Fig. 19–7 The Grand Stairway.

Remote Sensing

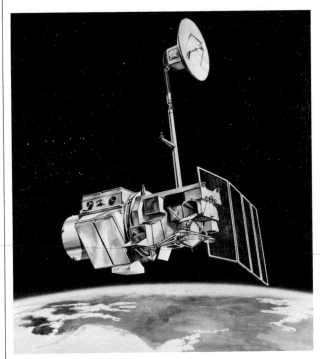

The innocent tourist viewing the Grand Canyon for the first time, finds that it is without question an impressive gulch, but nothing more than that. Somebody has to sit down and explain very carefully about geologic history and stripped domes before one becomes aware of the "big picture." Perhaps a view from a high-flying aircraft on a clear day would disclose immediately the surface geology of the entire four-state Colorado Plateau—or sending up a camera to take aerial photographs. What had been, at best, merely a conceptual or intellectual image at the local ground-level scale has suddenly been made lucid by viewing from afar. Now we are involved in *remote sensing*

The student reading this page is engaging in remote sensing. The eye, after which the camera is patterned, discerns from a distance a curious yet disciplined pattern of short black lines. This is recorded on the retina (the camera's sensitized film) and transmitted to the brain (the computer) where it is processed. The printed word is a code which the computer digests, unscrambles, sorts, analyzes, and stores.

Aerial cameras function, as does the eye, by sensing visible light and its reflectivity—so does the aerial TV camera. But the visible light portion of the total electromagnetic spectrum is only a tiny fraction of the whole. Cameras, lenses, filters, and films have improved immeasurably over the last few decades so that remarkable things can be accomplished, such as color photography which includes the near infra-red band just beyond what is apparent to the naked eye. But the real revolution in remote sensing has been in the areas of (1) sensors which function in an even greater range of wave-lengths and (2) platforms which carry those sensors—orbiting satellites such as Landsat 4 or Skylab.

All earthly surfaces absorb a percentage of solar radiation. There is a good deal of reflection involved, but those surfaces also become radiators themselves and emit energy back into space; however, diverse surfaces reflect, absorb, and radiate at varying rates of efficiency. It is these differences in energy reflection and radiation that the high altitude scanners record. Heat waves, for instance, are invisible yet we can feel them—so can the delicately calibrated sensor in space, zeroed in on the far infra-red band.

Data can stream back as rapidly as it is sensed or it may be stored on tape for later release. It can be translated into pictoral form or arrive in a raw digital state to be processed in the computer and spewed out as a coded pattern-map.

Radar is capable of going a step farther. Operating at radiowave frequencies, it is an *active* rather than a *passive* system for it sends out a signal which bounces off the earth and the intercepting sensor can give a reading as to both surface character and distance. One particular advantage over all other systems is that clouds, haze or night are of relatively minor significance in its operation.

Remotely sensed data have completely transformed the science of cartography, and each day we find additional relevant uses for this new and overwhelming horde of acquired information. Both daily and long-range weather forecasting are realities; so are things like forest management, agricultural monitoring, water pollution, resource inventories, archeological site identification, not to mention myriad military applications. The potential for use appears virtually unlimited.

Fig. 19–8 The Alps. It is not readily apparent from this general view but the Alps are folded mountains. Their great elevation is a result of massive and highly involuted folding of sedimentaries. Today's rugged detail, however, derives from ice erosion—both current and of the recent past—and it has made the reconstruction of the original flexures particularly difficult.

These rocks were caught in the vise of the African and Indian crustal plates drifting northward against the rigid shields of Europe and Asia. Mighty sinuosities resulted and not too long ago. They are today mountains of some magnitude and reflect in their involved, even recumbent folding, the great compressional forces that were brought to bear on original flat-lying sedimentary beds. That these great related ranges, extending half way around the world, are of fairly recent origin is evidenced by their extreme height. But this is being altered even now as erosion, especially intense glaciation, carves and hews their flanks, erasing some of the more obvious signs of original folding (Fig. 19-8).

Occasionally, in a road cut or on a cliff side, a perfect fold may be easily visible. Formed beneath the surface, erosion has not yet laid it bare, and its undisturbed outline is evidence of the manner in which folding occurs. Presumably, folding is going on in a great many places around the world today, but its progress is so slow that it is virtually impossible to measure. We know, however, from observing folds of all kinds in many places that folding has been a major force in the deformation of the earth's crust.

Occasionally, in a road cut or on a cliff side, a perfect fold may be easily visible. Formed beneath the surface, erosion has not yet laid it bare.

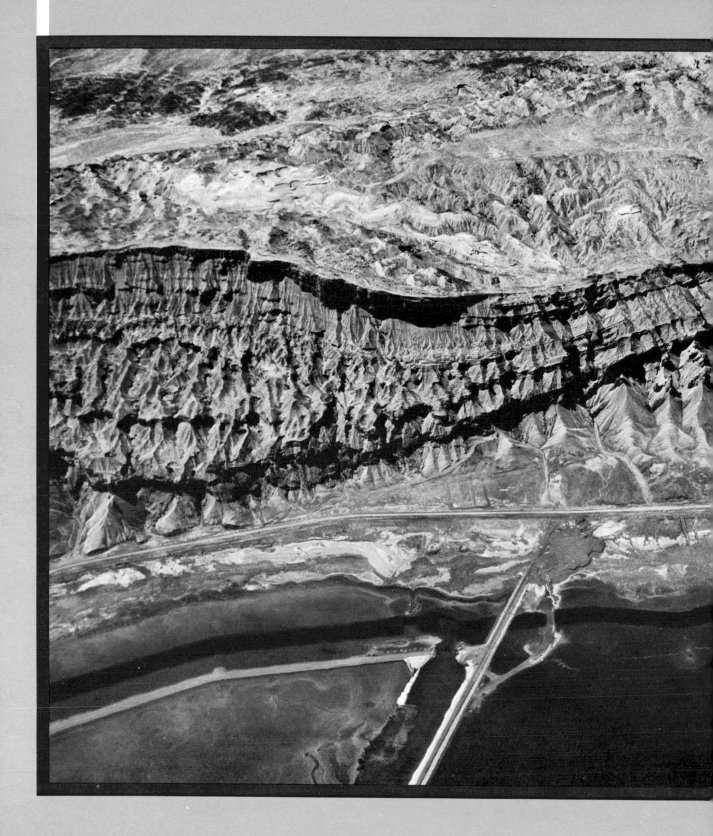

CHAPTER 20

FAULTING

Rising abruptly from the Dead Sea shore is the eroded, almost sheer west wall of the Jordan graben. The Dead Sea, occupying the lowest point in the valley floor, a record 1300 feet (396 m) below sea level at its surface, and with no outlet, is strongly saline, virtually bereft of life. Along the Sea margin are "mining" operations that extract potash and other useful minerals from the brine.

Large scale vertical displacement is not a rarity in this part of the world. This view is near Jebel Usdum (Sodom) at the extreme southern end of the graben.

INTRODUCTION

Faults are breaks or fractures in the earth's crust along which there is or has been movement. Nearly all rocks display cracks or joints resulting from their method of formation, such as the contraction of cooling lava or the alternating cooling and heating of surface rocks with the seasons. But if the fracturing is on a larger scale and there is readjustment along the plane of breakage, it is called a *fault*.

Faulting is a very common occurrence in the diastrophic episodes at the periphery of mobile plates.

The compressional forces at work in the earth's crust that cause folding also are responsible for faulting. But if the rock is too brittle or massive to relieve these pressures by folding or if the forces are exerted abruptly rather than slowly over a long period, then faulting will occur. Tensional dynamics too may pull the surface apart, opening up faults along which massive blocks can slip. Virtually all of this can be interpreted in terms of plate tectonics, either current or in the past; faulting is a very common occurrence in the diastrophic episodes at the periphery of mobile plates.

Much of this crustal faulting takes place beneath the surface of the earth or the sea and is traceable only through earthquake studies, but surface faulting causes certain characteristic landforms. Vertical displacement may be on a minor scale, forming escarpments or cliffs making the fault line easily visible, or massive blocks may be tilted up along the fault, resulting in high asymmetrical mountain ranges (Fig. 20-1). The Sierra Nevada in California, although slightly modified in places, is essentially a single tilted fault block. The sheer eastern scarp marks the fault line, and when viewed from Nevada or the Owens Valley is a spectacular sight in sharp contrast to the long gentle slope of the inclined western side of the block. It should not be inferred, however, that all or most of this displacement occurred at any one time. But the many lesser displacements must have been fairly recent to exhibit for us today such a sheer wall scarred by only moderate erosion.

SLUMP VALLEYS

We are probably a bit more accurate if we talk and think in terms of fault zones or slump valleys in many cases, for faults occur frequently in related clusters or series of parallel fractures. Rocks do not always break neatly along a single sharply defined plane; instead, like a green stick, there

Fig. 20–1 Elevated dramatically along an incisive fault line, the east wall of the Sierra Nevada towers above the Owens Valley.

fault plane

is a bit of mangling involved for some distance on either side of the less than scalpel-clean incision.

A selective erosional etching can enlarge the fault feature far beyond the magnitude of any original slippage.

Further, once any kind of vertical motion has occurred along a fault plane, soft strata previously protected underground, may be exposed to erosion. A selective etching can enlarge the fault feature far beyond the magnitude of any original slippage. Drainage too will be attracted to the newly made valley, and rivers are inclined to develop fault controlled patterns. It is certainly more than mere coincidence that the master streams of the California coastal range and most of their preponderate tributaries, follow southeast to northwest courses parallel to the San Andreas fault—the Santa Ynez, Cuyama, Salinas, San Benito, Russian, Eel, Mad, Klamath, and Smith. Many of these have opened up extensive flat-bottomed valleys, such as that of the Salinas; features that are fault dictated but not exclusively in and of themselves fault valleys.

When viewed from the air or on a map, a long fault zone may be an obvious landscape feature. Slim linear lakes or an invading finger of the sea may define the valley limits with great clarity. Volcanoes, furmaroles, and springs (many hot and mineralized) are surface evidence that a deep-seated fault has rather badly disrupted nature's underground plumbing system.

FAULT NOMENCLATURE

There is a certain basic nomenclature applied to the relative movements of elements on either side of a fault. If we are dealing with a line of fracture that is angled at something less than vertical, where one side, impelled by gravity, simply slides down the sloping fault plane, it is called a *normal fault*. But if compression forces it to skid up along the plane, the term *reverse fault* (or *thrust fault* when the angle of fracture is very low) is applied. Horizontal motion is also possible although seemingly less common, and this

NORMAL FAULT
A

REVERSE FAULT
B

STRIKE-SLIP OR WRENCH FAULT
C

Fig. 20–2 Fault nomenclature.

variety of fault is properly called a *strike-slip fault* (the British name *wrench fault* is perhaps more descriptive) (Fig. 20-2).

If two relatively vertical faults develop parallel to each other, then the block between may drop, causing a sheer-sided, flat-bottomed valley called a *graben* (German meaning literally grave or ditch). The Jordan River and the Dead Sea occupy the northern end of an extensive series of related grabens that include the Red Sea and the so-called Rift Valleys of East Africa (Figs. 20-3 and 20-4).

It is often difficult to determine whether a graben has

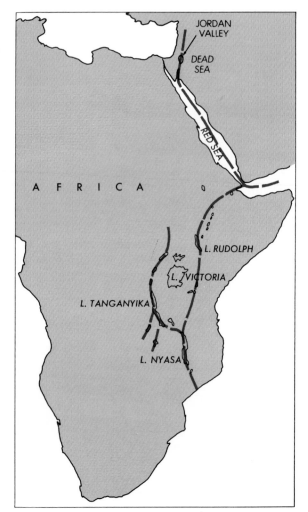

Fig. 20—3 Africa's rift valleys. Linear lakes and the Red Sea render visible on a map the extensive East African fault system.

Nevada is not a horst since it is merely tilted along a single fault plane, but the Great Basin, especially Nevada, displays many horsts. Here is a remarkable assemblage of faulted features extending for hundreds of miles over a wide area and all oriented in a roughly north/south lineation. Apparently tensional forces were applied in an east/west direction, and the resulting clusters of faults were formed at right angles to them. Movement along these fault planes over many years has caused innumerable horsts, grabens, tilted blocks, and enclosed basins at various levels. Continuing erosion has subdued the stark outlines of some of the uplands and partially filled the basins but in this arid region where erosion is somewhat retarded, the more recently formed features still retain much of their original character and are easily recognizable.

The name "playa", literally beach; these lakes feature more beach than water.

Often cut off one from the other and at different levels, each basin becomes a separate receptacle to collect the runoff from infrequent rains in its surrounding uplands. Periodically, broad sheets of water called *playa lakes* cover the basin floor, but they may attain only a foot or two maximum depth and evaporate away almost overnight. Hence the name "playa," literally beach; these lakes feature more beach than water. The normal aspect of the basin, except immediately following a rain, is a blinding, white, residual alkaline flat (Fig. 20-5).

EARTHQUAKES

There has been a popular image from as far back as the early Greeks, relating earthquakes and volcanoes. It has been observed that volcanic zones and seismic zones frequently show a good deal of coincidence. And, as we have learned from the constant press reports on Mt. St. Helens, "harmonic tremors" are a fact of life in the vicinity of live volcanoes, warning of new activity and impending eruptions. If this is the demonstrated case then it surely follows that similar low magnitude volcanic swarms anywhere in the

been formed by the middle block dropping or the blocks on either side being lifted, but the name graben still applies. In the upper Rhine Valley where the river is the boundary between Germany and France, it appears that although the Rhine flows northward in a graben, the Vosges on the west and the Black Forest on the east have been raised while at the same time the central block was slightly depressed.

If the central block is thrown up, resulting in an angular fault block upland, it is called a *horst*. The Sierra

Lake Baikal

One of the world's great grabens is that occupied by Lake Baikal in southern Siberia. The lake itself is 400 miles (744 km) long and 30 miles (48 km) wide, its surface 1495 feet (456 m) above sea level, and its bottom has been measured at 4250 feet (1295 m) below sea level. The towering sheer walls of the graben extend to 6600 feet (2012 m), displaying a total crustal relief of well over 2 vertical miles (3 km). So steep are its sides that for many years the Russian engineers constructing the trans-Siberian Railway were thwarted in every attempt at track construction along the western lake margin. Rather, they chose to approach the northern shore via the Angara River where it exited the lake and lay temporary tracks across the winter ice to the valley of the entering Selenga River. In the summer freight and passenger traffic reverted to barge and ferry. The western shore track was completed eventually, but only after great effort and expense.

Baikal's claim as the world's deepest lake at 5745 feet (1751 m) certainly stacks up impressively alongside those in the United States: Crater Lake (volcanic origin) 1932 feet (589 m), Lake Chelan (glacially scoured valley) 1419 feet (433 m).

Fig. 20–4 The distinctive Y-shaped head of the Red Sea graben with Sinai in the center. To the east, the Gulf of Akaba is a mere extension of the greater Jordan graben with the Dead Sea and Lake Tiberias/Sea of Galilee (almost obscured) both below sea level.

Fig. 20–5 Sevier Lake, Utah is a playa. In the spring, melting snows from the surrounding mountains will sometimes supply sufficient water to support a thin sheet in the basin bottom. But it will usually evaporate in mere days or weeks and the lake reverts to its normal character as an alkalai flat.

world are at the very least a tocsin—perhaps magma is in motion, working its way toward the surface. Such a contingency should not be ruled out.

As we have learned from constant press reports on Mt. St. Helens, 'harmonic tremors' are a fact of life.

However, for all that a rumbling volcano can shake up the immediate countryside, the *major* cause of seismic activity is faulting, and to a large degree faulting is also responsible for the occurrence of volcanoes. Diastrophism and convulsive mountain building are the resultants of crystal plate collision. A correlation of the plate tectonic map (Chap. 18) with that of volcanic and seismic activity (Chap. 21) will quickly demonstrate this predictable coincidence.

The margin of the Pacific Ocean in its present outline is a very unstable region marked by widespread faulting and continuing slippage and readjustment. From the Andes and the Sierra Nevada to Alaska, and from Kamchatka, Japan, and Indonesia to New Zealand and Antarctica, continuing seismic shocks must be accepted as an unalterable envi-

ronmental stringency even as tornadoes are in other parts of the world.

Crustal movement along existing faults or the formation of new faults is not a gradual thing. Pressures may build up slowly over many years, but when the rocks can no longer sustain that pressure, the break or the slippage comes with explosive suddenness. A part of the shock comes from what is called *elastic rebound* as rocks that have been somewhat deformed as they have attempted to sustain the increasing pressure suddenly snap back to their original shape. The result of this is an earthquake with the shock waves rolling out in all directions from the epicenter.

Displacement of mass may occur vertically or horizontally or both. If the fault is at the surface, vertical displacement is sometimes visible in the sudden formation of low cliffs, or if it is horizontal, fences, streams, and roads crossing the fault will be forced out of line. The infamous San Francisco earthquake of 1906 involved horizotal movement along the *San Andreas* fault that extends over half the length of California and well out to sea north of San Francisco (Figs. 20–6 and 20–7).

The west side of this deep-seated fault is the Pacific plate moving northward as it apparently has been for an extended period of time. A glance at Fig. 20–6 will show that Baja California, separated from the coast of western

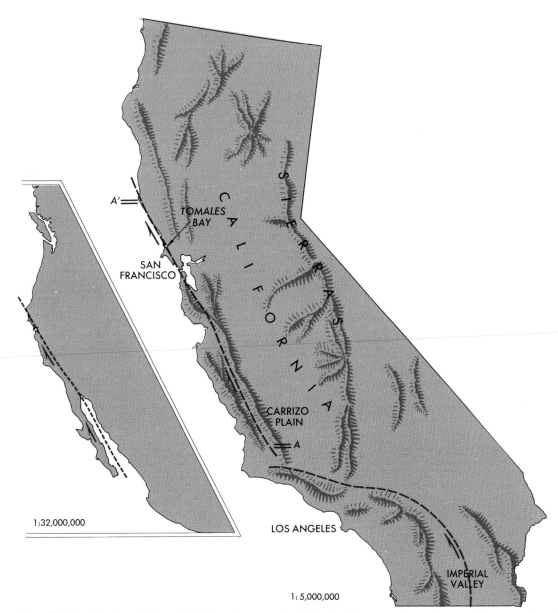

Fig. 20–6 The San Andreas fault. That the west side of the fault is progressing to the north is unquestioned. Reasonable correlation of the same rock structure has been made at A and A'.

Mexico proper by the narrow Gulf of California, would fit rather precisely into the configuration of that coast if it were shifted 200 to 300 miles (322 to 483 km) southeastward. The Gulf and its northward extension, the Imperial Valley, are both a part of the greater San Andreas fault sys-

tem, and it would appear that Baja and the entire western margin of California have been drifting northward in abrupt spurts along the line of the fault zone. Several attempts have been made to establish the rate of movement by matching rock strata on either side of the fault. To cite two fairly

Earthquake Swarms

What business does a tiny area in north central Arkansas have experiencing well over 19,000 earth tremors since January 12, 1982? So far only about 100 have been sensed by the inhabitants (one on January 21, 1983 reached a Richter magnitude of 4.5), the rest have been so minimal that only the seismometer could pick them up. But earthquakes are supposed to be a product of crustal plate margins and here we are about as far from the margin as one can get.

Moodus, Connecticut, is a long way too from where one would anticipate a trembling of the earth, but the good folks there have sensed, one way or another, over 500 of these same type of miniquakes since June 1982. And with their temblors comes "the Moodus noise" described as resembling far-off thunder or muted cannon fire.

There was a damaging earthquake in central Connecticut on May 16, 1791, long before any method of measuring energy quotient or epicenter was known. And only about 20 miles (32 km) northwest of the Arkansas tremors is New Madrid, Missouri where another unmeasured quake in 1811 may have been the most violent to ever afflict North America. So midcontinental plate earthquakes are not unknown—Charleston, South Carolina was wiped out in 1886. But does this persistent crustal trembling portend major disaster? We just don't know.

In the Mammoth Lake region of California over 3000 temblors were measured in one three week period and larger seismic jolts periodically for two years. The educated guess here is that since Mammoth is located on the rumpled edge of a crustal plate and has a long geologic history of sporadic active vulcanism, that the earthquake swarms signal rising magma from below. The same interpretation has been applied at Matsushino, Japan where 700,000 quakes were felt between 1965 and 1967.

The point that is obvious is that multitudes of miniquakes are not our standard everyday earthquake caused by sudden slippage along a fault plane. If this kind of symptom does indeed presage the abrupt movement of magma toward the surface then should we anticipate a Kilauea in Arkansas or an Old Faithful at Moodus? There are still a plethora of unknowns regarding the variables of crustal palpitation; and the earthquake geologists whom we entrust to make official diagnoses and, if not prescribe cures at least to sooth us appropriately, are the first to admit exactly that.

Fig. 20–7   From the air it is easy to trace the San Andreas fault across the Carizzo Plain.

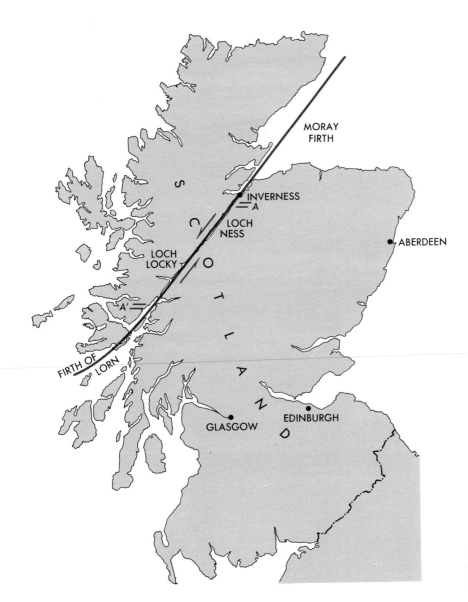

Fig. 20–8 The Glenmore fault. At one time strata A and A′ were contiguous across the fault.

well-substantiated studies: one indicates 350 miles (563 m) in 150 million years and the other about 220 miles (354 km) in 50 million years or an average speed of something on the order of 3 miles (5 km) per million years.

Several other large-scale wrench faults have been recognized and measured. One in the Philippines and another in New Zealand are currently active, but perhaps the best studied of them all, the great *Glenmore* of Scotland, seems to be quiescent, although statements of this kind

should always be made with caution. A cursory look at the map of Scotland will draw one's attention to the virtual cutting off of the northern highlands by a deep trench. Occupying the distinctive northeast to southwest linear valley are Loch Ness and Loch Locky, connected with each other and the sea at either end by the Caledonian Canal to form a navigable route across the heart of the highlands. The trench, although rasped out deeply by moving tongues of ice, is a shear zone, and proof of extreme horizontal movement along its plane has been demonstrated in many locations (Fig. 20–8).

Tidal Waves

Coastal areas, even those well outside of seismic zones, are subject to the effects of earthquakes at the sea bottom. Their shock waves are not only transmitted through the earth but through the overlying water as well. Such ocean waves, emanating from the sea depths, travel outward in all directions from the earthquake epicenter at speeds of hundreds of miles per hour and may devastate coastlines 1000 miles (1609 km) away.

One of the major problems is the detection and warning. Ships at sea are completely unware of a passing *tsunami*,[2] for the wave is only a few feet high and 100 or more miles (161 km) long. It also may be traveling at great speed (the deeper the water, the greater its velocity), yet the ship has no sense of abnormal sea behavior. However, as the shore is approached and water shoals, the leading edge of the tsunami slows while the mass behind runs up its back to build a high offshore swell. Moments later it breaks on the beach and surrounding ocean frontage to heights of 50 or 100 feet (15 or 30 m).

The Pacific with its great deeps and plethora of submarine volcanoes is the most active tidal wave ocean—at least one per year has been detected since 1800. On the average, a major tsunami can be expected about once every 10 years, but Hawaii, in its vulnerable midocean position, has experienced 37 in the past 125 years.

Sirens warned the city residents to evacuate. Some did and some didn't.

In April of 1946 an Aleutian trench earthquake triggered a wave that five hours later broke over Hawaii centering on the city of Hilo. Such was the damage and loss of life that Hilo set up a warning system essentially monitering all seismic activity around the Pacific basin. As a re-

[2] "Tidal waves" is something of a misnomer, and the Japanese term "tsunami" is coming into more frequent use. These may also be caused by volcanic explosions beneath the sea.

sult, within two hours after the great Chilean earthquake of 1960 sirens warned the city residents to evacuate. Some did and some didn't. When the tsunami arrived it came in three echelons, about 3 hours apart. The first two were 4 and 9 feet (1–3 m) high, the last a 20-foot (7 m) killer. Of the 61 dead, most had rushed downtown to view the excitement following the first waves.

Twenty-four to 36 hours after this same Chilean quake, coastal areas as far distant as New Zealand and Japan had experienced destructive high seas. The Anchorage quake in March 1964, similarly caused damage along the Oregon and California coasts.

In August 1977 a submarine earthquake measuring 8.3 on the Richter scale triggered a huge wave that swept the Indonesian island of Sumbawa. In July 1979 Lomblen, also in Indonesia, lost four villages and 700 of their inhabitants; and only two months later a tsunami resulting from an 8.0 earthquake northwest of New Guinea all but wiped out the little island of Yapen. Indonesia is immediately astride the Pacific marginal seismic/volcanic zone and tiny islands with coastally oriented fishing villages have proven to be particularly susceptible to seaborne disaster.

Building Construction

Buildings of reinforced concrete, of moderate height, and constructed on bed rock withstand the whipping, rolling action of earthquakes fairly well. But brick and masonry, especially if built on a loose fill base, are likely to be totally destroyed. Frame construction stands up nicely because it is flexible, although chimneys, plaster, and plumbing are usually damaged (Fig. 20–9).

One reason for the great loss of life in such quake-prone regions as Turkey and Iran is the almost universal use of unreinforced stone and baked mud as building materials for homes in the rural villages. But the record all-time earthquake catastrophe, in terms of resulting casualties, occurred in the Chinese province of Shenxi—it too was related to the use of fallible house structures.

The people lived like rabbits in a warren, their snug cave homes warm in winter, cool in summer, and eminently collapsible in an earthquake.

---

in an earthquake. In 1556, 830,000 folks learned about that the hard way.

Land Slides

Landslides too are often a destructive result of earthquakes. In the Hebgen Lake, Montana, quake of August 1959, landslides blocked a sizable river to form a lake, and in the process, a number of people were killed (Fig. 20–10). Or, in another variation, landslides may thunder down steep mountain slopes and into narrow arms of the sea or lakes behind confining dams. The damage is not the immediate result of the moving earth but rather the secondary effect of the resulting wave (called a *seiche*). Many Alaskan fjords exhibit a sharp line well above the water's edge below which no trees are visible, the tree line marking the highest point reached by a recent devastating wave. In the case of a dam-

Fig. 20–9   Most of the city of Popayan, Colombia was destroyed by a violent earthquake, April 2, 1983. But even cathedrals are not always immune to "acts of God."

Shenxi is the heart of north China's loess region where the powdery wind-blown soil has piled up deeply and compacted over the centuries. Caves can be cut into the valley sides with the simplest of implements and here the people lived like rabbits in a warren, their snug cave homes warm in winter, cool in summer and eminently collapsible

Fig. 20–10   Just to the west of Yellowstone National Park a 1959 quake induced a mammoth slide into the Madison River canyon. A new lake (left) was impounded behind the sudden dam, imaginatively named Earthquake Lake. Over in the park, the jolt cleared up some underground plumbing clogs and new and dormant geysers gushed to life.

formed lake, the waves may either rupture the dam or overrun it. In either situation settlements in the valley below are swept away in minutes.

By far the greatest disaster of this sort was experienced in the epochal Peruvian earthquake of May 31, 1970. Of the 50,000 estimated deaths from all quake causes, 20,000 were in the two mountain resort villages of Yungay and Ranrahirca. A U.S. Geological Survey report tells the story:

"In the beginning the avalanche, triggered by the quake, started with a sliding mass of glacial ice and rock about 3000 feet (914 m) wide and about a mile long on the nearby sheer slopes of Nevadas Huascaran (at 21,800 feet (6645 m), Peru's highest mountain).

It swept downward, dropping 12,000 feet (3658 m) vertically in a distance of 9 miles (14 km) and hit the town of Yungay at speeds up to 248 miles (399 km) per hour. As the ice and rock fragments rushed down the mountain, frictional heat changed the ice to water and the debris to a muddy mixture. By the time the mass reached Yungay it is estimated to have consisted of about 80 million cubic feet (2265,340m³) of water,

mud, and rock. A mud flow of such proportion originating from an ice mass, indicates a geologic process never before recorded."

EARTHQUAKE FORECASTING

Wouldn't it be helpful if we could monitor some kind of physical happening or measure something concrete that would give us a handle on the continuing question of "what is the likelihood of an earthquake occurring here?" (Fig. 20-11). In the past some California old-timers talked knowingly about earthquake weather; or a more bonafide expert would break into print with a learned generalization that "pressures have been building up ever since the big Frisco shake in '06 so another is likely anytime now—maybe." But in the last several years there have been rather sudden advances in the recognition of measurable phenomena. With the Russians, Japanese, and Chinese leading the way, researchers are now examining a number of promising lines of investigation.

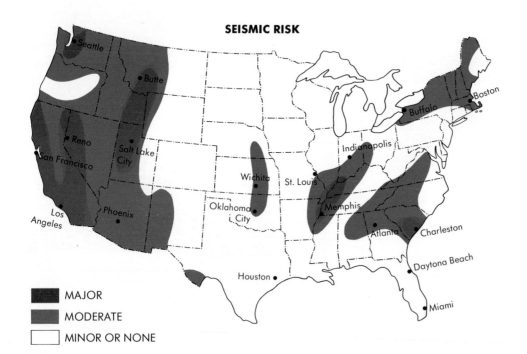

Fig. 20–11   Seismic risk.

"What is the likelihood of an earthquake occurring here?"

One approach is in the area of *dilatancy*, tiny cracks opening up in rocks under extreme pressure. This is quantifiable in itself but so are related mechanisms such as changes in rock volume, magnetism, electrical resistivity, and the speed of artificially induced seismic waves. Delicate tilt meters are now used to detect even minor changes in the level of the land or well water, and sudden increases in trace amounts of radon (a gas resulting from radioactive decay) in groundwater appears also to be an indicator of impending earth movement. The Chinese speak very seriously about abnormal animal behavior prior to a seismic episode—jumpy pandas, nervous yaks, and chows without appetites. But the world took a second look at Chinese methods on February 4, 1975 when they predicted a major earthquake at Haicheng (southern Manchuria), evacuated the city and saved an estimated 100,000 lives as a 7.3 intensity shock destroyed it. That forecasting is not yet a precise science was demonstrated about a year later when a quake of magnitude 8 shook Tangshan [150 miles (241 km) west of Peking] to the ground killing 655,000 people with no perceived warning at all.

The Chinese speak very seriously about abnormal animal behavior prior to a seismic episode, jumpy pandas, nervous yaks and chows without appetites.

An old theory that has achieved some immediacy with a 1983 report by two respected geologists, deals with the relationship between moon, tide and earthquakes. Given a fault line with built-in weaknesses, wouldn't it seem logical that the gravitational forces of moon and tides might set it off? However, studies attempting to correlate all of this on a worldwide basis have proved inconclusive at best. The recent investigation hedges its concerns by concentrating only within a limited area in southern California,

and looking especially at the San Andreas fault and related others oriented roughly north-south. In reading back along the historical trail of known quakes in this region the researchers found a remarkable coincidence. They are not to the point yet of predicting the next earthquake, merely likelihood. But the report goes so far as to point to November 1987 as a period of "extremely high potential."

But it does appear that we are on the right track. There are now things to measure that seem to be valid indicators and others will undoubtedly surface. The U.S. Geological Survey along with several universities and research institutes have set up seismic monitoring grids utilizing networks of automated sensors. There are currently two in California, one in New York, and one in South Carolina. Money is a continuing constraint, however, on both their efficient operation and projected expansion nationwide.

One provocative phenomenon that could probably stand a bit of close monitoring is the Palmdale bulge. Palmdale is a sleepy little community in the southern Mojave desert about 35 miles (56 km) northeast of Los Angeles, unremarkable except for the fact that it is situated directly atop the San Andreas fault. But in the early 1960s it was discovered that dramatic vertical movement was occurring here—a displacement of as much as 10 inches (25 cm) in Palmdale itself, which proved to be the center of a 120-mile (193 km) long oval-shaped bulge (Fig. 20-

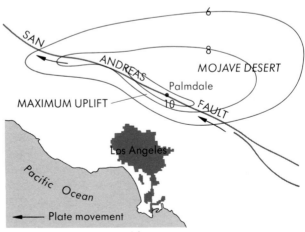

Contours represent elevation change in inches
Fig. 20-12   The Palmdale Bulge.

12). Could this be a symptom of an impending San Andreas paroxysm? Certainly, there is dilatancy and earth tilt here. Or is this merely a "false pregnancy"? At the very least we should be wary. A major earthquake this close to heavy population centers would inevitably result in hundreds of thousands of casualties and billions of dollars in property damage. But, so far, southern Californians remain casual and the city of Los Angeles continues to plan for a super jetport to be built soon in the southern Mojave.

Controlling Shocks

If we want to look even farther down the road, there is the intriguing thought that perhaps we can work our way out ahead of the game by *controlling* quakes rather than by merely warning of the inevitable. People have been thinking constructively about just that. One idea is to set off deliberately a series of minor tremors along a fault to relieve pressure and thus nullify a serious quake. It has been noted that the high-pressure injection of steam into oil wells or of noxious waste materials into underground bores has accidentally set off a whole series of barely perceptible shocks. So if we were to drill hundreds of 3 mile (5 km) deep bores along both sides of the San Andreas fault and then inject water into them alternately under great pressure, something interesting might happen. We are not yet prepared to find out exactly what. But in the interim we should try to keep the farm animals happy and tranquil in the barn so that they won't spook readily, and maybe we can forestall serious tremors in that fashion until other more scientific methods can be perfected.

A symptom of an impending San Andreas paroxysm or merely a "false pregnancy"?

Living in a seismic region can be a dangerous adventure, for earthquakes still occur suddenly with no advance warning; and although many are harmless, it is likely, even inevitable, that sooner or later a killer will come.

No part of the world is immune to earthquakes, but these are two great zones of almost constant activity. One

Year	Site of Quake	Richter Energy Quotient
1556	Shensi, China	-
1755	Lisbon, Port.	-
1811	New Madrid, Mo.	-
1857	Southern California	-
1872	Owens Valley, Calif.	-
1886	Charleston, S.C.	-
1899	Yakutat, Alaska	-
1906	Andes of Columbia and Ecuador	8.6
1906	Valparaiso, Chile	8.4
1906	San Francisco, Calif.	8.3
1911	Tien Shan, Sinkiang, China	8.4
1920	Kansu, China	8.5
1923	Sagami Bay, Japan	8.2
1927	Nan - Shan, China	8.3
1933	Japanese Trench	8.5
1934	India, Bihar - Nepal	8.4
1939	Chillan, Chile	8.3
1946	Honshu, Japan	8.4
1950	North Assam, India	8.6
1960	Central Chile - 3 major shocks	8.3, 8.4, 8.9
1964	Anchorage, Alaska	8.6
1970	Yungay, Peru	8.4
1976	Tangshan, China	8.0
1977	Northwest Argentina	8.2
1977	Sumbawa, Indonesia	8.3
1979	Pacific, N.W. of New Guinea	8.0

Fig. 20–13 Magnitudes of great earthquakes.

is the aforementioned Pacific littoral and by extension the West Indies and the largely submarine Atlantic ridge, and the second extends east/west through the high mountains of the northern Mediterranean, Asia Minor, the Caucasus, and the Himalayas.

A numerical scale devised by C. F. Richter in 1935 has become the definitive measure of earthquake magnitude. It expresses as closely as possible the energy released as deformed rocks, no longer able to withstand the strain, rebound. Some of the world's greatest earthquakes, in terms of energy quotient, are listed in Fig. 20-13. Many of those prior to 1935 were measured by more primitive means and have been assigned an estimated Richter magnitude. The scale is open-ended, but no known quake has yet reached 9.

CHAPTER 21

VULCANISM

The Basin of Mexico is almost surrounded by impressive snow-capped volcanic peaks, and the highest of these is Popocatépetl at 17,887 feet (5451 m), here viewed from Puebla. The concave flanks and pointed crest are classic signs that the great mountain is a composite cone built from alternating explosive and quiet eruptions. However, it remains sporadically active and the murcurial personality of this type of volcano is of some concern to the residents of both Puebla and Mexico City.

INTRODUCTION

The term "vulcanism" implies volcanoes—and they are, to be sure, a spectacular product of it. But vulcanism involves much more than mere volcanoes. *Any invasion of the earth's crustal zone by magma from below is properly called vulcanism.* If the magma pushes its way far up into the crustal strata but does not reach the surface, it is termed *intrusive* vulcanism, but if the magma flows out onto the surface via volcanic or other vents, it is called *extrusive* vulcanism.

The question may well be asked, "Why does molten material from the earth's interior force its way into and through the hard rock of the crust?" This is one of the questions that has vexed geologists for many years—not only "Why does magma invade the crust?" but "Where precisely does the magma originate?" It seems reasonably certain that release of pressure on the interior of the earth by deformation and fracturing of the crust triggers the outward movement of magma, but whether it originates initially at the earth's core, the mantle, or even the crust itself is difficult to determine. The core is the least likely in this speculation. Probably radioactive heat in the upper mantle or the lower crust is the magma source for most volcanoes.

Nearly all major mountain series are underlain by intrusive masses.

Thus nearly all major mountain masses are underlain by intrusive masses, and frequently there is some sort of extrusive vulcanism in evidence as well. Mountain-building mechanisms such as folding and faulting cause vulcanism, but vulcanism, once set in motion, can be a potent mountain-building mechanism in its own right. Massive intrusions may lift and warp the surface layers, and extrusive magma, finding its way to the surface through faults, can build huge piles of volcanic rock. Also, intrusions may be exposed by the wearing away of the softer overlying strata by erosion, and the hard igneous structure will stand up as highlands (Fig. 21-1). The Adirondacks and mountains of New England are of this type.

Fig. 21-1   In the massif central of France, ancient volcanoes have eroded away completely leaving behind only their former vents filled with solidified hard lava. These volcanic plugs or necks now stand dramatically above the current landscape. At LePuy, the St. Michel d'Aigulhe monastery has taken advantage of this prominent locale—but there are 265 stairs in the ascent.

INTRUSIVE VULCANISM

Batholiths and Stocks

The largest of the intrusive masses is the *batholith*, found to underlie every large mountain system and frequently exposed at the surface as the roots or cores of ancient mountains that have eroded away. But the batholith is not merely the upper margin of the monolithic lower crust. It is a lobe that has forced its way well beyond that and into the sedimentary (or altered rock) layers that form the continental epidermis, probably because of lessened pressure as a result

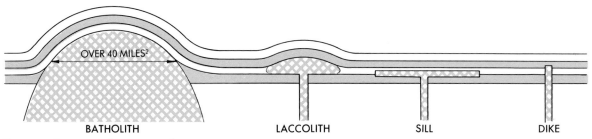

Fig. 21–2 Types of intrusive vulcanism.

of diastrophism (Fig. 21-2). Somebody has picked 40 square miles (104 km^2) out of a hat and proclaimed that anything smaller, although precisely the same kind of form, shall be called a *stock*. At least this gives us some idea of minimal size, for most batholiths are considerably larger.

The batholith and stock are coarse-grained igneous or plutonic rock that has obviously cooled slowly deep underground. At their point of contact with the stratum above is a metamorphosed aureole or *baked zone* where the combination of heat, pressure, magmatic fluids, and gases have commonly given rise to a concentration of useful minerals. Much of our knowledge of the batholithic periphery has been gained through the experience of mining these zones.

But nobody has ever seen the underside of what must be regarded as an essentially bottomless structure, nor has intrusive rock been seen to actually crystallize as extrusive lavas have. We are, as usual, dealing again with conjecture when we theorize as to the ultimate reason for batholith intrusion as well as the whereabouts of the rock replaced by the intrusion.

Laccoliths

On a much smaller scale than the batholith are several other kinds of magmatic intrusions. One of these, the *laccolith*, has some of the character of a small batholith or stock except that it is fed from below by a tube cutting across the overlying strata like the conduit of a volcano. However, where the volcanic bore gets into the surface rocks, magma rising up the laccolith feedpipe loses some of its impetus and spreads out between two flat-lying rock layers, eventually forming a lens-shaped mass that lifts and bows the surface into a dome (Fig. 21-2).

Sills

Similar to the laccolith but much more common is the *sill*. In this case the invading magma spreads out as a flat sheet, often for many miles, but does not throw up the overlying layers (Fig. 21-2). Sills come to our attention when erosion exposes an edge as a sheer wall, often darker and harder than the surrounding rock. The downcutting Hudson River has revealed a sill in the Palisades that is particularly striking because of the columnar jointing of the cooling magma. Another major feature standing out in a moderately subdued landscape is the Great Whin Sill of Northumberland, England. The black escarpment describes a 100-mile (1061 km) crescent from the Pennines to the sea, and since its steep side faces northward, it was seized on by the Romans as the foundation for Hadrian's wall (Fig. 21-3).

The Great Whin Sill describes a 100 mile (1061 km) crescent from the Pennines to the sea, and was seized on by the Romans as the foundation for Hadrian's wall.

Dikes

If the invasion of the surface layers is in the form of a thin sheet cutting sharply across all horizontal strata but falling short of intersecting the surface, then we have a *dike*. Visualize a sill turned on edge, except in the case of a dike, it draws magma from its entire lower edge rather than from a bore or conduit. Dikes often occur in swarms, most frequently radiating from a volcanic vent. If the resulting igneous rock is harder than that which surrounds it, erosion

Fig. 21–3   The Great Whin sill.

may lay bare the solidified magma in the volcano conduit as a *neck* or *plug* and the radiating dikes as long narrow ridges. Ship Rock, New Mexico, is an outstanding example (Fig. 21-4).

Pegmatites

It should be noted that although the originally molten rocks involved in intrusives are properly classified as plutonics and should therefore display a coarse granularity, only the batholith or thick sill is highly typical. Generally, the smaller intrusives isolate the magmatic melt into such thin or limited quantities that cooling is quite rapid and crystal formation is impaired. However, it is possible that a long intrusive finger may keep its contact with the molten mass below and selected gases and fluids will find their way into this backwater *(pegmatite)*. Occasionally, crystals will grow to huge size under these unusual conditions. In South Af-

Fig. 21–4   An igneous dike once occupied this slot. But it had the fatal flaw of being strongly jointed, and frost wedging ripped it out of the wall while ignoring the solid rock. Along the left side a portion of the dike remains displaying its distinctive horizontal jointing.

rica single mica crystals have been encountered measuring 10 to 15 feet (3 to 5 m) in width.

EXTRUSIVE VULCANISM

Volcanoes

Volcanoes are, of course, of particular interest not only because of their unpredictable explosive habits but because of their impressive structure. They tend to follow fault lines, and thus the world's great volcanic zones are coincident with the world's great seismic zones. Faulting, earthquakes, and volcanoes go together, and faulting is the ba-

sic cause for the other two. The Pacific's unstable margin is often popularly referred to as the *"Pacific Ring of Fire"* because of its extensive volcanic activity. Everywhere cones of all sizes, both active and dormant, are evidence of continuing vulcanism (Fig. 21-5).

It is never really safe to refer to a volcano as "dead." *Dormant* is the better term and should be used if there has been no eruptive activity in historical time. But even historic time is not always sufficient. In 79 A.D., Vesuvius, a not very imposing landform half hidden inside the broken remnant of the older Monte Somma, had never erupted in the memory of humans. There were not even any fanciful legends of vulcanism, so long ago had been its most recent activity. Yet in that fateful year Pompeii was buried in a

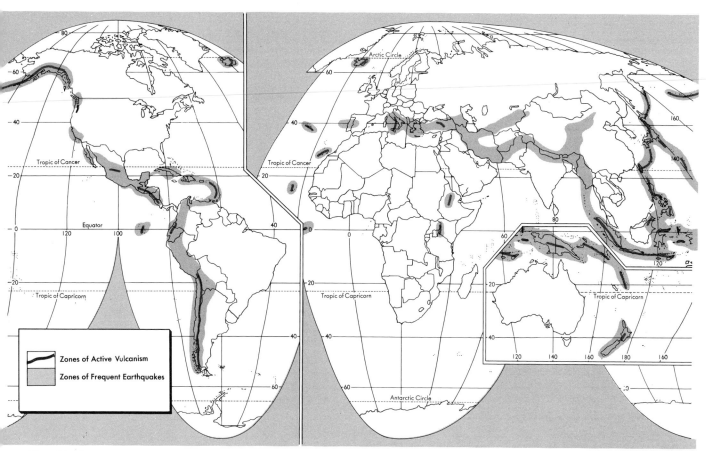

Fig. 21–5 The close coincidence of volcanic and earthquake zones.

great blast of fiery ash and Herculaneum was overrun by repeated flows of hot mud. Then for sixteen centuries there were only ten additional major eruptions. But in 1631, after 130 years of quiescence, Vesuvius began its modern more frequent eruptive cycle marked by increased lava flows that had been of little consequence earlier. So today Vesuvius is by any measurement active, its personality changed considerably from its former self; and if Somma is any evidence, there have been many times in the distant past when Vesuvius and its precursors have surprised the folks along the Bay of Naples.

There have been many times in the distant past when Vesuvius and its precursors have surprised the folks along the Bay of Naples.

The Cascade range, paralleling the U.S. west coast from northern California to British Columbia, is pimpled with a host of alabaster volcanic cones brooding in silent majesty. Mt. Rainier is the mightiest of the lot at 14,410 feet (4392 m) but there are other sizable glistening peaks, visible in good weather from hundreds of miles away: Shasta, 14,161 feet (4366 m); Adams, 12,307 feet (3750 m); Hood, 11,245 ft. (3427 m); Baker; 10,750 feet (3277 m); Jefferson, 10,495 feet (3199 m); Lassen, 10,457 feet (3187 m); 3 Sisters, 10,453 feet (3186 m); Glacier 10,436 feet (3181 m). Mt. St. Helens in the southwest corner of Washington was the smallest, only 9,671 feet (2947 m). As of May 18th, 1980, it abruptly became an 8,400 feet (2560 m) peak with the north side blown completely away and a yawning new crater 3000 feet (914 m) deep (Fig. 21-6).

St. Helens is a hot-blooded youngster that has slumbered beneath a mantle of snow with uncharacteristic calm for the past 127 years. It is so new by geologic measures that it displayed an almost perfectly symmetrical form, not even sculpted by the glaciers that disfigure its Cascade neighbors. But this perfection of shape tells the vulcanologist that there has been relatively recent activity. The evidences of chronic volatility are many—ancient cinder layers 37,500 years old, mudflows and lava eruptions repeatedly

Fig. 21–6   May 18, 1980—Mt. St. Helens belches her ash plume into the sky.

over the last 18,000 years. The modern cone is probably about 2500 years old built of alternating depositions of cinder, ash and lava. John C. Fremont witnessed an 1843 eruption and that was followed by over a decade of annual effervescence. Its last recorded outburst was in 1857.

St. Helens is a hot-blooded youngster that has slumbered for the past 127 years.

The 1980 St. Helens' eruption was a pyroclastic event, meaning that there was no liquid lava flows but rather the release of immense gaseous pressures out the side of the mountain. The result was solid ejecta of all sizes, from dust to jagged boulders, blown downslope. Some have estimated that the roughly 1.45 cubic miles (4.1 km^3) of solid debris, blasted to the north at upwards of 250 miles per hour (402 km per hour), was the largest avalanche in recorded geologic history. By carefully calling it an avalanche, they point up the fact that relatively little dust got up into the atmosphere because the "barrel of the gun", as it were, was pointed horizontally. There *were* huge clouds of dust in the atmosphere. It *did* cause trouble downwind. But it was not nearly what it might have been. In an area extending northward 5 miles (8 km) from the site of the explosion nothing of the virgin forest remained; for 7 more miles (11 km) trees were down, all pointing north; and the avalanche mass had pushed out as far as 17 miles (27 km) down the Toutle Valley.

There have been warnings all along the Cascade Range that these noble cones are not mere spectacular scenery. The very fact that they are so geologically modern with only moderate erosion scarring their classic outlines, should inspire vigilance. And close examination reveals clues in abundance—hot spots high on Mt. Baker, steam caves and a warm lake in Rainier's crater, little Wizard Island in Crater Lake and similar very recent cinder cones all around the base of Shasta. Mt. Lassen, of course, blew out its side in a manner similar to that of St. Helens, although at a far lesser scale, in 1914–15. If we had to guess "where next"? Lassen and St. Helens rank number one of course, but many volcano experts point to Shasta (probably active as late as 1786) and some pick Mts. Hood or Adams as possibles. The point is that every Cascade mountain is dormant, not dead, and dormant for only a relative short time—*all* are suspect.

Variations in Eruption.   No two volcanoes are alike in their eruptive habits and as we have seen from the Vesuvian example, an individual volcano is often capricious. Some are spewers of ash, like Irazu in Costa Rica, which erupted continuously for almost two years (1966–1968), subjecting the residents of the capital city, San José, to a seemingly endless rain of thick dust (Fig. 21-7). Others eject lava but with variations: Vesuvius goes off violently, with loud explosions and gushes of lava streaming down its flanks; Kilauea, the Hawaiian crater, quietly spills lava over its lip with filmy curtains of fire along its radiating fissures; Strombolian lava, cooling and crusting over lightly in the

Fig. 21–7   A black volcanic ash fall that is difficult to ignore. Here is the fishing village of Vestmannayjar on the Icelandic island Heimaey, 1973. A volcano simply blew up in their backyard. Dug out now and back to fishing, the weary residents have no guarantee that there will not be a new volcanic visitation at any time.

crater, traps gases beneath it, which explode every half-hour throwing incandescent clots high into the air. Then there are the spasmodic eruptors whose conduit becomes clogged with hardened igneous rock between eruptions. With a cork in the bottle it is not uncommon for the increasing pres-

There have been warnings all along the Cascade Range that these noble cones are not mere spectacular scenery.

sures to blow out suddenly through a weak spot on the side of the mountain. This is what happened to Mt. Pele on Martinique in 1902. Internal pressure started the cork, pushing it out of the vent in the form of a high spire, but before it could wholly clear the passage the side blew out. A dense cloud of intensely hot gases and self-exploding lava mist swept down the slope to the sea, completely wiping out the town of St. Pierre in an instant. This type of explosive debris, both superheated gas and solid particles, is airborne but heavy enough to respond to gravity following the surface terrain. It is called *nuée ardente* (glowing cloud).

Calderas.   There is yet another kind of happening that involves the reverse of eruption, that is, a sudden emptying of the magma chamber. When this occurs, the entire top of the mountain, lacking support, collapses and is engulfed into itself. The end result is a huge craterlike pit called a *caldera.* Crater Lake, Oregon, is one of these (Fig. 21-8). We know it is not a true crater because glacial evidences on the contemporary slopes indicate that a peak of over 12,000 feet (3658 meters) must have existed to support full-blown glaciers. This reconstructed volcano, dating back probably some 8000 years, has been named posthumously Mt. Mazama. But volcanoes can blow their tops—how do we know that it collapsed rather than blew up? We know because the entire surrounding countryside would have to be covered with debris from the old peak. There is some but not enough to support a theory of major explosion. So now we are faced with the question, What causes the abrupt evacuation of a magma chamber? And we have no sure answer.

Generations of crustal ulcers appear to be healing but the infection distressingly reasserts itself again and again when least expected.

Of more recent origin than Crater Lake is the caldera of Krakatoa, an island just off western Java. On August 27, 1883 there were four tremendous explosions, the last of

Fig. 21–8   Despite its name, Crater Lake is cradled in a caldera. Little Wizard Island is a recent cinder cone signalling new life centuries after the great summit collapse.

Lava

Newly formed Hawaiian basalt called *pahoehoe*.

The most common component of all lavas is silica but the range involved in amount is significant—from as little as 45% to as much as 75%. Silica-poor lavas (although they may average 50%) are referred to as *mafic* and they and their igneous end-products display a relatively high content of magnesium, iron and calcium oxides (hence mafic: ma-magnesium, fic-ferric). These are also fairly alkaline lavas/igneous rocks which break down into a particularly fertile soil. The silica-rich, acid lavas (65% or more silica) are called *felsic* and they cool into rocks which are rich in quartz as well as aluminum, sodium and potassium oxides (felsic: fel-feldspars, sic-silica).

Mafic rocks and lavas are dark because of their manganese and iron and the felsics are lighter in color—mafic basalt is black to dark grey, felsic rhyolite or dacite is buff to yellow or even pink. In between the two extremes are the intermediates, greyish andesite being the typical example.

The geographic significance of all of this, putting aside for the moment the parent rock's contribution to soil fertility, is that the amount of silica in a lava essentially controls its viscosity—the lesser the silica the more fluid the lava. So basalt is extremely fluid, andesite only moderately so, and dacite barely. And to follow one more step along this avenue of cause and effect, magmatic gasses attempting to dissipate into the atmosphere encounter no great difficulty as fluid lava wells out onto the ground, but increasing viscosity magnifies the problem of gaseous diffusion. Volcanoes featuring mafic lava are not nearly as prone to destructive explosion as those with felsic characteristics.

When Kilauea on the island of Hawaii erupts people rush to the crater to view the fireworks. "Curtains of fire" along radiating fissures are not really dangerous if one stands back a few hundred yards with a camera. And it is always prudent procedure not to stand in the path of rapidly flowing rivers of molten basalt. There is even a tourist hotel, Volcano House, on the rim of the crater where one can sit in the lounge and safely enjoy Pele's fury through the picture window.

There aren't any hotels at Mt. St. Helens and the curious with cameras are not welcome in the crater. Pre-1980 St. Helens was an explosive composite cone made-up of alternating cinder and andesite layers. But the fact that the 1980 paroxysm resulted in the whole side of the mountain being blown away, points to an increase in silica content in its magma supply. Support for this notion is the growth in the newly enlarged crater of a rapidly growing plug—a volcano within a volcano. And plugs are produced by dacite which almost wholly contains the trapped gasses within allowing the pressure to build to the danger point.

which was heard 3000 miles (4828 km) away in Australia. A towering dust cloud was thrown up, noticeable around the world for two years, and a great tidal wave drowned 36,000 people. Subsequent investigation revealed that a deep pit had replaced two-thirds of the island but that despite the dust and ash ejected, the bulk of the island had collapsed and the sea had rushed into the resultant caldera. The area is still active. A new volcanic island is building called Anak Krakatoa (child of Krakatoa). But even the original Krakatoa was merely a remnant of an earlier, much larger island. Like Vesuvius, standing in the breached caldera of Somma, generations of crustal ulcers appear on occasion to be healing but the infection distressingly reasserts itself again and again when least expected.

Agricultural people are drawn to frequently fertile lava soils, willing to take an unknown risk to derive a certain livelihood.

On recounting the loss of human life from volcanic disaster through the centuries, one might be inclined to wonder why anyone would choose to live with volcanoes as neighbors. But if a given cone can be classified conveniently as extinct, local residents come to love it, are in-

spired by its majesty, or even regard it as divine—that "the mountain" could be venomous and turn on its friends is unthinkable, until it does. Agricultural people are drawn to the frequently fertile lava soils, willing to take an unknown risk to derive a certain livelihood. And if the giant should cough up rejuvenating showers of ash periodically, so much the better. Anyway, absolute security is not a guarantee of living anywhere. Why would anyone choose to live in earthquake country, tornado alley, or along hurricane coasts? The law calls these "acts of God," and God seems, in the short run, less lethal than midtown or freeway traffic, which threatens us all on a daily basis.

Classification by Shape of Cone. The simplest classification of volcanoes is on the basis of the shape of the cone. Most of them fall generally into one of three or four categories that are easily recognizable, and the cone shape also gives some indication of the type of ejecta and the eruptive habits of the volcano.

First, there is the *cinder cone*. This is the product of a violently explosive volcano where the lava has solidified in the vent, forming a plug. The accumulation of steam and magmatic gases gradually develops sufficient pressure to blow the plug with such force as to shatter it into tiny fragments (called *cinders* and *ash*), and these are deposited in a symmetrical pile around the vent, the larger particles

lava flow

Fig. 21–9　There has been a lava spill in the foreground not too long ago, but the most recent event has been the formation of a perfectly shaped cinder cone.

nearest the vent and the finer ones farther away. The resulting cinder cone is steep sided (up to 37°, the maximum angle of repose of unconsolidated material) and usually symmetrical (Fig. 21-9). Although evident in many parts of the world, the cinder cone seldom achieves any great size, as erosion rapidly wears away what is essentially merely a pile of loose material. The disappearing islands of the Pacific are often mere cinder cones, where an eruption may throw up a pile of solid ejecta above the ocean level to be visible for a few days or months until wave action removes it.

The second type of volcanic cone is called a *shield* or *dome*. Here a quiet flow of fairly fluid lava issues from the vent, forming a vast low-angle cone as it cools. Mauna

A

B

Fig. 21-10 (A) A view across the Grand Coulee in eastern Washington reveals the multiple fissure flows that built up the Columbia plateau. (B) As new lava cools and begins to weather pioneer plants invade the little crannies that hold some water.

Loa/Mauna Kea whose 13,000 foot (3962 m) tips form the island of Hawaii is an excellent example of this type of cone, and if the low angle of the island's slope is traced to the sea bottom some five miles (8 km) deep, the true size of this multi-cone becomes apparent.

As often as not, however, a volcano passes through several stages in its history, alternating between explosive eruptions and lava flows. The resulting cone develops a combination of the low-angle shield and the high-angle cinder cone. These are *composite or strato cones* and display concave slopes with a sharp peak. Imagine first a cinder cone, but before it can erode away, it is overlain by lava flows. Then built on top of this another cinder cone, followed again by lava. The end product looks lke Fujiyama, Shasta, Rainier, or Egmont, the world-famous volcanoes that inspire poets and legends. These four are now all dormant, but El Misti in Peru is equally as impressive and still active, as is Mayon in the Philippines. Mayon, despite its generally low elevation and lack of picturesque snowcap, is reputed to be the most perfectly shaped of the world's composite volcanoes.

Mt. Lassen is a plug, a second generation volcano in the corner of a huge caldera.

The composite forms can build to great height but it requires millions of years of diastrophism to achieve its impressive size. *Plug* volcanoes on the other hand, may develop very quickly although their propensity for self-destruction is so great that they seldom survive for long. Constructed of lava so thick that it is barely plastic at all, it pushes upward in a bluntly rounded form. As the plug slowly extrudes its surface cools, becomes greyish in color, and igneous boulders cascade down its steep slopes. The only visible indications that it is flowing lava are glimpses of a glowing red interior through occasional cracks, and its inexorable growth. But such thick lava is not at all an effective dissipator of trapped gasses, and its usual fate is a massive explosion, often out the side of the mountain. A plug volcano is a very dangerous neighbor.

Mt. Lassen is a plug, a second generation volcano in the corner of a huge caldera. Ancient Mt. Tehama was an 11,500 feet (3505 m) strato peak supporting large glaciers as recently as 20,000 years ago, but it collapsed much as did Mt. Mazama, into a caldera and Lassen is its successor. After a relatively short and violent life Lassen is far from finished although badly deformed as of the moment.

Fissure Flows.   Lava, in pouring out onto the earth's surface, does not always issue from volcanic vents; it may well out of faults or fissures many miles in length. If the lava is highly liquid and the terrain fairly subdued, such *fissure flows* have been known to cover thousands of square miles and build up extensive plateaus as in eastern Washington, eastern Oregon, southern Idaho, and northern California. The bulk of peninsular India is also of this origin. In Washington the Columbia River Gorge and the Grand Coulee reveal along their sides a banded layering of differing colors and textures, each of which represents a separate flow (Fig. 21-10). On occasion, the hot lava will be heavily charged with gases, and as the lava cools, the escaping gases leave holes in the rock, making it extremely porous. Groundwater will occasionally flow through the permeable (*scoriaceous*) rock as though it were a pipe. It is also possible that in a moderately viscous flow, with the surface cooling and congealing through contact with the atmosphere, the hot, still liquid lava underneath will run on and out leaving an igneous cave or tube. In Hawaii and California's Modoc County these are common.

CHAPTER 22

WEATHERING AND MASS WASTING

Weathering is a good term for the disintegration of rock from the moment it is formed. Water, ice, temperature variations: all are effective, singly or in concert, in weathering hard rock into a variety of forms that are susceptible to removal. Above the tree line mountain slopes and basins are deeply veneered with coarse broken rock fragments called felsenmeer. Here is the active arena of ice-wedging in rock cracks or fractures, and judging from the piled debris it is a highly effective weathering mechanism.

INTRODUCTION

In the breakdown of earth processes resulting in distinctive landforms, there has been specified two basic categories: the tectonic forces and the gradational forces. Weathering and mass wasting does not fit neatly into either one. It is close to being a part of the gradational process and admittedly there may be a bit of hair-splitting involved when the claim is made that it is not, but the crux of the matter revolves about the function of the agents of gradation: running water, moving ice, wind, and waves. Each of these is an entity unto itself, operating essentially under its own set of rules. However, in order to qualify as an element of gradation, *each must demonstrate the ability to pick up material, carry it off (sometimes long distances), and finally deposit it*. It is this transportational function that is the nexus of gradation.

Weathering as will be seen shortly, breaks down solid rock into a form in which it can be more easily transported and therefore acts primarily as a preliminary to gradation; it aids and abets but does not grade. *Mass wasting* is a step closer to gradation than weathering; there is movement here—that is what the term *wasting* means. But the energizing force is simply gravity and the distance that material moves is distinctly limited. One might argue that without gravity there would be no running water so where is the difference? But wind is a gradational process and gravity is not the whole story. Ice can flow uphill and ocean waves perform nongravitational rites.

> *V*irtually any rock is susceptible to a degree of disintegration from the moment of its formation

The word *erosion* should be introduced here, a general term signifying all aspects of the breakdown and removal of earth materials. But in this organizational system erosion is too general, so both weathering and mass wasting will be regarded as prefatory and helpful to the agents of gradation. The graders might very well get the job done without these complementary processes, but if weathering disintegrates the rocks and mass wasting impels the resul-

tant debris into place for easy pickup and removal, they undoubtedly aid and speed the entire operation.

WEATHERING

Virtually any rock is susceptible to a degree of disintegration from the moment of its formation (or at least from the moment it appears on or near the surface), for although massive stone seems indestructible, every rock has flaws that can be exploited by one or a combination of environmen-

Fig. 22–1 Columnar jointing, formed as igneous rock cooled, has allowed access to water and ice. As the tall columns are pried loose their size is further diminished by free-fall impact at the cliff base.

tal influences. For instance, most rocks are cracked. If they are of igneous origin, the contraction of cooling in nearly every case results in some kind of jointing, often very pronounced (Fig. 22–1). Other varieties of rock succumb to even minor forces of uplift and pressure by breaking in some fashion, from out-and-out faults to minuscule stress and fatigue fractures. And once there are cracks, the rock has been weakened and is on its way to eventual destruction. Not only do breaks in basic structure lessen the ability of a rock to sustain its own weight and that of strata above it, but they also allow the intrusion of outside forces whose effect is to widen the crack and increasingly attack the interior. Tree roots, water, acids—each is effective in its own way and they frequently operate in concert.

The mechanical structure of certain rocks leads to distinctive modes of disintegration via crumbling, scission, or decay. Sedimentaries display a strong tendency to separate back into their original units. If sandstone or conglomerate is nothing more than chunks of rock stuck to-

gether, the urge to come unstuck is built in. And sedimentaries too are almost surely stacked up in flat layers, one atop the other, which rather automatically makes every bedding plane a zone of weakness (Fig. 22–2). Formations of igneous derivation might be considered less susceptible to this because of their massive character, but the size of crystal is frequently the critical factor. Obsidian is reasonably resistant to weathering, granite is not. The larger crystals of granite make it liable to selective attack, some crystals proving much less resistant than others. But how about the metamorphics, which are by definition denser and harder than most rocks? As hard as it is, slate foliates into thin sheets cleanly and easily, and the banded schists are inclined to follow suit.

Given the multitudinous assemblages of rocks that comprise the surface of the earth's crust, with their strengths and their flaws, weathering as a process must function in a variety of ways. Standard pedagogy has been to simplify this variety by separating all weathering into two very dif-

Fig. 22–2   Bedding plane weakness in sedimentary layers.

ferent kinds—*chemical* versus *mechanical*. Such a dissociation is artificial to the extent that chemical and mechanical weathering usually operate in conjunction. But there are extensive regions where one strongly dominates the other. Generally speaking, since chemical weathering requires water and is most effective under conditions of high temperature, the warm moist climate areas are especially permissive, while the deserts, high altitudes, and high latitudes exhibit the decisive effect of mechanical processes. A thumbnail visual clue as to which is dominant is an assessment of the general harshness of the landscape. Rounded, muted, subdued outlines denote chemical weathering; sharp-edged and jagged landforms, reminiscent of lunar panoramas, are of mechanical origin.

If sandstone or conglomerate is nothing more than chunks of rocks stuck together, the urge to come unstuck is built in.

Chemical Weathering

Chemical weathering might be characterized simply as the rotting, mouldering, or decay of rock materials—chemicals are reacting with a wide variety of minerals to alter their form and in so doing they weaken or destroy the rock. The most familiar example of the chemical decomposition of a seemingly indestructible material is the rusting of steel. In an unbelievably short time, handsome shiny steel, immune to abrasion and other mechanical wear, can be reduced to a reddish powder and simply blown away. All that is required is the presence of water. The processes are *oxidation* (chemical union with oxygen) and *hydration* (chemical union with water). Copper is a little more resistant than steel but *carbonation* (reaction with carbon dioxide in air or water) turns it soft and green on the surface.

Most minerals have a chemical Achilles heel.

Rocks are simply combinations of minerals, and most minerals have a chemical Achilles heel. Each can be at-

tacked and broken down by a different chemical process. If clear water is not effective, then perhaps the acids contributed by vegetative decomposition will be. If these fail, those minerals that do react may add to the groundwater still other corrosive properties. Rare is the mineral that can stand up to an all-out chemical attack (Fig. 22–3). One that does much of the time is glasslike silica, a widely occurring rock mineral. The chemistry professor always makes his or her little speech at the beginning of each term beseeching the students to keep their acids in glass containers because they will eat their way out of anything else. So it is with natural acids—silica resists. Silica's hard unaltered crystals appear in great profusion in beach sands as they are released from their rock bond by the decomposition of neighboring minerals and carried off to sea.

Another common mineral, calcium, appears widely as calcium carbonate (limestone). In arid country this is considered to be a highly resistant rock and often stands above the surrounding countryside as mesas or hogback ridges. It is not entirely the lack of moisture in the desert that makes limestone difficult to break down, for it is only moderately soluble in clear water—it is the lack of vegetation. Once groundwater filters through a carbon dioxide-rich mat of decaying organic matter, it becomes carbonic acid; and carbonic acid destroys limestone in short order by readily taking it into solution. Virtually all of the world's great underground caverns have been etched out through the dissolution of limestone by cold groundwater tinged with carbon dioxide (see Chapter 27).

The widespread evaporatives are particularly susceptible to being taken right back into the solution from whence they came by just plain water.

The feldspar family appears as an ingredient in a great array of mineral complexes (making up approximately 60 percent of all igneous rocks), and it too breaks down easily in the presence of even very dilute carbonic acid. The resultant is soft flaky clay. And the widespread evaporatives (table salt is an example) are particularly susceptible to being taken right back into the solution form whence they came

Fig. 22–3   Selective solution
weakens the entire structure.

by just plain water. These kinds of reactions weaken a rock by creating soft spots and cavities. The surface crumbles or sometimes peels away in rounded layers *(spalling)*, and the underlying fresh surface is then exposed to the weather for its round of treatments.

Mechanical Weathering

One may try to sort out the purely mechanical or physical influences that pry away chunks of rock and shatter them. But it is not always easy to perform such an act of isolation for, as often as not, chemical processes are involved at the same time. Surely the wedging effect of ice in a rock crack is mechanical: ice has a greater volume than water and we are witnessing the same tremendous force being exerted as when the solid metal of an engine block is ruptured after the antifreeze has been forgotten. But ice is water and where

there is water, chemical change is also present. How much of the rock's vulnerability that allowed the ice to pry away a segment must be attributed to chemical disintegration? The steel waterpipe suffers from rusting as well as freezing. The point is that a part, and often a very large part, of the total weathering is unquestionably mechanical. In a closed system it is estimated that at 32°F (0°C) ice will exert at least 2000 pounds per square inch (141 kg per cm^2) and this pressure can build to 3000 pounds per square inch (211 kg per cm^2) at a few degrees below 0°F (-18°C). Of course, a closed system does not exist in nature, but water in a crack freezes first at the top, which does form a quasi-seal. Even if we deal with only a very conservative 1000 pounds per square inch (71 kg/cm^2), few rocks, with the water-filled crack a point of established weakness, can withstand this kind of pressure. Any concomitant chemical reaction is essentially unimportant. Scattered at the base of the weathered rock will be an accumulation of coarse sharp-edged debris, evidence that the rock's disintegration is being accomplished largely by mechanical means.

---

*H*igh altitudes are the special operating arena for maximum efficiency in frost wedging.

High altitudes are the special operating arena for maximum efficiency in frost wedging. The frequent freezing and thawing on a daily rather than seasonal basis tremendously speeds up the process of rock destruction. It is not at all uncommon for the ground at tree-line to be absolutely covered with angular rock fragments of many sizes. The Swiss have given this the highly descriptive name *felsenmeer* (rock sea). Ice can be an effective weathering mechanism in the high latitudes and deserts too, except that near the poles alternate freezing and thawing is limited to a short fall and spring, and in the middle latitude desert where winters are severe enough for ice, water is not always present even in the form of dew. But the desert displays an alternate type of crystalline wedging that sometimes goes unrecognized. This is the formation of salt crystals. The action is much like that of ice and similarly requires water. As the water evaporates away at the surface of a porous rock, moisture from deeper down is drawn outward by capillary action until finally there is a great concentration of ever-larger saline crystals built up in the surface openings—intergrain and shallow crack rupturing results.

---

*A*luminum barn siding pulling away from the walls because it was affixed with steel nails expanding and contracting at a different rate.

Also in the desert, where most of the year the diurnal temperature contrast is of great severity, it is possible that rock surfaces can be made to crumble away by rapid alternation of intense heat and cold. Each mineral exhibits its own individual coefficient of expansion, and since every rock is a mélange of minerals, radical differences in rates of expansion and contraction can presumably tear the surface of a coarse-grained rock apart (Fig. 22–4). Although there are those who would play down the influences of temperature on rock disintegration, anyone who has witnessed aluminum barn siding pulling away from the walls because it was affixed with steel nails expanding and contracting at a different rate cannot be a total unbeliever.

Fig. 22–4 Desert boulders wearing away before our eyes. Many processes are at work here: wind scouring, spalling caused by salt crystals in intergrain apertures, and strongly contrasting day and night temperatures. This was once a single solid rock. Notice the reduction in size from top to bottom.

Fig. 22–5 Massive granite released from a great compressing overburden of rocks by erosion (ice has been the latest eroder and a considerable weight in its own right) expands most rapidly at the surface and peels away in layers. The Yosemite domes are formed in this manner—here is Liberty Cap.

Earlier the concept of unloading was introduced, the gradual uplifting or rebounding of deeply buried rocks as millions of tons of overburden are removed by erosion or the melting of continental ice sheets. Those rocks are massively compressed in their original state and as the overlying pressure is gradually eliminated, not only does the entire mass drift upward, but the surface rock layers tend to pull away from those below. This scaling off of the exterior in broad flexures is called *exfoliation* and is beautifully displayed in the granitic Yosemite domes (Fig. 22–5).

Hordes of hungry chipmunks gnashing their tiny incisors.

Tree roots too, inserting themselves in any sort of available rock crack, can exert inexorable pressures. We are tempted to feel sorry for the gaunt gnarled pine perched atop a rocky battlement, its roots forced to search for soil and water in seemingly sterile rock apertures. But the Bristlecone pine, inhabiting just such a difficult mountain environment in eastern California and Nevada has survived for perhaps 5000 years and in that length of time we might transfer our sorrow to the rock, for measurable wedging and breakage has occurred along the course of every

Fig. 22–6 This sea cliff was already jointed both vertically and horizontally and doomed to eventual breakdown, but probing tree roots and groundwater acids from mats of vegetation are aiding greatly in the fragmentation process.

Avalanche

Summer visitors to scenic Swiss mountain villages are enthralled by the rustic beauty and bucolic tranquility of the Alpine countryside. They are not there when the fearsome avalanches roar. In every Swiss valley these winter scourges are a fact of life and simple survival requires a critical knowledge of their behavior. What might appear to be a random pattern of town and farm building location is in fact a very carefully arranged strategic plan based on centuries of experience with the character of local avalanches. Certain ridges act as avalanche "splitters" and carefully protected forest plots on the hillsides, known as *bannwalds* (forbidden forests) impede, if not completely halt, runaway snow slides. The villages are placed in lee positions, protected from the well defined avalanche tracks and much of the valley floor and lower slopes is allocated to permanent fields and pasture.

Given a mountain grade well above the snow line, whose degree of slope is something less than sheer so that it can accumulate a depth of snow, a potential avalanche is in the making. Piled up snow over a winter season is laminated in character. At the bottom is old snow, near solid ice from percolating melt water and great pressure from above. Interspersed throughout the depths are hard slick sun-crusts that were subjected to freezing and thawing on the surface over periodic weeks of good

weather—then light fluffy layers from recent storms that form an unstable base for heavy wet new snow on top. And projecting, over-balanced above all of this, are massive snow cornices fashioned by strong winds whipping across the ridge top.

At a certain point the equilibrium becomes so delicate that one more snowy day or a single skier can trigger an avalanche. And once the snow mass breaks loose it picks up speed, quickly skidding along an icy interior plane. Rising above and ahead is a great cloud of billowing ice particles that spill down the mountainside like a chilly nuée ardente. Essentially frictionless, some dry powdery snow clouds have been clocked at 200 miles per hour (322 km per hour) followed by the avalanche proper at over 100 miles per hour (161 km per hour). And propelled out ahead is an invisible blast wave which of itself can be very destructive.

Deep-seated avalanches, scouring at their base to bare rock, move more slowly with friction supplying a great deal of meltwater at the point of contact. During the summer in high mountain basins avalanche tracks are visible on every slope where they have completely stripped away the trees and moved great quantities of rock and debris, sometimes clear across the valley and up the opposite side.

In 218 B.C. the story goes that half of Hannibal's 38,000 troops and elephants were killed by avalanches as they attempted a winter crossing of the Alps. The worst U.S. disaster was the complete burying of two trains in the Cascades in 1910 with the loss of 96 lives. Today the danger is greater than ever before as snowmobiles and skiers invade the mountains in ever increasing numbers, and prevention teams are out in populated mountain regions, utilizing carefully placed explosives to set off little avalanches before big ones can develop. But the system is far from foolproof. Winter vacations in the mountains carry with them a certain calculated risk; while those who choose to spend a lifetime in a Swiss village or a place like Juneau, Alaska, at the base of sheer mountain walls, are involving themselves in a continuing adventure.

probing root. That roots have lifting power is beyond question for we find street trees in every city easily dislodging ponderous slabs of sidewalk concrete (Fig. 22–6).

Even burrowing animals contribute their bit. This is

not to say that hordes of hungry chipmunks, gnashing their tiny incisors, masticate impervious stone, but in the construction of their tunnels or warrens they do remove support for rocks on slopes, causing them to tumble, and these same burrows open up channels of attack for roots and groundwater.

MASS WASTING

All loose particles on the earth's surface are attracted by gravity and each makes every attempt to respond, so that if we but look around us as we roam the earth, we will find a great deal of movement going on. Everything is headed in the same relative direction—downhill—some in a slow, pedestrian manner, some pell-mell with a maximum of sound and fury, some at a steady pace, some in a highly spasmodic and unpredictable fashion—but all materials progress downslope as best they can. Chunks of bedrock break off from undercut cliffs or are pried away by ice; water-lubricated volcanic ash and fine dust may swirl by as a muddy matrix carrying a cargo of large boulders; soil inexorably creeps downward causing terracing and wrinkling of the sod that tries to hold it in place; or a single bounding pebble may upset the delicate equilibrium of a piled apron of debris and the entire accumulation rumbles off to find a more stable angle of repose. Even snow reacts violently as an unconsolidated mass in the form of avalanches, carrying with it tons of rock debris.

All of this motion is part of the universal cycle of denudation wherein the high places are whittled away and the accumulated residue transported to the low spots. But this drift of loose materials downslope—this creep, slide, flow, glide, skid, slip, fall of all kinds and sizes of rock detritus—is only the first and somewhat limited part of the transportation system that ultimately moves each particle to the sea. Mass wasting, in impelling the end products of weathering to the bottom of the slope, simply places them in an advantageous position for the final agents of gradation to carry them off.

Slow Movement

The slow, steady movement of materials downslope (as opposed to the sudden runaway slide or fall) is everywhere, but the actual motion is usually difficult to detect. On a clean tilled field in rolling terrain it is clearly evident that the soils upslope are increasingly thinner, even to the point of the hilltop being bald. But when a slope is wooded or supports a brush or grass cover, soil creep must be inferred from the cracking of retaining walls and the downhill lean of fence posts or trees. If there is winter freezing, then ice crystals replace the soil water and an upward swelling action called *heave* takes place. Melting, then, completely collapses the soil into a mass of extremely fluid mud, which sags a step downward. Every freeze and melt (and there can be several each year) alternately lifts and drops the soil with a net movement down the slope.

The Role of Water

Water is a major ally in the efforts of masses of earth to relocate themselves at a lower level, but it is not indis-

Fig. 22–7 In the building and maintenance of mountain roads there is always a certain risk. The forest on the steep slopes are slide stabilizers but they are not infallible if the earth movement is deep-seated. Repairs here may consume six months and the work crews are fairly safe in assuming more trouble before long.

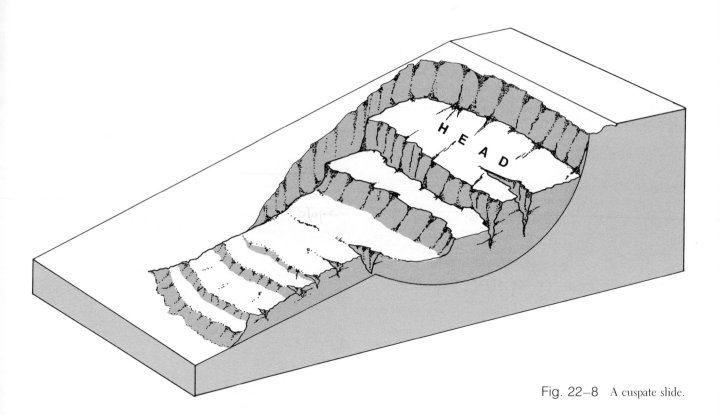

Fig. 22–8   A cuspate slide.

pensable. There are landslides without water, as where earthquakes trigger the action. But water, whether released from ice and snow, precipitating as rain, or introduced as groundwater, is an excellent lubricant and in saturating the soil adds tremendously to its weight. After every rainstorm, the highway department in mountainous country is a busy agency clearing away slides of all sizes. The simple arrival of spring in many parts of the world, with its sustained melting snow and ice, will bring the same result (Fig. 22–7).

Water is an excellent lubricant and in saturating the soil also adds tremendously to its weight.

Landslides and Earthflows

Landslides or earthflows, although technically not exactly the same, are descriptively named and are simply the rel-

atively abrupt movement of large masses of soil and rock downslope, with water as an abetting agent more often than not.

Landslides are subject to many variables in their susceptibility to movement. A common one is the existence of an inclined hard-rock or impervious clay stratum on which soil and debris have piled up. With groundwater penetrating only to this level, it becomes a line of moisture accumulation and an obvious glide or slip plane for the sudden movement of surface materials.

When soils are deep and no slip plane exists, (1) degree of slope, (2) soil character and depth, (3) vegetative cover, and (4) water are all a part of the answer to "why did the hill come down into my backyard?" As often as not the initial motion in deep soils is a sudden slump in a concave or cup-shaped pattern (Fig. 22–8). The "head" pulls away and drops but the mass movement is a rotation that causes the soil to lift farther down the slope. Both the deepseated surface cracks and the holding-basin formation allow the retention and penetration of surface water and

Fig. 22–9 The twin and related evils of fire and mud.

this adds greatly to both ease of further slippage and total weight.

A cuspate slide, though, tends to move sporadically and there is time for attempts at remedial action, the chief one being to drain away as much ground water as possible. Retaining walls, plastic covers to repel the rain, or piling driven into the slide proper may all help a bit, but the slumping mass has now developed as a major zone of weakness and seldom can it be controlled permanently. Next year's rainy season or next year's snow melt may set it off again.

With increasing urbanization and the urge to seek the view site (or get above the smog), contractors, architects, and unhappy homeowners have all become intimately involved with landslides. In the residentially exclusive canyons surrounding the Los Angeles basin the story goes, "If

Fig. 22–10 Free-fall debris from the high cliff face piles up in coalescing *talus* slopes at the base.

the summer brush fires don't get you, the winter mud slides will." Fire clears away the vegetation on the surrounding slopes, exposing the loose soil to occasional torrential winter rains. Runoff, concentrated in the canyons, can bring with it such a heavy sedimentary load that the consistency is that of thick gruel and it is called a mud slide rather than a flood (Fig. 22–9). The moral: do not build in a canyon or the mud will come and engulf you. And one probably should add, don't build at the top of a scenic sea cliff or the waves undercutting your shelf will suddenly plunge you into the surf, and don't build in a mountain valley because all sorts of things keep dropping in.

*D*on't build in a mountain valley because all sorts of things keep dropping in.

Rock Avalanches

In the mountain valley, if we ignore the snow avalanche, the chief villain is the rock slide or rock avalanche. We are dealing here with felsenmeer except that the angular rock debris often builds to a considerable depth in steep draws for lack of any flat surface. Here is a potential disaster for the region below, for broken rock fragments have little cohesion with the ground (often bare rock) or each other and the slope is likely to be excessively steep. There develops, then, a loosely organized and extremely heavy mass hanging poised above the mountain basin. It takes very little to set such an avalanche in motion—one more bounding rock from above, one whistling marmot scrambling across its face and off it goes thundering into the abyss.

Free Rock Fall

Free rock fall is common from high-standing crags and cliff faces. Big chunks pried free by mechanical weathering either smash themselves into tiny pieces on impact (actually a process of mechanical weathering itself) or go bouncing, rolling, and skidding well beyond the cliff base. Lesser pieces, all the way down to fine powder, arrange themselves at the foot of the escarpment in cone-shaped aprons according to size: the finest at the top and the coarsest at the lower margin. This *talus* or *scree* slope contains no binding material of any kind and is totally unconsolidated, a situation that dictates a maximum slope of about 37°. But such a tenuous equilibrium is easily disturbed, certainly by additional rockfall, and a talus environment is a restless one, constantly shifting and readjusting and endlessly rejuvenated from above (Fig. 22–10).

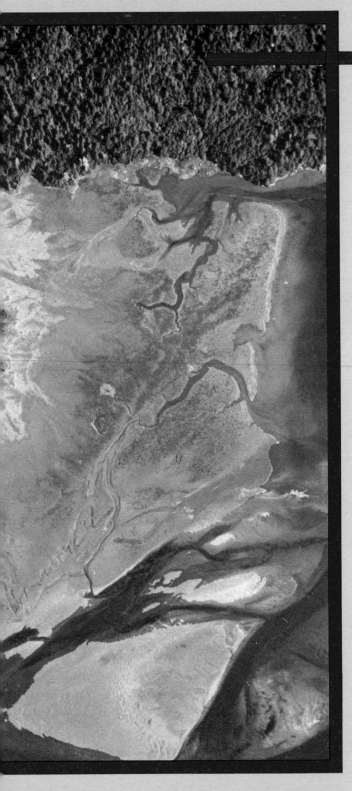

CHAPTER 23

RUNNING
WATER

When a river runs to the sea it immediately
encounters difficulties of its own making. Abruptly
diminished velocity necessitates deposition of the silt
load that was transported with such ease upstream,
and this results in accelerated fill and the
development of new land at the river's mouth. But
the new land is almost dead flat, so that as the
stream must run across it, to the now more distant
sea, it fouls its channel with further deposits. In
constantly seeking a less impeded course the stream
mouth is forced to occilate and form a swampy fan-
shaped delta—vegetation establishing itself on the
new land compounds the difficulty of easy stream
discharge.

INTRODUCTION

By all odds, the most important of the gradational agents is running water. Without question, wave erosion of shorelines is very effective and widespread, but waves merely nibble at the periphery of continents, while running water operates over their entire surface. Even in the desert where surface stream courses carry water only now and again after infrequent storms, they may move an enormous amount of material in a very short time. The hard baked earth and the violence of the typical desert downpour mean maximum surface runoff, leaving the land scarred and eroded into numerous deeply cut channels and arroyos (Fig. 23–1). Elsewhere, every little rivulet and creek is a gradational agent delivering its load of material to the master stream to carry away, and gullying as it goes.

RUNNING WATER AS A GRADER

Slope, of course, is a major factor in stream erosion because without slope the stream ceases to flow. But remember, all gradational agents must accomplish three functions: they (1) *pick up materials*, (2) *transport them*, and (3) *ultimately deposit them*. When streams encounter a low gradient, which is frequent in their lower courses as they debouch into the sea, they deposit their load and thus fulfill the last part of their gradational function. It is sometimes useful to break *gradation* down into its components of *degradation* or cutting and eroding, and *aggradation* or depositing and building up. *Transportation* comes in between these two, and the final result is to carve away the high spots and fill up the low ones.

All streams carry sediments. There is no such thing

Fig. 23–1 There is not a drop of water anywhere in this Imperial Valley desert landscape where less than 3 inches (8 cm) of rain falls annually, yet the marks of riverine erosion are incised on the land. Obviously, stream runoff is very effective following the occasional torrential downpour. This type of symmetrical drainage pattern is called dendritic because of its supposed visual similarity to the ribs and veins on a leaf surface.

as absolutely clear running water on the surface of the earth. Obviously, a big muddy-appearing river such as the Hwang Ho or the Missouri is moving a great deal of soil in suspension, but even sparkling mountain cascades are transporting surprisingly large amounts of material. It is simply that their capacity is so great relative to their load that they appear clear.

Every stream, large or small, transports three categories of load: (1) the *solution load*, those invisible components that water dissolves from the rocks over which it passes; (2) the *suspension load*, all of the fine material easily carried by running water, usually well up in the main current; and (3) the *bed load*, that segment of the solid flow that is heaviest and impelled along in contact with the stream bed. Swirling sands and skipping pebbles are in constant motion, but even sizable boulders advance spasmodically during the highest water period when increased velocity and buoyancy are temporary aids.

There is no such thing as absolutely clear running water on the surface of the earth.

The chief control of a stream's ability to transport material is its velocity. Volume too affects this but the importance of velocity is overwhelming; thus high-velocity mountain streams exhibit a huge load-carrying capacity, while sluggish rivers barely moving across a gentle gradient are very ineffective transporters of sediment. It is on this basis that an attempt is made to classify running water into general types.

Try to visualize streams as living entities endeavoring to reach an equilibrium between actual sediment load and their capacity. Such a balanced stream where the load is exactly equal to the amount it is equipped to carry is called a *graded stream*. Although this perfect equilibrium is virtually impossible to achieve in nature, many streams come fairly close and all strive toward this graded ideal. This means that a high-velocity stream with tremendous energy but underloaded will work hard to find more materials to carry. Such a water course is called *youthful*. On the other hand, a placid low-velocity stream may find itself heavily overloaded, especially if it is fed by vigorous tributaries. Its in-

creased volume fails to offset its low velocity, and this type of stream attempts to rid itself of its surplus load. It is classed as *old age*. In between these two extremes are many streams with median velocity and median load that approach the graded ideal. On occasion from season to season, such streams may exhibit youthful or old-age tendencies, but since they normally show a reasonable balance between load and velocity, they are called *mature*. The vigor of youth, senility of old age, and balance of maturity are all represented here, and the simile aids in understanding stream behavior.

It does not necessarily follow that youthful streams must flow on youthfully eroded landforms.

Caution, although this same classification system and terminology of youth, maturity, and old age was used earlier to characterize the various stages of the geomorphic cycle, it does not necessarily follow that youthful streams must flow on youthfully eroded landforms, mature streams on maturely eroded landforms, etc.

YOUTHFUL STREAMS

The high-velocity water course with its tremendous energy is attempting to avail itself of a sediment load commensurate with its velocity. It has only one arena of operation and that is its channel, so it works with great vigor on that channel. But the friction of running water over rocks is scarcely capable of eroding those rocks at all effectively. Certainly, water is able to take into solution some of the minerals, thereby weakening the rock and allowing pieces to tear away and be carried off, but these rock fragments and others delivered from the valley slopes, varying in size from boulders to sands and silts, become the "teeth" of the stream and give it great cutting ability. Especially effective are the hard, sharp-edged quartz fragments in sands that allow a youthful stream to abrade and corrade its channel downward at a very rapid rate. In doing so, it begins to

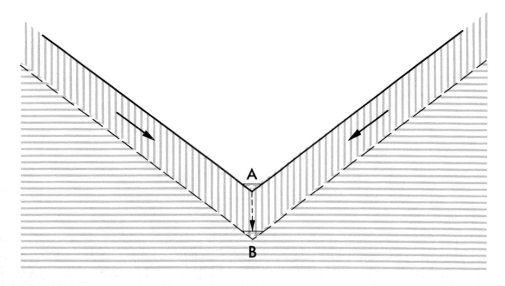

Fig. 23–2 V-shaped valley and youthful stream erosion. As the stream works its valley down from A to B, the material on the valley sides, the slopes steepened now, is delivered to the stream via gravity.

appease its appetite for sediments by cutting them out of the channel while at the same time gradually lowering its gradient (and velocity), thus requiring an increasingly smaller amount of load to balance velocity.

If it is cutting through soft rock or soil, the youthful stream typically forms a V-shaped valley and occupies the entire narrow valley bottom (Fig. 23–2). The slope of the V is determined by the steepest angle of repose of loose soil. But as the river works its way rapidly downward, that angle is increased and great volumes of soil and rock slide into the valley and supply the stream with added material. Also, any tributary creeks and rivulets find their gradients becoming steeper as they attempt to maintain their mouths at the level of the constantly lower master stream, and thus they become more effective eroders and deliver increased quantities of sediment.

Where the youthful stream flows across hard-rock strata, it cuts a vertical-sided gorge or slot rather than a V, since the rock is capable of maintaining sheer cliffs without sliding. However, as soft layers that have been protected previously from the atmosphere by overlying hard rock are exposed along the sides of the gorge, weathering begins its work immediately and further weakens them, so that eventually the harder stratum is undercut to the point where it collapses and all this debris is delivered to the stream via gravity. Many of the spectacular gorges in the Colorado

Plateau have been developed in this manner. They are commonly a half-mile or more in width across the top, narrowing via terraces toward the bottom. Each terrace surface represents a hard-rock layer overlying softer strata, while far below in the bottom of the canyon, the river occupies a narrow slot as it busily cuts its way downward (Fig. 23–3). This type of canyon development, of course, depends on the exposure of beds with differing resistances to weathering as in the Grand Canyon and many others. The Snake and Columbia rivers, on the other hand, have worked their way down through hard rocks of similar resistance, and their gorges tend to be narrow and sheer-sided.

As youthful streams charge across the countryside, they have a marked tendency to maintain relatively straight courses, for it is characteristic of them to overrun and cut away obstacles rather than to detour around them. Slides may choke the channel for short periods, but the energetic stream will rapidly chew its way through. Lakes, falls, and rapids are common features in early youth as a drainage pattern establishes itself over newly uplifted terrain, but these are transitory. The upper end of a lake will silt up rapidly, while at the same time the downward erosion at its outlet will cut away the lip that maintains the water level and the lake will gradually drain away (Fig. 23–4).

Similarly, falls and rapids will be erased within a short time. Close to home, we have the examples of the St.

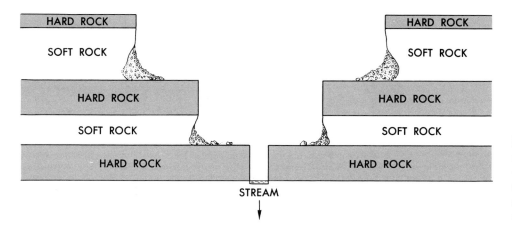

Fig. 23–3   Youthful stream erosion—canyon. By cutting its way across alternating hard and soft strata, the stream exposes soft rock to weathering and erosion.

Lawrence with many rapids, Niagara Falls, and the Great Lakes in its drainage system. The rapids are being smoothed out, Niagara Falls is retreating rapidly and left to itself (which is unlikely in the light of its economic benefits), the river will drain the lakes within a very short geologic span.

Like people, streams are only as old as they act.

Do not equate the length of time that a stream has been in existence with its category of youth, maturity, or old age. *Its behavior is what determines its classification.* Like people, streams are only as old as they act. Obviously, a stream attempting to erode a hard-rock channel will maintain its youthful characteristics much longer than one cutting through soft strata. But given sufficient time with no tectonic interruptions, in the normal course of events, all youthful streams will lower their gradient to the point where they can achieve maturity.

Fig. 23–4   Youthful stream. As it hurries by, often over falls and rapids desperately searching for added materials to transport, the river scours, abrades and consumes its channel floor. Those big boulders in the stream bed, motivated by weathering, the continued downcutting of the racing water, and gravity came tumbling down from the valley sides. During high water some of them can be jostled along downstream—those that are too big are quickly rasped down to size by water-borne sand.

MATURE STREAMS

Mature streams come closest to the graded ideal. Remember that there is actually no such thing as a perfectly graded stream. Even one where velocity matches load most of the year may find itself with increased volume during springtime's melting snows or a sustained rainy season and will sweep its channel clean and cut into its bed in an attempt to find a larger load to match its greater capacity. In this case increased volume rather than velocity will allow it to carry somewhat more. Conversely, during a dry season its volume will be reduced so that the stream must drop a part of its load, thereby choking the channel with gravel and sand bars. These render it difficult for the main current to maintain itself, and the flow may break up into many lesser courses following tortuous routes around the barriers (*braided stream*) (Fig. 23–5). But nonetheless, any stream that normally exhibits a reasonable balance between load and carrying capacity is called *mature*.

Typically, the mature stream, having lost some of the ability of the youthful stream to overrun obstacles in its path, has developed moderate bends or meanders in its course.

For instance, if a slide were to block its course in early maturity, the stream would detour around the slide. In doing so, the current that attempts to run in straight lines would be diverted against the bank. Here it would cut horizontally on the outside of the bend, expanding that bend farther to the side. Then bouncing off this side, it would be diverted again to the other side and cut again laterally to form another bend, and so on downstream. Thus once these bends or meanders are established, the main current, moving alternately back and forth across the river, cuts at the outside of bends and enlarges the meanders. Here we have something quite different from the youthful stream where all of its great erosional energy is expended downward; the mature stream that cuts moderately downward during high water normally erodes laterally, forming meanders and widening its valley floor. To illuminate the full picture, we should divide the flowing water of the mature stream into two segments: first, the main current that has the greatest velocity and some youthful characteristics as it cuts away at the bank of the outside bends and keeps its channel clean and deep; and second, the more slowly moving bulk of the stream that flows along the inside bends. It displays some old-age characteristics because it cannot carry the entire load

Fig. 23–5 A braided stream.

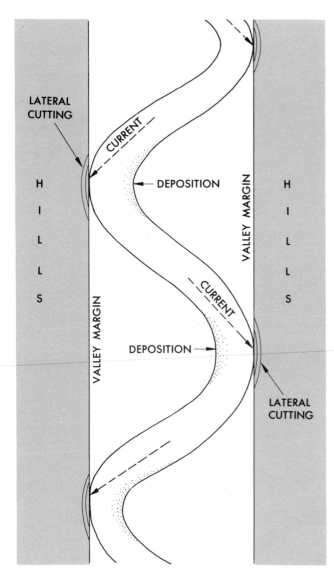

Fig. 23–6   Mature stream behavior.

supplied it and drops the load along the inside bend, thus building up these areas to become a portion of the valley bottom. Inside bends then display gently shelving *slip-off slopes* built up by deposition, while outside bends will be deep with a cliffed bank (Fig. 23–6).

This lateral cutting combined with matching deposition across the river means that a mature stream occupies a broad valley floor as opposed to the youthful stream where

the channel was the valley floor itself. The outside bends touch and cut away the sides of the valley on both sides, continually expanding the width of the flat bottom. The proper way to express this is to say that the meander belt (the width from outside bend to outside bend) is the exact width of the valley floor (Fig. 23–7). Any road attempting to take advantage of such a valley through rough country must bridge the same stream repeatedly. Also, because of the flat valley, flooding may well be a hazard during high-water periods.

OLD-AGE STREAMS

Any water course flowing across a low-gradient plain and attempting to transport a load of sediment beyond its capacity is properly called an *old-age stream*, irrespective of how long that stream has been in existence, for under these circumstances it rapidly acquires a fundamental set of senile characteristics. Basically, it is an overloaded stream and therefore an active depositor as it attempts, like all other streams, to balance load and velocity. The place of obvious deposition is the stream channel itself, and typically it is choked with islands and bars and is very shallow. However, since most old-age streams have developed from the mature stage, the already formed moderate meanders are becoming enlarged, and again we must divide the stream flow up into two parts: the swifter-flowing main current attacking the outer bends versus the slow-moving portion of the inside bends. Despite the fact that an old-age stream is overloaded, sufficient deposition occurs at the inner bends to allow active lateral erosion on the outside bends, and thus the meander pattern becomes extremely sinuous and involved. When this looping course has reached its ultimate development, it becomes increasingly frequent for two outside bends to work their way laterally toward each other until they coalesce and the stream flow cuts across, establishing a new channel and leaving an abandoned meander called an *oxbow lake* (Fig. 23–8). Unlike the mature stream, the meander belt occupies only a small part of the wide old-age stream valley. Scars on the bluffs on both sides of the valley (*cusps*) indicate that the stream has been at work, widening its valley, and give evidence that the meander belt shifts back and forth across the valley from time to time.

Fig. 23–7 The modest meanders signal maturity. So does the steep bank at the outside bend balanced by a reciprocal shelving beach on the inside. And the valley width has expanded a great deal from the earlier youthful "V".

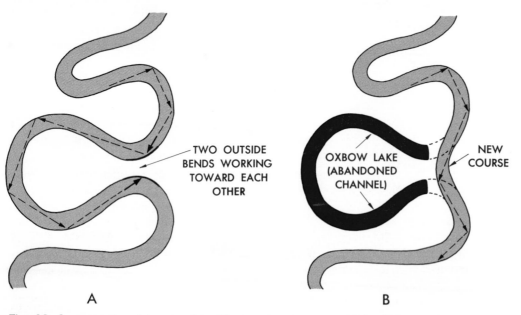

TWO OUTSIDE BENDS WORKING TOWARD EACH OTHER

OXBOW LAKE (ABANDONED CHANNEL)

NEW COURSE

A

B

Fig. 23–8 Formation of an oxbow lake. The lateral cutting on outside bends is carried to its extreme when two involved meanders coalesce to form a new channel, leaving the old abandoned loop as a lake.

Fig. 23–9   Old-age stream valley. (A) Crescentic scars (cusps) on the bounding bluffs. (B) Abandoned channels. (C) Oxbow lakes. (D) Swamps. (E) A serpentine river course occupying only a part of a broad floodplain. All are distinctive features of an old-age stream valley.

Although the old-age stream carries all the sediment that it can to the sea and deposits it in the form of a delta, and further loses part of its excess load through channel deposition, this is inadequate for complete relief. Normal behavior is for the stream to overflow its banks during even relatively minor high-water periods. Several days of heavy rain in its drainage area are usually enough to cause some overflow, and during the spring, snow melt in the uplands drained by its tributaries often results in extensive flooding. The low banks that suffice to keep the stream in its channel during normal times are easily overrun during high water. These banks, called *natural levees*, exist because as the river abandons its channel, its velocity is suddenly checked and heavy deposition occurs here along the channel margin. Then the water spreads out for miles across the entire valley floor as a still lake and drops its entire load of sediments. In this way the old-age stream gets rid of its overload and at the same time builds up its valley higher above the sea in an attempt to increase the gradient and velocity to enable the stream to carry its load.[1]

Once the old-age stream has flooded, the former channel may be abandoned, for usually it is slightly higher than the general level of the valley because of constant channel deposition. So an entirely new drainage system may establish itself, or more than one, only to repeat the overflow and abandoning again and again. Thus the broad, flat valley of an old-age stream presents an amphibious appearance with permanent swamps, oxbow lakes and constantly changing channels (Figs. 23–9 and 23–10).

The valley of the lower Mississippi was like this before

---

[1] In lower courses of rivers debouching into the sea, this is seldom successful, as delta building at the stream mouth offsets the added height of the valley floor.

The Yellow River

The Yellow Plain in Flood.

Cold winter monsoon winds sweeping out of Mongolia and the Gobi carry with them a cargo of fine, amber-hued, desert dust. Particle size is on the order of flour, small enough to be easily transported by the wind but of sufficient weight to drift like dry powder snow against the slopes of broken terrain. In the provinces of Shenxi and Shanxi in eastern China the depositional process has been unremitting until great piles have accumulated, some to as much as 450 feet (137 m) deep. This is loess, wind transported soil.

The Yellow River on its long southerly reach through the Loess Plateau, experiences no difficulty at all in incising a deep gorge in the soft compacted soil, acquiring as it goes a huge sedimentary load—several large tributaries add their contributions along the route. Here is the fine yellow desert dust on the move again, this time as *alluvium*, water transported soil. And the running water, even though the gradient is fairly steep at about 4 feet (1.31 m) to the mile, is muddy in appearance and silted to near capacity. At this point the river course shifts sharply to the east and heads out on a 500 mile (805 km) run across the North China Plain to the sea, but simultaneously its gradient drops abruptly to no more than 10 inches (.3 m) to the mile. The river finds itself attempting to transport a substantially larger sedimentary load than the laws of physics will allow—that it cleanse itself promptly becomes a pressing urgency. Three-quarters of the solids in suspension must either settle out in the channel itself or be distributed over thousands of square miles by repeated large-scale flooding.

Coveting the inherent fertility of the twice transported alluvium, Chinese agriculturalists moved early out onto the swampy North China Plain with their first order of business to make the river "behave". They acquired a great deal of experience over the long centuries, in their little exercise of building the dikes ever higher. But in the process they have forced the Yellow River to flow along a sinuous ridge of its own making, inevitably magnifying the enormity of flooding in the event of a single break anywhere along the hundreds of miles of levees. They call the river "China's Sorrow" for there have been innumerable floods with huge loss of life and crops. And still they struggle to force an old-age stream to comport itself abnormally so that they might cultivate the soil born of the flood.

A major problem for the levee menders after each serious break has always been how to get the escaped river back up onto its channelled ridge when water's normal inclination is to flow downhill. Sometimes they succeed and sometimes they don't. In the last 3,500 years there have been 26 major channel changes, a discomfitting situation for those whose fortunes are directly tied to stream behavior.

Since 1960 the situation has been significantly altered for the first time in history. A massive dam completed at San-men Gorge near the Yellow River exit onto the Plain, has blocked the loessal sediments and they are piling up behind it. Out on the North China Plain the river with no suspended load is scouring its channel and the danger of flooding has been greatly reduced. All is well for a few years. The official estimate had been that the San-men dam was good for 50 years before it completely silted up. However, measurements have shown that the accumulation behind the dam is far ahead of schedule and that by 1990 it will be San-men waterfall. Once again the Yellow River downstream will be filled with substantially more solids than it can handle.

The government plan is for a system of as many as 46 dams on the master stream and its tributaries, and during the short time gained before they all silt up, the effort will be made to terrace, irrigate and reforest most of the semi-arid loess watershed. Not much time to accomplish a huge project. It will be interesting to watch.

Fig. 23–10 Old-age stream. Meanders don't come any more involuted than these. A careful examination of the wide valley bottom will reveal oxbow lakes and abandoned looping channels.

it was occupied. De Soto found it to be a 300-mile (483 km) wide malarial swamp. But such valleys have a certain attractiveness to an agricultural civilization if only the river can be tamed. The deep and constantly rejuvenated soils and dead-flat terrain make exceptionally productive farmland. So, clear the tangled vegetation and drain the swamps and then above all make the river stay in its channel where all self-respecting rivers are supposed to flow. The most

obvious and simplest way to do this is to build up the natural levees so that at next year's high water, they will be just high enough to contain it. This works nicely except that all the sediments that used to be deposited on the floodplain must now be dropped in the channel bottom. A few years of this and the channel has built itself up to the point where high water comes dangerously close to the levee top, and the levee must be extended a little higher (Fig. 23–11).

A stream flowing 50 feet (15 m) above the roofs of their houses is a dangerous neighbor.

Perhaps people should not attempt to farm old-age stream valleys, for they are trying to have their cake and eat it too. The deep soils are there because old-age streams are supposed to flood as a mechanism to get rid of surplus load. In making the river stay in its channel, humans are creating an artificiality, and a stream flowing along a ridgetop 30 to 50 to 100 feet (9 to 15 to 30 m) above the roofs of their houses is a dangerous neighbor (Fig. 23–12). There may be other ways to control the river and many have been tried, but none with complete success as yet.

DELTAS AND ALLUVIAL FANS

In order to be classified as full-fledged gradational agents, streams must pick up, transport, and deposit materials. Youthful streams, aided and abetted by weathering, are es-

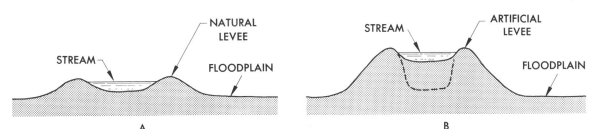

Fig. 23–11 Levees. The natural levee (A) serves to contain the river during normal water periods, but artificial levees (B) that hold the water in its channel, even during high water, must be built constantly higher to compensate for increased stream bottom deposition.

Fig. 23–12 Life can be a damp experience when one opts to reside in the Mississippi floodplain. All of the artificial constraints in the world are no guarantee that the great river will not revert to norm.

pecially effective in picking up sediments of all sizes, but even mature and old-age streams can erode if often only laterally. And then these sediments must be deposited elsewhere. Here the old-age stream is most efficient both in its channel and its floodplain, with the mature stream accomplishing the same objective to a lesser degree. But any river can be made to deposit simply by lowering its velocity. For instance, when a stream discharges into the sea or lake, no matter whether it is youthful, mature, or old-age, its velocity is suddenly checked, and it begins to deposit its load. First, the larger and heavier materials are dropped near the shore, and then, as its forward impetus declines, the progressively smaller sediments precipitate. This sorting action of stream deposits is typical. Such a depositional feature is called a *delta*, and it eventually builds up to sea level or slightly above, thus extending the shoreline. Often the delta

is much wider than the water course itself, as the offshore currents distribute the loose deposits extensively, or the stream chokes its own channel as the delta reaches sea level and constantly switches to alternate lower channels.

Along the western front of the California Sierra Nevada are a series of overlapping alluvial fans (alluvial piedmont) extending for hundreds of miles with their lower gentle slopes utilized for orchard plantings.

A very similar situation occurs when a high-velocity river suddenly runs out of a mountain canyon and onto a

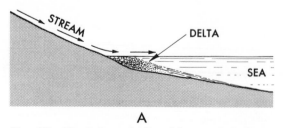

A

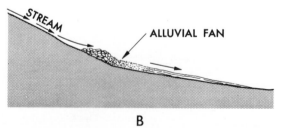

B

Fig. 23–13 Delta and Alluvial fans. Both display the sorting by size that is mandatory in stream deposition.

Fig. 23—14   Formed in much the same fashion as a delta, the fan-shaped deposit *(alluvial fan)* at the canyon mouth is a function of an abrupt lowering of stream velocity as it suddenly flows out onto a plain.

plain. Here again, an abrupt checking of velocity results in deposition along the mountain front and the dry-land equivalent of a delta is formed. Coarse materials are piled up into a steep slope at the valley mouth, succeeded by finer sediments and lesser slope (Fig. 23–13). Finally, the stream flows on across the plain carrying what it can. This type of deposit is called an *alluvial fan* (alluvial means water deposited) and is fan shaped because all deposition must be in the stream channel. As this becomes rapidly choked, the stream establishes a new channel to one side that in turn is fouled with deposits. The net result is that the stream oscillates from side to side, spreading its deposits in a triangle or fan, with its apex at the mouth of the mountain valley (Fig. 23–14). During drier periods, much of the water soaks into the coarse gravel at the head of the fan to percolate downslope underground and reappear as springs on the lower fan. Along the western front of the California Sierra Nevada are a series of overlapping alluvial fans *(alluvial piedmont)* extending for hundreds of miles with their lower gentle slopes utilized for orchard plantings.

CHAPTER 24

GLACIATION

Columbia glacier, Valdez, Alaska. A *rumpled icy river, flow lines clearly evident, appears to be in motion even in a still photo. Of course it is in motion—it's just that it progresses so slowly that an act of faith is required akin to believing that the hour hand on a clock really moves. Great icebergs periodically break off the leading edge and drift out into the nearby shipping channels endangering oil tankers as they service the Valdez Alaskan pipeline terminal.*

INTRODUCTION

Brittle ice like solid rock can be made plastic through pressure; however, ice requires a great deal less pressure. Usually, the simple accumulation of deep deposits of snow in a region where snow receipt is in excess of the rate of melt and where the terrain is subdued enough to retain the snow depth without loss via avalanche, is sufficient to start the process. The weight of the total snow mass plus alternating seasonal freezing and thawing will change the lower elements from snow crystals into ice pellets, which eventually compact into an ice sheet. Further weight from above will be translated into slow outward flow along the ground. Normally ice, like water, will find it easiest to move down whatever slope exists, but unlike water, if sufficient pressure is exerted from behind, ice can run up over moderate elevations in its path. Such a moving mass of ice can obviously accomplish a good deal of gradation.

Moving ice operates very similarly regardless of scale so that accessible valley glaciers are effective little laboratories.

Essentially, we recognize two types of glaciers: (1) *valley or alpine glaciers*, which are relatively small and restricted to a single valley or series of valleys in the mountains, and (2) *continental glaciers*, which move across whole continents or large parts of them. Valley glaciers are widespread in nearly all high-mountain ranges, and many of them are fairly easy of access (Fig. 24–1). But there are only two continental glaciers in existence today: the great ice sheet covering Antarctica and the smaller Greenland icecap (Fig. 24–2). Both are remote, and it becomes difficult if not impossible for the average researcher to observe continental ice sheets in action. Luckily, moving ice operates very similarly regardless of scale so that the accessible valley glaciers are effective little laboratories where we can study the principles of glacial behavior.

VALLEY GLACIERS

By observing the behavior of moving ice in all types of valley situations, that is, valleys filled with ice clear to the sea, receding glaciers that are no longer as far advanced as they have previously been, and abandoned valleys with the evi-

Fig. 24–1   Valley or Alpine glacier. Confined between limiting valey walls, the tongue of ice snakes its way downslope eroding as it goes.

enue of egress. In the bottom of the basin where the ice is thickest and the pressure greatest, a good deal of gouging and scraping will take place, thus deepening the basin floor. Also, large blocks of rock will be frozen into the underside of the ice, and as it begins to move, these will be carried away (*quarrying*), acting at the same time as tools or teeth to further cut and polish the floor. While the bottom of the basin of accumulation is being eroded in this manner, the edges of the ice around the perimeter of the basin will similarly take hold of rocks and soil by freezing them into the ice mass and pulling them away from the sides. This process is called *plucking* and acts to steepen the sides of the basin into cliffs. The action is continuous; new ice is immediately formed in contact with the cliffs as the plucked materials are carried off with the general flow. Gradually, constant plucking will enlarge the basin and cut it farther and farther back toward the mountain peak. The resulting roughly circular amphitheater with steep cliffs all around except the downslope side where the ice is spilling out is called a *cirque* (Fig. 24–3). In order for ice to develop cirque cliffs hundreds of feet in height, it need not fill the cirque to the top, but by simply being in contact with all cirque margins and continuously plucking at their base, the cliffs are undercut to the point where great masses of rock slide and tumble onto the ice surface from above and are carried off. If enough of these cirques are gnawing at a mountain's flanks, each working its way headward, they are capable of eroding away an entire peak.

If enough cirques are gnawing at a mountain's flanks they are capable of eroding away an entire peak.

Thus right here at the headwaters, so to speak, of a valley glacier, the ice is capable of removing large quantities of material (1) by freezing rocks into its underside which in turn cut and grind the cirque floor, (2) by plucking and freezing material into its sides wherever it is in contact, which again supplies it with tools for additional cutting, and (3) by carrying on its surface rocks and soil that have fallen from above.

As the tongues of ice push out of the cirque and down the original stream valleys, they continue to erode and

Fig. 24–2   Continental glacier. Towering cliffs of pristine blue-white ice push out to the sea along the Greenland shore. Huge chunks are snapped off periodically by sea-water buoyancy to form majestic bergs, a process called *calving*.

dence of ice erosion exposed to full view, we have a very clear picture of how moving ice functions and what kind of work it can accomplish.

Erosion

First, ice has a very considerable ability to erode and pick up material from the valley floor and sides. Let us begin with a basin of snow accumulation high in the mountains. Here the vertical pressure of deep snow is causing the bottom layers of ice to move horizontally. Obviously such a basin will have been drained of its annual meltwater by a stream flowing out of its downslope margin, and such a stream will have cut a valley. So when the ice attempts to move outward, it will follow this already existing easy av-

Fig. 24–3 This naked cirque once cradled an ice mass which nourished a sizable glacier. Now it has been abandoned as the ice melted back, and its stark outline is muted by rock slides from above.

modify the shape of the valley. A youthful stream valley in mountainous regions is normally V-shaped, but as the blunt-nosed ice lobe pushes down, it deepens and widens it into a U-shape (Fig. 24–4). The plowing action of the inexorable ice advance is not to be ignored, but the same types of freezing at the bottom and plucking at the sides as occurred in the cirque, continuing for as long as the ice occupies the valley, are immeasurably more effective.

The surface of a typical valley glacier is scarred with deep *crevasses,* usually crescentic in shape with their concavities presented upslope. Such ice chasms are a result of the frictional drag of the glacier along its sides and bottom, thus allowing the top central portion to move more rapidly (Fig. 24–5). Frequently, tributaries of the original stream that occupied the valley floor flow down the overhanging mountain slopes and across the surface of the ice, only to plunge down the crevasses and eventually find their way out via under-ice tunnels. Such streams usually deposit part of their sedimentary load on the ice as their velocity is suddenly lessened, and it is mingled here with piles of loose

material that slide down the continually undercut valley sides. Many glaciers eventually become so mantled with rock debris that they appear as massive mud slides rather than as ice.

Once the ice has melted away the real effects of ice erosion become apparent in the abandoned valley. Outstanding is Yosemite Valley in the California Sierras (Fig. 24–6) where many thousands of years ago, temperatures must have been sufficiently cooler than they are today to support a sizable glacier. Now it is gone, but the ice-modified valley remains with its evidences of former ice occupation visible to even the casual tourist. The Merced River flows peacefully along the broad valley floor that is somewhat flattened today by post-ice alluvial deposits. The sheer valley sides are polished and scarred with *striations,* deep grooves cut by the rocks frozen into the side of the ice as it advanced (Fig. 24–7). Near the top of these cliffs, above the level of the ice surface, are seen the sloping remnants of the old V-shaped stream valley; and the original tributaries to the pre-ice Merced River, which at one time flowed

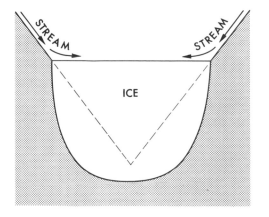

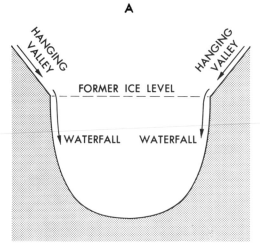

Fig. 24—4   Ice-modified youthful stream valley. (A) The dashed line traces the outline of the former river valley now occupied by ice, deepened and widened from V to U. (B) Tributaries of the original master stream fall freely as waterfalls from the mouths of *hanging valleys* after the ice has disappeared.

Fig. 24—5   Mer de Glace (sea of ice), Chamonix, in the French Alps. The glacier surface is far from smooth. Shear is inevitable as what is basically a brittle substance attempts to flow at several rates of speed simultaneously.

down these slopes and onto the glacier top, now occupy what are called *hanging valleys* and drop off the cliff top in scenic waterfalls. Far up at the head of the valley is a cirque with its deepened bottom occupied by a *tarn lake.*

At higher latitudes than California, where temperatures were cool enough to support ice clear to sea level, glaciers ran down valleys to the sea. Many of these are abandoned now, and the sea has pushed up the valleys in long fingers for many miles. These are called *fjords,* the Norwegian term being applied because of its wide occurrence along the Norwegian coast, but they are also found in large numbers in southeastern Alaska, British Columbia, southern Chile, and New Zealand. Aside from the fact that they are now arms of the sea, fjords exhibit features identical to those of Yosemite and other similar ice-carved valleys well above sea level. The sea has been able to invade fjords for great distances for two reasons. First, ice deepens the original stream valley, and since the stream

Fig. 24–6   There is no more classic example of the broad "U" shape and hanging valley waterfall than Yosemite. The Merced River, once a normal mountain stream course, has been strongly modified to Yosemite's mouth, but from that point on downslope it still occupies a "V".

reached the sea at sea level, the ice cutting has deepened the lower valley below sea level. Second, during the period when major ice sheets covered extensive land areas, the normal runoff of continental precipitation was held up. Thus the sea, from which all precipitation ultimately derives via

Fig. 24–7   The glacier has passed this way! It's gone now but its "footprints" remain. Both the straiae and glossy polish are evidences that a mass of ice with rocks and fine sand imbedded in it ground across this granite surface.

evaporation from its surface, must have been considerably lower. As the water level rose with general ice melting, the invasion of deepened valleys was inevitable. However, seldom does the sea push inland as far as the cirque, so that streams normally now flow down the old ice-glaciated valley into the head of the fjord. In many cases at these relatively high latitudes, the cirques and upper valleys still support small glaciers.

Deposition

Moving ice with its great erosional and transportational capacity must eventually deposit all of its accumulated load at a lower level. This is done in several ways, but first we must recognize that *as long as a glacier exists, it continues to move forward.* The so-called receding glacier is flowing forward and carrying its load of debris downslope despite the fact that the tip of the glacier is measurably retreating. This simply means that the rate of melt at its lowest extremity is more rapid than its rate of advance. Once the glacier has melted back to a high enough elevation for melt to decrease so that it equals advance, the nose of the ice becomes static. This is the normal situation, when melt and advance reach an equilibrium, but once again, whether the nose is advancing or retreating, the ice mass is moving forward. This is important to remember because this forward flow allows the glacier to function as an endless belt

Fig. 24–8 Terminal moraine. Here is the point where melting exceeds advance and whatever the glacier is carrying is dumped. Everything from monstrous boulders, to fine silt, to chunks of rotten ice are mixed helter skelter. Athabaska glacier, Jasper National Park.

and constantly to deliver its load of debris to the ice margin and deposit it there.

Glacial meltwater characteristically appears muddy or milky, as it carries a heavy load of sediment prepared for it by the abrasive grinding action of the glacial mill.

The chunks of rock frozen into the glacier and carried on its surface vary in size from many tons to fine ground powder. When they come to the end of the line, the ice melts out from under them, and they are dumped helter-skelter into an unsorted heap at the ice margin. Mixed in with the boulders and silt may well be chunks of rotten ice breaking off the end of the glacier. These melt away before too long leaving a cavity, which when filled with rainwater, is called a *kettle lake*. When the glacier nose is static (neither advancing nor retreating), this heap of loose material builds up to a considerable height and is called a *terminal moraine*. Its upslope side next to the ice is typically quite steep, but the downslope gradient is modified by rocks

sliding and tumbling as they are released by the ice (Fig. 24–8).

Occasionally, a terminal moraine ranging from valley wall to valley wall will act as a dam to impound the glacial meltwater, and as the glacier eventually melts away, long *finger lakes* will appear, as in the Italian Alps (Fig. 24–9). More commonly, however, the streams issuing from under the melting ice will break through the moraine here and there and flow off down the valley. This meltwater characteristically appears muddy or milky, as it carries a heavy load of sediment prepared for it by the abrasive grinding action of the glacial mill, and it deposits the surplus as soon as possible on leaving the glacier. The usual point of deposition is at the downslope margin of the moraine. Here glacio-fluvial deposits take the form of alluvial fans as the streams constantly choke their channels and seek new ones, eventually forming convex aprons of sorted alluvium in advance of the moraine stretching far down the valley. This feature is called a *valley train* and may be separated from the moraine to which it is attached by noting the absence of sorting of materials in the ice-deposited moraine as opposed to the careful sorting according to size of the alluvium.

There is one other area of deposition in conjunction

Fig. 24—9 The glacier that once moved down-valley from the snowy peaks in the distance deposited its morainal debris in the direct foreground of the photo, forming a dam. Roads and houses now take advantage of the modest elevation. Lake Wanaka, New Zealand.

with the lower end of the glacier and that is under the ice itself. Near the nose of the glacier, not only is the underside of the ice heavily charged with the debris that it has picked up in its journey, but the holding capacity of the ice is becoming less as the rate of melt increases. Thus, like an overloaded old-age stream, the glacier must drop some of its load. This deposition of all sizes of materials is under the ice, the weight of which compacts it into what is usually described as boulder-studded clay. It is called *till*, or *ground moraine* (Fig. 24–10).

So there are three types of glacial deposition near the nose of the ice: (1) the till deposited as a compacted flat mass under the ice; (2) the terminal moraine as an asymmetrical crescent-shaped ridge deposited at the ice margin; and (3) the valley train as a convex-shaped, sorted alluvial apron deposited by meltwater at the outer edge of the moraine.

Ice does not appear to flow and must be measured carefully over a period of years to ascertain even slight movement.

One word of caution relative to the speed of these operations: the entire concept of moving ice as an agent of gradation is dependent on an understanding of the time involved. Even running water, which requires considerable time to accomplish significant gradation, is more obvious than ice, for water moves rapidly, can be observed carrying and depositing materials and, on a small scale as in the formation of gullies, can demonstrate its cutting power before ones very eyes. But ice does not appear to flow and must be measured carefully over a period of years to ascertain even slight movement. The best way to appreciate what it is capable of doing is through a before-and-after comparison of mountain valleys. Here we can find ice in all its phases and fully appreciate its capabilities.

CONTINENTAL GLACIERS

It is one thing to point to the continental glaciers of Greenland and Antarctica (Fig. 24–11) and classify them as the only ones in the world, but it is quite another to convince oneself of the radical theory that much more extensive ice sheets covered large parts of North America and Europe in the none too distant past. This is a difficult proposition to swallow and a good deal of very convincing evidence had to be presented before the unbelievers came around. Today we not only have conclusive proof that such

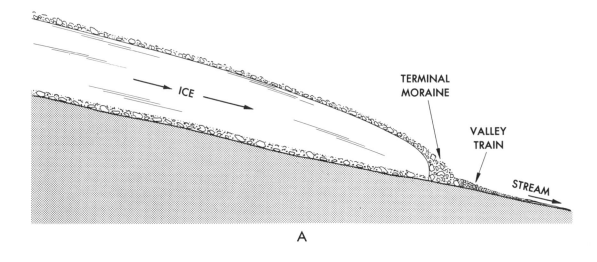

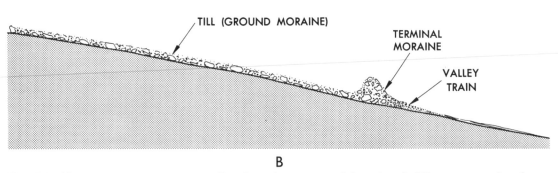

Fig. 24–10 Glacial deposition. (A) A valley glacier brings material downslope building up a complex of depositional forms at its terminus. (B) After the ice has melted back.

continental glaciers did indeed exist, but we know all the details of multiple advances and retreats. The history of the ice is written on the land for all to see, but the observers were blind until they could imagine that such a thing was possible. It remained for Louis Agassiz, a Swiss naturalist, to first suggest in 1834 that ice-abandoned valleys in the Alps must have at one time supported glaciers. He based his reasoning on a comparison of the similarities of deposits between the empty valley and one in which an active glacier was present. In other words, ice left evidences of its former existence. Some of these evidences had been puzzling geologists in both Europe and the United States for many years. Why were large boulders present in Iowa that had no relationship to the local bedrock but did resemble

that of Canada? Some researchers had suggested answers such as Noah's flood. And what about those deep grooves in solid rock, as on Kelley's Island, Ohio? Probably made by dinosaurs in the rutting season? Agassiz's theories answered these questions and many more very logically, and further investigation gradually built up a body of irrefutable proof that continental glaciers must have existed. It is entirely fitting that Louis Agassiz, the champion of continental glaciers that no one ever witnessed, should have his memory perpetuated in the name of a lake that must have existed once but vanished thousands of years ago. *Lake Agassiz* at one time covered a large part of North Dakota and adjacent Manitoba.

About a million years ago ice began to form in Can-

Louis Agassiz

The problem was *erratics*, great boulders out of place in England when it could be proved that they were of Scandanavian origin. Early 19th century experts on matters of these kinds were in fairly unanimous agreement that Noah's flood had washed them over there, or as one of their member, the pioneer British geologist Charles Lyell suggested with unknowing prescience, perhaps icebergs afloat on the great floodwater could have carried a cargo of rocks frozen into them to be released upon melting. The fanciful notion that immense sheets of ice now long gone had the capacity to transport loads of debris and that Noah was not in the act, simply didn't enter anyone's mind.

The Swiss farmer lives and functions daily cheek-by-jowl with the glaciers. To him it had never been a novel idea that ice moves forward and then retreats leaving as evidence of its passing erratics, piles of unsorted rocks and soil, scratched and polished valley walls. But simple everyday peasant wisdom had somehow eluded the famous scientists who were only now beginning to free themselves from the long ingrained reflex that one consults the Bible, not necessarily the facts, for answers to perplexing problems in nature.

The first truly scientific postulation that a continental ice sheet had overrun much of Europe was presented to the Swiss Society of Natural Science in 1829 by Ignatz Venetz. He was a Swiss bridge engineer who had been in the mountains and talked with the inhabitants of glacial valleys. His own follow-up research convinced him that they knew whereof they spoke, but the Swiss Society was not impressed. Neither were they five years later when amateur naturalist Jean de Charpentier put forth a

somewhat refined version of the Venetz hypothesis. Non-professional apostates always find the "anointed" a difficult audience to convince.

Nonetheless in that audience in 1834 was 27-year-old Louis Agassiz (1807–1873), professor of natural history at the University of Neuchatel, and a well-regarded expert on fossil fish and marine biology. He was intrigued and accepted an invitation from Charpentier to spend the summer at Bex in the Rhone Valley. The first-hand evidences that he observed of every type of glacial activity thoroughly convinced him of the righteousness of the cause and he became a vigorous researcher, writer and apostle.

Agassiz proved to be supremely adept in pursuing research grants so that his year-round projects were well-funded and staffed. His two books, *Studies in Glaciers* and *Glacial Systems*, were handsomely printed and distributed widely sparking great interest and considerable controversy. But his theories were supported by a huge volume of impressive research too—for several years his home base was a permanent research station on the edge of Unteraar glacier. And the doubters were coming around, even Lyell eventually. It should be remembered that in this era, talk of continental glaciers had to be a radical theory since Antarctica remained yet to be discovered; it was not even recognized that Greenland was capped by a single huge ice sheet.

In 1846, Agassiz arrived in America as a lecturer in the new science of glaciology. He found evidences of continental glaciation everywhere he looked. At Harvard where he filled a specially created chair, he was much respected and his courses and lectures were in great demand. But ultimately, far from his beloved Unteraar, his interest in ice behavior waned. Now a permanent American resident, Agassiz continued his scientific research but reverted to his old biological interests. During the last decade of his life, still exhibiting great enthusiasm for a cause, he gleefully attacked the new theories of evolution being espoused by Charles Darwin. Ironic that the early scoffer at the constraints of biblical dogma on true science should end his career as the defender of creationism against new and enlightened ideas.

After Louis Agassiz' death in Boston at the age of 66, a 2,500 ton (1,859 t) erratic, striated and polished by glacial ice, was brought from the Swiss Alps to mark his grave.

Fig. 24–11 The margin of the Antarctic ice sheet advances into a chilly sea. Icebergs here are relatively thin and tabular in contrast to the impressive craggy Greenland variety.

ada and northern Europe in much the same manner as has been described in mountain basins, that is, a continuing accumulation of snow gradually changing its lower elements into ice over a very extensive area. As the depth and pressure increased, the ice was forced outward in all directions. In some cases this was downslope or, as in much of North America, across generally subdued terrain, but in Europe the pressure was sufficient to cause ice, centered in the Karelia/Finnish region, to surmount and overrun such major obstructions as the Scandinavian mountains and Scottish highlands. Even in North America, the Adirondacks, New England mountains, and northern Appalachians were overwhelmed by an extensive sheet of continental ice. Four or five major advances and retreats have been traced, extending roughly as far south as the Ohio and Missouri rivers in the United States (Fig. 24–12) and southern England to central Germany and the Ukraine in Europe, a result of fluctuating temperatures over the centuries. But on the whole, the behavior of these continental glaciers and the gradational work they accomplished were almost identical to those of ice in Alpine glaciers except that not being confined to limited valleys, the scope of their influence was very much larger.

Erosion

By carefully examining the countryside over which the continental glaciers must have swept again and again, we may establish two somewhat differing zones. First, there was an inner zone extending a few hundred miles out in all directions from the center or centers of ice origin. Here the ice was thickest as it passed over the land and therefore exerted a great deal of vertical pressure. Also, the underside was fresh, clean, and relatively unclogged with soils and rocks, so that as a youthful stream, it had a great capacity for freezing materials into it. This might be called the *zone of active erosion*, for as we observe the surface today, it is obvious that the glacial effect was not only to remove all soil right down to bedrock but to gouge and abrade the rock itself, cutting out in many places deep cavities that have subsequently become lakes (the Great Lakes, for example). There is a great deal of similarity in the general aspect of landscape in northern Minnesota, Michigan, Wisconsin, and eastern Canada and that of Finland and northern Russia. The thousands of lakes at varying levels, scrambled drainage, and great exposure of bedrock are typical everywhere. And the uplands overrun by such ice sheets are rounded, striated, and polished (Fig. 24–13).

There is a great deal of similarity in the general aspect of landscape in Northern Minnesota and Finland.

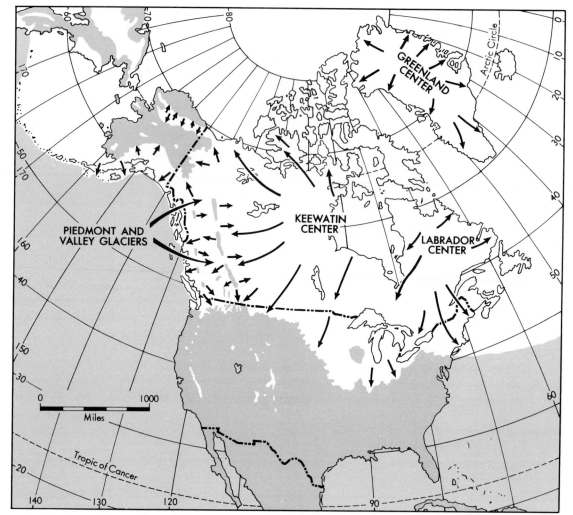

Fig. 24–12 Extent of continental glaciation in North America.

Deposition

The second zone is beyond the first, far out from the center of activation, and especially near the southern edge of the glacier. Here the ice was thinner, melting much more rapidly, and heavily choked with the waste it had accumulated in passing over the erosional zone. This is the *zone of active deposition*. Like the old-age stream, the ice was not only overloaded but demonstrated increasingly less capacity for picking up and carrying material. Here too the leading edge of the glacier had reached a point of equilibrium between melt and advance, and even minor temperature changes could cause violent fluctuations in the position of the ice front. Extensive moraine belts, great loops extending for hundreds of miles, occur throughout this zone and indicate temporary pauses and readjustments of the glacial margin. In front of each moraine is the alluvial apron laid down by meltwater, here called *outwash plain* (identical to the valley train except that it is much more exten-

Fig. 24–13 Parts of coastal Greenland are no longer under the continental ice, but the glacial grinding, cutting and polishing of the recent past is mutely evident in today's countryside.

sive), and throughout the region are great depths of ice-compacted till effectively masking the original terrain details. This is the history of the North German Plain and much of the Middle West.

The driftless area of *southwestern Wisconsin. Here is an example of what the Middle West might have been like without glaciation.*

The question is often asked, "From the standpoint of humans' utilization of the land, has continental glaciation been advantageous?" To those who occupy the zone of deposition, the answer is probably "Yes". Deep new soils and generally subdued landscape have helped to make this region one of productive farms. It is possible that the soils are too sandy or gravelly in places and clayey in others, and disrupted drainage has caused considerable swamp development, but on the whole, such soils can be made productive with proper management. An excellent comparison with "what might have been" is provided by the *driftless area* of southwestern Wisconsin where for unknown reasons the glaciers completely surrounded but did not move over several thousand square miles.[2] Here is a sample of what the Middle West might have been like without glaciation—rough hill country with thin soil, except in the river bottoms, and generally an unproductive region.

In the zone of erosion, glacial action in removing the soil has ruled out agriculture, but at these northerly latitudes, agriculture could only be marginal at best. So in laying bare the bedrock, a good many mineral deposits have been exposed at the surface, the disorganized drainage has produced hydroelectric sites, and the thousands of lakes make a fisherman's paradise. Even here glaciation might be interpreted as a blessing.

SECONDARY EFFECTS OF GLACIATION

Continental glaciation also has some secondary effects beyond the direct effects of the passage of ice over land. The general lowering of the world's sea level has already been mentioned, and it is obvious that coastlines everywhere have felt the repercussions of such an effect. For instance, the movements of animals and humans across land bridges that must have existed in Southeast Asia and the Bering Strait are indirectly related to glaciation through the lowering of the sea.

And then there is the great volume of meltwater released on the land. Existing rivers must have carried more water than they do today and were more effective eroders. In many places, enclosed basins such as the one occupied by the Great Salt Lake were filled to the brim with huge

[2] Drift is a general term referring to all types of glacial deposition; therefore "driftless area" is a negative terminology denoting no drift and thus no glaciation.

Fig. 24—14   Ancient Lake Agassiz once covered all of this country near Portage LaPrairie, Manitoba, and much more. But one must conjure up considerable imagination to visualize this flat and fertile plain at the bottom of a lake of glacial meltwater.

lakes. Old beach lines of ancient *Lake Bonneville* are easily visible along the Wasatch slopes near Salt Lake City, and the salt deposits of the now evaporated lake are evident in the well-known Bonneville salt flats. Lake Agassiz was formed when normal drainage to the north was blocked by the ice, and until the glacier disappeared, the lake maintained itself. Today only Lake Winnipeg remains as a tiny remnant of the original (Fig. 24–14). Ice also blocked the northeast drainage of the Great Lakes via the St. Lawrence River and caused them first to drain southward via the Chicago River to the Mississippi and later to drain eastward through the Mohawk gap and the Hudson River. The only good break through the Appalachians occurs here, formed by a tremendous stream of meltwater. The Erie Canal, so influential in our history, was virtually foreordained by this great channel. Equally important in the Pacific Northwest is the Grand Coulee, a drainage channel dug out by the ice-diverted Columbia River carrying many times the volume of water it does today. After the ice melted back, the Columbia abandoned the Coulee and resumed its normal course, but the Grand Coulee Dam at the head of the Coulee now can direct Columbia water once again down this ready-made channel to irrigate the desert (Fig. 24–15).

The movements of animals and humans across land bridges that must have existed in Southeast Asia and the Bering Strait are indirectly related to glaciation.

These are the gradational effects of moving ice. What caused the sporadic cooling and warming that allowed the continental glaciers to advance and retreat and caused the development of extensive valley glaciers in the high mountains all over the world? Nobody knows—speculation is rife and there are theories without end. We do know that the overall long-term climatic trend today is warming, and

Fig. 24–15   Looking south down the Grand Coulee. The contemporary Columbia flows away to the right, but an ice-blocked, melt-water-swollen Columbia once surged southward cutting a massive channel. Water in the Coulee bottom today is stored for irrigation.

nearly all existing glaciers are in retreat. How long will this last? It is perhaps provocative to note that the time lapse between the last retreat of the continental glaciers and the present is not so long as some of the inter-ice periods between advance and retreat. Maybe we are currently in an inter-ice period and there are more advances to come.

CHAPTER 25

WIND

The desert and the beach are the great arenas for wind motivated sand, restless dunes constantly in motion. Wind-carried sand, moreover, is an effective cutter and erodes although each grain is too heavy to stay airborne for very long.

INTRODUCTION

The dusty streamer trailing out behind a car driven along an unpaved road, the rising cloud above a tractor plowing a field, or the swirl of dust across the skinned portion of a baseball infield, momentarily disrupting the game—all are everyday evidences that wind transports minuscule bits of rock and soil. But can this activity be translated into significant quantities? After all, it is obvious that the huge boulder, which the glacier, the raging torrent, or the storm-driven wave can handle with impressive ease, is well beyond the capacity of even tornadic winds to so much as jostle. Small pebbles edged along the ground are probably the absolute maximum size to which strong winds can impart any motion.

Small pebbles edged along the ground are probably the absolute maximum size to which strong winds can impart any motion.

One generally accepted estimate by A. K. Lobeck gives us a statistical base with which to work. He calculated that the average dust storm suspended one ounce (28.35 g) of solids in each 10 cubic feet (.28 m^3) of air, not a very imposing figure. But take some of these 10-foot (.28-m) cubes and stack them up for several hundred feet (meters), then extend this along a front of 300 miles (483 km), and finally pile them in depth for another couple of hundred miles (322 km) to cover the area involved in a normal dust storm. In other words, form a new three-dimensional air mass 300 feet (91 m) high, 300 miles (483 km) long, and 200 miles (322 km) wide. Add a shallow bottom stratum of much heavier sand so that now we are talking in terms of tens or even hundreds of millions of tons (90,718,500 t) of suspended solids moving along at a rapid rate. This is one average dust storm. Multiply this many times and pump into the equation an element of time, and we have an impressive operation (Fig. 25–1). Any medium that can accomplish transfer of earth materials of this magnitude simply has to be classed as a gradational agent.

However, although not proscribed by gravity in the same way as are rivers and glaciers, there are limitations

Fig. 25–1   This photo was taken in semiarid Western Queensland, Australia, in 1965 during one of its periodic droughts. Government soil conservationists, running tests during this particular dust storm, estimated that 50,000 square miles (80,465 km^2) of countryside were involved and that over a billion tons (907,185,000 t) of soil were taken aloft.

beyond mere size of particle to what moving air can accomplish in the realm of gradation. In order for wind to be an effective agent, loose dry sands and soils must be present at the surface protected by a minimal vegetative cover. This means that to a large extent, wind erosion is relegated to the arid regions of the world where vegetation is scarce to nonexistent and the lack of moisture in the soil keeps the particles from clodding and adhering one to the other. However, it is possible for beach sands, even in humid areas, to dry out rapidly above the high-tide line and thus to be subject to wind attack. Also, it has been demonstrated in several parts of the world that glacial debris or alluvial deposits may be removed by wind if its velocity and the vegetation and moisture conditions allow. But all things considered, of the several gradational forces with which we are dealing, wind is probably one of the least widespread and effective.

TRANSPORTATION AND EROSION

The carrying capacity of wind and its ability to pick up materials is, like running water, determined by its velocity. Almost any vagrant breeze can sweep up dust and fine particles, but it takes a steady wind of at least moderate veloc-ity to move sand. Once these sharp-edged particles are in suspension, the moving air has armed itself with effective cutting tools to wear away whatever is in its path, thereby supplying itself with still more materials. The original load may have derived from a mantle of mechanically weathered rock or the loosely deposited alluvium of an infrequent desert stream. Their removal lowers the surface, often exposing the bedrock, which is polished and abraded as the wind-blown materials scour it. If bedrock is hard and massive, it resists abrasion and will acquire a high polish, but if it is of softer material, it will be progressively cut away. Frequently, sandstone is exposed on the desert floor or on cliff faces, the hard sand grains cemented by softer shales that can be rapidly eroded, freeing the sharp-edged quartz to aid in further cutting. On the other hand, when sizable stones and pebbles are mixed with finer material, the erosive action of the wind is selective and only particles up to sand size are removed, the desert floor eventually becoming paved with a mosaic of smoothed cobbles called *desert pavement* that protects the underlying strata.

Any surface exposed to the constant sand-blast effect of wind-carried debris, especially sand, will inevitably be cut, chewed, and polished (Fig. 25–2). Even automobiles caught in sandstorms have their paint removed, metal surfaces pitted, and glass frosted in a remarkably short time. Rocky cliffs facing prevailing winds in the desert display

Fig. 25–2 Pitting, polishing and abrading of rock strata of varying resistance are clearly evident in the formation of Angel Arch in Utah. The arch itself is the result of crossbedding and vertical faulting, but its current smoothed outline has almost certainly been influenced by a natural sandblasting.

Deflation in the Sahara

Egyptian basins and oases.

Below Sea Level

Above Sea Level

Perhaps the largest deflation basin in the world is the Qattara depression in the Egyptian Sahara west of the Nile—185 miles (298 km) long, 75 to 100 miles (121 to 161 km) wide, and 420 feet (128 m) below sea level. Nobody suggests that the wind alone cut out this huge cavity. Surface runoff during desert storms probably had a hand in originally breaching the rock, but the fact that the bottom is below sea level is significant. It is definitely not a structural feature, and intermittent streams flowing to the sea cannot bite below that base level. The wind is patently the basic factor and it has worked the surface downward until it reached the groundwater table. The entire bottom of the basin is an extensive salt marsh and has stabilized at this level since the wind cannot remove the damp soil. But it is a repellent region to humans; the marsh is so treacherous that the Qattara was an effective defensive barrier employed by the British against Rommel's Afrika Corps in World War II.

Today there is a tentative plan (remaining to be imple-

mented by the Egyptian authorities) to channel Mediterranean water from El Alamein, less than 50 miles (80 km) away, into the depression. In theory the plunging stream would activate turbines for power generation and the huge new inland sea would raise water tables, change the local climate, and in general rejuvenate its Saharan neighborhood.

The Qattara is not the only wind-excavated depression in this part of the Sahara. There is a whole family of lesser ones in the immediate vicinity. Most of them are oases, some below sea level and some above, but nearly all are at a groundwater base level, which, by allowing flowing springs, seepages, and wells, both nourishes life and limits further deepening by wind action. To the south and west of this extensive region of deflation is a large area of sand dunes and ridges, which are rather obviously at least a partial product of the material blown out of the basins upwind.

tremendous honeycombing as the softer elements are etched away, leaving the hard materials in high relief. One of the peculiar features of some desert landscapes often attributed to wind erosion is top-heavy balanced pinnacles. Whether wind is responsible for the original pinnacle is questionable, but the more rapid cutting at the bottom is typical of wind action, for sand, the most effective cutting tool, is relatively heavy and seldom is suspended more than a foot off the ground. Observation of telephone poles when one is driving through dry country will reveal either metal shields attached at ground level or a pile of rocks to protect them from being cut down by the wind—and the wood of the lower portion of the pole will display an etched and polished surface.

Although hard and generally resistant to wear, the individual sand grains employed in this continuous sand blasting eventually lose their sharp corners and become rounded. Beach and riverine sands, although utilized similarly as cutting tools, do not display quite this degree of wear, so the geologist can often make a reasonable distinction between ancient desert sand deposits and those of the beach, giving an important clue in efforts to reconstruct the evolution of a given region.

As a transporting agent, wind may carry tiny dust particles high into the atmosphere and transport them great distances. Colored sunsets and muddy rainfalls are often experienced thousands of miles from major dust storms or explosive volcanic eruptions. The plowing of the buffalo grass along the eastern Rocky Mountain front followed by an extended drought in the thirties gave rise to an extensive "dust bowl" when the prevailing westerly winds carried the fine-grained topsoil eastward in great clouds. Much the same experience has been endured along the southeastern semiarid margins of the Australian desert and in the so-called "virgin lands" of south-central Siberia.

Automobiles caught in sandstorms have their paint removed, metal surfaces pitted, and glass frosted in a remarkably short time.

Deflation

Wind does not always move over the surface of these arid regions from a single direction or at a constant rate. Often it is gusty and capricious, and the something less than flat surface terrain sets in motion channeled air streams and eddy currents. For instance, a swirling eddy in the lee of a minor ridge may cut downward and dig out a sizable basin. This general digging ability of swirling wind currents is called *deflation*. Once the basin is begun, its conformation encourages further eddy currents and stronger ones, and if the subsurface material being worked is loose soil or easily eroded rock strata, huge basins of great depth may be excavated.

Canyons, slots, and narrow gorges act to capture moving air from any of several directions and direct it, funnel-like, down the chute.

Channeled Air Currents

When wind moves across broken terrain, the eddy current directed downward on the lee side of ridges is not the only effect. Canyons, slots, and narrow gorges act to capture the moving air from any of several directions and direct it, funnel-like, down the chute. Such channeled winds not only blow constantly from the same direction but are squeezed through a restricted orifice and their velocity is greatly magnified. The accelerated polishing and scouring of both the floor and walls of constricted defiles by debris-carrying high-velocity winds are strikingly evident.

DEPOSITION

Extremely fine dust carried from the desert by prevailing air currents, such as Asia's monsoon, is inclined to drift like so much snow against impediments in its path and to fill in the low spots where terrain is broken. Air is further cleansed of solid suspended particles by increasing precipitation beyond the desert margin, as each tiny droplet forms about a dust nucleus. All of this deposited material is called *loess* (see Chapter 15).

Larger particles than those that go to make up simple dust and/or loess, particularly sand, require higher velocity winds than are normal to be taken into suspension. These are commonly carried short distances just above the surface and pushed along the ground, piling up and drifting

Fig. 25–3 On the surface of every massive, wind-driven dune are myriad ripples, each a tiny dune unto itself.

against all sorts of minor obstacles. The result is the formation of dunes along beaches and in those portions of the desert where sand is plentiful. If the wind is rather constant from one direction, the dunes are likely to take on a crescentic shape with the convex exposure upwind and of rather gentle slope, while the concave inner slope is very steep *(barchans)*. As long as the sand on the windward slope is not anchored by vegetation growing on it and is free to move, the entire dune will migrate slowly forward as the wind erodes the front of the dune and dumps it forward grain by grain over the crest. As each grain falls into the lee of the dune and out of the wind, it drops down the abrupt backside, maintaining this steep slope. If the wind displays somewhat less constancy of direction, dunes will take on less classic, merely ridgelike forms. However, even these normally show a gentle windward slope versus a steeper lee. And on the surfaces of most dunes, close examination will reveal tiny rows of ripple marks at right angles to the wind, which are miniature dunes in themselves, moving

Fig. 25–4 Wind billowed sand at the Souf Oasis in Algeria. Groundwater is at the base of the sand so that each date grove must be excavated to this point. Sand is not necessarily beneficial to palms as the usual desert photo might imply—given no frost, great heat at the crown, and copious water at the root, date palms will prosper nicely almost anywhere.

But danger lurks out there in the dunes; they advance relentlessly and constantly threaten to overwhelm the farmer's artificialities. Many little men with many little shovels have kept the enemy at bay through the centuries, but the battle is endless.

The Waikato Dunes

Monterey pine on the dunes.

At the mouth of the Waikato River, 25 miles (40 km) south of Auckland, New Zealand are the billowing Waiuku sands. Energized by boisterous winds off the stormy Tasman Sea, these actively ambulating beach dunes have threatened to engulf the adjacent countryside as they march steadily inland. But successful efforts at controlling dune migration here date back only to about 1940 when experiments with planting the California Monterey pine (pinus radiata) proved successful. Through the years a system has been perfected which culminates in a mature radiata forest on the beach dunes, a forest which: (1) stabilizes sand movement, (2) produces a continuing supply of merchantable saw timber, and (3) as a happy by-product, provides for the harried urban vacationer a pleasant quasi-wilderness recreation site.

Currently the Waiuku sands support a sizable forest of mature trees and they are being cut and replanted in a controlled sustained-yield system. But today the planners are accommodating their forestry and dune stabilizing efforts to yet another intriguing complication—the sands of Waiuku are titanium-rich ironsands suitable for use as commercial ore. With the establishment a few years ago of an integrated steel mill just 8 miles (13 km) distant, the forest is in the midst of a mineral as well as a wood-products development.

In order to utilize efficiently this dual resource, the ore-rich region has been divided into 250-acre (101 ha) blocks for clearing and mining one at a time. Mature timber goes to the mill, the ironsand to a separator for preparation of blast furnace charges where the process is a simple scrubbing with water followed by magnetic separation of the useful titano-magnetite. All residue, which makes up a very high proportion of the initial product, is returned to the quarry site and a new forest begun. So, a sustained-yield, block-cutting, lumbering regime has been initiated complicated only slightly by the short-term borrowing of the soil so that some of its useful elements may be removed. Curiously, preliminary research tends to indicate that the trees grow slightly better on the processed sands than the original.

forward and exhibiting the typical asymmetrical outline (Fig. 25–3).

Along many beaches and desert margins, these moving dunes have been known to march inexorably across virtually anything in their path. Roads, oases, even woods and buildings have been overrun (Fig. 25–4). They can be stopped if some kind of vegetation is encouraged to grow on their surfaces, but sterile sand is not an ideal bedding ground for plants, and new dunes are constantly being formed (Fig. 25–5). In the Landes district of southern

Fig. 25–5 Planting marram grass (introduced from the North Sea coast of Europe) on a beach dune surface to slow its migration.

Fig. 25–6 In the distance typical flat-lying sedimentary beds, but those sandstones in the foreground are sedimentaries too and their bedding planes are at odd oblique angles. This is called *crossbedding*, clear evidence of dune origin rather than deposition in shallow water. In addition, each little sand grain displays a muting of its sharp edges, unlike those of the beach.

Contemporary erosion is working here on the softer cementing materials putting the resistant grains back into circulation after a lengthy masquerade as sedimentary rock.

France, dunes from the Bay of Biscay shore have eventually been fixed after overrunning hundreds of square miles of territory and even a sizable village. Today, planted pine trees in the old dune region produce valuable timber and turpentine.

Probably not more than 20% of the world's deserts are sandy.

There seems to be a commonly held misconception that most of the world's deserts are made up of endless miles of billowing sand. Such deserts are to be found and admittedly they are spectacular, but they are not so common as to be typical. In order for sand dunes to form, the supply of sand must develop more rapidly than it can be blown away: a seashore or a soft sandstone that erodes readily (Fig. 25–6). This occurs only now and again so that rocky deserts of one kind or another are more common than the sand dune. Probably not more than 20% of the world's deserts are sandy. In the Sahara, where research into wind erosion has been carried on for a much longer time than in other regions, three types of deserts are recognized: *hamada* or polished bed-rock desert, *reg* or pebble desert (including desert pavement), and *erg* or sandy desert.

CHAPTER 26

WAVES

The active margin of a restless sea. Waves are most effective during storms when they assail the base of each sea cliff with a maximum of fury, but even on a quiet day they work and rework beach materials to ever smaller size. The action is uneven. Harder rock strata are bypassed and left as miniature islands to be subdued later. Rock debris, reduced to sand size, forms shelving beaches which force the waves to sacrifice energy running up them. And those same sandy beaches, constantly rearranged by tide and surf, block the easy egress of the little creek.

So the beach zone is an interesting theatre. Overall though, the aim of a relentless sea is to reduce the dimensions of the world's landmass.

INTRODUCTION

Unlike clear running water that has little erosive capacity unless it carries sharp-edged tools, waves can throw themselves against an unprotected shoreline in such massive assaults that the sheer weight of the water is a powerful agent of gradation. No one who has not seen at first hand storm waves battering a cliff front or beach can fully appreciate the violence attained. Tons of water repeatedly dashed at high velocity against even solid rock can accomplish a tremendous amount of erosive work in only a few days or hours, and loose sand offers virtually no resistance. Not only the weight of the water itself, but its effect of compressing air in cracks, vents, or caves in a cliff helps to tear loose great chunks of rock. And like running water, seawater can take into solution certain minerals, thereby removing them from the rocks and weakening the entire structure. We must not forget the buoyancy of water, which allows the waves to lift massive rocks up to several tons in weight, once they have been loosened, and if not carry them away immediately, jostle them until they gradually work free.

All of this is accomplished without the aid of sands, gravels, and boulders as grinding and cutting agents. But waves do have cutting teeth of this kind, some very large, derived from the wearing away of shorelines, and their effect is to multiply many times the erosive effectiveness of waves.

The weaker or softer elements are incised first, somewhat like the selective honeycombing of a cliff face by blowing sand except that ocean waves, armed with an array of tools from huge boulders to tiny sharp-edged fragments of silica, can destroy at an immeasurably faster rate. Caves, arches, and deeply undercut bluffs result, while harder strata are bypassed and left behind as isolated islands and *stacks* by the retreating shore (Fig. 26–1). However, these latter hard-rock residuals, although a common feature of eroding coasts, are themselves transitional, for their resistance is merely relative and they too are amputated at their roots by vigorous wave action in a short time.

The buoyancy of water allows the waves to lift massive rocks up to several tons in weight.

Although every day is not a stormy day and waves do the bulk of their work during storms, they are continuously active, nonetheless, attempting to lower any land that stands

Fig. 26–1   Stacks, a result of selective wave erosion.

above sea level. Even on a calm day this work is visible, although at a much subdued rate. Island inhabitants know all about wave action, for it threatens them from all sides and they are frequently forced into a variety of attempts to thwart the encroachment of the sea.

Island inhabitants know all about wave action for it threatens them from all sides.

A strange example is the case of Falcon Island in the Southwest Pacific Tonga archipelago. First discovered as a tiny islet in 1885, it disappeared in 1894 only to reappear successively in 1896, 1909, and 1927. Each time it reared its head above the sea, wave planation reduced it to a mere reef, the last time in 1949. Born of vulcanism, destroyed by the surf, this tectonic/gradational standoff, if not typical, is certainly illustrative. Other classic cases are those of northwest Europe where the measured retreat of inhabited island coastlines has been carefully recorded from Roman times. The north York coast of England has worn back as

much as three miles (5 km), inundating several towns and villages and destroying many square miles of productive farmland. Out in the stormy North Sea is the island of Heligoland. A tiny fraction of its former self today, it would have long since disappeared except for its military value and the subsequent reinforcement of its margins by cement breakwaters and seawalls (Fig. 26–2).

WAVE ACTION AT SEA

The waters of the sea respond to the gravitational pull of sun and moon (tides), salinity and temperature differences (upwelling and convectional currents), Coriolis force (horizontal current patterns), sea-bottom seismic and volcanic activity (tidal waves or tsunamis), and wind (waves). This last, the normally observable ocean waves, is merely the wind ruffling the surface of the water. The depth at which they can be felt is a function of their size at the surface, but in no case is it measured at more than a few score of feet. So waves are extremely shallow, but because we are

Fig. 26–2   Known since the early Roman times, Heligoland was once a sizable island. We have a long record of its accelerated decline to the current less than 42 acres (17 ha). Today, its tortured red sandstone cliffs continue to be assailed by North Sea waves in the few places where protective devices have yet to be installed.

working with a huge homogeneous liquid mass, energy applied at the surface at one side of the ocean can be readily transmitted via waves many thousands of miles to the other. The high wind energy of a Caribbean hurricane is translated into a frenzied sea, and a few days to a week later is experienced as crashing surf in such diverse locations as the beaches of Scotland, Morocco, and Brazil.

The wave certainly gives an illusion of horizontal motion but the water is merely sloshing around in the container.

To generalize a bit, there are two distinctly differing types of waves: (1) the ponderous and predictable great ocean rollers with their periodicity and general parallelism and (2) the many variations on the totally disorganized, choppy, or broken sea. The rollers are the basic waves produced by major storms somewhere in the ocean basin and carrying that energy basin-wide; the others are simply kicked up by local breezes and temporarily superimposed atop the rollers. In either case, the water is not going anywhere—the wave certainly gives an illusion of horizontal motion but the water is merely sloshing around in the container. An excellent simile, frequently cited, is the rippling of tall grass in the wind. There is motion but no real forward progress. To put it another way, the *wave form* moves swiftly but not the *wave medium*. Watch a cork, a flat chip of wood, or a jellyfish. They bob up and down as each wave passes under them but there is no forward progress. Now substitute a Portuguese Man of War, a gaudy purple jellyfish with a cartilaginous vertical sail, for the "straight" jellyfish and he will move forward as the wind catches the sail, but out in front will be little visible ripples and behind a tiny wake indicating movement through the water—the water itself is not advancing at all.

WAVE ACTION ON THE SHORE

The actual motion then is entirely vertical and any given particle of water will follow a circular pattern, oscillating

roughly as deeply below a median sea level as the waves are above it, and eventually returning to its original point of departure. But when wave forms encounter a shoaling bottom, this easy circulation is impeded and a frictional drag sets in. Surface evidence that the waves are "feeling bottom" is a shortening up or crowding of the rollers as the forward ones are slowed and those coming up behind step on their heels. There is also a strong tendency for bottom drag to pull the wave fronts around to something approximating a parallel approach to the shoreline, no matter how broken that shore may be. This is called *wave refraction* (as with light through a prism, refraction simply means to deviate from a straight line) (Fig. 26–3).

It is at this point of friction with the ever shoaling bottom that the wave is forced to abandon its vertical circulation, and as the wave fronts abruptly steepen, the entire mass of water is thrown forward in a phalanx of breakers and rushes up the beach. Thus the line of breakers is some indication as to the shallowness of water. Generally speaking, a swimmer can wade out to the breakers even if they are far off shore, but if they are in close to the beach, it means a rapid dropoff instead of a gently shelving bottom.

At this critical location somewhat offshore where the drag begins, the first bit of shoreline erosion is accomplished as the water, attempting to complete its' vertical oscillation, bites into the bottom. Just beyond the breaker line, the bottom is lowered sufficiently to allow the water to achieve its circulatory cycle. Once this is accomplished, friction ceases and the line of breakers moves closer to the shore with the bottom erosion continuing just behind it. Thus a flat bench (*wave-cut bench*) is formed at a specific depth, sometimes cut into solid rock, at other times excavated easily from loose sands and gravels (Fig. 26–4). The materials derived from this operation are carried forward by the breaking wave, cutting and scouring as they go, to aid in the formation of the beach. If the bottom is not too shelving and the line of breakers is fairly close to shore, the forward rushing water will attack the shore and break away quantities of debris, especially if a high cliff is being undercut, and slides from above deliver added material periodically. All of this is worked over, broken, smoothed, and reduced in size by continued wave action, and then most of it is removed so that the waves can get back to work on the cliff itself.

Fig. 26–3 As waves approach the shore and begin to "feel bottom," their crests close in on each other and there is an effort by the whole wave system to swing around and roughly parallel the beach. That distant plunging breaker is additionally signaling the initial point of bottom erosion.

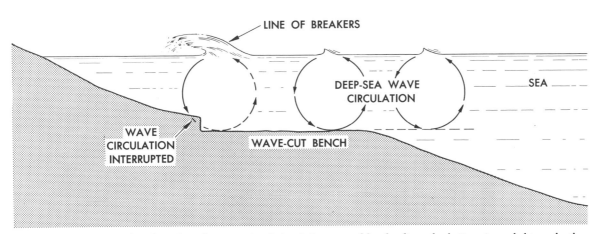

LINE OF BREAKERS

DEEP-SEA WAVE CIRCULATION

SEA

WAVE CIRCULATION INTERRUPTED

WAVE-CUT BENCH

Fig. 26–4 Wave-cut bench. When deep sea circulation is interrupted by shoaling, the bottom is eroded to a depth sufficient to allow resumption of normal circulation.

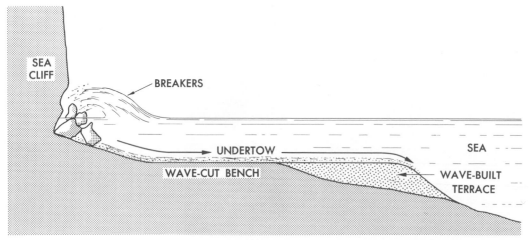

Fig. 26–5 Wave-built terrace. Fine materials, derived from the wearing away of bottom and sea cliff, eventually find their way to the outer edge of the wave-cut bench to be deposited as a terrace.

TRANSPORTATION AND DEPOSITION

The removal is accomplished in two ways. First, there is the *undertow* or the water retreating down the beach after each wave (often channeled into *rip currents* rather than a sheet of water). It carries much of the eroded debris with it. Some of this is moved back up the beach by the next wave, but the finer particles eventually drift out into deeper water and are deposited beyond the wave-cut bench, building up a depositional terrace (*wave-built terrace*) at the rear of

the bench (Fig. 26–5). The rocks and sand that repeatedly advance and retreat with each wave aid in smoothing and abrading the beach, while at the same time they are being reduced in size so that gradually they too may be carried off by the rip currents to be added to the wave-built terrace.

The second means by which material may be transported from its original site of erosion is via *longshore drift*. Very seldom does a wave strike the shore precisely parallel, despite the general effects of refraction, so that if the winds that form the waves are prevailing from a particular direc-

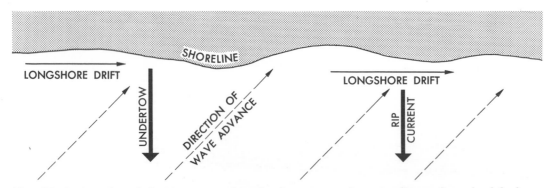

Fig. 26–6 Longshore drift. Waves approaching the shore at an angle carry sediments forward, while the undertow or rip currents return them directly downslope. The net movement of any given particle exposed to these alternating influences is lateral.

Fig. 26–7 A sandy beach is developing in each little cove. It is built of wave degraded materials, swept clean of the cliff bases, and sediments delivered by rivers. All is transported longshore to these quiet sanctuaries.

tion for any considerable length of time, there will be a movement of sediments and debris along the shore as well as onshore and offshore (Fig. 26–6). The undertow tends to move its load directly away from the shore downslope in answer to gravity, but the succeeding wave will carry the same material back up the beach at an angle, so that gradually beach-forming sands will be transported great distances longshore. Often they will come to rest in quiet coves, while cliff headlands, which are producing much of the material as the waves beat against them, will have their bases swept clear (Fig. 26–7). *Spits* and *bars*, generally parallel to the shore, are also products of longshore drift. Of course, the winds may shift, often seasonally or even daily, and reverse the direction of drift. Locally, tidal currents set in motion by some peculiarity in shoreline configuration may complicate the pattern. And storms will periodically disrupt with great violence the formation of drift features (Fig. 26–8).

Fig. 26–8 The little volcanic cone in the foreground, Mt. Maunganui, was captured by a spit working its way out from the mainland forming a *tombolo*. Inside the entrance is the commodious joint harbors of Tauranga and Mt. Maunganui, New Zealand.

The residents were greatly chagrined, living in a city named Long Beach with no beach at all.

During the construction of the breakwater enclosing the Los Angeles/Long Beach harbor, the famous beach for which Long Beach was named was almost entirely removed by new currents set in motion by the partially completed breakwater. For a number of years, the residents and city officials were greatly chagrined, living in a city named Long Beach with no beach at all. However, with the completion of the breakwater, the beach returned and all was well. This is not a rare occurrence, for longshore currents are fickle and subject to change as normal erosion alters the coastal outline. *Groins* or *jetties* at right angles to the shore are commonly seen as beach-front residents attempt to catch and hold the sand moving longshore. They are sometimes the subject of lawsuits by their neighbors downshore whose beaches would be the recipients of those sands if they were allowed to move normally.

SHORELINE CLASSIFICATION

The traditional method of organizing all of the world's diverse shorelines into a small number of simple categories has been to recognize those which are (1) *submerged* or *drowned*, (2) *emerged* or *uplifted*. No matter what the local detail, any coast can be assigned one or the other of these labels and reasonable generalizations can be made regarding the action of ocean waves on it. To illustrate, let us consider for the moment bodies of water smaller than the ocean.

Submerged Shorelines

Imagine a dam built across the course of a stream and backing up the water to form a lake, which, in finding its own level in an eroded terrain, develops a sharply crenate shore featuring deep embayments and prominent points of land. This is a drowned or submerged shoreline, and water will react in the same fashion whether it is on a large or small scale, salt or fresh (Fig. 26–9). The coast of Norway is an example of the invasion of the land by the sea advancing up the valleys while the ridges stand out as bold peninsular headlands—so are the coasts of southeastern Alaska, Newfoundland, the Mediterranean, and our own mid-Atlantic states. Perhaps the land is sinking rather than the water rising. It does not matter. Anytime we encounter this relative movement of land and water, whichever is the movee (even if both are), the shoreline of even a moderately eroded land surface will show this broken character.

Emerged Shorelines

The emergent shore, on the other hand, is theoretically smoothly shelved and featureless, the kind of shore we find along the margins of evaporating lakes such as the Great Salt Lake or the Caspian Sea. Here the former bottom is exposed, smoothed, and planed by wave action and often floored with water-laid sediments, a direct contrast to the steep and jagged character of the drowned coast.

Classification should therefore be easy: any variation on a smooth and gentle shore derives from emergence, and the rough and broken coast must result from submergence. However, from what we are learning of the recent history of the earth, especially events that would affect variation of sea level, it begins to appear that pure emergent shorelines are moderately rare—much more so than was at one time supposed. Given the melting of continental ice sheets and the general rise of sea level from 200 to 400 feet (61 to 122 m) in the last 20,000 or so years, it is unlikely that many coasts have escaped some sort of modern drowning.

Far upon a mountain slope we may encounter a series of marine terraces appearing as giant steps. Each platform, backed by a steep wall, can be demonstrated to be the product of past surf planation and sea cliff erosion (Fig. 26–10). This is good evidence of coastal uplift, as are elevated coral remnants that must, by definition, have been nourished during their formation at or below sea level. But even these observations do not constitute absolute proof that the most recent movement has been a relative lowering of the sea, although it certainly makes it appear likely.

The emerged shoreline does exist. The Baltic Sea and Hudson Bay are becoming shallower as the land responds to the unloading of its massive load of ice by drifting upward at a measurable rate; lava cascading down the slopes of Hawaii forms a new shoreline; live coral and mangrove extend the coast out to sea in the tropics and so does the active river delta. But this all means that we·must be a lit-

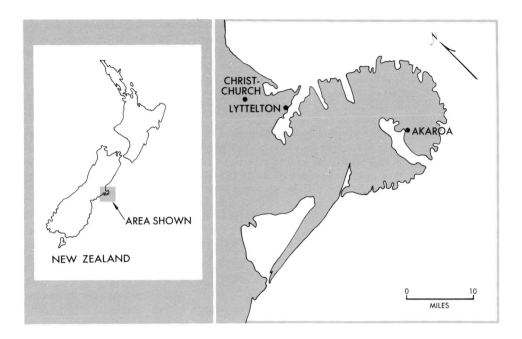

Fig. 26–9 Submerged shoreline. The sketch map indicates the location of Akaroa on Bank's peninsula, a partially foundered volcano. Originally it stood well offshore, a perfect cone complete with crater and symmetrical drainage pattern. But as it slowly sank (or the water rose), fingers of the sea pushed their way up each stream valley, even to the point of breaching and flooding the crater at Akaroa. The photo is taken from the hills behind the village with the Pacific to the left.

Holland's Delta Project

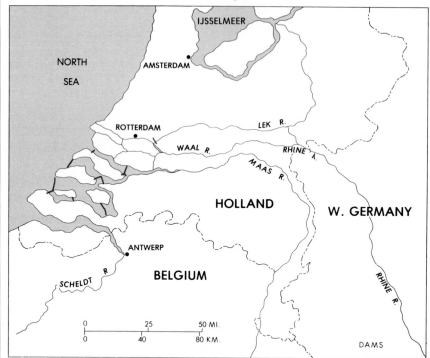

The great involved marshy delta of the Rhine and a bevy of lesser streams, are of major significance to the Netherlands. Not only does the modern port of Rotterdam at its northern periphery control the chief navigable access to the upper Rhine (W. Germany, France and Switzerland), but the drained and reclaimed agricultural land of Zeeland makes up more than one third of the nation's productive polders. Yet the threat of vicious North Sea gales and flooding rivers is a constant worry—at best, with 434 miles (698 km) of vulnerable dikes, a delicate equilibrium has been achieved.

The disastrous storm of February 1953 severely damaged nearly all of the diked land in Holland, and disrupted the delta's complex balance almost as thoroughly as did the German bombs of WWII. This was not the first seaborne cataclysm, they extend far back into history, but in the wake of 1953's events it became obvious that if this strategic and heavily populated region was to escape permanently recurring disaster, something major would have to be done. Merely shoring up the dikes once again was not the long-term answer.

The plan that emerged proposed 4 major dams across the delta mouths, in essence amputating 440 miles (708 km) of estuarian coast line; these to be backed up by a series of seven secondary dams upstream to regulate river outflow. Converted now to fresh water lakes, the old estuary channels would function as a buffer against tidal salt water that threatened to poison arable land. At the north the canalized channel to the sea from Rotterdam would be left open to serve Rhine traffic, and similarly to the south the Scheldt River mouth, Antwerp's deepwater access, would remain undammed.

The Delta Project has been a mammoth effort. Not only huge expenditures of capital and manpower but whole new processes in design and engineering were developed. Its scale is so large that the machines to make it work had to be built on the spot and new ingenious methods invented along the way. It is almost in place now and scheduled for completion in the late 1980's—perhaps the world's largest project to stabilize a coastline whose natural inclination is to maintain a state of constant flux.

Fig. 26–10   Facing the sea with a steeply cliffed frontage is a well defined marine terrace. Its flattish surface was once the smooth offshore sea bottom and its inner margin a wave-cut bluff.

tle more careful than we have been in the past in assigning the emergent tag.

GRADATION OF THE SUBMERGED SHORELINE

Wave activity along the typically ragged and broken submerged shore is consistent in its general outline everywhere. First the indentations are removed by both cutting and filling, and then the entire coast is driven back relentlessly. The exposed headlands are attacked with great vigor, especially if the sea is deep immediately offshore and the full force of the waves can be concentrated directly on the land without the impediment of friction with the bottom. Under the effect of this attack, the point of land is cliffed and pushed back rapidly. The great mass of material resulting from this erosion is partially shifted longshore, the longshore currents depositing much of their load as *bay-mouth spits* **and** *bars* across the mouths of the adjacent coastal indentations and as beaches at their heads. Typically, these spits curve inland at their tips away from the waves, which break on their seaward side, and except for minor breaks to accommodate tidal currents, they effec-

tively cut off the bay from the ocean waves (Fig. 26–11). Essentially, then, the bay becomes a quiet lagoon. Streams emptying into it from the land build undisturbed deltas, which with the tangled roots of their vegetative cover, gradually transform the lagoon into land—low-lying and swampy but land. So as the waves continue to push back the headlands, the bays build out to meet the bars at their mouths and the contour of the coastline changes from one of involved crenulations to a smooth straight line. Now the sea can attack the shoreline impartially and does so, gradually eroding it back at a much retarded rate, for the wave-cut debris, some of which moved longshore to form spits and bars, has also formed beaches and a generally more shelving approach to the land so that much of the energy of the waves is expended in running up the beach and erosion is slowed. During storms, however, the sea revives much of its old vigor and rapidly pushes the coastline back again (Fig. 26–12).

Wave activity along the typically ragged and broken submerged shore is consistent in its general outline everywhere.

Fig. 26–11 Two examples of spit formation. (top) The mouth of the Russian River north of San Francisco. Longshore drift from the right, reinforced by stream sediments, threatens to seal off the river from the sea. Regular dredging is required to allow normal drainage. (Bottom) Port Angeles, Washington, on the featureless north coast of the Olympic peninsula. The open Pacific is 50 miles (80 km) to the right, and if it were not for the spit (Ediz Hook) built by Pacific-activated longshore drift, the port could not exist.

GRADATION OF THE EMERGED SHORELINE

The newly formed coastline is a portion of the old sea bottom and is likely to be flat to shelving and generally fea-tureless, reflecting its formation as wave-cut bench or wave-deposited terrace. This means that on a newly emerged shoreline, the line of breakers is well offshore, and at this point the bottom is being eroded away to a depth sufficient for adequate vertical circulation of the water. The debris from the erosional work is thrown forward by the breakers

Fig. 26–12 Seaside homes are in great demand because of their recreational opportunities and ocean view, but with the view come certain risks. When the waves kick up a little too high for invigorating surfing it is eminently possible to be swept right into that liquid scenery. January, 1983 along the California coast.

and commonly piles up at this point into a sandy bar called an *offshore* or *barrier bar* (Fig. 26–13). These may extend for great distances—off the Texas coast they are as much as a mile offshore—and may be broken only occasionally by tidal channels.[1] Once again as in the case of the bay-mouth bar, a quiet lagoon is established that rapidly fills with shore-derived sediments and becomes a marshy ex-

tension of the mainland. Meanwhile, the big ocean waves are lashing the seaward side of the bar and as rapidly as the wave-cut bench can be expanded, are pushing the bar inland. The bar is moving onshore, the land is moving offshore, and when they meet a new shoreline has been established. Wave action now continues to wear away the coast and move it back at a more modest rate (Fig. 26–14).

Deltas are another situation. Far from passive, merely submitting to wave maltreatment, they continue to build, often more rapidly than the sea can cut them back, although longshore and tidal currents may control sedimen-

[1]The experts tend to disagree as to whether waves are capable of building up such a bar above sea level, some contending that offshore bars must have a basement of an older beach now drowned by a recent minor rise of the sea.

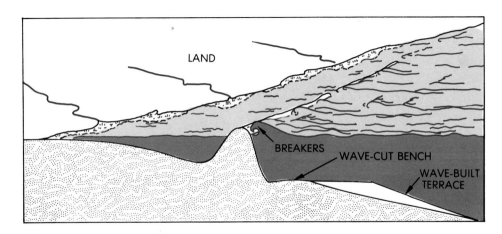

LAND

BREAKERS

WAVE-CUT BENCH

WAVE-BUILT TERRACE

Fig. 26–13 Offshore bar. The bar advances landward as rapidly as the waves can cut away the bottom. At the same time, stream deposits build up the shore and it advances seaward.

Fig. 26–14 Already a brackish marsh is building in the lagoon behind a longshore bar. The break is kept open reasonably well by the tide, but eventually new land will extend all the way to the bar.

tary deposition to the degree that the delta develops some form of asymmetrical shape. Coral reefs too, alive and growing, often regenerate as rapidly as battering ocean waves can deform them.

But these are the exceptions and given time, even they can be overcome. In general, we can say that irrespective of type of shoreline, ocean waves are involved in one fashion or another in wearing away the land. Theoretically, left to themselves, waves, with their ability to erode the wave-cut bench below sea level, could eventually remove the last vestiges of dry land and we might be faced with attempting to conduct our business on an all-water earth. There is plenty of water to cover a smoothed-off globe and plenty of deep places in the ocean to accommodate all of the continental sediments. But happily, so far at least, we have had the tectonic forces to counteract this catastrophe.

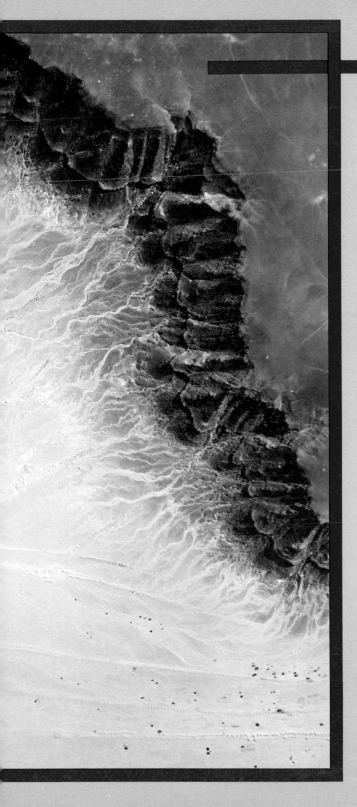

CHAPTER 27

GROUNDWATER AND RELATED WATER PROBLEMS

Draining from the dark escarpment is a network of surface watercourses which, here in Libya, do not flow very many days each decade. But the coarse gravel of the upper alluvial fans is highly porous and there is groundwater below each wash. This is critical to the sustenance of life. Careful examination of the slopes will reveal line-ups of tiny circles. These are qanats, a series of opening for maintenance of underground conduits which bring water from the upper fan to the village and its irrigated plots. The low-odds risk of a flash flood is a cheap price to pay for an assured water supply.

INTRODUCTION

The subject of groundwater is pursued at this particular juncture because in some ways it can be regarded as a legitimate agent of gradation, along with streams, ice, wind, and waves. Moving groundwater can pick up rock materials, transport them long distances, and finally deposit them. And when groundwater performs these functions, it automatically classifies as a grader. The fact that groundwater's mineral load is restricted to those that can be carried in solution is something of a limiting factor, but wind is a certified member of the gradational fraternity and cannot take anything into solution (nor essentially can ice). So every agent has its peculiarities, groundwater among them.

Groundwater's mineral load is restricted to those that can be carried in solution.

The erosive or gradational province is not the only one involving groundwater. Probably of more importance to humans is its control of stream discharge and of the water level of lakes and swamps, as well as a source of irrigation and potable water for a large part of the world's population. However, when we delve into matters of this kind we open up a whole Pandora's box of interrelated water problems, for it becomes apparent immediately that groundwater and surface water share a high degree of mutuality and complete divorce of one from the other is not only difficult to attain but perhaps not wholly desirable. Also, the student must be aware of the human's role regarding water resources and the significant interrelationships of all the earth's water.

THE HYDROLOGIC CYCLE

As a point of departure we must remember that there is only so much moisture on this earth. None is lost or gained from outer space, but at any given moment some of it is in the sea, some on the land in the form of water or ice, some in the crust itself as groundwater, some as suspended water droplets and gaseous water vapor in the atmosphere. The cycle of change in both form and place is called the *hydrologic cycle* (Fig. 27–1).

All of the world's water derives from the oceans, and most of it returns ultimately to this source. Evaporation from

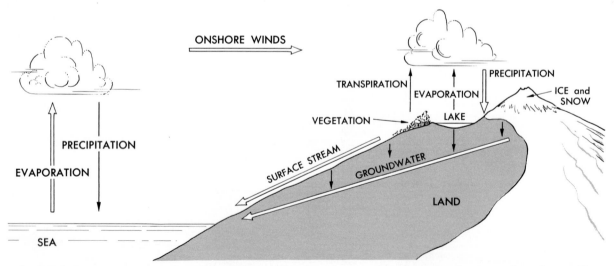

Fig. 27–1   The hydrologic cycle. The basic cycle involves four elements: (1) evaporation of moisture from the sea, (2) movement of water vapor from over the sea to over land, (3) precipitation of rain or snow onto the land, and (4) the return of this precipitation via gravity drainage.

the sea provides the water vapor for rain and snow. If it condenses and precipitates over the oceans, then its return is immediate, but if it is carried over the land, a number of things may happen to it before finding its way back. Some moisture falls as snow and remains on the land until the spring melt before flowing back to the sea via rivers. Or it may become part of a semipermanent ice mass and be detained many hundreds or thousands of years. Much of the water vapor will fall as rain, some flowing off immediately on the surface, some remaining in lakes and swamps, some evaporating directly or via plants to be precipitated again, and some soaking deep into the ground to percolate gradually downslope to the sea or to appear at a lower level at the surface as springs and seepages.[1] This last is groundwater.

POROSITY VERSUS PERMEABILITY

How much groundwater occurs in any given place and its character and movement depend largely on two variables: (1) climate, which controls precipitation, evaporation, and vegetation and (2) peculiarities of the soil and subsurface rock. The great bulk of ground water has simply seeped down from above, but certain rocks and rock materials have a much greater water-holding capacity or *porosity* than others. Loose sands and gravels, with their irregular particles allowing a maximum pore space, may contain up to 50% water, as can some clays. Sandstones too, although part of their intergrain areas are filled with cement, may be highly porous. On the other hand, dense shales and igneous rocks usually exhibit a very low porosity. But an ig-

Fig. 27–2   A stratum of vesicular lava is not only porous but high permeable as well. The frothy appearance results from escaping gases as the lava congealed.

neous rock in cooling often develops fissures, joints, or cavities from escaping gases and then, although the rock itself is dense, may be capable of containing large quantities of groundwater (Fig. 27–2).

Clay, with its frequent sponge-like ability to soak up water, is not generally permeable.

But we should make a distinction here between *porosity* and *permeability*. It does not always follow that a material capable of holding large quantities of groundwater will allow that water to move through it, and it is this freedom of movement that is the essence of *permeability*. Clay, with its frequent sponge-like ability to soak up water, is not generally permeable, for the tiny size of each clay particle holds the groundwater through molecular attraction. Fissures too, if formed in dense rock with an opening only at the top, may collect water and hold it but horizontal percolation or flow will be impossible. On the other hand, coarse granular layers, whether unconsolidated or indurated, normally allow ready movement of water through them, as do rocks whose fissures, joints, or cracks are in-

---

[1] This theoretical system is not absolutely perfect, although given time it comes very close. The lag in time before moisture is returned to the ocean can be very long so that water or ice held in lakes or glaciers may appear to be permanently lost to the dynamics of circulation. But ice both melts and evaporates and lakes are surprisingly transitory; even the great Antarctic ice cap flows away from the pole and breaks off into icebergs. The closest to permanency in water loss is that fraction tied up as a part of the chemical composition of rocks and magma, or in an underground reservoir completely lacking in permeability, which has trapped ancient surface waters. But even here, humans have found it advantageous on occasion to tap and draw on these reservoirs, and vulcanism, metamorphism, chemical weathering, etc., can and do release water through chemical reaction.

terconnected throughout an entire strata. Permeability, then, is obviously an important factor in the completion of the hydrologic cycle and in the feeding of wells, springs, and seepages.

THE WATER TABLE

Assuming a considerable depth of homogeneous permeable material, *the groundwater table is the top of the saturated soil.* In a humid region this may be quite near the surface or even coinciding with it after a particularly heavy rain. But normally it is at least a few feet down, overlain by a layer of loose soil where water is rapidly lost through evaporation and transpiration. This zone of soil water or *vadose water* is generally within the reach of plant roots and reflects immediately day-to-day or even hour-to-hour weather fluctuations, so that characteristically it is saturated for only short periods of time and more often than not holds as much air as water in its pores.

A water table will roughly approximate the contours of the terrain that it underlies.

Under usual conditions a water table will roughly approximate the contours of the terrain that it underlies, higher under hills and declining under valleys (Fig. 27–3). The groundwater is percolating downslope and its upper margin thus assumes the gradient of a surface stream; like a surface stream, it discharges into the master drainage artery in the valley. The river at the valley bottom may receive the groundwater discharge directly via springs as it intersects the water table, or if the river itself is slightly above the water table, the coarse sands and gravels of its bed will intercept the flow and direct it down the valley under the surface waters of the river. Swamps, lakes, rivers, or the sea, which occupy the lowest point of a given region, act as regulators of the slope of that region's water table. Loss of water supply, as during an extended dry period, will cause the water table to flatten out and be found at an increasingly greater depth beneath the hills, but it will continue to intersect

the surface in the valleys; as long as any gradient at all is maintained, groundwater will percolate downslope.

Wells bored below the water table will produce water, but how rapidly they will refill themselves after emptying is determined by the permeability of the rock. No matter how permeable the rock or how humid the climate, too many wells will lower the water table, and increasingly deeper bores will be required to tap groundwater. Obviously, an inventory of the groundwater resources should be undertaken before uncontrolled well boring takes place—a procedure that has seldom been followed anywhere in the world until the water table has been lowered drastically and disaster is imminent.

It is very embarrassing to drown some winter after going bankrupt in a farming venture the previous summer for lack of moisture.

The Santa Clara Valley in California, a region of irrigated orchards and, more recently, burgeoning urbanization, is a case in point, illustrating the results of continued overdraft on the ground water supplies. The increasing depths of wells through the years are shown in Fig. 27–4.

Reservoirs as Rechargers

Groundwater levels can go both ways—water can be put back into the ground as well as taken out[2] —and it is not absolutely necessary to stand around thirsty for 40 years waiting for the natural rainfall to do the job. Look at the peak in the middle of the well depth chart. This is very directly the result of building retaining reservoirs. The climate did not change dramatically and suddenly the rainfall doubled, and no new exotic water was brought in. It became a matter of driving a maximum of the surface water downward into underground strata instead of letting it run back to the sea unused. The special rainfall peculiarity of the Santa Clara Valley (and much of California) is that all

---

[2] There are a number of places, the Santa Clara Valley being one of them, where removal of groundwater has been so severe and the underground structure so weak that measurable sinking of the land has occurred. In such a circumstance, compression of the water-holding strata permanently lessens their capacity. This situation also strengthens the argument against massive overdraft without rejuvenation.

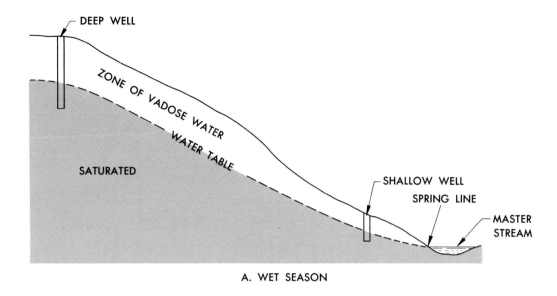

A. WET SEASON

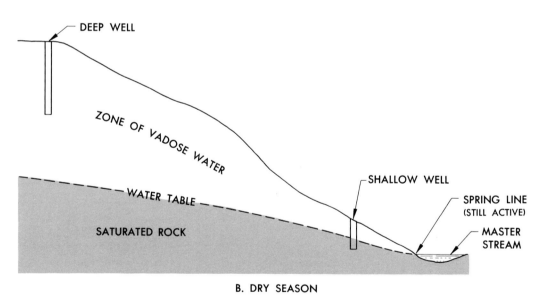

B. DRY SEASON

Fig. 27–3 The groundwater table. (A) The water table tends to approximate the surface contour. (B) But during an extended dry period, it flattens out considerably, although still intersecting the surface at the spring line. As a consequence, the well at the top of the hill, despite its great depth, is in a less advantageous position than the shallow one on the lower slope.

the rain arrives in a few winter months and the long-empty stream channels, frequently grown up with brush and weeds, simply cannot handle the runoff. The result is serious flooding as well as the loss of valuable fresh water that would have been so welcome the summer before (Fig. 27–5). It is very embarrassing to drown some winter after going bankrupt in a farming venture the previous summer for lack of moisture.

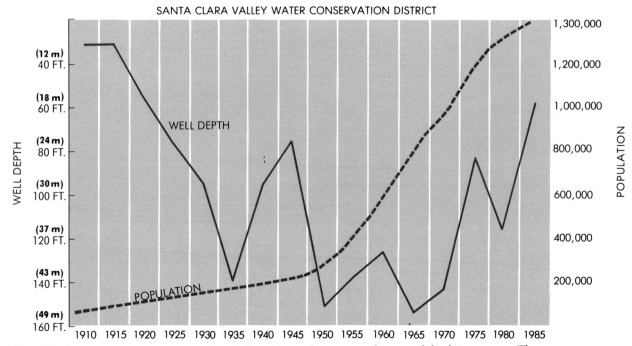

SANTA CLARA VALLEY WATER CONSERVATION DISTRICT

Fig. 27–4   Well depth in the Santa Clara Valley. A general trend toward increased depth is apparent. The fluctuations are, to a degree, a result of unreliable rainfall, but of greater significance has been the construction of reservoirs to recharge underground strata, as in the 1940s. The sharp upturn in 1965–1970 reflects the introduction of exotic Central Valley water.

Fig. 27–5   Too much of a good thing at the wrong season, Marin County, California. Perhaps everyone's morale could be improved if boating in the city streets were officially classified as recreational use of the water resource.

Fig. 27–6   Anderson reservoir (dam in the left foreground) in the hilly country bounding the Santa Clara Valley. The recreational feature of the artificial lake, so obvious here, is merely one of several in this type of multipurpose development.

A series of reservoirs to impound the seasonal surplus solves a number of problems:

1.   It is a flood control measure, saving for beneficial use winter flood water that would have, at best, been lost to the sea, or at worst, drowned people and destroyed property.
2.   It provides accessible lakes for recreational use.
3.   It recharges the underground water system.

But why worry about this last? Let's just use the water right from the reservoir. Water *can* be diverted directly from a reservoir, but if the bottom of the holding basin is at all porous, a certain amount of water can be added to the underground supply at the same time and a great deal more can be forced downward if it is drawn off and spread out in shallow percolating beds below the dam (Fig. 27–6).

Evaporation from Reservoirs

One of the beauties of storing water underground rather than on the surface is to negate almost completely the loss through evaporation. Unhappily, the places in the world that find themselves in need of water storage are frequently those with long hot summers and low humidity. Under these conditions the effects of evaporation can be enormous. No two reservoirs evaporate at precisely the same rate, variations depending on such things as climate, dimensions of surface exposure, percolation, etc., but recent experiments in the northern Negev region of Israel indicated a loss in excess of 75% from evaporation alone. Obviously, any method of cutting down on such a loss rate must receive serious consideration. Some interesting work with the use of nontoxic chemical films to inhibit evaporation appears to hold real promise for both reservoirs and percolating beds, especially those that are not to be utilized for recreation. Water skiers and motor boats are not wholly compatible with an effective film.

Sea Water Incursion

In regions draining to the sea the groundwater approximation of the surface flow delivers fresh water via springs or seepages at about sea level. If well overdraft occurs upslope in the system, hydraulic pressure is diminished and salty seawater can invade the porous underground strata. This means that the normally fresh-water coastal-plain wells will be "salted out," often a local disaster, unless a certain minimum groundwater table can be maintained.

Water skiers and motor boats are not wholly compatible with an effective chemical film to inhibit evaporation.

Further, there is the readily observable reciprocal relationship between the level of lake, pond, swamp, and streams and adequate groundwater. Since much of the surface water feeds the underground strata, a lowered water table increases the withdrawal through percolation, thus lowering their level. But most of these surface water features, because of their location at valley bottoms or local terrain sags, also receive groundwater percolation from upslope. So in essence, *each sustains the other*. Farm ponds, reservoirs, and flooded percolation beds are generally recommended by the regional planner as effective aids in stabilizing stream flow and well and lake levels.

WATER QUALITY

Of equal importance to the mere availability of water is its quality. There is no use having water if you can't use it. Not too many years ago it was a virtual article of faith that natural filtration would rejuvenate any water supply no matter how badly it had been polluted. Wasn't even raw sewage, the epitome of effluvial pollution, susceptible to cleansing by filtration? Thus the whole process of forcing surface waters underground to be withdrawn via springs and wells boasted the great merit of purification. But increasingly the variety of chemical waste products that contaminate our streams includes elements that defy the simple purge of filtration. Especially notable are insidious pesticides/insecticides, chemical detergents, and a number of industrial wastes.

Increasingly the variety of chemical waste products that contaminate our streams includes elements that defy the simple purge of filtration.

There is the additional problem that groundwater, although often less polluted than surface water, is frequently "hard." This simply means that in its passage through porous strata it has taken salts into solution, a part of its normal function as an agent of gradation, but one which affects its utility to humans. The degree of hardness varies greatly depending on the character of underground rock and the water's temperature (heat as an industrial discharge into a stream can contribute to hard groundwater as well as destroying the immediate surface ecology). Often hard water merely means that soap lathers with difficulty and moderate salt deposits build up in household plumbing (Fig. 27–7), but hardness can be so severe as to attain the saltiness

Fig. 27–7   Hot water pipes are particularly susceptible to the accumulation of salt deposits in regions utilizing hard water for the household. This one was replaced after only two years.

of seawater and the water becomes totally unusable for either irrigation or drinking. All of this is not to imply that *only* groundwater can be hard, but it is much more common than in freely moving rivers.

SOLVING THE WATER PROBLEM

How do we attempt to solve our dual-faceted water problem: (1) sufficient water in the right places at the right time and (2) water of an acceptable quality for general utility? There is little question that a problem exists, even if we were to assume a static world population—an unlikely situation for some time to come—and to ignore for the moment the dubious probability of desalinizing the ocean (see Chapter 30). Simple drinking water for the world's hordes involves a huge volume in itself, compounded many times as humans agglomerate into an urban environment of flush toilets and green lawns. Agriculture dependent on the ir-

rigation ditch is even more demanding of water on a per-acre basis than the housing tract, while heavy industry, always carefully sited on a freshwater lake or large river, uses the most water of all.

Transportation

We have already considered the principle of conserving wet-season surplus moisture in reservoirs and underground holding basins for use during the dry season. But let us apply this same kind of thinking on a regional basis. Now we get into the concept of transporting water from surplus to deficit regions. This is old stuff. The pipe was invented to take water out of the river and into the house, or to move it clear into the next county, or to transfer it from the wet side of the Rockies in Colorado *under* the continental divide to the dry side, or to impel it 300 miles (483 km) out into the Australian desert to the gold mines of Kalgoorlie, or to transport it 1000 miles (1609 km) from northern California to the south. Why not transport Columbia River water to the arid southwest, or Mackenzie River water to

Fig. 27–8 Feather and Sacramento River water flowing uphill (south toward the camera) on its way to southern California. Those along the route partake as need requires. Here in the Kettleman Hills district on the west side of the San Joaquin Valley, longstanding desert has abruptly turned green. But the entire concept of large-scale water transportation has been under increasing fire from state voters. To generalize a bit, the pros tend to be the southern multitudes with their huge water deficit, and the cons the much smaller numbers from the green north. All are apprehensive about escalating costs.

Augmenting the Volga

The Volga at Volgograd (Stalingrad).

There was a day in the distant past when the "Mother Volga" flowed freely to the Caspian Sea, gathering the waters from its tributaries as it went and delivering a huge quantity of fresh water at its mouth. But the modernization of Russia has placed a major emphasis on the utilization of the river as a reliable navigation artery, as a generator of hydro power, and as a seemingly endless source of industrial and agricultural water. The major stricture on the Volga's free flow has been a series of dams which have transformed a majestic rolling stream into a series of quiet lakes, and ever increasing water withdrawal has resulted in the "Caspian Problem".

During the last 40 years the level of the Caspian Sea has gone down dramatically (over 9 feet [2.7 m] currently and continuing to drop) diminishing fisheries, shoaling harbors, and increasing salinity. If this trend can be alleviated, or even balanced at today's level, it would be a major accomplishment. Is it possible to somehow inject more water into the Volga system to solve the Caspian Problem, at the same time increase power production, and supply additional water to irrigate the arid wastes? Soviet engineers and planners have inquired into just this question from as far back as 1932 and an intriguing plan has gradually evolved.

At about 61°N, just west of the Ural Mountains, the headwaters of three rivers are only a few miles apart. Two of these, the Pechora and Vychegda, flow north to Barents Sea discharging quantities of unused water. The third, the Kama, is a major tributary to the Volga, its confluence at the big bend near Kazan about 600 miles (966 km) upstream from the Caspian Sea. The proposed project would dam all three rivers near their headwaters to form a reservoir at the low interstream divide. A large part of this impounded water would then be released into Kama drainage to find its way directly to the Volga—in essence reversing the flow of the Pechora and Vychegda. In the process of adding water to the Volga, the new dams would produce power for the far north, and the Kama and its reservoir would be opened up to extensive navigation.

Critics of the project have been quick to point out the loss of agricultural, pastural and forest land through flooding—the huge reservoir would cover 6000 square miles (15,500 km²) almost as much as Lake Ladoga, European Russia's largest freshwater body. Probably 60,000 people would have to be resettled. Migratory fish running upstream from the Arctic Ocean would find the depleted and dammed rivers almost impossible to negotiate, although this loss could be more than offset by planting millions of other useful fish fingerlings in the reservoir.

Are these kinds of efforts at reengineering nature really worth the price (estimated at something in excess of 1 billion rubles)? And even if money costs balance out through gains in agriculture and power generation, etc., how do we measure the costs of relocated people and damage to fragile environments? Engineers tend to ask the question, "*can* it be done?" Geographers and planners ask, "*should* it be done?" At this point the answer to the first question is yes. Soviet officials are still mulling over the second.

the Great Basin, or even the Ob or Yenesei River water to Kazakhstan? It seems that the only limit is need coupled with engineering and cost feasibility. But Washingtonians are not likely to react with great enthusiasm to a redirected Columbia, or Canadians to the use of Mackenzie water to irrigate the orange groves. Neither were northern Californians and Mexicans particularly thrilled when their water went elsewhere, but standing up and saying so proved to be no deterrent to "progress" (Fig. 27–8). The problem comes in agreeing on a point where local regional interests and human aesthetic values become as important as the efficient use of resources. We *must* utilize our resources, but perhaps when we come to the point of substituting the word "rational" for "efficient" (and then define rational to the satisfaction of one and all), we will be on the right track.

The problem comes in agreeing on a point where local regional interests and human aesthetic values become as important as the efficient use of resources.

Recycling

All of this leads to an emphasis on more efficacious use of the water we now have, and the key term here is *recycling.* Water is never used up. If we drink it, flush it, wash or irrigate with it, the water is returned sooner or later somewhat changed—we call it polluted. The trick is to clean it up. This operation can be two or even threefold in its benefits. First, if we can eliminate all of the harmful side effects of pollution or contamination such as water-borne human diseases, destruction of wild life, stench, etc., it is well worth almost any cost involved. Second, the by-products may aid in cutting down that cost, as Milwaukee discovered many years ago when its sewage treatment operation yielded a profitable commercial fertilizer. Salts from distillation and productive use of energy derived from cooling hot-water pollution are further examples. Finally, of course, there is the reason behind the original effort: clean water to be used again. If drinking recycled sewage water is shocking to your sensibilities, be assured that you have probably been doing exactly that all these years. Most towns on rivers draw their drinking water from upstream and dis-

charge their sewage downstream. If there were only one town per river this system might work, but there are usually a lot of people upstream and a little filtration and a dash of chlorine your only protection. Actually, carefully monitored recycling, operating under an enforced master code, ensures much higher quality water than most of us usually imbibe.

·

AQUIFERS

So far, we have been discussing rock with a certain homogeneity of permeability, but this is not always the case, for many different strata may be present underlying a given region, each with its own characteristics. An important occurrence is a permeable layer, such as sandstone, confined between impervious strata and tilted so that it intersects the surface. Water entering the exposed sandstone outcrop percolates along it as if in a pipe. But normal groundwater will be limited to the zone above this confined *aquifer*, for the impervious layer immediately atop the sandstone stops its downward movement and also acts as a barrier between free association of the water in the aquifer and the groundwater. Shallow wells plumb only the local groundwater that is subject to the vagaries of seasonal rainfall and if the region is arid, is easily exhausted. But wells tapping the deeper aquifer assure themselves of a more continuous supply, especially if the intake is in a better watered area. Also, since the permeable surface outcrop is often the highest point in the aquifer, wells drilled to this lower level are usually *artesian* because of the hydrostatic pressure within the closed system. As long as the outlet is lower than the intake, water will rise in the well. Some will flow; others merely show a rise of level above that of the general water table. In either case the term "artesian" is applied (taken from the first recognized well of this type in the French province of Artois).

One of the largest artesian systems in the world underlies much of interior Queensland and northern New South Wales in Australia (Fig. 27–9). Here a confined and highly permeable gravel bed receives water at its highest point in the humid Great Dividing Range and carries it for almost 1000 miles (1609 km) west beneath an arid to semiarid country where water is at a premium. Many hundreds of bores tap this aquifer, supplying life to what would oth-

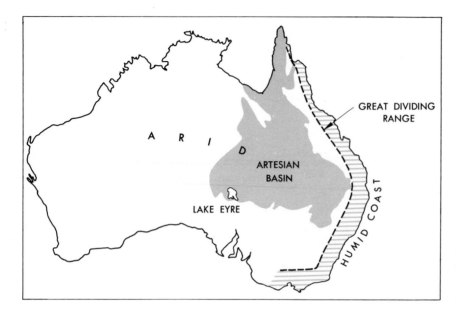

Fig. 27–9 Australia's Artesian Basin. A part of the heavy rainfall along the Great Dividing Range finds its way into the exposed aquifer that carries it downslope under the semiarid interior, to regain finally the surface in the greater Lake Eyre Basin below sea level.

erwise be desolate country. But inevitably, too much water has been withdrawn, and most wells that at one time flowed freely at the surface now must be pumped. And too, the quality of the water deteriorates with the distance it moves through the aquifer so that only in the eastern half of the basin is salinity low enough for reasonable utility. Even here the water is too brackish for general irrigation and is used only for watering stock.

If drinking recycled sewage water is shocking to your sensibilities, be assured that you have been doing exactly that all these years.

In the Lake Eyre Basin, below sea level although in the heart of the continent, the lower end of the aquifer reaches the surface. The highly saline water flows out here as artesian springs, evaporating almost as rapidly as it appears and depositing salts around each outlet. These are called *mound springs*, building up to a foot or two in height, and are a result of artesian water under great pressure, high salinity, and an extreme rate of evaporation (Fig. 27–10).

Similar artesian systems are found in many other parts of the world, although none is as large as that in Australia.

The Dakota sandstone, underlying the northern Great Plains, has its intake in the Rockies and is a major source of water for an extensive semiarid region. North of the Po River in Italy is a line of flowing springs, the *Fontanili* Fed by melting snows in the Alps, the underground water moves down through an aquifer of coarse glacial and alluvial material. Many of the north Saharan oases exist because aquifers transport Atlas Mountain waters far out into the desert to reappear as springs.

HOT GROUNDWATER

The *heat gradient* of the earth's crust, that is, the increase of heat with increase of depth, may affect groundwater moving along a deeply buried aquifer. Theoretically, there is a depth limit to rock that can contain water, for at several miles down, the pressures exerted by overlying strata are such that rock pore space becomes inadequate to accommodate a liquid. Water may be present but it is chemically combined in the rocks. But above this limit, sufficient heat still exists to raise the temperature of moving groundwater, and when it discharges, the water is in the form of a thermal spring. This type of spring is not rare,

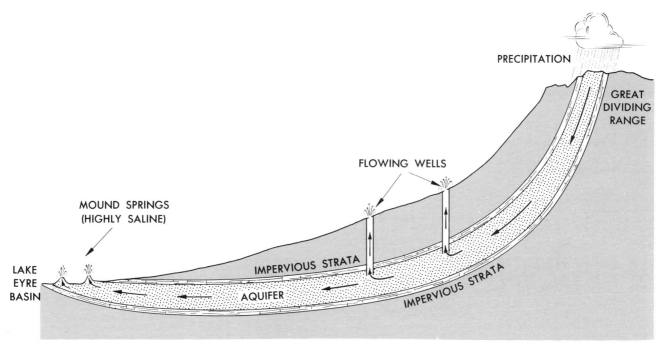

Fig. 27–10 Generalized cross section of Australia's Artesian Basin. Since the outlet of the wells and the mound springs are below the level of intake in the mountains, they are artesian in character. Salinity increases as the groundwater moves along the aquifer.

but much more common are the hot springs, geysers, and steam vents associated with regions of recently active vulcanism. Here the heat of the earth's interior has been brought well up into the crust where it is easily encountered by even very shallow groundwater. Such well-known tourist attractions as Yellowstone Park, the Rotorua district of New Zealand, and Iceland are fantastic assemblages of every conceivable type of thermal phenomenon, and each is underlain by hot volcanic rocks (Fig. 27–11). To be sure, cooling magma emits hot water, steam, and other gases so that all of this activity is not wholly a result of heated groundwater, but surface waters percolating downward to contact the hot rocks below are without doubt an important element.

Hot springs originating in this manner may achieve very high temperatures indeed, many reaching the boiling point. If the heat is even greater, then the water becomes steam and issues through vents in the rock as jets. The term *fumerole* or *solfatara* is usually applied if the pure steam is contaminated with magmatic gases and emits an odor.

Geysers, which are merely intermittently eruptive thermal springs, are of this same origin. They require a long tube reaching from the surface to the heated rocks below. As the tube fills, the superheated water at the bottom is under too much pressure from that above to turn into steam, but when all of the water reaches the boiling point and some bubbles out at the top, pressure is released and the water at the bottom flashes into steam causing an eruption. The great volume thrown out by many geysers indicates that an underground reservoir must also be present, interconnected with the tube.

Geothermal Power

Violent forces such as geysers and steam jets are not merely so much spectacular scenery but have occasionally been put to productive use. A pioneer venture in the volcanic sub-Appenines of Italy was producing practical electricity from

Thousand Springs

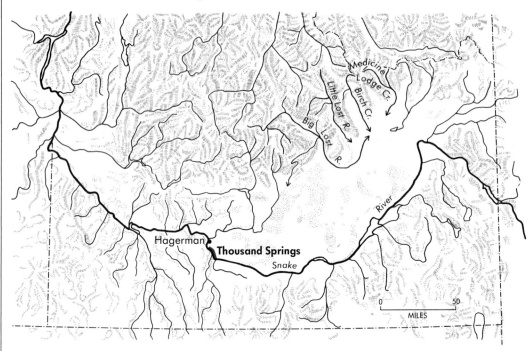

If we measure simply total volume of water delivered per hour, perhaps the most remarkable series of springs anywhere in the world occur in the Snake River canyon of southern Idaho. Here for 10 miles (16 km) upstream from the town of Hagerman literally thousands of vigorously flowing springs issue from a porous stratum exposed along the north wall of the defile. Nobody is absolutely certain where this huge volume of water originates, but two bits of evidence point to a likely source: (1) there are no tributaries of any kind entering the Snake from the north for a 200-mile (322 km) stretch short of Hagerman; this despite the high Rockies and continental divide not too far off and (2)

there are several streams draining southward from the Rockies, notably the sizable Lost Rivers, which disappear abruptly into porous lava beds.

If this is the source of the Snake River springs, then we must visualize an aquifer, probably a buried *vesicular* lava sheet connecting the Lost River sinks with canyon wall flows. Vesicular lava, when molten, is highly charged with gases and as it cools and congeals, the tiny apertures created by the escaping gases result in an exceptionally permeable layer. The underground water does not percolate sedately through this kind of aquifer but flows virtually unimpeded as in a pipe.

natural steam as early as 1913, and Luther Burbank was involved in drilling shallow steam wells at The Geysers near Santa Rosa, California, in the 1920s. These two sites, under new management and much expanded, are among the

world's largest today. In addition, important projects are operative in New Zealand, Japan, Iceland, Mexico, and Chile, and experimental work is being done in many nations (Fig. 27–12).

Fig. 27–11 When a hot spring is fouled by viscous mud (much of it a result of chemical decomposition of rock under the influence of hot water) it finds its way to the surface via these little mud volcanos. They swell and then collapse one after the other with obscene "plopping" sounds. Yellowstone mud volcanos are sited directly over a thermal plume that brings heat from the mantle close to the surface in the middle of the North American plate. The entire geologic history of the Park region has been one of active vulcanism—two calderas one within the other, massive lava flows, and periodic explosions. Little mud volcanos and related thermal phenomenon are signals that the entire district is far from dormant.

Fig. 27–12 Groundwater flashing into steam as it contacts hot subterranean rocks has the potential for cheap electric power production for it obviates the orthodox and costly requirement of both fuel and boiler to activate generator turbines. Here the powerhouse is at the riverside for ease of condensing and the steam is brought to it by a phalanx of huge pipes. Steam plumes mark well-head sites.

Electrical power derived from the geothermal resource is not really much different from that of a normal coal or oil-fired thermal plant in its basic scheme. Coal burned under a boiler produces steam, which is directed to the turbine blades of an electrical generator. But in the geothermal plant, nature supplies the fuel and the boiler and there is a huge theoretical cost advantage not to mention lesser air pollution. Natural steam and hot water, however, are always, to some degree, contaminated by dissolved salts and gases and consequently corrosion and scaling of the equipment results. For example, near California's Salton Sea is an immense underground hot water reservoir, but up until now its utilization has been impeded by an equally immense salt content. Further, hot waste water and brine discharges are sometimes difficult to deal with.

In the geothermal plant nature supplies the fuel and the boiler and there is a huge theoretical cost advantage.

We probably should not regard geothermal power as the universal substitute for traditional energy sources any more than we regard sun or wind, although in terms of practical development, it is much more advanced today. But, given the right location and a fortuitous assemblage of physical characteristics, the development of both cheap and efficient power from geothermal sources can add significantly to our total energy package.

SOLUTION

The very fact that a large proportion of our groundwater is hard or even brackish is an indication that in all of its movements groundwater is taking into solution and removing mineral matter from the rocks through which it passes. Even perfectly clear cold water has the ability to dissolve some minerals or to combine chemically with the rock to form new minerals. But groundwater is seldom clear, since it must pass through overlying material as it percolates downward to become groundwater. Decomposing vegetation on the surface and the soil below it adds elements that give the water certain solvent characteristics and then, depending on the type of rock it encounters, the water reacts chemically with those rocks. If the water is heated, it frequently becomes a much more efficient solvent, and in the process of being heated will often pick up various new elements from magmatic gases.

Karst

An excellent example of the sort of solution activity in which groundwater can engage is the removal of calcium carbonate by either hot or cold water with a high carbon dioxide content. Percolating through the cracks in a massive underground limestone stratum, such groundwater can etch out sizable cavities in a relatively short time. Given a longer time, such huge caverns as Carlsbad and Mammoth Cave result. Eventually, the terrain above these caves can be affected as the roof supports are removed and slumping and cave-ins manifest themselves at the surface in deep holes called *sinks* or *dolines*. Normal surface streams may disappear into such sinks only to reappear many miles away, having flowed as an underground river through the cave. In central Florida where the groundwater level is high, these sinks become lakes. Further solution within the cave will cause increased slumping until the original surface all but disappears and the landscape takes on a fantastic lunar aspect of sharp eroded spires. Eventually, even these become subdued through the activity of wind and rain, erosion, and weathering. Such a landscape, deriving its character from the continued expansion of solution caverns beneath it, is called *karst*, after the region in Yugoslavia where this type of landform was first described. Following an earlier discussion of the geomorphic cycle, karst landscapes display, like all others, the typical sequence of youth, maturity, and old age. Scattered sink holes only slightly deforming the original surface are youth; major collapse resulting in maximum local relief is maturity; and final virtual peneplanation at a lower level is old age (Fig. 27–13).

DEPOSITION

Cave Deposits

All solution is admittedly not this scenic, but it is a constant process and, over the centuries, an effective one. And

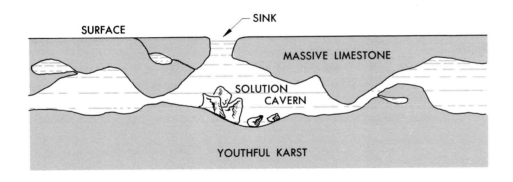

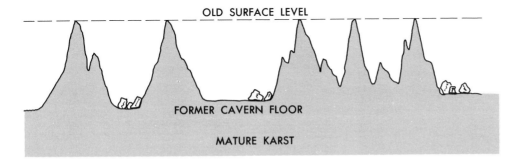

Fig. 27–13 The developmental stages of karst landscape.

if there is solution and removal of materials, there must be deposition. Sometimes deposition is long delayed as when the charged water finds its way to the sea, adding its bit to the salt content there, and it remains for the sea to accomplish the deposition. Frequently, however, some deposition occurs along the route to the sea. Coarse aquifers become indurated, mineral layers build up around spring outlets, and colorful crystals develop in rock cavities. If a cave is open to the atmosphere, lime-rich water dripping from the ceiling partially evaporates and deposits hanging needles (*stalactites*) to be matched by counterneedles on the floor (*stalagmites*) where each drop evaporates further (Fig. 27–14).

Each cellulose molecule is faithfully replaced by silicon to give us petrified wood.

Petrified Wood

Even such elements as silicon, normally strongly resistant to solution, may be taken up by certain types of warm groundwater and redeposited. Where organic remains such as tree trunks gradually decompose, each cellulose molecule is faithfully replaced by silicon to give us petrified wood (Fig. 27–15). Other minerals too, many of them valuable economically, are concentrated by this process of *groundwater replacement*. What causes groundwater to give up its minerals? Cooling of hot water may account for some deposition; evaporation and chemical change as new elements are taken into solution, or the activity of specialized plants such as those that give color to some hot springs, are responsible elsewhere. So groundwater constantly dissolves, transports, and deposits and thus is deserving of recognition as at least a moderate gradational agent along with wind, waves, ice, and running water.

Fig. 27–14   Moving groundwater transports minerals in solution, its solvent limits set only by temperature and acidity. But once it comes into contact with freely moving air, as in this cave environment, evaporation commences and the minerals crystallize into an unreal fairyland of needles and pillars.

Fig. 27–15   This is not wood, these seeming massive logs; all of the cellulose is gone. Warm groundwater percolating through volcanic ash has substituted dissolved silicon for cellulose by a process called groundwater replacement.

PART SIX

THE OCEANS

CHAPTER 28

THE BOTTOM OF THE SEA

All through the oceanic depths are whole ranges of volcanic mountains with the occasional high point breaking the surface. The Galapagos Archipelago is one of these, set off slightly from the East Pacific Rise. When extruded basalt meets the sea it steams mightily, cools abruptly, crystalizes its silica, and forms black puffy billows called pillow lava. Here along the shore of Fernandina Island in the Galapagos this intimate relationship is consummated.

INTRODUCTION

Well over two-thirds of the surface of the earth is covered by the sea at the present moment—29% land versus 71% water. This has not always been the case in the past nor will it be in the future, for although the great ocean basins have tended to display a certain basic stability through the ages, the sea margins have fluctuated widely. On a great many occasions, large segments of the continents that are high and dry today have been invaded by shallow seas, while at other times, notably the Pleistocene period about 1,000,000 years ago, sea levels were greatly lowered as continental glaciers impounded an exorbitant fraction of the earth's total water supply.

This process of continually changing sea levels is operating even now. Recent measurable uplift has occurred in the high middle latitudes merely because of the release of the great weight of continental ice, and frequent more dramatic volcanic eruptions and faulting are accomplishing rapid changes of level in many other parts of the earth. At the same time, erosion inexorably carries sediments into the sea and builds new land along its margins. Yet, if the Antarctic ice should suddenly melt through some small gain in atmospheric temperature, all of the world's coastal cities would be engulfed.

The ocean has dominated the earth's surface ever since primordial rains first filled the crustal cavities.

Nonetheless, despite the transitory nature of precise coastlines at any given moment, the ocean has dominated the earth's surface ever since primordial rains first filled the crustal cavities. But we know very little about it, especially the character of the land beneath the waters.

SOUNDING THE SEA

Except for shallow inshore fishing banks and well-frequented harbors, the sounding of the ocean bottom has been for centuries only a sometime thing. Even official pilot directions and hydrographic charts of coastal waters abound with notations such as "depth unknown" or "probably shoal." Out in the open ocean, where the bottom is a long way from the top, only specially equipped oceanographic vessels have found it worthwhile to even attempt to measure depth with long non-stretching wire lines maneuvered by high-speed winches. But still the difficulties of lead and line soundings have been such that by 1923 only some 15,000 reasonably accurate deep ocean measurements had been made.

It remained for the development in 1940 of sonic gear, which utilizes the principle of echo timing, to revolutionize and speed the process immeasurably. Increasing numbers of ships are being equipped with these instruments and sonar profiles are becoming more or less routine; yet even sonar has its limitations. Echo soundings before 1950 are all suspect because of the lack of recognition that a frequency-regulated power source was an absolute must; and even now there still exists a critical depth limit of about 2000 fathoms [1 fathom equals 6 feet (1.8 m)].

Another problem has been seawater strata of variable temperatures and/or salinity, transmitting sound at differential rates. These normally occur at no great depth, for it is the surface waters that react to daily and seasonal heating and cooling. During World War II, submarines discovered that they could remain undetected beneath a lens of warm water that developed during sunny afternoons and distorted the sonic waves of searching vessels. This was called the *"afternoon effect."*

Representative of some of the new and imaginative approaches to an old problem is the Navy's FISH, a submersible device towed along the bottom by a ship. With both vertical sonar and side-scanning [up to 600 feet (183 m)] capabilities, it has the great advantage of functioning well below the troublesome layers of surface water. But FISH too has its problems, being hampered by a depth limit of less than 300 fathoms (5486 m).

We have by now evolved at least a general outline of the underwater topography of most of the world's seas, and although we are still a long way from being able to construct a detailed contour map of the sea floor, the present knowledge is a great advance from that of Magellan who, in 1521, hung a standard inshore sounding line over the gunwale as he crossed the South Pacific, and when it failed to touch bottom proclaimed that here was "the deepest spot in the ocean" (Fig. 28–1).

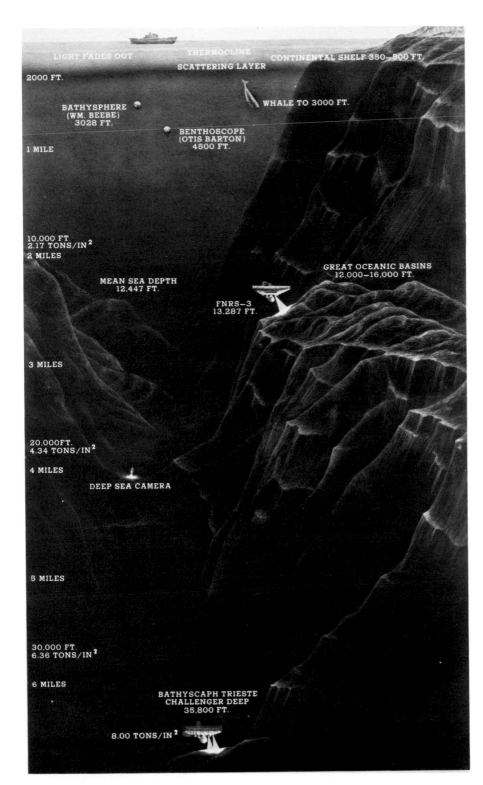

LIGHT FADES OUT

THERMOCLINE

SCATTERING LAYER

CONTINENTAL SHELF 350–600 FT.

2000 FT.

BATHYSPHERE
(WM. BEEBE)
3028 FT.

WHALE TO 3000 FT.

BENTHOSCOPE
(OTIS BARTON)
4500 FT.

1 MILE

10,000 FT.
2.17 TONS/IN^2
2 MILES

GREAT OCEANIC BASINS
12,000–16,000 FT.

MEAN SEA DEPTH
12,447 FT.

FNRS–3
13,287 FT.

3 MILES

20,000FT.
4.34 TONS/IN^2

4 MILES

DEEP SEA CAMERA

5 MILES

30,000 FT.
6.36 TONS/IN^2

6 MILES

BATHYSCAPH TRIESTE
CHALLENGER DEEP
35,800 FT.

8.00 TONS/IN^2

Fig. 28–1   Some appreciation of the great depths involved in the world's oceans may be gained from this artist's rendering. There is ample room in the sea for all of the continents to be submerged with water to spare. The Pacific alone encompasses fully one third of the earth's surface.

THE CONTINENTAL SHELF

The major ocean basins are deep—there is no question about that. Average depths run in the neighborhood of 13,000 to 14,000 feet (3962 to 4267 m) or almost 3 miles (5 km), but an important part of each ocean overlies the subdued margins of the continents. In some cases this *continental shelf* extends for many miles offshore beneath its epicontinental sea; elsewhere it may be very narrow. But where it is existent, the continental shelf is regarded as a legitimate part of the continent, albeit temporarily flooded (Fig. 28–2). Even minor lowering in sea level would expose extensive tracts of this sea bottom, as it has undoubtedly in the past. Dogger Bank, a particularly shallow fishing grounds in the North Sea, is a part of the continental shelf today, yet trawlers have frequently brought up bits of wood and manufactured artifacts indicating that this was recently a forested inhabited land. And off the U.S. Atlantic Coast mammoth teeth, giant sloth bones, and bits of peat containing grasses and twigs have been dredged up from a bottom now 50 fathoms (91 m) below the sea.

The actual break-off point between the continent and the ocean depths is at the outer edge of the continental shelf, wherever that might be. The 100 fathom (183 m) line was quite arbitrarily selected as the critical point for many years, but with increasing knowledge of the character of undersea topography, the International Committee on the Nomenclature of Ocean Bottom Features redefined, in 1953, the continental shelf as "the zone around the continents, ex-

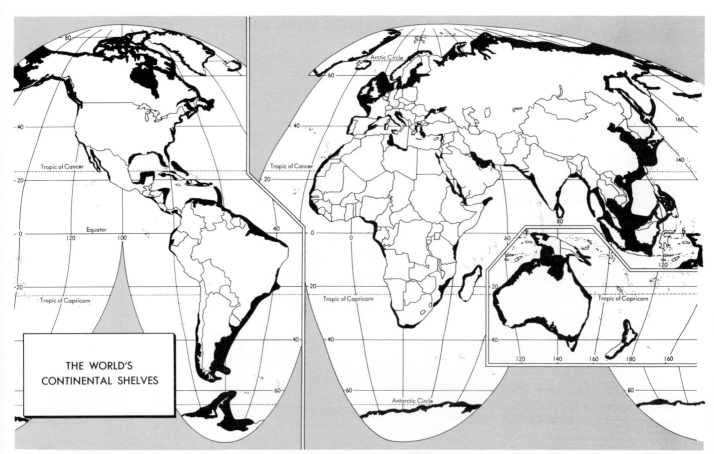

Fig. 28–2   The world's continental shelves. The great ocean basins lie beyond the margins of these shelves.

tending from low-water line to the depth at which there is a marked increase of slope to greater depth"; in other words, no specific depth, for shelves vary in their shoalness, but rather where a sharp break in topography occurs. If every continental shelf were exposed and only the true ocean basins contained the seas, the land surface of the earth would measure about 35% and the oceans would recede to 65%.

The continental shelf is regarded as a legitimate part of the continent, albeit temporarily flooded.

The continental shelves are that part of the sea bottom most like land—life abounds. The critical element is light, penetrating in sufficient quantities to support photosynthesis. In this lambent blue-green world vegetation growth is explosive; not merely seaweed anchored to the bottom rocks (Fig. 28–3), but myriad swarms of phytoplankton—those tiniest of all microscopic sea life which form the absolute base of the marine food chain. They are eaten by larger zooplankton, they in turn by small fish and crustacea, and they by still larger predators—"little fish have bigger fish lurking near to bite 'em, and they in turn have bigger fish, and on ad infinitum."

In this lambent blue-green world vegetation growth is explosive;

The fertilizer to sustain this frenzied pullulation is washed down by continental rivers, imported by sweeping ocean currents, delivered from the deep by coastal upwelling. And contributing to a self-regenerative cycle of life, there is on the shelf, a constant soft precipitation of organic remains as the phytoplankton expire after a life span of a few days or even hours. Like the farmer who plows in a field of clover as green manure, one season's crop residue nourishes the next. Many of the world's great fisheries are in these epicontinental seas, especially those based on

Fig. 28–3 Edible seaweeds of many kinds are harvested from the continental shelf in Japan. But this green vegetation responding to light, heat, and nutrition availability is only one link in a luxuriant and involved marine biosphere.

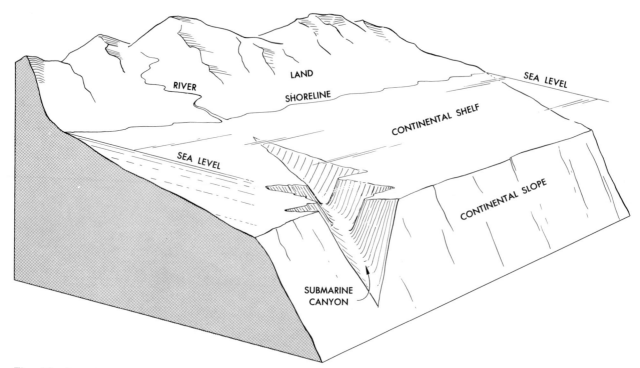

Fig. 28–4   Submarine canyon (highly diagrammatic). These canyons may reach depths of over 1 mile (1.61 km) and are often situated immediately opposite the mouth of a major river.

demersal, or bottom-dwelling species, such as cod or halibut.

It would appear at first glance that the continental shelves should be smooth-surfaced because of the subduing effects of wave action and continuous sedimentation, but this is not necessarily the case. The floor is frequently irregular, in particular where continental glaciers have run far out to sea and not only scarred and gouged the surface but also dumped huge quantities of debris that stand up today as *banks*.

Much more dramatic than these relatively minor irregularities are the widespread canyons scoring the continental shelves to great depth. At their deepest points, the outer edge of the shelf, some of these canyons can reach a mile (1.6 km) below sea level and have frequently been compared with the Grand Canyon of Colorado (Fig. 28–4). The deepest appears to be Great Bahama Canyon with a maximum wall height of well over 2 miles (3 km), but a

Russian survey of Zhemchug Canyon in the Bering Sea indicates by far the greatest volume of any known extant.

Their origin remains a mystery and a source of heated controversy among submarine geologists, but several of these canyons have been studied rather carefully and their basic characteristics are known. They appear to be remarkably like stream-carved canyons on the land, having steep v-shaped contours with tributary valleys along the sides. Since many of the submarine canyons are off the mouths of present-day continental rivers, or where former rivers have flowed to the sea, it has been postulated that they were incised during a low-water period as normal steam valleys and subsequently drowned. The canyons of the Hudson and Congo rivers seem to be simple extensions of those river courses, and the Monterey canyon is immediately opposite the ancient mouth of the Salinas River. But this type of thinking assumes that during fairly recent times the sea level was lowered to a much greater depth than many researchers are willing to accept. These people talk in terms of turbidity currents, heavy mud-saturated currents perhaps triggered by earthquakes, and then, impelled by gravity down the steep wall at the outer edge of the continental slope, scouring and cutting the canyon contours. Whatever the cause, the canyons of the shelf are a major and highly dramatic feature.

THE CONTINENTAL SLOPE

Where the continental shelf sharply falls away to the great depths of the ocean proper, there occurs the most imposing escarpment in all the world. It is called the *Continental Slope* (Fig. 28–4). Varying in height from place to place, it probably averages 12,000 feet (3658 m), which is in itself impressive as a single steep slope. But when continental slopes of almost 30,000 feet (9144 m) are encountered in a number of places, we begin to realize the enormity of this undersea feature. Undoubtedly, the south wall of the Himalayas is the greatest single slope visible to us on the land, but if we combine the continental slope, which begins almost at the beach off the Peruvian coast, with the Andean heights immediately inland, we find what is virtually a continuous slope of about 42,000 feet (12,802 m), or almost double that of the Himalayas. This is where the ocean begins. These deep basins beyond the continental slope have always contained the sea, in sharp contrast to their shallow and transitory extensions onto the margins of the continents.

OCEANIC RIDGES

Apparently, the bottom of the sea has been subjected to most of the tectonic forces that have affected the rest of the earth's crust. No longer do we regard the ocean floor as essentially a vast level plain with only occasional ridges or deeps, for as soundings and samplings have become more frequent, they have disclosed evidences of a very rugged bottom topography indeed, resulting from faulting and vulcanism on a scale comparable to that of the continents (Fig. 28–5). It now appears that extensive plains, such as the one discovered in 1947 by a Swedish expedition southeast of Ceylon in the Indian Ocean, are something of an exception. These are called *abyssal plains* and there are others: the Sohn (off New York), the Hatteras, the Bering, and the Argentine, to name some of the larger ones. But

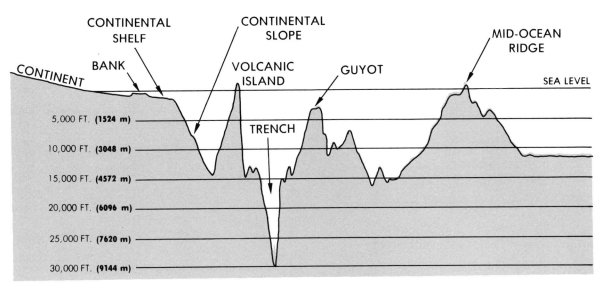

Fig. 28–5 Idealized sea-bottom configuration (great vertical exaggeration).

Fig. 28–6   The island of Santa Maria in the Azores, one of the few peaks in the largely submerged mid-Atlantic ridge high enough at 27,000 feet (8229 m) above the seafloor, to break the surface. Lava cliffs, thin soils, and painfully terraced slopes give evidence of its recent volcanic origin.

the remarkable thing is that the deep-sea topography is not even more rugged since the compensatory erosional forces of the land are lacking. Only when an oceanic mountain rears itself above the surface do waves, rain, and wind attack it, and because of this it is generally believed that many undersea landforms are of extreme age.

The earliest of the major undersea topographic features to be studied and plotted was the Atlantic Ridge, for it was directly athwart the trans-Atlantic cable route between Europe and North America. Subsequent investigation has revealed the outlines of a substantial mountain range running the full length of the Atlantic (with only one

Only when an oceanic mountain rears itself above the surface do waves, rain, and wind attack it.

narrow break near the equator called the *Romanche Trench*) from beyond Iceland in the north to roughly 50°S latitude where it veers off to the east to pass south of Africa. This great cordillera, several times as wide as the Appalachians, reaches its greatest height of 27,000 feet (8230 m) above

the seafloor in the Azores, one of only a few places where it breaks the surface (Fig. 28–6). Ascension, Tristan da Cunha and St. Peter and St. Paul's rocks are also limited islands, but for the most part the ridge is submerged a mile or more although it averages 5000 to 10,000 feet (1524 to 3048 m) in elevation (Map, 1).

In Iceland, active volcanoes, geysers, and seismic tremors are simple facts of life.

That the volcanic Atlantic Ridge is active is readily observable where its peaks break the sea. The tiny isolated colony on Tristan da Cunha was forced to flee its island home in 1961 by threatening lava flows.[1] The people of the Azores have become conditioned to ignore frequent minor earthquakes. In Iceland, active volcanoes, geysers, and seismic tremors are simple facts of life. But perhaps

[1] Most have happily returned to their rocky potato fields after a hiatus of several years in Britain where they experienced modern civilization's benefits of the common cold, television, city traffic, and the income tax.

the most spectacular recent happening was the 1963 birth of a new volcanic island, named *Surtsey* after a legendary Norse giant, a few miles off the south coast of Iceland. It was the child of a sudden fissure near the crest of the Atlantic Ridge some 425 feet (130 m) below sea level and the violent explosions and lava emanations quickly built it into a legitimate island [200 feet (61 m) high and 2000 feet (610 m) long by the fourth day]. Surtsey remains active today although its early explosive character has diminished somewhat; it has built to over one square mile (1.6 km²) in area and 600 feet (183 m) in height.

Rachel Carson's description of the Atlantic Ridge in her classic little volume *The Sea Around Us* is almost poetic.

". . . most of the Ridge lies forever hidden from human eyes. Its contours have been made out only indirectly by the marvelous probings of sound waves; bits of its substance have been brought up to us by corers and dredges; and some details of its landscape have been photographed with deep-sea cameras. With these aids our imaginations can picture the grandeur of the undersea mountains, with their sheer cliffs and rocky terraces, their deep valleys and towering peaks. If we are to compare the ocean's mountains with anything on the continents, we must think of terrestrial mountains far above the timber line, with their silent snow-filled valleys and their naked rocks swept by the winds. For the sea, too, has its inverted "timber line" or plant line, below which no vegetation can grow. The slopes of the undersea mountains are far beyond the reach of the sun's rays, and there are only the bare rocks, and, in the valleys, the deep drifts of sediments that have been silently piling up through the millions upon millions of years."[2]

In the short time since Rachel Carson composed her lyric description of the Atlantic Ridge much has transpired in our understanding of undersea mountain-building processes. 1962 marked the first proposal that drifting continents might well be related to midocean crustal convection expressing itself in subterranean volcanic ranges and sea-floor spreading. Follow-up research has revealed much new information.

For instance, every ocean basin has a midocean ridge

[2]Rachel Carson, *The Sea Around Us* (Fair Lawn, NJ: Oxford University Press, 1951), pp. 68–69. Quoted by permission, Oxford University Press.

system, no two exactly alike, and all connecting one to the other in at least a tenuous fashion. Beyond Iceland is an extension of the Atlantic Ridge into the Arctic Sea, swinging north of Greenland and on well past the pole. At its southern extremity the ridge curves sharply southeastward around the Cape of Africa. In the Indian Ocean there is a double chain, the western one appearing to be the master ridge with the Maldives and Laccadives as emergent peaks. They coalesce in the south to form an inverted Y shape. One arm veers southwest to connect with the Atlantic system while the other swings on around to the southeast between Australia and Antarctica. A lesser branch in the north angles off to insert itself into the Red Sea/Northeast African rift complex (Map 16).

Least typical of the midocean ridges is that in the Pacific.

Least typical of the midocean ridges is that in the Pacific—it is neither high and imposing nor truly midocean, but is set off eccentrically to the east. Called the East Pacific Rise, this ridge roughly parallels the South American coast about 2000 miles (3219 km) offshore for most of its course, but north of the equator it intersects Mexico and seems to enter the Gulf of California, thereby getting itself involved in the greater San Andreas fault system. Despite the occasionally high peaks visible as small islands—Easter, Clipperton, Revilla Gigedo—the average height of this range is only about 4000 feet (1219 m) above a seafloor 10,000 to 12,000 feet (3048 to 3658 m) deep. Its continuation southwestward to unite with the Indian Ocean ridge south of New Zealand is relatively easy to trace. However, a possible congress via Cape Horn with the Atlantic Ridge is evanescent at best.

A peculiarity of midocean ridges that makes them very distinctive relative to their assumed continental counterparts is a great linear rift valley where the crest should be—like a long crater running the full length of the range, which may very well be the case. Ridge water temperatures are higher than the surrounding sea, gravity anomalies are typical (indicating possible rising magma), and bottom samples indicate a great deal of freshly solidified igneous rock. The Atlantic Ridge displays this feature beautifully

Seabottom Oases

Giant tube worms at almost 2000 fathoms (3650 m).

Everywhere in the ocean basin there is life, even in the stygian depths where pressures are equal to several tons per square inch and temperatures close to the freezing point. But this is an immensely difficult environment and only a few exotic and grotesque creatures have managed to adjust, nourished by organic debris that precipitates from the sunlit strata far above. Here is the conventional life zone, the thin layer near the surface where light can penetrate and photosynthesis is possible. The sun is the energy source for all life in the oceans—or so we thought until 1977 when seabottom exploration just north of the Galapagos Islands, at depths of 11,000 feet (3353 m), revealed a whole new world of vibrant life functioning entirely independent of the sun and photosynthesis.

The Galapagos exploration site, and a subsequent one near the tip of Baja California in 1979, are along the crest of the East Pacific Rise where the seafloor is pulling apart at a measurable rate and salt water is coming into contact with superheated magma. The result has been highly mineralized hot water jets spurting up from the ocean bottom. As they cool abruptly in the 36° (3°C) seawater the minerals precipitate out to form chimneys and blanket the slopes at their base. But the most surprising feature of the underwater hot spring phenomena is the abundance of life surrounding each jet: extensive beds of foot-long (.3 m) clams with huge blind white crabs scrambling over them, whole swaying forests of 8 foot (2.4 m) long blood-red tube worms, dandelions (a yellow relative of the Portuguese man-of-war) attached to the bottom by a long filament, smaller worms whose tubes are formed of the minerals in the warm water, whelks, barnacles, giant mussels, leeches, and bristle worms.

It is not heat that is the attraction for life around the demersal jets, although the 854°F (350°C) temperatures do indeed heat the surrounding water for a limited distance, but rather the abundance of mineral sustenance. Hot springs eject hydrogen sulfide which is susceptible to metabolization by commonly occurring bacteria. Then rapidly multiplying bacteria become the primary food source for higher organisms, and this leads to lush little oases of life spotted along the crustal plate interface—an ecosystem based entirely on chemical synthesis. This is remarkable, to put it mildly. To the marine biologist vent communities are as exciting and provocative as the sudden discovery of a lost valley of prehistoric dinosaurs, and we presume that there are many more in every ocean.

but the Indian Ocean ridges are wider, and feature up to six deep parallel rifts, and the East Pacific Rise exhibits only modest allusions here and there.

Somehow in the seafloor spreading process, stresses are set up, which strongly fracture the newly formed crustal material at right angles to the ridge. These *fracture zones* or *transform faults* are so numerous as to dominate the ocean bottom completely for hundreds of miles on either side of the central ridge; and very commonly, if not al-

ways, there is lateral movement along the fault lines offsetting the ridge and its rifted crest. Someone has said that the Atlantic Ridge appears as an 8000 mile (12,675 km) long sliced loaf of French bread, the slices slipping a bit in both directions from center. Slippage can be on a very large scale. The Romanche Trench is 26,000 feet (7925 m) deep and has offset the Atlantic Ridge over 600 miles (966 km) to the east, while south of Iceland the ridge crest shifts abruptly 300 miles (483 km) westward.

Ridge-related volcanic islands and seamounts are rarely encountered at the ridge crest but rather along the flank. Whether this is wholly a result of transverse faulting has not been determined. The Azores, Jan Mayen, and Ascension in the Atlantic are each located very close to a fracture line; Iceland, Ascension, and St. Peter and St. Paul's Rocks are near the top of the ridge; the St. Helena, and Tristan da Cunha have developed far down the eastern slope.

TRENCHES

Much more prominent in the Pacific than its limited mid-ocean ridge are its *"deeps."* These long trenches are a great deal deeper than the general ocean floor, and by far the largest number of them and those with greatest depth are found in the Pacific. They are ranged around the margin of the ocean basin rather than in its center and always in conjunction with active vulcanism and frequent seismic activity. Thus it follows, virtually without question, that they are related to the very faulting and vulcanism that raises the high mountains along their margins, and this, almost certainly, is a product of perambulating crustal plates and an expanding seafloor. In the western Pacific where the chains of mountain peaks form island arcs along offshore fault lines, the trenches align themselves parallel and immediately adjacent to the convex side of most arcs. These are the greatest known depths anywhere on earth.

Until a relatively few years ago, it was thought that Mindanao Deep in the Philippine Trench was the deepest of all at 34,428 feet (10,494 m) below sea level, but the British found Challenger Deep in the Mariana Trench to be 35,700 feet (10,881 m) deep, and a further sounding just a few miles away by the Russian oceanographic vessel "Vityaz" in 1957 led to a claim of a 36,000 foot (10,973 m) depth. Simply measuring these depths has proved to be a very difficult operation through the years, even with specially equipped ships and trained personnel, yet the U.S. Navy astounded the world in January 1960, when two men in a bathyscaph were successfully lowered to the bottom of the Mariana Trench and brought back alive and well. It may be difficult to comprehend just how far down 36,000 feet (10,973 m) is unless we compare it with, say, the elevation of Mt. Everest, the highest known point above sea level at 29,028 feet (8848 m), or visualize Mt. Rainier on top of Mt. Whitney, the two together failing by over a mile (1.6 km) to equal the depth of several of these oceanic trenches. Others of similar depth include the Kurile Trench, 34,020 feet (10,369 m); Ramapo Deep off Japan, 33,000 feet (10,058 m) and the Tonga Trench, 34,860 feet (11,601 m).

Although the total area of ocean bottom involved in these narrow deeps is minuscule, some of them extend for great distances. Off the west coast of South America, for example, where the volcanic activity is on the mainland and there is virtually no continental shelf, the 25,000 foot-deep (7620 m) Chile/Peru Trench parallels the coast for over 2000 miles (8219 km). And the Aleutian Trench on the south side of that island chain is almost as long. In the other oceans, marginal trenches are present but in limited numbers; however, they once again occur adjacent to volcanic seismic regions. The West Indies and the island archipelago south of Cape Horn are accompanied by great deeps, as are the south coasts of Java and Sumatra in the Indian Ocean.

SEAMOUNTS AND CORAL

Charles Darwin on the celebrated voyage of the *Beagle* in 1835 was the first to suggest that the Pacific *atoll*, with its ringing coral reef and shallow enclosed lagoon, might be related to undersea volcanic mountains. He envisioned a slowly sinking volcanic peak with coral growing upward around its margin at a rate rapid enough to match subsidence. At that time, not enough was known about either coral or Pacific vulcanism to prove or disprove such a theory, but today it appears that Darwin was essentially correct. The Pacific floor is liberally peppered with steep-sided volcanic cones or *seamounts*, often arranged along fault lines. And there is a considerable body of evidence pointing to a sinking of the seafloor or a rise in sea level or both. Flat topped seamounts or *guyots*, chiefly in the Pacific but also in the other oceans, suggest wave planation of the tops of volcanoes once at sea level but now drowned. And the thousands of atolls, so typical of the tropical Pacific, are evidence in themselves, once the peculiarities of coral growth are understood (Fig. 28–7).

The coral polyp is a living creature that forms about

Fig. 28–7   Swain's Island. This is not a true atoll because no sealevel connection with the brackish lagoons exists (there is probably a subterranean one); but it was formed in the manner of all atolls. Only 1.5 x 1 miles (2.4 x 1.6 km) in area, and with no elevation over 20 feet (6 m) high, tiny Swain's Island has been American territory since 1925. It is governed from Eastern Samoa 170 miles (274 km) to the south and the 100 or so inhabitants fish and manufacture copra for their livelihood.

Fig. 28–8   It's easy to see why this is called brain coral. Only one of many reef-forming corals, its polyps are microscopic in size. Colors vary with variety.

itself a calcium carbonate chamber from which it can reach out and feed off the zooplankton in seawater. Although there are many different kinds of coral growing under a wide variety of conditions, the several types of colony-dwelling corals, whose social tendencies lead them to construct massive calcium condominiums or reefs, require a fairly limited set of physical circumstances in order to flourish (Fig. 28–8). First, they cannot survive if water temperatures fall below 65° to 70°F (18° to 21°C), so that if we draw a set of lines across the world's oceans at about 30°N and S, we will have for the most part enclosed all those tropical seas that will permit their growth.[3] Second, coral requires a reasonably saline water and does not do well, for instance, where rivers discharge large quantities of fresh waters into the sea. Third, the muddy turbid waters of many beaches and river mouths have a deleterious effect on coral.

[3]Warm currents moving poleward may push this limit somewhat beyond 30°F (−1°C) in places, and conversely cold currents moving equatorward inhibit reef development at rather low latitudes. Also, temperatures much above 90° (32°C), such as are experienced in the enclosed Red Sea and Persian Gulf, often affect coral growth.

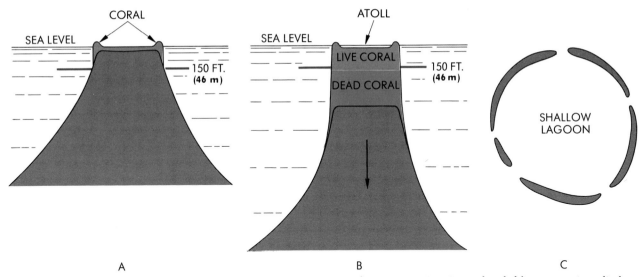

Fig. 28–9   The development of an atoll. (A) An oceanic mountain peak in cross section, its top beveled by wave action, displays coral reef formation. (B) Again in cross section, the continued development of coral at or near the surface as sea level slowly rises (or the mountain sinks or both). (C) A plain view of a circular atoll with its shallow lagoon.

And finally, coral requires light and feeds on plankton with the same requirement, so there is a depth limit, generally described as about 150 feet (46 m), below which reefs cannot form.

Colony-dwelling corals whose social tendencies lead them to construct massive calcium condominiums or reefs.

With respect to this last condition, borings in 1954 on Eniwetok, an atoll in the Marshalls, penetrated coral for 4500 feet (1372 m) before encountering a volcanic rock basement, thus rendering virtually mandatory an explanation of sinking seamounts. The reef below 150 feet (46 m) is, of course, no longer living, but atop this base, coral has been able to maintain itself, suggesting that the sinking was a slow process. The outer edge of the reef, supplied with fresh seawater and the food within it, grows most vigorously and may rise above sea level if waves constantly break over it (Fig. 28–9). We see this same sort of development in the offshore or *barrier reefs*. The Great Barrier Reef off the east coast of tropical Australia is 1500 miles (2414 km) long and occupies the outer edge of the continental shelf, which has been depressed slowly (Fig. 28–10). *Fringing reefs* are found immediately adjacent to many tropical coasts, but slow sinking can cause a fringing reef to become a barrier reef as the land is inundated and the former beach line, marked by growing coral, remains visible some distance offshore. Coralline limestone frequently encountered at some distance up a mountain slope or a former atoll with its lagoon high and dry is evidence that uplift may also occur.

But for every coralline atoll or volcanic island in the Pacific there are probably an equal number of totally submerged peaks, many of which we do not even know exist. Each time an oceanographic expedition ventures out, a hatful of new seamounts is discovered and the supply seems endless. It would appear that a 12,000 foot (3658 m) mountain might be difficult to hide anywhere on this earth, but if we were seeking the best possible place, we could not go too far wrong if we selected the bottom of the Pacific. For instance, as recently as 1967 the Scripps Institution of Oceanography ship *Argo* stumbled across two peaks of this size southwest of Hawaii during routine soundings. Only 58 miles (93 km) apart, both are guyots with flat

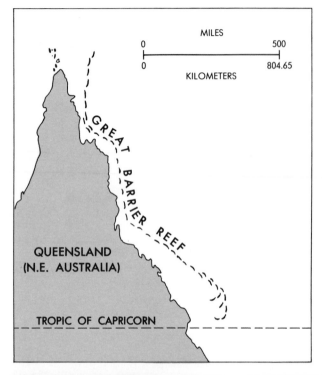

summits more than 30 miles (48 km) wide, and both rise over 12,000 feet (3658 m) from a bottom depth of 17,000 feet (5182 m).

An ominous deep growling against the usual background din of scolding porpoises and nattering squid.

New discoveries are made occasionally by some method other than sounding. Scientists monitoring underwater noises north of New Zealand in 1964 identified an ominous deep growling against the usual background din of scolding porpoises and nattering squid. It led them to initiate a detailed sonar check near Great Barrier Island where they discovered six previously unsuspected 8000 foot (2438 m) submarine peaks, three of which proved to be active vol-

Fig. 28–10   Australia's Great Barrier Reef. The most extensive barrier reef in the world accurately marks the outer margin of the drowned continental shelf.

Fig. 28–11   A new little peak that just surfaced along the Pacific Ring of Fire. Despite its size, it is a furious fumer and could very well build to larger size although the big ocean waves are already nibbling at its flanks. If we could drain away the ocean, we would find that this is no little peak at all but a sizable volcano rising many thousand feet above the ocean bottom. It just happens that at this moment the crater is virtually coincident with sea level.

canoes tentatively named Rumble I, II, and III (Fig. 28–11).[1]

Of some interest in this connection is Cobb's seamount just a few hundred miles off the Washington coast. A 10,000 foot (3048 m) guyot, its remarkable feature is that its 80 acre (32 ha) summit is only 120 feet (37 m) below the surface. But shallow as that may seem, it had been too deep for ships to run afoul of and hence had remained unknown until 1950. In that year, government researchers studying fish migrations were struck by the great concentrations of certain species in this one limited area. The at-

traction, as it developed, was the shallow guyot plateau within reach of sunlight and the subsequent marine growth for fish food. Divers have now explored the surface of Cobb's seamount, and plans are afoot for the University of Washington to sponsor a permanent structure extending above sea level, which will be utilized for meteorological, biological, and oceanographic data gathering.

So we can see that the sea bottom is far from uniform. At least the gross outline is known to us today, although we should expect constant new discoveries and refinements of theories. The shallow continental shelf with its minor irregularities and spectacular canyons is quite a different world from the tremendous continental slopes and deep ocean basins featuring whole mountain ranges, great linear deeps, and isolated seamounts.

---

[1] An inspired bit of originality in the nomenclature department has been demonstrated in the West Indies, just north of Grenada, where a spasmodically erupting submarine volcano has been christened Kick-em-Jenny.

CHAPTER 29

CURRENTS AND TIDES

The stately medieval cathedral of Mount St. Michel crowns a rock islet in Normandy's Bay St. Michel. Until recently high tides regularly surrounded it with water, imposing a splendid, if temporary isolation from the mainland. Its only connection has been a .25 mile (.4 km) long causeway across the shallows. But the bay has become increasingly silted and now only one high tide in ten inundates all of the mudflats. The French Ministry of the Environment has exhibited alarm and is searching for means to reverse nature's evolution.

INTRODUCTION

Within the seemingly constant and random motion of the waters of the sea a fundamental sense of order is inherent. Two basically predictable mobility patterns become apparent immediately; the first is currents and the second, tides.

Whatever sets the surface currents in motion initially (probably a combination of prevailing winds, temperature variations, salinity differences, and bottom configuration), once they begin to move they respond to the deflective force of Coriolis. Hemispheric gyrals result, always clockwise in the north and counterclockwise in the south. Along east coastal frontages there are inevitably warm currents, poleward flowing while along their counter west coasts (at least in the tropics and subtropics) cold currents move equatorward. This is the identical stream of water whether classified as warm or cold because we do not specify any particular temperature. Warm or cold is relative, so that a current flowing poleward from the tropics into higher altitudes is warmer than the water it is flowing through, hence a warm current. But that same current, if having traversed the high middle latitude, is then encountered moving equatorward invading tropical waters, is called cold, for a large part of its original heat has been dissipated and it is now colder than the water through which it is flowing. This is the standard pattern irrespective of ocean basin or hemisphere (Fig. 29–1, and Map V).

The tides are no less regular in their basic pattern. A direct product of the relative positions of moon and sun, their pulsations are totally predictable in general outline even if the details are myriad in their variations.

OCEAN CURRENT PATTERNS

Atlantic Currents

By all odds, the most powerful of the poleward-flowing warm currents is the *Atlantic Gulf Stream*, a veritable river in the sea, carrying huge volumes of indigo tropical water far to the north at speeds of up to 3 knots. Because of its distinctive color, the current is readily visible to the observer crossing it off the coast of the United States, and the balmy temperatures and occasional cumulus clouds and storms associated with it are reminiscent of the tropics. But even this great current must respond to the influence of the earth's rotation. As it moves past Cape Hatteras it veers away from the coast toward the northeast, pulling in behind it a cold water eddy current from the Arctic called the *Labrador Current.* As these two, the brilliant blue tropical waters and the icy green current from the Arctic, merge over the Outer Banks, the cold runs under the warm. Virtually permanent fog results at this juncture, as warm air is abruptly cooled and the roiling waters mix oceanic ingredients ideal for propagation of plankton and fish life.

Tidal pulsations are totally predictable in their general outline even if the details are myriad in their variation.

Now the Gulf Stream trends eastward, impelled by the Westerlies, losing some of its vigor as it widens out, and spilling minor eddy currents both north and south as it goes. Upon reaching the coast of western Europe in the vicinity of the British Isles, once again, the tendency to curve to the right is instinctive, the bulk of the *West Wind Drift* flows southward much cooler and more subdued. But a lesser branch, split off by Ireland, swings northward, hugging the Norwegian coast past North Cape and eventually fading out along the Russian north coast. Cool as it has become by now, this branch exerts a warming influence at these high latitudes and is responsible for ice-free harbors far north of the Arctic Circle.

The main current, cool now as it moves to the south and into tropical water off the coast of Portugal and Morocco, is called the *Canaries Current.* Its generally low temperature is reinforced by the upwelling of cold waters inshore as it pulls to the right and once again, the mixing of cold and warm water sets up conditions highly favorable for oceanic microorganisms, which are basic fish food (Fig. 29–2). As the southerly reach is concluded, the *Canaries Current* becomes tropical and, warming rapidly, continues clockwise to flow due west north of the equator. Here it is termed the *North Atlantic Equatorial Current.*

At the center of this great North Atlantic whirlpool is a region peculiar unto itself in all the world; this is the

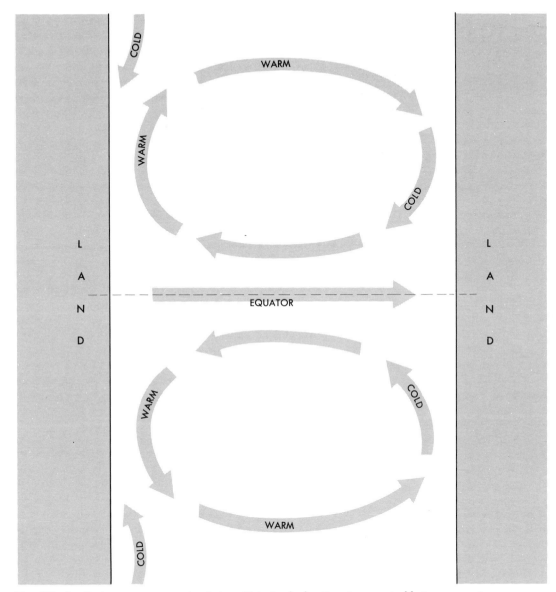

Fig. 29–1   Basic ocean current circulation. This standard pattern is recognizable in every major ocean basin.

Sargasso Sea. Each ocean has the identical gyral characteristics, and in every case its center coincides with the generally calm Horse Latitudes' permanent high-pressure cell. But only the North Atlantic combines this quiet, currentless, windless sea with the great accumulation of sar-

gassum weed. Seemingly endless miles of ocean contain the weed brought in originally by the Gulf Stream from the West Indies where storms have torn it loose from its rocky coastal habitat. And through countless thousands of years, the weed has evolved the ability to maintain and re-

Fig. 29–2   Every modest cove in Japan, it seems, supports a little fishing community. The broad *Kuro Shio (Japanese current)* sweeping along their coast encounters the cold, southward probing *Oya Shio (Okhotsk current)* just offshore, and the mingling of the waters encourages copious marine life.

produce itself here in midocean. The old legend of the Sargasso Sea as a graveyard of ships has some small basis of fact in that, given enough time, any floating debris is likely to find its way into the heart of the spiraling current circulation. And the Bermuda High is a permanent area of calms that makes sailing difficult. But the sargassum is scarcely thick enough to entrap any seaworthy craft, and the actual facts of the matter have been greatly exaggerated and embroidered on through the years.

In the South Atlantic a mirror image of the North Atlantic current circulation prevails: the poleward flowing *Brazilian Current*, following the South American coast, is warm; the broad and still warm *West Wind Drift* flows to the east; the large and vigorous *Benguela Current*, with its associated upwelling, draws Antarctic water equatorward off of West Africa; and finally, the westward-moving *South Atlantic Equatorial Current* completes the circulation. Even the cold Labrador eddy current finds a counterpart in the *Falkland Current* of the South Atlantic.

The old legend of the Sargasso Sea as a graveyard of ships has some small basis of fact.

Between the two equatorial currents running to the west is an apparently out-of-place *Equatorial Countercurrent* exhibiting a diametrically opposed course. This is explained by the piling up of surface water on the west side of the Atlantic by the constantly blowing Trades and the easterly Equatorial Currents. It is common knowledge that the water level at the Atlantic end of the Panama Canal is several feet higher than at the Pacific end—this despite the fact that sea level is supposed to be the same the world over. But the winds are onshore on the east side of the isthmus and offshore on the west. Here is the phenomenon that causes the *Equatorial Countercurrent*. As the major warm currents of each hemisphere curve away to flow poleward taking some of this water surplus with them, the remainder flows "downhill" to the east in the windless Doldrums. This same situation is supposedly accountable for the high current velocities of the *Florida Current* and other elements contributing to the Gulf Stream as they flow east to northeast through the restricted channels of the West Indies.

Pacific Currents

The Pacific displays the same basic pattern as the Atlantic. The *North Pacific Equatorial Current* curves northward off the coast of Southeast Asia and becomes the *Japanese Current (Kuro Shio* or *Black Current* because of its dark blue coloration). This warm flow in turn curves right when it reaches central Japan to become the *West Wind Drift,* and in so doing sucks in behind it a southward-probing finger of cold water from the Arctic called the *Okhotsk Current*

The Gulf Stream

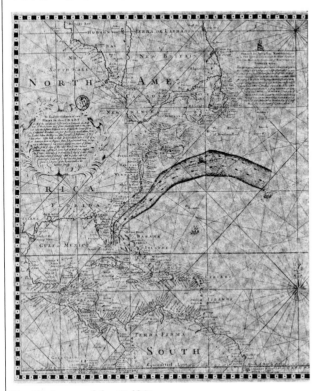

Franklin's map of the Gulf Stream.

Benjamin Franklin (1706–1790) in 1769, then Postmaster General of the United Colonies, ordered the first chart of the Gulf Stream to be drawn as an aid for British mail packets. New England captains had long known something of the character of the great eastward flowing current through their close contact with fishermen working the outer banks, and they would use it to their advantage heading east while carefully shunning it on the westward passage. But Englishmen "had been too wise to be counselled by simple American fishermen" and consistently took two weeks longer on the Liverpool to Boston run than did their American counterparts. They learned reluctantly.

But having cleared up that practical problem, there still remained a tremendous dearth of knowledge regarding the overall character of the Gulf Stream. Organized scientific investigation into the realm of oceanography had never really been at-

tempted—yet there was an untapped source of information on ocean and wind currents that no one had considered seriously, ship's logs. How about collecting and amassing all of that cumulative data? Young Lieutenant Matthew Fontaine Maury (1806–1873) became an early champion of this approach, reasoning that a significant overall picture of ocean conditions would reveal itself in the aggregate where the single log was seldom read by anyone beyond the ship's master and the owner.

Maury was a scientist and congenitally curious scholar disguised in a naval uniform. Throughout his long career he constantly quested for and championed the accumulation of knowledge in all things nautical. So in 1842 when he was put in charge of the newly organized Depot of Charts and Instruments (later to become the U.S. Naval Observatory and Hydrographic Office), he quickly embarked on a ship's logs collection program. From that emerged a folio of *Wind Currents and Charts*, adopted internationally and constantly brought up-to-date with the accumulation of new observations.

Through the years such a great store of data on the oceans accrued that Maury felt he should disseminate his material in a book, *The Physical Geography of the Sea*, published in 1855. It eventually ran to 19 editions and was translated into five foreign languages. One large chapter dealt with the Gulf Stream: it began,

"There is a river in the ocean. In severest droughts it never fails, and in the mightiest flood it never overflows. Its banks and bottom are of cold water while its current is warm. The Gulf of Mexico is its fountain and its mouth is the Arctic Ocean. It is the Gulf Stream. There is in the world no other such majestic flow of waters. Its current is more rapid than the Mississippi or the Amazon and its volume is a thousand times greater."
(Actually its volume is three thousand times greater than that of the Mississippi.)

Today a building at the Naval Academy bears Maury's name—he has been called the Father of the Academy. And still at the top of every chart issued by the U.S. Hydrographic Office is the notation "founded upon the researches made and the data collected by Lt. M.F. Maury, U.S.N."

(Oya Shio). Here again, the cold current underruns the warm, developing a region of fog and often turbulent seas. Unlike the North Atlantic, the Aleutian chain and Gulf of Alaska do not allow more than a very minor part of the *West Wind Drift* to invade Arctic seas, and essentially the entire current turns south to parallel the North American coast and become the cold *California Current*.

The penguin, classic symbol of Antarctica, has no business way up north at the equator, but there he is, portly and dignified, on the Galapagos beaches.

In the eastern Pacific the *South Equatorial Current* is a powerful stream as is the *Equatorial Countercurrent* to the north of it, but although the *Countercurrent* maintains itself across the entire ocean basin, the *South Equatorial Current* runs into difficulties as it progresses westward into the island world of the East Indies and Australia. The general tendency is to curve south, but each island group and narrow strait forces adjustments and deflection of some part of the stream, and in the absence of a definitive continental wall to force its poleward curvature, it tends to become badly disorganized. One fraction, called the *East Australia Current*, moves along the coast of the southern continent and is eventually picked up by the Westerlies and swung east. And here, under the influence of these strong winds that roar around the world uninterrupted by land, a sizable *West Wind Drift* is established. On reaching the far tip of South America, this current turns northward, pulling with it large quantities of Antarctic water, and the world's greatest cold current comes into being. This is the *Humboldt Current*, which constantly maintaining its low temperature through massive upwelling, flows along the coast of South America to the equator. The penguin, classic symbol of Antarctica, has no business way up north at the equator, but there he is, portly and dignified, on the Galapagos beaches, nourished by cold imported Antarctic waters. Here again, but on a larger scale than in other oceans, the exchange of salts through overturning and the invasion of tropical seas by icy water stimulates the active growth of zooplankton and produces a great concentration of fish.

Indian Ocean Currents

The Indian Ocean is somewhat different from the others. South of the equator a reasonably typical gyral exists, but the limited sea of the Northern Hemisphere is strongly affected by the seasonal monsoon winds so that no permanent currents exist and the surface waters tend to move in opposite directions in winter and summer.

DEEP SEA CURRENTS AND WAVES

The currents that have been described so far are all surface currents, propelled in large part by winds. On occasion, over continental shelves, they scour the bottom, but at most they represent only the movement of surface waters to a depth of a few hundred fathoms. What about the great mass of the sea? Does it move in any coordinated predictable manner? Very little is known of the ocean depths, but in theory the warm and thus lighter waters of the tropics should drift poleward at the surface, while cold Arctic and Antarctic waters subside and move equatorward along the bottom. A few experimental testings beneath the Gulf Stream have indicated that there is indeed a current of cold water moving in opposition to that at the surface, but no widespread definable pattern has yet been found. In addition, deep-sea photographs have shown well-defined ripples in the bottom sediments, very similar to those formed by waves near the shore. This surely suggests the presence of waves or current in the ocean depths.

The tides, of course, activate all the ocean waters in their daily rhythms, as do the sporadic tsunamis. There is also some cause to believe that deep waves, quite unlike the normal surface waves, are widely occurring phenomena. But at the moment, we have only the vaguest hints concerning deep-sea movement, and it will undoubtedly require many years of work before these riddles are solved.

TIDES

Basic Causes

Theoretically, the oceanic tides are a simple and highly predictable phenomenon. We know their fundamental cause—the gravitational attraction of the moon and sun,

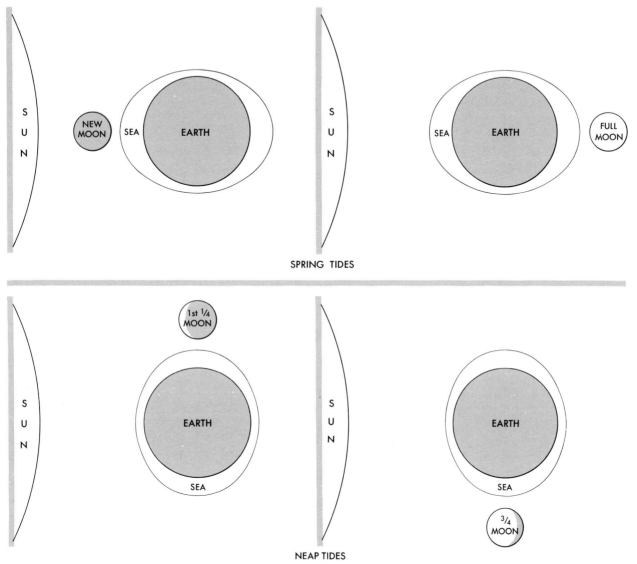

SPRING TIDES

NEAP TIDES

Fig. 29–3   Idealized relationship of tides to phases of the moon. The basic cause of tides everywhere is illustrated here, although variations are infinite, the result of strictly local factors.

but the moon, although many times smaller than the sun, is so much nearer to the earth that it exerts roughly double the force of the sun. Even the ancients were somewhat aware of this relationship between the moon and the tides, for in most places on the earth the time of high and low tide progresses by 50 minutes each day as the moon rises 50 min-

utes later. And the height of each tide varies regularly each month as the moon waxes and wanes (Fig. 29–3).

When the moon is new (dark), it is between the sun and the earth, and the combined forces of both act to pull the mobile waters away from the earth into a bulge. At the same time, a similar oceanic bulge appears exactly opposite as a result of centrifugal force.[1] Now as the earth rotates on its axis, every coastline is exposed to each bulge 12 hours apart. A roughly comparable situation occurs once again during the full moon when the earth, moon, and sun are all in line, but this time the earth is between moon and sun. Two sizable oceanic bulges are once more in evidence and each location experiences two high tides. These are the highest tides of each month, a result of moon and sun acting in concert, and are called *spring tides* (an unfortunate term for they have nothing to do with season). Subsequent tides exhibit the same sequence of daily occurrence, but the high tides become progressively lower until at the quarter phases of the moon, the relative positions of sun and moon are at right angles and their forces oppose one another. Now the tidal bulges are much less exaggerated and the mildest high tides, or *neap tides*, occur. So each day there are two high tides and two low tides six hours apart, but the greatest of these are at the new and full moon every month and the least are at the quarter.

Variations

Once again, however, it should be emphasized that this is theoretical, for although the attractive force of moon and sun is most certainly the basic motive force, the actual character of the tides at any given place varies widely. Puget Sound has 12 foot (4 m) spring tides, Tahiti only 1 foot (0.3 m), and parts of Korea, Alaska, and the Siberian Pacific coast have 30 to 40 foot (9 to 12 m) tides. Yet all face out on the same ocean. The tides at opposite ends of the Panama Canal vary over 15 feet (5 m). Furthermore, the typical cycle of two high and two low tides advancing 50 minutes each day is found to occur less than unanimously. The Gulf of Mexico has only one rise and fall; the Pacific Coast of the United States has two but of differing heights; and Tahiti does not have the 50-minute advance each day of its mild but otherwise normal tides (Fig. 29–4).

These great variations, sometimes occurring only a few miles from each other, are the result of both bottom to-

---

[1]The earth rotates daily about an axis through its own center. However, the earth/moon system also rotates monthly about its center of gravity, which is not at the center of the earth. In effect, the earth wobbles about this eccentric point, causing a centrifugal force in the opposite direction from the moon.

In addition, the high tide opposite the moon is, to a degree, the moon pulling the earth's solid mass away from the fluid water.

Fig. 29–4 A 20 feet (6 m) tide in New Brunswick. Did all of those people drown?

pography and local coastline configuration. The head of the Bay of Fundy in Canada's Maritime Provinces has, at 50 feet (15 m), the greatest tidal variation known. But it is only partially explainable by its long funnel shape and shallowing bottom—surely some peculiarity of the continental shelf contributes to this phenomenal tidal range that is not shared by other similarly shaped bays. A joint United States/Canadian venture to harness the energy of Fundy tides was begun in 1935 at Passamaquoddy Bay (an indentation of the Bay of Fundy), but was later abandoned. However, the idea of utilizing tidal power has been around a long time, and in Brittany's Rance River estuary it has finally

Fig. 29–5   The world's largest tidal power plant at the mouth of the Rance River estuary, St. Malo, France.

been realized. Here the confined tidal surge four times each 24 hours has been producing 544 million kilowatts annually since 1967 (Fig. 29–5).

In inshore waters, tidal currents may commonly reach high velocities, and shipping is advised to avoid certain restricted channels during flood tides. The Norwegian Mael-

Fig. 29–6   Narrow defiles not only confine and magnify normal tidal currents but frequently spawn a dangerous savage churning called rip tides. Small craft are advised to stay clear of such sites when the tides are turning.

strom in the Lofoten Islands creates devastating whirlpools. And much-used Seymour Narrows between Vancouver Island and the mainland, a part of the Inside Passage to Alaska, has proved to be so dangerous (15-knot currents) that demolition of shoal rocks has been carried out in an attempt to tame its fury. Tide rips too, along a broken coastline, where retreating tides from opposite shores meet in frothing eddies, are both common and dangerous (Fig. 29–6).

The idea of utilizing tidal power has been around a long time.

Many river ports, such as Calcutta on the Hooghly, are dependent on *tidal bores* to keep their deep-sea channels open, but here and there a bore becomes so exaggerated that all navigation must be suspended temporarily. A virtual wall of water rushes up the river channel, sometimes for many miles. To produce a bore, a river mouth must not only experience a high tidal range, but some sort of obstruction such as a bar or spit that functions to hold the rising water back until finally it is overcome with almost explosive force. Then if the estuary is funnel-shaped, as is the case at Hanchow Bay, China, where perhaps the world's best known bore is encountered, the height of the advancing wave builds up rapidly. Most rivers discharging into the sea do not experience a bore, but nearly all are troubled by tides backing up normal stream flow, causing upstream flooding and allowing saline water to invade irrigation outlets and the like.

CHAPTER 30

RESOURCES FROM THE SEA

Solar salt evaporation is such a simple matter that nearly everywhere along the world's coasts, if the sun shines at all, small-scale operations are carried on. But this is no small-scale operation near Redwood City on San Francisco Bay—it has been billed as the largest solar salt recovery enterprise in the world. In total, over 50,000 acres (20,234 ha) of tidal flat are involved (some of the shallow pans visible at the top of the photo). The stockpile is served by a dredged deepwater channel.

INTRODUCTION

Considering the fact that the world ocean completely dominates all other earth features and that it is undoubtedly the fundamental cradle of life, we are stupendously uninformed as to its character. Deep-water sailors, shore dwellers, and fishermen have had to amass a considerable body of practical knowledge to assure their livelihood, and this has been invaluable to those who would know the sea. But compared with our understanding of the land, or even the phenomena of the atmosphere, the total accumulation of oceanographic information from the earliest of times is relatively sparse and rudimentary. Only in very recent years has the science of oceanography been recognized and sophisticated instruments put to use to delve into the mysteries of the sea. Already, many long-held misconceptions have been corrected and old generalities hedged and qualified, and we may expect this process to continue for some time, as it is always the case during the pioneering stages of a science.

Perhaps the sea is the ultimate resource if we but learn to use it, but the corollary of utility is knowledge.

It behooves us to learn more about the oceans—and soon. In a world rapidly depleting its mineral resources, increasing its population beyond the capacity of the land to feed it, and constantly requiring larger quantities of fresh water than seem to be available for more than the near future, the sea offers some respite. Perhaps it is the ultimate resource if we but learn to use it, but the corollary of utility is knowledge.

MINERALS

Presumably, all the known elements of the earth's crust are represented in the salts of the sea, for it is the erosion of the land that is largely responsible for the ocean's salinity. No sooner is a landmass upraised above sea level than it is attacked by wind, rivers, ice, and waves, and the entire area with all of its mineral constituents is whittled away to be returned to the sea (Fig. 30–1). But the problems of recovering selected salts for human use are manifold. Often it is the chemistry of the matter that defies solution, but even when the chemistry is possible, the economics of cost feasibility as often as not defeats such efforts. At present, only a relatively few salts are removed directly from seawater in commercial quantities. The age-old evaporation of seawater by the sun still furnishes sodium chloride to many coastal districts, and more recently magnesium and bromine have been successfully isolated on a large scale at relatively low cost.

On numerous occasions in the past the oceans have cooperated in making salts available by intruding into shallow embayments far inland where they have been trapped and slowly evaporated away, depositing their salts in thick beds. These may be mined today, sometimes at great depth, by sinking shafts or forcing hot water into them to form brines that can be piped to the surface for processing and recovery. Some of the largest underground caverns ever made by humans underlie the city of Detroit where thick salt strata have been mined for many years. They date back to Devonian times when tropical temperatures evaporated a sea that overlay much of the northeastern United States. Aiding further mining efforts is the fact that salts precipitate selectively, thus concentrating various differing and reasonably pure layers (Fig. 30–2).

On other occasions, portions of ancient seas have been trapped in underground cavities or porous rock, and there concentrated brines are drawn off today by bores or wells for processing. The Dead Sea, so saline that its waters sustain no life, is the remnant of a former larger sea, in this case at the surface and therefore easy to "mine" for its minerals.

Where humans cannot cope with the chemical problems of isolating certain salts, plant and animal life in the sea may act as intermediaries. Many are equipped with the ability to select out and concentrate a particularly valuable mineral, and if large numbers of their remains are deposited along with the sediments of old evaporating seas, there may be formed a rich stratum susceptible to recovery by mining. Our familiar commercial petroleum is of organic origin too, and forms as a result of the precipitation of dead plants and animals to the bottom of shallow seas. The ex-

Element	Tonnes per Cubic Kilometer	Tons per Cubic Mile	Element	Tonnes per Cubic Kilometer	Tons per Cubic Mile
Chlorine	12,137,853.95	89,500,000	Nickel	1.22	9
Sodium	6,713,114.75	49,500,000	Vanadium	1.22	9
Magnesium	867,958.27	6,400,000	Manganese	1.22	9
Sulphur	568,597.62	4,200,000	Titanium	0.68	5
Calcium	257,675.11	1,900,000	Antimony	0.27	2
Potassium	244,113.26	1,800,000	Cobalt	0.27	2
Bromine	41,499.25	306,000	Cesium	0.27	2
Carbon	17,901.64	132,000	Cerium	0.27	2
Strontium	5,153.50	38,000	Yttrium	0.14	1
Boron	3,119.23	23,000	Silver	0.14	1
Silicon	1,898.66	14,000	Lanthanum	0.14	1
Fluorine	827.27	6,100	Krypton	0.14	1
Argon	379.73	2,800	Neon	0.07	0.5
Nitrogen	325.48	2,400	Cadmium	0.07	0.5
Lithium	108.49	800	Tungsten	0.07	0.5
Rubidium	77.09	570	Xenon	0.07	0.5
Phosphorus	44.63	330	Germanium	0.04	0.3
Iodine	37.87	280	Chromium	0.03	0.2
Barium	18.83	140	Thorium	0.03	0.2
Indium	12.75	94	Scandium	0.03	0.2
Zinc	6.37	47	Lead	0.01	0.1
Iron	6.37	47	Mercury	0.01	0.1
Aluminum	6.37	47	Gallium	0.01	0.1
Molybdenum	6.37	47	Bismuth	0.01	0.1
Selenium	2.58	19	Niobium	0.007	0.05
Tin	1.90	14	Thallium	0.007	0.05
Copper	1.90	14	Helium	0.004	0.03
Arsenic	1.90	14	Gold	0.003	0.02
Uranium	1.90	14			

Fig. 30–1   Sea-water concentration of elements.

Fig. 30–2   Early 19th century Midwestern pioneers were strongly dependent on "salt licks" for their domestic salt. These were naturally occuring springs with a high salt content, the water laboriously boiled down to its essential crystals. But what the lick signaled was thick strata of underground salt left behind by evaporating Devonian seas. Today they are being tapped by deep mines. Almost .25 of a mile (.4 km) below Detroit are these massive caverns hewn out of solid salt and traversed by giant trailer trucks.

Metals From the Sea Floor

Manganese nodules.

Scattered widely over the ocean bottom are accumulations of nodules, up to 3 feet (1 m) in size, which have proven to be rich in manganese, cobalt, nickel and copper. They are of such recent discovery that little is known about their mode of formation or distribution, but nonetheless they have excited great interest among the industrial nations of the world who see them as an answer to declining and/or politically unreliable sources on the land.

Although most commonly found in 5000 to 20,000 feet (1524 to 6096 m) of water, they have been encountered at depths of as little as 500 feet (152 m). Almost every dredge sample of ocean bottom seems to contain a few nodules, and in some areas they form a virtual cobblestone pavement. Manganese content is about 25% with 3% nickel, copper, and cobalt. In 1982 the United States imported more than 75% of its nickel requirements, 98% of its cobalt and manganese, and 15% of its copper—and the experts guess that nodules are being formed on the sea bottom at a more rapid rate than manganese is being consumed worldwide. Practical mining techniques have already been partially worked out with a combination dredge and vacuum system pumping up potato-sized nodules off Hawaii. But large-scale economic mining remains well off in the future.

Even more daunting than the task of nodule recovery would be the mining of speculative massive pure ore-bodies deposited by the action of seabottom hot water jets along crustal plate margins. These are the same demersal hot springs that have spawned extraordinary life-forms in underseas oases along the East Pacific rift zone. Supposedly iron, copper and zinc along with cobalt, gold, molybdenum, mercury, vanadium and manganese precipitate to the ocean floor when the superheated jet encounters near freezing sea water. The Gorda Rise, a broken ridge off the Northern California/Oregon coast has been targeted as a possible ore source within official U.S. waters and is scheduled to be offered for lease soon by the Department of the Interior.

act chemistry of decomposition and the effects of heat developed by subsequent rock deformation to form a liquid hydrocarbon are not fully understood; but without the sea and the life that inhabits it, we could have no oil, for it cannot be made economically in the laboratory.

By far the mineral of greatest value drawn from the sea today is this petroleum. It occupies a category slightly different from oceanic salts extracted directly from sea water in that oil is sucked up from porous rocks in the continental shelf in much the same manner as an oil well operates on land. The fact that the shelf is moderately flooded merely leads to the use of floating platforms as a work base (Fig. 30–3). The Gulf Coast, California, Alaska, Indonesia, Australia, and New Zealand are all involved in exploiting their offshore reserves and many other nations are avidly exploring. The proposed Mohole project goes a step further and envisions the use of a more complicated platform far out at sea, which opens up a whole new range of possibilities. Once this kind of platform is proved practical, any mineral that can be made to flow through a pipe will be obtainable from beneath every part of the ocean floor. Sulphur deposits, for instance, are now tapped from some distance under the continental shelf by drill holes introducing superheated water to melt the sulphur so that it can be drawn off in liquid form. Other soluble minerals have this same potential.

Fig. 30–3   Mobil Oil's largest offshore rig (Transocean I) puts down a gas well on Leman Bank in the North Sea. In the adjudication of national mineral rights, this sector 30 miles (48 km) northeast of Great Yarmouth, has been allocated to Britain. Other nations involved in the North Sea are Norway, Denmark, West Germany, and Holland.

Sulphur deposits are tapped by drill holes introducing superheated water to melt the sulphur so that it can be drawn off in liquid form.

To exploit the resources hidden away in submarine strata by old-fashioned hardrock methods is a good deal more difficult to imagine, yet most minerals are susceptible to removal only in this manner. *Adits* (horizontal shafts) do extend from mine heads on the shore out under the sea in a number of places, but again we are involved only with

the limited continental shelf. To tap mineral deposits locked in a rock bond below 10,000 to 15,000 feet (3048 to 4573 m) of seawater will surely require some wholly new concept in mining technology.

But there is still a third method, a simple sweeping, scraping, or vacuuming of loose materials from the seafloor. Sand and gravel, sometimes by-products of deep-channel dredging, are prosaic nonminerals in most lexicons, but they are required in huge quantities and may often be acquired more cheaply offshore than on land. Shell as a source of basic lime is dredged up in a similar fashion. But diamonds, lying about mixed with gravel on the continental shelf, are far from prosaic, and off the coast of Southwest Africa (Namibia) they are being retrieved by the use of giant vacuum hoses. This is a relatively new technique and one that hopefully can be applied to a broader range of minerals at greater ocean depths.

DESALINIZATION

The sea as a source of potable water has always been an intriguing idea, especially to those inhabitants of the Mediterranean and Tropical Dry Climate regions where rainfall is limited to nonexistent, yet the ocean is at their front door. In total there are some 18,500 miles (29,773 km) of such dry-land seacoast where the only barrier to an unlimited supply of fresh water is a modest salt content. Surely there must be a simple way to solve this little conundrum. But obviously, it is not simple at all or these extensive regions would be blooming instead of continuing to suffer their water shortage.

Many nations are becoming concerned today and are looking critically at the sea, for we are all suddenly discovering that the underground water sources are exhausted, surface supplies are not only inadequate but polluted, and even traditionally pure and sweet rainwater must transit an increasingly contaminated atmosphere before arriving in some sort of altered state. In the United States, for example, we currently use over 500 billion gallons (1 billion m³) of water per day (almost three times the rate of 30 years ago) and with normal population increase, expect to require nearly 600 billion gallons (2 billion m³) by 1990. Yet our sources are essentially the same ones, perhaps used a little more efficiently under the pressure of need, but at the same time deteriorating in quality and increasing in cost of delivery. Conservation can help—fewer green lawns, less frequent showers and toilet flushings—but these are palliatives, not cures, for a very obviously escalating problem that just will not go away.

The Bureau of Reclamation, Department of the Interior, and similar bureaus in other countries are beginning to attack systematically the seawater problem (as well as brackish groundwater and polluted surface sources) and are in a position to bring to bear, for the first time, sophisticated technology backed by national treasuries. The effort is truly international. But given today's world of runaway inflation, governments pressed to establish spending priorities have as often as not scaled down research and development funds for a whole array of promising ventures, desalinization among them.

The basic mechanics are not the difficulty—simple distillation has been with us for a long time and remains in modern plants the most widely utilized process. A variety of refinements are employed to increase efficiency in a volume production effort; systems with names like *long-tube vertical multiple-effect, multistage flash distillation*, and *forced circulation vapor-compression* sound impressive, but it is still old-fashioned distillation.

There are 18,500 (29,773 km) miles of dry-land seacoast where the only barrier to an unlimited supply of fresh water is a modest salt content.

There is also freezing as a means of separation. A saline solution cooled to its freezing point will deposit crystals of sweet water: frozen sea ice in the Arctic is not salty. Theoretically such a phase-change system will require lesser energy than other methods and because of minimal corrosion, can take advantage of cheaper materials for plant construction. Also there are variations in utilizing the selective passage of seawater through several kinds of plastic membranes— *electrodialysis, reverse osmosis, transport depletion* and *piezodialysis.* However, no one is too sure just how efficiently any of these will operate on a truly large scale, and there are always economic matters to consider,

such as the high cost of construction where equipment must stand up to corrosion and scaling and the equal or higher cost of power.

We may gain a bit of insight into just how well large volume reverse osmosis works when the world's largest plant goes into operation at Yuma in the late 1980's. Cost has not been a factor here because under a U.S. agreement with Mexico we are obliged to guarantee pure Colorado River flow across the border, water currently so fouled by U.S. irrigation waste that it is too salty either to drink or use for irrigation. The $150 million Yuma facility is scheduled to clean up all of that at a rate of 100 million gallons (0.4 million m^3) per day.

Frozen sea ice in the Arctic is not salty.

Not long ago purified seawater produced from small experimental plants retailed at well over $1.00 per acre-foot versus a national average of less than 70 cents per acre-foot for water from conventional sources. We still do not have a large-volume producer of converted water in this country but the efficiencies of better designed plants over the last few years, combined with somewhat larger volume than before, has already brought prices per acre-foot down in some cases to 95 cents to $1.00. When we realize that conventional water costs are rising and that a quoted municipal price of approximately 70 to 80 cents per acre-foot does not include the high cost of distribution paid for by consumers out of their other pockets, it can be seen that at least in limited situations, practical converted seawater is close to reality.

It is important to realize that no single desalinization process can be applied universally. Types of fuel available, size of plant, slightly brackish groundwater versus seawater—all of these are elements to consider. For instance, the tiny oil-rich country of Kuwait on the shore of the Persian Gulf has the highest per-capita income in the world. Its fresh water requirements are met entirely by conversion units, but since the population is so small and cost is no object, we can scarcely point to the Kuwait experience as typical. In most places the equation that must balance out is need versus fuel cost.

Shevchenko

The Mangyshlak Peninsula juts out into the Caspian Sea along one of the world's most repellent shores. At 92 feet (28 m) below oceanic sea-level the land-locked sea itself is a turgid saline solution capable of supporting fish life but several times saltier than normal sea water—and the climate, featuring great temperature extremes and an average rainfall of 5 inches (13 cm) per year, is less than attractive for any kind of permanent population. Yet, here at the southern base of the peninsula has developed a flourishing modern town, Shevchenko, dating back to about 1955.

The magnet that attracts large numbers of people to this forbidding littoral is oil, but to sustain them they must have a reliable source of fresh water. So they drink Caspian Sea water, not directly of course, but after it has been processed. Shevchenko is the only community in the USSR that utilizes artificially prepared potable water exclusively, and its desalinization plant is the largest in the country. Additionally, of particular interest to observers from the international community, is that water purification is the deliberate by-product of a thermonuclear fast-reactor electrical power plant.

At the First World Congress on Desalinization and Water Re-use in Florence, Italy, July, 1983, the Soviets reported on the current nuclear power/desalinization complex operation and on some of the lessons learned in over 20 years of on-line experience. The original installation in 1961 experimented with a 3-stage evapo-distillation unit, a 2-stage flash unit, and both electrodialysis and ion exchange membranes, all of which produced only about 18,500 to 31,500 gallons (70 to 120 m^3) per day. But by 1982 fresh water output had escalated to 34,500,000 gallons (130,000 m^3) per day with plans for continuing expansion, and a 10-stage concurrant system had emerged as the reliable and efficient heart of the operation.

The ever present scourges of scaling and metal corrosion throughout the plant had also been addressed and largely corrected by working with various metal alloys and water additives. Carbon dioxide added to the initial sea-water and crushed chalk introduced into preheaters proved to be the critical amalgam that dramatically cut down on scaling in the evaporators. In the heat exchange tubes it was found that an aluminum-brass alloy stood up best against corrosion; in the high temperature stages stainless clad steel; and in the low temperature stage carbon and stainless steel.

All of this is of great significance to those who would emulate the entire system or any part of it. Shevchenko is a pioneer in both long-time/large-scale water purification and the utilization of waste heat from a nuclear power plant as its sole energizer. Techniques pioneered here will prove invaluable to many other nations.

Why not solar energy? Locations experiencing desert or near-desert conditions are blessed with almost constant sunshine, especially just a few miles back from the foggy coast. The potential is there but we are still lagging a bit in the practical application aspect. However, increasingly efficient solar cells and heat collectors are in the developmental stages while at the same time there is belated official recognition that free, nonpolluting solar power offers some hope.

That the sun can do the job is demonstrated by a simple little clear plastic tent fitted over a black-bottomed salt-water container. This solar-activated distilling device operates much as a greenhouse and is used widely where only limited amounts of water and portability are required. The evaporating water in the pan, condensing on the underside of the sloping tent, drains into gutters at the side (Fig. 30–4).

Nuclear power appears also to hold much promise, especially for dual-purpose electrical generating/water conversion plants. A major problem of the nuclear-powered generator has always been its tremendous discharge of heated water that has been used in cooling. The beauty of the water

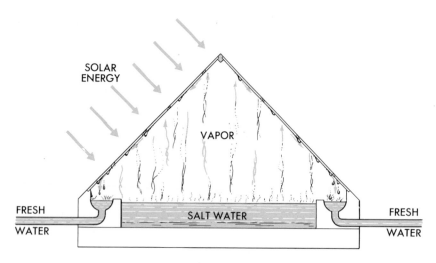

SOLAR
ENERGY

VAPOR

FRESH
WATER

SALT WATER

FRESH
WATER

Fig. 30–4   Simplified solar desalinization.

conversion function is that it can utilize this by-product (heat) to produce a second useful product, thereby cutting both costs and pollution.

A major facility of this kind [150 million gallons (0.6 million m³) per day fresh water capacity] was scheduled to be constructed on an island off the southern California coast, but after much publicity and controversy it has been abandoned. However, a joint Mexican/United States study team has recommended several sites for a similar venture in the general Gulf of California/Colorado Delta region. They talk of an eventual two billion gallons (0.008 billion m³) per day of fresh water. There remains the frequently voiced objection that *any* nuclear reactor leaks small amounts of radiation and surely earthquakes or other natural catastrophes could loose a radiation holocaust. (The Gulf of California is an extension of the San Andreas fault.)

In 1982 the United States fresh water production from desalinization was less than the current needs of a city the size of Honolulu.

If we could go one step further and think in terms of a tri-purpose operation, still greater economies might be achieved. Visualize a powerhouse (nuclear or conventional) whose surplus heat desalts brine, thereby producing both fresh water and useful salts for sale. The Bureau of Reclamation estimates that a single seawater desalinization plant producing 50 million gallons (11,000 m³) per day could also yield 20% of the U.S. annual salt requirements, 25% of the magnesia and 1% of the potash.

Are we making progress or is most of this effort still merely experimental? To put it in perspective, in the United States and its territories (chiefly the Virgin Islands and Puerto Rico), the 1982 fresh water production from all desalinization sources was about 100 million gallons (22,000 m³) per day, less than the current needs of a city the size of Honolulu. But if we compare how far we have advanced from a 1956 production figure of 2 million gallons (400 m³) per day, it is apparent that we are not standing still either. There are so many variables involved that any generalized prognostication at this point becomes more than a little murky.

Another thought has been injected recently into this idea of dual-purpose facilities—utilize the waste cold derived from processing liquefied natural gas. There is a good deal of current argument revolving about both the economics and the safety of transporting liquefied gas at −260°F (−162°C), but some is already moving by ocean freighter and being landed on both coasts and tentative plans call for much expanded shore facilities. To change the product from a liquid back to gas produces cold as a by-product, which can be applied to seawater. The resulting slush of ice crystals and concentrated brine can be easily separated, thereby producing cheap pure water.

FOOD FROM THE SEA

Fish

The present worldwide oceanic food harvest measures out at roughly 100 million tons (91 million tonnes) per annum and is increasing at a fairly constant rate of 5% each year. Of this, over 90% of the total catch is fish, about half being consumed directly while half is converted into fish meal (Fig. 30–5). But this is merely a fraction of the sea's latent productivity. As of now, only a relatively small number of species are considered worthy of the chase, certain limited regions are badly overfished (Puget Sound, the North Sea, etc.), and ancient methods are faithfully adhered to by even some of the so-called advanced nations. A conservative projection points to at least 300 million tons (272 million tonnes) per annum from a well-managed world fishery, the key word here being "managed."

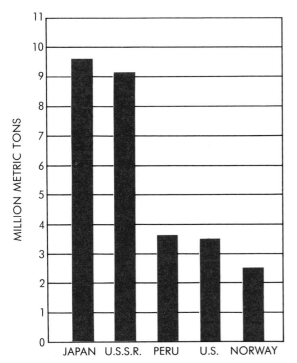

Fig. 30–5   Annual catch of chief fishing nations—1983. (Source: F.A.O.—United Nations.)

We are dealing in this instance with a renewable resource and the aim of good management is exploitation to the point of maximum yield within the limits of maintaining or even improving the resource—while always keeping a weather eye cocked to the total interrelationships of oceanic ecology. This sounds easy until we try to implement it. A major preliminary research project lies ahead, for before any sort of long-range plan can be formulated some knowledge of the life cycles of every variety of fish anticipated in the ultimate catch must be gained. Once a plan is effected, everyone has to pull together and obey the rules. All of this is a very "iffy" proposition if past performances of even the 14 major fishing nations (not to mention several hundred lesser ones where fish may be literally the staff of life) with regard to agreement and enforcement of international codicils are any measure of success. (Fig. 30–6)

Take the simple case of jurisdiction over coastal waters (of which more will be said shortly). It does not seem to matter who is involved; everybody on the beach gets excited when a fleet of foreign ships pulls up a few miles offshore and wants to argue about national versus international waters while they catch your fish.

Then there is the technology of the matter. Surely the New England doryman sculling about in the freezing fog with a few hooks on a line is doing a serious disservice to the great American image of an efficient technocracy. That he has managed to overfish the offshore banks does not constitute proof of the method. A modern fleet might sweep through the same waters one month a year as part of a master plan, take the maximum compatible with species recovery, and move on to other fisheries. There must be a major revolution in both equipment and techniques. The kinds of gear that will be needed to fish the ocean depths of 10,000 to 15,000 feet (3048 to 4572 m) have not been invented yet.

Other Foods

The suggestion has been made that in the interest of maximum oceanic food yield we should back off down the food chain and instead of eating the fish, eat what the fish eat. This is akin to the Chinese eating rice, a direct product of the soil, instead of feeding it to animals and then eating them. In terms of pure calories much is gained by leaving the animals out of the system. So let's eat *plankton*

Fig. 30–6 In some societies fish constitutes the major source of protein in the national diet. And even though the numbers of vessels in the fishing fleet and the magnitude of the catch do not loom large in international fishery statistics, the industry is of extreme local significance.

*F*ish congregate to graze these lush oceanic pastures and fishermen follow to round up their livestock.

The great plankton "gardens" occur in those parts of each ocean basin where cold currents invade warmer waters (or vice versa), thereby initiating vertical turbulence on a large scale. This bringing up of salts from the deep ferti- lizes the surface waters where, in the shallow stratum subject to atmospheric light and heat, a massive spontaneous flowering takes place. Fish congregate to graze these lush oceanic pastures and the fishermen follow to round up their livestock. Farther out in the ocean basins, in the middle of each hemispheric gyral, there is a strong tendency for stratification of water layers not too different from the atmospheric stratification of a stable air mass. Here plankton growth is significantly less than in the regions that enjoy the current activator. It has been fashionable on occasion to speak of "ocean deserts," but this terminology implies that broad regions exist without *any* plankton or fish and that simply is not true. Perhaps we could add the appropriate salts (derived as by-products of desalinization plants) via aircraft and thereby short-circuit the need for upwelling and overturn. If we are going to farm the sea it should be perfectly legitimate to fertilize it for greater yield.

But how about the harvest? A sea-going combine sounds a little "way-out." Problems of processing and the aesthetics of established food habits have yet to be addressed seriously. As an interesting aside, Peru, which has emerged as a major fishing nation only in the last few decades, processes almost 100% of its catch into fish meal, and over half of that is sent abroad (chiefly to Europe and the United States) as poultry and animal feed. So here we are going *up* the food chain instead of *down*.

We do farm the oceans today but only on a micro scale. Oysters and seaweeds are grown on racks or grates in selected coastal environments (Fig. 30–7), and fish corrals in many parts of the Orient impound certain saltwater species whose natural habitat is the shallows. Such activities could, and probably should, be expanded, but since the areas involved are usually mere hundreds of yards offshore, we are dealing here with the thinnest of ocean margins—scarcely an all-out effort to exploit the food potential of the sea.

NATIONAL/INTERNATIONAL JURISDICTION

So we have decided to go to the oceans for food, fresh water, and minerals. Fine, but who owns the sea? As long as transportation was the chief use that was made of the oceans,

Fig. 30–7   Aquaculture. Oyster rafts off the coast of Kyushu, Japan.

the concept of international waters seemed reasonable. Most nations made the exception of landlocked bays and harbors and a general extension of 3 miles (5 km) offshore because that was about as far as old-fashioned cannon balls could be fired effectively. The rest was free to any and all. However, what worked for transportation has not worked for fisheries. Certain species of fish are pelagic; that is, they swim freely in schools, throughout the ocean basin without being tied to the shallow bottom and food resources of the continental shelf. These fish can be caught by anyone willing to go after them, and the rich, energetic, and efficient usually bring in more than their fair share, at least in the eyes of the rest.

Most countries can live with this, especially after they expand their sovereignty to 12 miles (19 km) offshore so as to be sure of encompassing most of the continental shelf and the attendant varieties of fish that contribute to their critical coastal fisheries. The kind of thing that has led to great controversy has been when a 12 mile (19 km) limit does not properly protect those fish that a nation regards as its own. The Pacific salmon suddenly abandons its' pelagic character and heads back to the freshwater streams to spawn. If great numbers in their run to the Frazier River of Canada are caught by Americans standing 13 miles (21 km) off the river mouth, no laws have been violated but there are a lot of unhappy Canadians. And unhappy Ecuadorians and Peruvians have been much in the news when they have sent warships 200 miles (322 km) offshore to arrest U.S. tuna vessels peacefully fishing out of sight of land in what they view as international waters. However, it is the Peruvian/Ecuadorian contention that the largesse of the cold green Humboldt, sweeping northward along their shores, is their national treasure, and if it takes a 200 mile (322 km) offshore limit to ensure it, that is the limit they will claim.

This 200 mile (322 km) fisheries extension has sud-

Fig. 30–8   Soviet fishing vessels, often a "mother" ship with a brood of dependent trawlers, have not been particularly welcome working the American continental shelf. The U.S. reaction has been to extend national jurisdiction to 200 miles (322 km). This Russian trawler was intercepted and brought into the port of Boston for flouting these limits, where she faces confiscation and/or a fine.

denly become very popular and much of the world seems to be going along. In March of 1977 American tuna vessels off Peru abruptly lost their support from home when the United States announced a 200 mile (322 km) limit of its own. In that one month of March, U.S. citations were issued for fishing violations to 53 Soviet ships, 33 Spanish, 10 Japanese, and 1 South Korean. And in April 2 Soviet vessels were seized near Boston, their illegal cargoes confiscated and a fine of $240,000 imposed. In addition, one of the captains was fined $10,000, given a nine-month suspended sentence and put on probation for a year. So obviously, we mean business (Fig. 30–8). If all nations go to 200 miles (322 km) and each seriously polices its claim, there can be major repercussions among the leading fishing nations—Japan and the USSR, in particular, who range the world in search of their catches. Reportedly, the USSR has obtained 60% of its catch from waters that are now under the jurisdiction of nations who have extended their fishing limits to 200 miles (322 km).

It is the Peruvian/Ecuadorian contention that the largesse of the cold green Humbolt is their national treasure.

With regard to minerals from the bottom of the sea, it has been conceded generally that a country is entitled to

all the wealth of the adjacent continental shelf (to date there has been no serious mining beyond the shelves in deep ocean basins), irrespective of the limits of national waters. Surely we would be concerned if the USSR decided to dip into our Gulf Coast offshore oil reserves with a string of wells on the continental shelf. But even accepting this rather generous assignment of rights, snags can easily occur. The North Sea is epicontinental in that it overlies in its entirety a very shallow shelf, and it is fronted by eight nations. They have had their arguments over fisheries in the past, but when oil and gas were discovered a few years ago in the rocks on the continental shelf, all eight felt that they had equal claim. In this particular case a convention of claimants led to reasonable compromise and each nation now shares the mineral rights. But other, less readily solved jurisdictional problems are inevitable.

An opportunity for true internationalism of the earth's resources in which all shall share alike.

If we look into the not too distant future, to the possibilities of recovering, say, manganese nodules or of drilling for oil below two or three miles of ocean—who shall claim this? The affluent nations who can supply the capital and expertise and also accept the risk of failure? Or is this an opportunity for true internationalism of the earth's resources in which all shall share alike? Of course, the rich may not feel that the effort is worthwhile if only a tiny fragment of the product can be retained. Or they may choose to go ahead and then keep it all under a protective umbrella of their military. The world's nations have to make some decisions and formulate some rules.

In 1973 an International Law of the Sea Conference was convened to address a wide range of problems including those of deep sea-bottom mining. After eight years of sometimes rancorous debate, often matching Third World underdeveloped nations against the great powers, a formula was worked out that seemed to be acceptable to all— it turned out to be acceptable to all but the United States. The Reagan administration balked at a provision that established an International Seabed Authority authorized to determine who would be allowed to mine the ocean-bottom. Their contention is that unless American mining companies are guaranteed mining rights the treaty cannot be ratified. 130 nations voted in favor of the basic proposal, the U.S. voted against it and 17 nations (most of Western Europe and the USSR) finally opted to abstain. This action, at least for the moment, aborts a whole host of additional agreements on the use of the sea, such as fishing, navigation, pollution and scientific research. Individual nations are now free to each act on its own initiative in mining the sea-bottom and a number, including the U.S., Britain and the USSR, have already passed enabling legislation.

POLLUTION

As we have seen, the combined waters of all the oceans function in a variety of ways. Among others they act as a vast processing plant converting solar energy into protein, a warehouse of dissolved minerals and fresh water, a source of tidal energy, a definitive influence on the behavior of the lower atmosphere, and a receptacle for all the world's wastes. This last sounds reasonable enough—if we cannot throw our garbage into the deepest basins of the earth, where the miles of overlying water will dilute, disperse, and decompose it, what can we do with it? This is a normal, natural process. The entire earth utilizes the oceanic garbage can as each flowing stream carries its load of waste, dis-

solved or solid, to the sea. Burning of our garbage won't do as a universal alternative, even if everything would burn, because combustion is the chief villain in the fouling of the atmosphere; and surely we can't bury it all.

The entire earth utilizes the oceanic garbage can as each flowing stream carries its load of waste to the sea.

There is nothing inherently wrong with disposing of wastes in the oceans, but the kind, amount, and location of such disposal must be taken into consideration. Of particular concern are the coastal waters where both industrial complexes and great human agglomerations regard the neighboring sea as a convenient sewer. The sheer volume of garbage, raw sewage, and industrial waste discharged into the shallowest part of the ocean basin simply overwhelms the ability of seawater to accept and transform the effluent into an innocuous secondary product. These same coastal waters are subject to increasing demands of competitive use such as desalinization, cultivation of shellfish, and recreation.

To ruminate about the sea as the ultimate resource while at the same time destroying it simply does not make good sense.

Absolute bulk of discharged material is not the whole of the problem, bad as it may be; even worse are the non-degradable chemicals, many of them out-and-out poisons, that are finding their way to the sea. We have certainly witnessed the disastrous results of massive oil spills (Fig. 30–9) in recent years but, although less visible, the various chemical by-products of industry and agriculture may be more far-reaching in their effects. DDT will not degrade and appears to be capable of building up deadly concentration far back along the food chain—so does lead, mercury, and radioactive debris from nuclear fallout. And many of these contaminations are not limited to the coastal zone. Since tetraethyl lead was introduced into gasoline over 50

Fig. 30–9   An oil tanker on its way to Davey Jones' locker. This is what the world dreads, anywhere, in any sea.

years ago, studies in the Pacific have shown at least a ten-fold increase, while strontium 90 can be isolated from virtually any random sample of seawater. One qualified witness to ocean pollution is Thor Heyerdahl, the Norwegian anthropologist who, while crossing the Atlantic on the papyrus raft "Ra II," observed the water from only a few feet away during the leisurely cruise. He reported that besides the usual presence of old crates and grapefruit rinds far out in midocean, at no place was his craft ever free of clots of floating oil or tar.

And far down on the deep ocean-bottom are growing piles of odd receptacles full of radioactive waste materials, nerve gases, and other indestructibles, dumped by various agencies where hopefully they will do the least damage. One would think that people clever enough to perfect a nerve gas could also develop an antidote so that this toxic concoction would not have to be jettisoned into the sea with the thin walls of an untested container our only protection.

We may already have done irredeemable damage to our oceans. Certainly, we must find ways of stopping *immediately* those kinds of contamination that can make it worse. To ruminate about the sea as the ultimate resource while at the same time destroying it simply does not make good sense. If we cannot find a way to get at the heart of the matter soon, we may very well discover that we are engaged in long-range global suicide.

APPENDIX A

PRECIPITATION

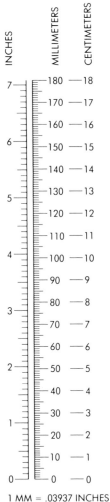

INCHES | MILLIMETERS | CENTIMETERS

7	180 — 18
	170 — 17
	160 — 16
6	150 — 15
	140 — 14
5	130 — 13
	120 — 12
	110 — 11
4	100 — 10
	90 — 9
	80 — 8
3	70 — 7
	60 — 6
2	50 — 5
	40 — 4
	30 — 3
1	20 — 2
	10 — 1
0	0 — 0

1 MM = .03937 INCHES
1 CM = .3937 INCHES

SELECTED METRIC EQUIVALENTS

TEMPERATURE

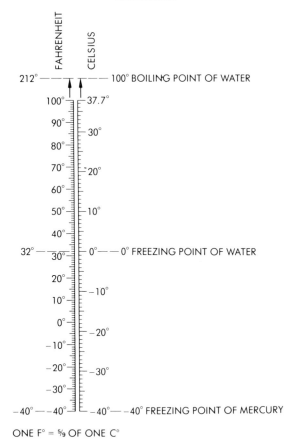

FAHRENHEIT | CELSIUS

212° —— 100° BOILING POINT OF WATER
100° — 37.7°
90° — 30°
80°
70° — 20°
60°
50° — 10°
40°
32° — 30° — 0° —— 0° FREEZING POINT OF WATER
20°
10° — −10°
0°
−10° — −20°
−20° — −30°
−30°
−40° —— −40° — −40° —— −40° FREEZING POINT OF MERCURY

ONE F° = ⁵⁄₉ OF ONE C°

LINEAR DISTANCE

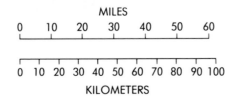

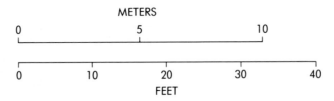

1 METER = 3.28 FT.
1 KM = .62137 ST. MILES

AREA

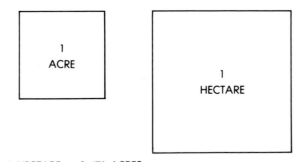

1 HECTARE = 2.471 ACRES

VOLUME

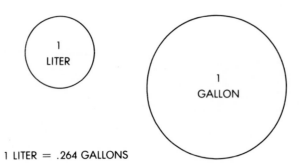

1 LITER = .264 GALLONS

THE KÖPPEN CLIMATIC CLASSIFICATION

There are five basic climatic categories in the Köppen system, symbolized as A, B, C, D, and E. These are further subdivided by adding lower-case letters to indicate lesser variations of temperature and moisture within the major groupings.

A. TROPICAL HUMID CLIMATES

Coolest month must be above 18°C (64.4°F).

Af—Tropical Rain Forest (f—feucht or moist) No dry season. Driest month must attain at least 6 cm (2.4 in.) of rainfall.

Aw—Tropical Savanna (w—winter) Winter dry season. At least one month must attain less than 6 cm (2.4 in.) of rainfall.

The following lower-case letters may be added for clarification in special situations:

m (monsoon)—despite a dry season, total rainfall is so heavy that rain forest vegetation is not impeded.

w′—autumn rainfall maximum.

w″—two dry seasons during a single year.

s (summer)—summer dry season.

i—annual temperature range must be less than 5° C (9°F).

g (Ganges)—hottest month occurs prior to summer solstice.

B. DRY CLIMATES

No specific amount of moisture makes a climate dry. Rather, the rate of evaporation (determined by temperature) relative to the amount of precipitation dictates how dry a climate is in terms of its ability to support plant growth. This is reckoned through the use of formulas that are not included here. See Selected Bibliography for Section 3, Köppen and Geiger.

BW (W—wuste or wasteland)—desert.

BS (S—steppe)—semiarid.

The following lower-case letters may be added for clarification in special situations:

h (heiss or hot)—average annual temperature must be above 18°C (64.4°F).

k (kalt or cold)—average annual temperature must be under 18°C (64.4°F).

k′—temperature of warmest month must be under 18°C (64.4°F).

s—summer dry season. At least three times as much precipitation in the wettest month as in the driest.

w—winter dry season. At least ten times as much precipitation in the wettest month as in the driest.

n (nebel or fog)—frequent fog.

C. TEMPERATE HUMID CLIMATES

Coldest month average must be below 18°C (64.4°F), but above −3°C (26.6°F). Warmest month average must be above 10°C (50°F).

Cf—no dry season. Driest month must attain at least 3 cm (1.2 in.) of precipitation.

Cw—winter dry season. At least ten times as much rain in the wettest month as in the driest.

Cs—summer dry season. At least three times as much rain in the wettest month as in the driest. Driest month must receive less than 3 cm (1.2 in.) of rainfall.

The following lower-case letters may be added for clarification in special situations:

a—hot summer. Warmest month must average above 22°C (71.6°F).

b—cool summer. Warmest month must average below 22°C (71.6°F). At least 4 months above 10°C (50°F).

c—short cool summer. Less than four months over 10°C (50°F).

i—see A climate.

g—see A climate.

n—see B climate.

x—maximum precipitation in late spring or early summer.

D. COLD HUMID CLIMATES

Coldest month average must be below −3°C (26.6°F). Warmest month average must be above 10°C (50°F).

Df—no dry season.

Dw—winter dry season.

The following lower-case letters may be added for clarification in special situations:

a—see C climate.

b—see C climate.

c—see C climate.

d—coldest month average must be below −38°C (−36.4°F).

f—see A climate.

s—see A climate.

w—see A climate.

E. POLAR CLIMATES

Warmest month average must be below 10°C (50°F).

ET (T—tundra)—Warmest month average must be above 0°C (32°F).

EF—All months must average below 0°C (32°F).

APPENDIX C

THE 7TH APPROXIMATION SOIL CLASSIFICATION*

Within this system are the following six subdivisions, beginning with the general and progressing to the finite:

Orders   Great Groups   Families

Suborders   Subgroups   Series

Orders

Order Name	Definition	Derivation of Formative Element	Approximate Equivalents in Old System
Entisols	Soils without pedogenic horizons		Azonal, some Low-Humic Gley soils
Vertisols	Cracking clay soils	L. *verto*, turn	Grumusols
Inceptisols	Soils with weakly differentiated horizons showing alteration of parent materials	L. *inceptum*, beginning	Ando, Sol Brun Acid, some Brown Forest, Low-Humic Gley, and Humic Gley soils
Aridisols	Soils with pedogenic horizons, low in organic matter, usually dry	L. *aridus*, dry	Desert, Reddish Desert, Sierozem, So-lonchak, some Brown and Reddish Brown soils and associated Solonetz

(continued)

*For greater detail and a full explanation of this soil classification, including subdivisions of families and series, see Soil Survey Staff, Soil Conservation Service, U.S. Dept. of Agriculture, *Soil Classification*; A *Comprehensive System (7th Approximation)*, U.S. Government Printing Office, Washington, D.C., 1960—with supplements.

Orders *(continued)*

Order Name	Definition	Derivation of Formative Element	Approximate Equivalents in Old System
Mollisols	Soils with nearly black, organic-rich surface horizon and high base supply	L. *mollis*, soft	Chestnut, Chernozem, Brunizem (Prairie), Rendzinas, some Brown, Brown Forest, and associated Solonetz, and Humic Gley soils
Spodosols	Soils have an accumulation of amorphous materials in subsurface horizons	Gk. *spodos*, wood ash	Podzols, Brown Podzolic soils, and Groundwater Podzols
Alfisols	Soils with gray to brown surface horizon, medium to high base supply, and subsurface horizons of clay accumulation		Gray-Brown Podzolic soils, Gray Wooded soils, Noncalcic Brown soils, Degraded Chernozems, and associated Planosols and some Half-Bog soils
Ultisols	Soils with horizons of clay accumulation and low base supply	L. *ultimus*, last	Red-Yellow Podzolic soils, Reddish-Brown Lateritic soils of the U.S., and associated Planosols and Half-Bog soils
Oxisols	Soils that have mixtures principally of kaolin, hydrated oxides, and quartz	F. *oxide*, oxide	Laterite soils, Latosols
Histosols	Organic (peat and muck) soils	GK. *histos*, tissue	Bog soils

Suborders

Formative Element	Derivation of Formative Element	Connotation of Formative Element
alb	L. *albus*, white	Presence of albic horizon (a bleached eluvial horizon)
and	Modified from *Ando*	Andolike
aqu	L. *aqua*, water	Characteristics associated with wetness
ar	L. *arare*, to plow	Mixed horizons
arg	L. *argilla*, white clay	Presence of argillic horizon
bor	Gr. *boreas*, northern	Cool
ferr	L. *ferrum*, iron	Presence of iron
fibr	L. *fibra*, fiber	Least decomposed stage
fluv	L. *fluvius*, river	Flood plains
hem	Gr. *hemi*, half	Intermediate stage of decomposition
hum	L. *humus*, earth	Presence of organic matter
lept	Gr. *leptos*, thin	Thin horizon
ochr	Gr. *ochros*, pale	A light-colored surface
orth	Gr. *orthos*, true	The common ones
plag	Ger. *plaggen*, sod	Presence of plaggen epipedon
psamm	Gr. *psammos*, sand	Sand textures
rend	Modified from Rendzina	Rendzina-like
sapr	Gr. *sapros*, rotten	Most decomposed stage
torr	L. *torridus*, hot and dry	Usually dry
trop	Modified from Gr. *tropikos*	Continually warm
ud	L. *udus*, humid	Of humid climates
umbr	L. *umbra*, shade	A dark-colored surface
ust	L. *ustus*, burnt	Of dry climates, usually hot in summer
xer	Gr. *xeros*, dry	Annual dry season

Great Groups

Formative Element	Derivation of Formative Element	Connotation of Formative Element
acr	Gr. *akros*, at the end	Extreme weathering
agr	L. *ager*, field	An agric horizon
alb	L. *albus*, white	An albic horizon
and	Modified from *Ando*	Andolike
anthr	Gr. *anthropos*, man	An anthropic epipedon
aqu	L. *aqua*, water	Characteristic associated with wetness
arg	L. *argilla*, white clay	A argillic horizon
calc	L. *calcis*, *lime*	A calcic horizon
camb	L. *cambiare*, to exchange	A cambic horizon
chrom	Gr. *chroma*, color	High chroma
cry	Gr. *kryos*, coldness	Cold
dur	L. *durus*, hard	A duripan
dystr, dys	Gr. dys, ill; *dystrophic*, infertile	Low base saturation
eutr, eu	Gr. *eu*, good; *eutrophic*, fertile	High base saturation
ferr	L. *ferrum*, iron	Presence of iron
frag	L. *fragilis*, brittle	Presence of fragipan
fragloss	Compound of *fra(g)* and *gloss*	See the formative elements *frag* and *gloss*
gibbs	Modified from *gibbsite*	Presence of gibbsite
gloss	Gr. *glossa*, tongue	Tongued
hal	Gr. *hals*, salt	Salty
hapl	Gr. *haplous*, simple	Minimum horizon
hum	L. *humus*, earth	Presence of humus
hydr	Gr. *hydor*, water	Presence of water
hyp	Gr. *hypnon*, moss	Presence of hypnum moss
luo, lu	Gr. *louo*, to wash	Illuvial
moll	L. *mollis*, soft	Presence of mollic epipedon
nadur	Compound of *na(tr)*, and *dur*	
natr	Modified from *natrium*, sodium	Presence of natric horizon
ochr	Gr. base of *ochros*, pale	A light-colored surface
pale	Gr. *paleos*, old	Old development
pell	Gr. *pellos*, dusky	Low chroma
plac	Gr. base of *plax*, flat stone	Presence of a thin pan
plag	Ger. *plaggen*, sod	Presence of plaggen horizon
plinth	Gr. *plinthos*, brick	Presence of plinthite
quartz	Ger. *quarz*, quartz	High quartz content
rend	Modified from Rendzina	Rendzina-like
rhod	Gr. base of *rhodon*, rose	Dark-red colors
sal	L. base of *sal*, salt	Presence of salic horizon
sider	Gr. *sideros*, iron	Presence of free iron oxides
sphagno	Gr. *sphagnos*, bog	Presence of sphagnum-moss
torr	L. *torridus*, hot and dry	Usually dry
trop	Gr. *tropikos*,	Continually warm
ud	L. *udus*, humid	Of humid climates
umbr	L. base of *umbra*, shade	A dark-colored surface
ust	L. base of *ustus*, burnt	Dry climate, usually hot in summer
verm	L. base of *vermes*, worm	Wormy, or mixed by animals
vitr	L. *vitrum*, glass	Presence of glass
xer	Gr. *xeros*, dry	Annual dry season
sombr	F. *sombre*, dark	A dark horizon

Subgroups

Formative Element	Derivation of Formative Element	Connotation of Formative Element
abruptic	L. *abruptum*, torn off	Abrupt textural change
allic	Modified from *aluminum*	Presence of extractable aluminum
arenic	L. *arena*, sand	Sandy texture
clastic	Gr. *klastos*, broken	High mineral content
cumulic	L. *cumulus*, heap	Thickened epipedon
glossic	Gr. *glossa*, tongue	Tongued
grossarenic	L. *grossus*, thick, and L. *arena*, sand	Thick sandy layer
limnic	Gr. *limn*, lake	Presence of a limnic layer
lithic	Gr. *lithos*, stone	Presence of a lithic contact
leptic	Gr. *leptos*, thin	A thin solum
pergelic	L. *per*, throughout in time and space, and L. *gelare*, to freeze	Permanently frozen or having permafrost
petrocalcic	Gr. *petra*, rock and calcic from calcium	Petrocalcic horizon
plinthic	Gr. *plinthos*, brick	Presence of plinthite
ruptic	L. *ruptum*, broken	Intermittent or broken horizons
stratic	L. *stratum*, a covering	Stratified layers
superic	L. *superare*, to overtop	Presence of plinthite in the surface
pachic	Gr. *pachys*, thick	A thick epipedon

THE GEOLOGIC TIME SCALE

Eras		Periods	Beginning of Each Period
Cenozoic	Quaternary	{ Recent (holcene) { Pleistocene (ice age)	15,000 years ago 1,500,000 years ago
	Tertiary	⌈ Pliocene | Miocene { Oligocene | Eocene ⌊ Paleocene	15,000,000 years ago 30,000,000 years ago 40,000,000 years ago 50,000,000 years ago 60,000,000 years ago
Mesozoic		Cretaceous Jurassic Triassic	120,000,000 years ago 160,000,000 years ago 180,000,000 years ago
Paleozoic		Permian Pennsylvanian ⎫ Carboniferous Mississippian ⎭ Devonian Silurian Ordovician Cambrian	225,000,000 years ago 270,000,000 years ago 300,000,000 years ago 345,000,000 years ago 375,000,000 years ago 435,000,000 years ago 540,000,000 years ago
Precambrian (Cryptozoic)	Proterozoic	Keweenawan Huronian	1,000,000,000 years ago
	Archeozoic	Timiskaming Keewatin Unknown	Possibly 3,350,000,000 years ago to 5,000,000,000 years ago

Glossary

A

absorption The change, on contact, of solar radiation to heat.

abyssal plain A deep sea-bottom plain.

accordance All ridge tops at the same height, an indicator of an earlier surface.

advection The horizontal transfer of an air mass.

Agassiz, Lake A huge ancient lake of glacial meltwater. Current Lake Winnipeg is a residual.

aggradation The depositing of transported material. A part of the gradational process.

agonic line A line on a map connecting all points of no compass variation.

air drainage Dense, cold surface air flowing downslope in response to gravity.

air mass A great body of air that displays a singular homogeneity of both temperature and moisture.

albedo The reflectivity of light waves from an unpolished surface.

Aleutian Low The tiny but intense (especially in winter) North Pacific Subpolar Low cell. It is a major generator of middle latitude storms.

alluvial fan A depositional feature at the mouth of a mountain canyon. As a river abruptly lessens its velocity on exiting the canyon onto a plain, it drops its suspended material in a sloping, progressively sorted sequence according to size.

alluvial piedmont A series of coalesced alluvial fans along a mountain front.

alluvium Stream transported and deposited silt.

Alpine glacier See valley glacier.

alto (clouds) Higher than normal.

andesite Greyish, moderately silica-rich lava and rock. A fine-grained relative of diorite.

Antarctic circle The northernmost latitude in the Southern Hemisphere experiencing one 24-hour period of light and one of dark each year—66.5°S. Lat.

ante meridiem (A.M.) Before the sun has transited a given meridian, hence morning.

anticline An upbowed rock strata.

Antitrades Poleward flowing upper air wind currents directly above the Trades. See Trade Winds.

aphelion The greatest distance of earth to sun each year—on or about July 3.

apparent motion (astronomy) The seeming motion of sun or fixed stars to those viewing them from the moving earth.

aquifer A confined and tilted permeable underground stratum exposed at one end to receive moisture. Water percolates down through the porous structure and can be tapped by well punctures from above.

arch Mildly upfolded strata, two-dimensional. Anticline.

Arctic, A (air mass) Extremely cold and extremely dry, reflecting its source region, Antarctica or the frozen north polar seas.

Arctic/Antarctic Front The somewhat theoretical line of contact, in the high latitudes, between Arctic and Polar air masses.

Arctic circle The southernmost latitude in the Northern Hemisphere experiencing one 24-hour period of light and one of dark each year—66.5° N. lat.

artesian well A flowing well. One that taps an aquifer whose water entry point is higher than the well head.

ash (volcanic) Fine flourlike bits of solid volcanic ejecta.

association, vegetative Mutually supportive plant communities. Pure stands of a single specie are virtually never encountered in nature.

atoll A semicircle of coral islets enclosing a shallow lagoon. The entire feature marks the crest of a submerged volcano.

azimuth True compass direction.

azimuthal projection Transference of a portion of the globe's network of latitude and longitude lines onto a tangent plane. True direction from its center.

azonal (immature) soils New or disturbed soils not yet in equilibrium with their environment, especially climate and vegetation.

Azores high See Bermuda high.

B

banks Shallow plateaus atop the continental shelves. Receptive of greater light and hence a larger plankton supply; they are unusually attractive to fish.

bannwald Carefully protected forest plots

on the slopes above Swiss village. They are maintained to impede the winter avalanches if not to entirely block them.

barchan A crescentic sand dune with a gentle slope upwind and a steep one to the lee.

barometer (aneroid) An instrument measuring air pressure. A metal diaphragm enclosing a partial vacuum compresses or expands in response to differing pressures.

barometer (mercurial) An instrument measuring air pressure. A column of mercury in a glass tube responds to differing pressures.

basalt A dark, silica-poor lava and rock. A fine-grained relative of gabbro.

base exchange (soil) Colloidal absorption of ions and their release in exchange for other ions.

batholith A large intrusive plutonic mass, at least 40 square miles (104 km^2) in size.

bed load (streams) Swirling sands and skipping pebbles impelled along the channel bottom by stream flow. These are the heaviest materials that the stream is capable of transporting.

Bermuda high The North Atlantic Horse Latitude cell featuring generally dry sunny weather, sinking air, and high pressure.

Bonneville, Lake A huge ancient lake of glacial meltwater. Current Great Salt Lake is a residual.

Bora A cold, high velocity gravity wind descending the Julian Alps at the head of the Adriatic.

boreal forest High latitude conifers. See taiga.

braided stream One in which a broad channel is choked, during low water, by deposits of coarse gravel and the stream breaks up into a tortuous array of mere trickles and rivulets.

C

caatinga A dense thorn thicket occurring in semiarid northeastern Brazil.

caldera An immense craterlike cavity at the crest of a volcano, formed by collapse as the magma chamber is abruptly evacuated.

calendar A timekeeping device that attempts to mesh the diverse elements of year, month, week and day. See Gregorian and Julian.

calorie The heat required to raise the temperature of one gram of water 1°C.

calving As continental ice sheets flow inexorably out to sea, seawater buoyancy snaps off huge chunks to form icebergs.

Canadian shield (Laurentian) The hardrock core of North America. Centered on Hudson Bay. It involves the eastern half of Canada and a bit of the adjacent United States.

canyon, submarine Deep "V"-shaped valleys, remarkably like stream-carved canyons on land, which often occur opposite the mouths of large rivers. However, their exact origin remains a subject of lively debate.

capillary action (soil) The blotterlike action of a dry soil particle adjacent to a moist one, thereby pulling groundwater in opposition to gravity.

carbon 14 A radioactive dating mechanism for organic remains but is distinctly limited in accuracy beyond about 40,000 years.

carbonation A chemical reaction with carbon dioxide.

cartography The art and science of mapping.

castellanus (clouds) Turreted in appearance.

Celsius scale (temperature) 0° equals the freezing point of water, 100° its steam point.

centigrade (temperature scale) See Celsius scale.

chaparral A low, woody brush characteristic of the Mediterranean climate region.

chernozem A black high-humus soil developed in the middle latitudes under 15 to 20 inches (38 to 51 cm) of annual precipitation and short grass cover. First identified and named in the Ukraine.

Chinook Warm dry wind descending lee mountain slopes.

cinder (volcanic) Coarse bits of solid volcanic ejecta deriving from a violent explosion as lava solidifies in the vent.

cinder cone A steep, usually symmetrical volcanic cone built of unconsolidated ejecta.

cirque A semicircular rocky amphitheatre formed at the head of a valley glacier. Plucking and quarrying enlarge the structure and advance it upslope.

cirrus clouds Formed of ice crystals, they are characteristically hazy or feathery in appearance.

climate Generally a longtime mean of momentary weather observations.

climax vegetation A natural vegetation association that is in equilibrium with the environment. Theoretically, it will continue unchanged infinitely.

climograph A graph on which can be plotted the average of longtime monthly temperatures and precipitation records.

clouds Tiny water or ice droplets in suspension, usually at a high level. Caused by adiabatic cooling.

cloud seeding Adding to the atmosphere substances such as silver iodide to provide freezing and/or condensation nuclei.

cold front A segment of the Polar Front, on the westerly side of Standard Middle Latitude cyclones, where local circulation causes cold air to advance against warm.

colloids (soil) Tiny soil particles that when combined with water, become gelatinous in consistency.

composite volcanic cone A pointed concave-shaped volcanic cone built by alternating cinder showers and lava flows.

condensation The change in form of water vapor to liquid water, usually as a result of lowered temperature.

conduction The transfer of energy through contact.

conformality (map) The quality of true shape on a map.

conglomerate A sedimentary rock. Pebble size or larger rocks cemented together by any of a variety of materials.

conic projection Transference of a portion of the globe's network of latitude and longitude lines onto a tangent cone.

coniferous Cone bearing trees. Usually needleleafed and evergreen.

continental drift The concept that continents can and do break apart from an original whole and slowly drift as individual entities.

continental glacier A massive sheet of ice overlying whole continents and flowing out in all directions from the center(s).

continental shelf The subdued margins of the continents overlaid by epicontinental seas.

continental slope The sharp drop-off to the deep sea-bottom that marks the outer edge of the continental shelf.

contour A line on a map connecting all points of equal elevation.

contour interval The vertical distance represented by each contour line on a map.

convection A rising, warm central column of matter and sinking, colder side columns triggered by heat at the base.

convectional precipitation See convection. Violently rising moist air, triggered by heat, expands and cools to a sub-dew point temperature.

core, earth's The extremely dense, superheated center of the earth, thought to be solid, despite great heat, because of the pressure exerted on it. An outer core may be molten.

Coriolis force The earth's rotation causes an apparent lateral drift of horizontally moving objects. The force increases with latitude from nil at the equator to maximum at the poles. See Ferrel's law.

craton See shield.

crescent moon Only a slim margin of the moon reflecting sunlight. Appears as a bright crescent from the earth.

crevasse (glacial) Deep stress fractures in flowing alpine ice as the top center, unimpeded by friction, advances more rapidly than the rest of the ice mass. Crevasses tend to be cresentic in shape with their concavity upslope.

crossbedding Sandstone formed from dune deposition displays a variety of obliquely aligned strata as opposed to the normal strict parallelism of most sedimentaries.

cumulous clouds Fleecy, fluffy, cottony appearance. Often a small fair weather cloud but can also include towering thunderheads.

cusp The scars of laterally cutting meanders along the marginal cliffs of an old-age floodplain.

cyclone Generally, any low pressure center. Specifically, in India, Australia, Pakistan, a violent hurricane/typhoon type storm.

cyclone, Standard middle latitude A large, slow moving, frequent and generally mild product of the Polar Front.

cyclone, weak tropical An apathetic, characterless tropical storm whose chief feature is cloudiness and heavy rain—probably a product of the Intertropical Front.

cyclonic (frontal) precipitation Light moist air ascending over a denser air mass, expands and cools to a sub-dew point temperature.

cylindrical projection Transference of a portion of the globe's network of latitude and longitude lines onto a tangent cylinder.

D

dacite Light colored, extremely viscous, silica-rich lava and rock.

Daylight Saving Time A summer adjustment of zone time so that a given zone keeps the time of the adjacent zone to the east.

deciduous A tree or bush dropping its leaves during one season of the year, usually in response to cold or aridity.

declination The number of degrees of error from true north in a magnetic compass reading.

deflation The ability of swirling, eddying wind currents, with their suspended sands, to cut downward and form depressions, particularly in the lee of a ridge.

degradation The cutting, abrading, or erosional aspect of the gradational process.

delta A depositional feature built at the mouths of rivers debouching into a sea or lake. The material is progressively sorted according to size as the stream loses its velocity.

desert pavement Pebbles and larger rocks cemented together by the saline remains of evaporating moisture. The finer material has been removed by wind.

dew Condensation on a cold surface above 32°F (0°C), a result of air cooling through contact.

dew point Air cooled to the point of saturation. Further cooling will produce condensation.

diastrophism Mountain building. See tectonic forces.

diffuse daylight Radiated short-wave solar energy, as a result of scattering, arrives at the earth in a soft light form.

dike A relatively small intrusive plutonic mass. It is in the form of a thin vertical sheet cutting across horizontal strata.

dilatancy Tiny cracks opening up in rocks under extreme pressure. A possible clue to earthquake occurrence.

diorite Coarse-grained plutonic relative of andesite. Grey colored and moderately silica-rich.

Doldrums An equatorial belt of high temperature, low pressure, heavy rainfall, and generally fickle breezes or calm.

doline See sink.

dome Mildly upfolded strata—three dimensional.

dome, breached A circular upfolded structure, the top removed via erosion with the concentric strata exposing themselves as inward facing cuestas.

dome, volcanic cone See shield volcano.

drift, glacial A general term referring to all types of glacial deposition.

driftless area Hilly country in southwest Wisconsin and adjacent Illinois and Iowa, which somehow escaped being overrun by continental glaciers.

dunes Asymmetrical sand ridges piled up by wind action. Often advancing downwind.

dust (atmospheric) Solid particles in atmospheric suspension (excluding ice).

dyne The force that will accelerate one gram of mass one centimeter in one second.

E

Easterlies (Polar) High latitude easterly winds. Encountered in both hemispheres at latitudes 60° to 70°.

ecliptic The plane of the earth's orbit about the sun. Also, the plane of the moon's orbit around the earth.

edaphic Strong control of vegetative character by soil.

elastic rebound Along an active fault, rocks deformed as they attempt to sustain increasing pressure, suddenly snap back to their original shape.

elements There are 92 naturally occurring elements, which are the basic building stones of all known matter.

El Niño A warm, southward flowing current that inserts itself periodically between the massive cold Humboldt current and the coast of Ecuador and Peru. Chaos ensues.

eluviation The removal of soluble minerals and small particles from a surface soil horizon via leaching.

emerged shoreline One in which the most recent relative movement of land and water has resulted in the exposure of the fringing sea-floor. Frequently a flattish, straight, shelving coastline.

epiphytes Plants which utilize a host tree as a perch to escape the dark forest floor. Unlike parasites they do not draw their nourishment from the host.

equal area (map) The quality of area equivalency on all parts of a map.

equator A great circle whose plane is at a right angle to the earth's axis, hence one which bisects the earth into the Northern and Southern Hemispheres.

Equatorial, E (air mass) Located in equatorial regions, it is a hot and very moist air mass.

equinox, fall (autumnal) Both hemispheres receive the same amount of solar energy; the midday sun is overhead at the equator, and day and night are equal everywhere. On or about September 21.

equinox, spring (vernal) Both hemispheres receive the same amount of solar energy; the midday sun is overhead at the equator, and day and night are equal everywhere. On or about March 21.

equiseta An ancient plant featuring seedless reproduction. Its upright tubular jointed stem is high in silica, hence the popular name "scouring rush" for its use as an abrasive polishing agent.

erg Sandy desert.

erosion A general term for the wearing away of rock. See gradational forces.

erratic (glacial) Huge boulders transported by now departed ice sheets that are out of place on the local geologic scene.

evaporation The change in form of liquid water to vapor, usually as a result of heat.

evapotranspiration Quantifying the evaporation from the soil plus transpiration from plants.

exfoliation Scaling of surface rock layers as a result of pressure release.

eye (hurricane) The warm, windless, cloudless center of a hurricane.

F

Fahrenheit scale (temperature) 32° is the freezing point of water, 212° its steam point.

fathom One fathom equals 6 feet (1.8 m).

fault Any deep-seated crustal fracture along which there is movement.

fault, normal An inclined crustal fracture with the upper block sinking and/or the lower block rising.

fault, reverse An inclined crustal fracture with the upper block rising and/or the lower block sinking.

fault, strike-slip (wrench, transverse) A crustal fracture along which there has been horizontal motion.

fault, thrust A very low angle crustal fracture along which the upper block has risen.

felsenmeer Extensive ground cover of sharp-edged angular rock fragments, usually at high altitudes.

felsic lava An acronym formed from fel-feldspar and sic-silica. A silica-rich lava, hence thick and viscous.

Fennoscandian Shield The hard-rock core of Europe. Centered on Finland, it involves much of Sweden and Karelia.

Ferrel's law Any horizontally moving object in the Northern Hemisphere will exhibit an apparent right-hand deflection. Left-hand in the Southern Hemisphere. See Coriolis force.

finger lake A "U"-shaped, ice-deepened valley dammed at its lower end by a moraine, thereby impounding water.

fissure lava flows Highly liquid flows of lava emanating from cracks or fissures in the surface. The resulting igneous rocks from such flows arrange themselves in flat-lying horizontal layers.

fjord A "U"-shaped, ice-deepened valley intruded by the sea as the glacier melts away.

foehn See Chinook.

fog (advection) A transported radiation fog, usually as a result of a gentle breeze.

fog (radiation) Tiny water particles in suspension, usually near the ground, a result of air cooling below the dew point via a temperature inversion.

folding The actual flexing of brittle rock over time into any of a variety of plications.

foliation (rocks) The quality of fracturing readily into flat, leaf-like planes.

fossils Plant and animal remains of earlier life preserved for observation and study today in sedimentary rocks.

fracto (clouds) Torn or rent.

fracture zones See fault, strike-slip.

freon (fluorocarbon) The common gaseous propellant in aerosol spray cans. It is possible that this gas will contribute to a breakdown of the ozone layer.

front The line of contact between two differing air masses.

frost See hoar frost. Also, any air temperature below 32°F (0°C).

full moon The moon's position relative to the earth exactly opposite the sun so that its light face appears as a shining perfect disc.

fumerole A natural steam jet contaminated with magmatic gases and emitting an odor.

fungus Any of a group of thallophytic plants comprising the molds, mildews, rusts, and smuts. They are without chlorophyll and do not involve themselves in photosynthesis.

G

gabbro Coarse-grained plutonic relative of basalt. Dark colored and silica-poor.

galeria forest A narrow riverine forest almost completely dependent for its moisture on the watercourse.

geanticline A particularly large-scale up-bowed rock strata.

geologic time The recognition of the immense time span of the earth's existence (perhaps as long as 5 billion years) and that all geological phenomena must be measured relative to this scale.

geomorphic cycle The progressive erosion of a landform through youthful, mature, and old-age stages from its inception to its removal.

geosyncline A particularly large-scale down-bowed rock strata.

geothermal power Tapping natural underground steam via wells and directing it to turbine powered generators.

geyser A spasmodic erupter of hot water and steam.

gibbous moon Over half of the moon reflecting sunlight so that it appears from the earth as a slightly bloated half moon.

glacial polish A mass of ice with fine abrasive sand embedded in it imparts a satiny gloss to the rocks that it overrides.

gneiss A metamorphic rock featuring wide, sharply defined banding of alternating dark and light-colored minerals.

Gondwanaland The speculative reassembly of Southern Hemisphere drifting continents into a single landmass.

graben Two vertical crustal fractures between which the central block has slipped downward producing a steep-sided, flat bottomed valley.

gradational forces Those forces that cut, dig, and whittle away at the end products of diastrophism. They further transport the accumulated materials for some distance and deposit them.

graded stream A theoretical stream where velocity and sedimentary load are in precise equilibrium.

Grand Coulee An abandoned watercourse carved by the ice age, meltwater swollen Columbia River in eastern Washington.

granite A commonly occurring, coarse-grained igneous (plutonic) rock.

granitization The theory that granite may be an end-product of the metamorphosis of a wide range of rocks.

graphic scale (map) A graduated line showing distances on a map.

Great circle The largest circle that can be drawn on a globe; a circle that bisects a globe, or the line described on the surface of a globe by a plane passed through its center.

greenhouse effect Short-wave solar energy, allowed easy ingress through the glass walls, is transformed into long heat waves by absorption. The glass is no longer transparent to the heat waves, which are trapped within.

green manure One season's crop, usually a legume such as clover, plowed under to act as organic fertilizer for the next crop.

Greenwich The prime meridian. One-half of a great circle passing through both poles and Greenwich, England. Designated 0° longitude, it is the line from which all longitude is measured.

Gregorian calendar Our current everyday calendar dating from 1582.

groundwater Relatively shallow subterranean water in soil and rock. It exhibits normally, a close relationship to precipitation and surface water sources.

groundwater replacement Molecule-by-molecule replacement of buried organic material by minerals dissolved in circulating groundwater.

guyot A flat-topped submarine volcano. Probably the result of wave planation when sea level was lower, the mountain higher, or both.

H

hachures (contour map) Short lines at right angles to a closed contour line indicating a depression.

hail Solid ice balls precipitating frequently from violent convectional storms. They are formed as raindrops but have been elevated repeatedly into the freezing zone.

half moon Half of the moon reflects sunlight so that it appears from the earth as a bright half sphere.

halophyte Plants with the capacity to utilize groundwater with a high content of salt.

hamada Polished bedrock desert.

hanging valley Ice-truncated tributary valleys that once intersected a master stream whose subsequent deepened, widened valley has been occupied by an alpine glacier. Once the ice has melted back, the tributary valley displays a waterfall at its mouth.

Hawaiian high The eastern end of the

North Pacific Horse Latitude belt. Typically dry sunny weather, sinking air, and high pressure.

Hawaiian volcanic cone See shield volcano.

heave, soil Freezing groundwater in surface soils swells it upward. Subsequent melting allows collapse and a perceptible sag downslope.

hoar frost Ice crystals forming on cold surfaces as moist air is cooled below the freezing point by contact.

horizon, soil Succeeding parallel strata in soil with increasing depth, each distinctive from the other by color or texture.

Horse latitudes A subtropical region of high temperatures, high pressure, minimal rainfall, and generally calm conditions. Encountered especially at sea in both hemispheres at approximate latitudes 30°.

horst Two vertical crustal fractures between which the central block has been thrown up.

humidity A measure of water vapor in the atmosphere.

humilus (clouds) Flattened cumulus.

humus Partially decomposed organic matter in soil, imparting a dark color as carbon is released. Also significant in soil texture contributing to both friability and moisture retention.

hurricane An intense summer/fall storm of the tropical seas—spawned by the Intertropic Front. Normally encountered in the southwest corners (Northern Hemisphere) and northeast corners (Southern Hemisphere) of major ocean basins. The South Atlantic is the exception.

hydration A chemical union with water.

hydrologic cycle The loss of oceanic water via evaporation; the return of that moisture to the earth's surface via precipitation; and finally, the restoration of surface water to the sea via rivers, springs, and icebergs.

hygroscopic The ability to act as a condensation nucleus, such as dust particles in the atmosphere.

I

Icelandic Low The tiny but intense North Atlantic Subpolar Low cell. A major generator of middle latitude storms.

igneous rock Any rock that has cooled from an original molten state.

illuviation The accumulation of soluble minerals and small particles in a subterranean soil horizon; *a* result of leaching from above.

insolation The total solar energy received at any given point on the earth's surface.

International Date Line Roughly the 180th meridian. A day is lost in crossing from east to west, gained from west to east.

Intertropical Front (Equatorial Convergence Zone) The somewhat theoretical line of contact in the Tropics, between Equatorial and Tropical air masses.

intrazonal soils Soils reflecting some local peculiarity of relief, parent material, etc. Not in equilibrium with their environment, especially climate and vegetation.

inversion (temperature) A gain in altitude results in an increase in temperature. An anomaly. The opposite of lapse rate.

ion microprobe A dating device (ancient rocks) that involves a tightly focused beam of high energy ions.

ionosphere See thermosphere.

isagon A line on a map connecting all points of equal compass variation.

isarithm See isopleth.

isobar A line on a map connecting all points of equal barometric pressure.

isohyet A line on a map connecting all points of equal precipitation.

isopleth A line on a map connecting all points of equal value.

isostasy The theory that uplift in one part of the earth must be compensated for by equal depression elsewhere.

isotherm A line on a map connecting all points of equal temperature.

J

jet stream (polar night) High latitude, high velocity air currents active only during the six months of polar darkness.

jet stream (upper Westerly) High velocity, narrow, westerly currents near the top of the troposphere. They are roughly above the Polar Front and both appear to act as middle latitude storm tracks.

Julian calendar The first modern western calendar in use from 46 B.C. to 1582.

K

karst Landscape resulting from the collapse of subterranean limestone caverns.

katabatic (wind) Cold air flowing downhill at high velocity responding to gravity. An extreme form of simple air drainage.

Kelvin scale (temperature) 0°K equals absolute zero; 273°K the freezing point of water; and 373°K its steam point.

kettle lake Morainal depression caused by the melting away of a large chunk of broken glacial ice.

kipuka The Hawaiian term for an island of vegetation surrounded by barren recent lava flows.

Köppen system (climate) A quantitative system of climatic classification. First introduced in 1918, it has been modified many times.

L

laccolith A relatively small intrusive igneous mass. It is fed from below by a conduit cutting across strata, but the bulk inserts itself between layers as a lens-shaped mass and lifts the surface above into a dome.

land and sea breeze The gentle daytime onshore, and evening offshore breezes, that result from the differential heating of land versus water over a 24 hour period.

lapse rate (adiabatic-dry) A mass of air impelled aloft will experience a loss of 5.5°F per 1000 feet (3°C per 305 m) as a result of expansion.

lapse rate (adiabatic-wet) If condensation is occurring within it, a mass of air impelled aloft will experience a loss of less than 5.5°F (3°C) per 1000 feet (305 m). 5.5 degrees (3°C) will be lost due to expansion but a modicum will be returned through condensation.

lapse rate (normal) Assuming relatively still air, a gain of altitude will cause a loss of 3.6°F (2°C) for every 1000 feet (305 m).

latent heat of condensation Evaporation requires heat, which means that each molecule of water vapor in the atmosphere has an increment of latent heat. Condensation, the reverse of evaporation, releases this heat.

laterite (latosol, ferrallite) A leached forest soil of the tropics and subtropics featuring an A horizon low in silicon and high in oxidized iron and aluminum. Generally, infertile for cultivation without careful management.

latitude Distance north or south of the equator. Measured in degrees from 0° (equator) to 90°N or S (poles).

Laurasia The speculative reassembly of Northern Hemisphere drifting continents into a single landmass.

lava Molten rock on the surface of the ground.

leaching The selective removal of soluble minerals from the topsoil by percolating groundwater.

legume Plants with the ability to fixate large amounts of nitrogen from the air in nodules on their roots.

levee The natural, but quite low, containing bank of an old-age stream. The levee is often artificially built up to many times its normal height, to keep the river in its channel during high-water periods.

libration The apparent minor wobbling of the moon as it rotates, allowing a view of slightly more than one-half of its surface. Caused by the moon's elliptical orbit and an ecliptic not quite parallel to that of the earth.

lichen A composite vegetative organism featuring a symbiotic relationship between a fungus and an algae.

lightning Electrical discharges between storm clouds and the ground, with other clouds, or within the same cloud. Common in violent convectional disturbances.

limb (of a meridianal great circle) One-half of a meridianal great circle, from pole to pole, assigned a distinctive degree number east or west of Greenwich.

limestone A sedimentary rock normally formed by the compression of lime-rich sediments accumulated on the sea-floor. Also includes coral and sea shells cemented into a coarse rock.

loess Powdery wind-transported soils.

longitude Distance east or west of the prime meridian (Greenwich). Measured in degrees from 0° (prime meridian) to 180° (its opposing limb).

longshore drift The transport of beach materials laterally along the coast.

M

mafic lava An acronym formed from mamagnesium and fic-ferric. A silica-poor lava, hence highly liquid.

magma Molten rock underground.

magnetic north The attractive point for magnetic compass needles located over 1000 miles (1609 km) from the North Pole.

magnetosphere The influence of the earth's magnetic field far beyond the outer margin of the atmosphere. See Van Allen radiation belts.

malee Scrub eucalyptus thickets in Australia.

mangrove A low tropical maritime tree whose habitat is shallow sea water necessitating "knees" on its roots for aeration. It throws out aerial roots and develops extensive thickets.

mantle, earth's The bulk of the earth's mass located between core and crust. Thought to be made up of a dense basic mineral like olivine, and capable of plastic flow, under pressure.

maquis See chaparral.

marble The metamorphic rock formed from limestone.

marine terrace A wave-abraded platform, formerly submarine but now exposed well above sea level as a terrace. Evidence of either a lowered sea or uplifted coast.

marsupial An ancient mammalian family whose young are born in an advanced embryonic stage and then transferred to the mother's pouch. Generally superceded by placentals.

mass wasting The drift or slide of accumulated debris on slopes. The end product of weathering.

mature landform One in which erosion has altered the original form and slope to a maximum of local relief.

mature stream One in which the sedimentary load-carrying capacity (determined chiefly by velocity) and the load it attempts to carry are in rough equilibrium.

Mercator projection A conformal projection constructed from a cylindrical base—great navigational utility.

meridian The preferred definition: one-half (or a limb) of a great circle passing through both poles designated by a distinctive longitudinal number. It may also be defined as simply a great circle passing through both poles.

mesopause The line separating the mesosphere from the thermosphere (ionosphere)—60 miles (97 km) above the earth.

mesosphere That atmospheric layer extending outward from the stratosphere to about 60 miles (97 km).

metamorphic rocks Rocks transformed by the application of great pressure and heat to a thoroughly different end-product. They are normally much harder and denser than the original and display a complete molecular restructuring.

midocean ridge Every ocean displays a

ridge (not always exactly midocean), which is interpreted as the active element in sea-bottom spreading, and ambulatory crustal plates. Vulcanism along this seam not only produces new crust but forces existing rocks to move laterally.

millibar A force of 1000 dynes per square centimeter.

minerals Naturally occurring combinations of elements.

Moho (Mohorovičić) The sharp line of discontinuity between the earth's mantle and crust.

Mohole The projected boring through the crust and into the mantle. Abandoned in 1966.

monadnock A hard-rock erosional remnant standing above the general level of the surrounding countryside.

monotremes Extremely rare and archaic egg-laying mammals.

monsoon The alternating summer onshore winds (rainy season) and winter offshore winds (dry season), basically a result of differential seasonal heating of land versus water.

moraine, ground See till.

moraine, terminal A ridge of ice-deposited material marking the end of a glacier where it has remained static for some time. Unlike water deposition, the ice-deposited debris is unsorted.

mound springs Flowing springs that discharge in a region of high evaporation. If the water is of great salinity, deposits will build up rapidly around the vent.

N

neck (volcanic) See plug.

new moon The moon is between the earth and the sun and no reflected light is discernible. Appears dark from the earth.

nimbo, nimbus (clouds) Dark, with rain either occurring or obviously imminent.

North Pole The northern end of the earth's axis—90°N. lat.

nuclear fission A neutron is used to split the nucleus of a heavy atom (e.g., uranium 235) into two lighter fragments, thereby producing energy.

nuclear fusion The fusion of two nuclei of light atoms, thereby releasing energy. Deuterium and tritium are attractive possibilities.

nueé ardent Literally "glowing cloud". An incandescent mixture of superheated gases and cinders blown explosively out the side of a volcano. It moves at great speed, yet is heavy enough to respond to gravity and flow downslope.

O

obsidian Volcanic glass formed upon the abrupt cooling of lava.

occlusion (occluded front) The rapidly advancing Cold Front in a Standard Middle Latitude cyclone overtakes the Warm Front, leaving only cold air at the ground.

offshore bar A long, slim, sandy feature paralleling the coast at some distance offshore. Barely above the level of the sea, it is built, theoretically, by sand thrown up as ocean breakers work the seaward side.

old-age landform One in which the original feature has been almost totally removed by erosion, worn away to a near plane at a lower level.

old-age stream One in which a low velocity impedes the stream's capacity to carry its sedimentary load—overloaded. As a result, the stream is an active depositor in its channel and floodplain.

orographic precipitation Wind forcing a moist air mass up a mountain slope, thereby causing sub-dew point cooling via expansion and condensation.

outwash plain Sorted glaciofluvial deposits beyond the outer edge of a continental glacier's terminal moraine. Similar to a valley train.

oxbow lake A water-filled abandoned meander in the floodplain of an old-age stream.

oxidation Chemical union with oxygen.

P

paleomagnetism At the time of igneous rock formation, ferrous minerals align themselves with the prevailing earth's magnetic field. For unknown reasons the magnetic field changes from time to time, thereby providing a dating device.

paleontology The science that deals with life of past geologic periods.

Pangaea The speculative reassembly of all the world's drifting continents into a single landmass.

parallel A circle of latitude whose plane is at right angles to the axis, hence parallel to the equator.

parallelism The inclination of the earth's axis always in the same direction.

parasites (plant) Plants which live on a host and draw their nourishment from it by tapping into its vital systems.

patterned earth (polygons) Surface hexagons in the tundra, probably a result of expansion as soil water above the permafrost layer freezes with the onset of winter.

pedalfer An acronym referring to the leached soils of humid climate regions.

pedocal An acronym referring to the unleached soils of dry climate regions.

pedology The study of soils.

pegmatite A long intrusive finger of plutonic rock maintaining contact with a batholith below. It has been the recipient of selected gases and fluids. Occassionally, huge crystals will develop.

peneplain A gently rolling erosional plain, the quintessential old-age landform.

perihelion The nearest approach of earth to sun each year—on or about January 3.

permafrost Permanently frozen subsoil. Even in the summer it thaws to no more than a foot or so below the surface.

permeability The free movement of a liquid through porous rocks or soil.

petrified wood Circulating groundwater has replaced the cellulose of a buried tree, molecule by molecule, with dissolved silica.

phase change The theory that dense rocks increase their volume when heated, thereby rupturing surrounding rocks.

photochemical A reaction wherein particle structure changes are a result of radiant heat absorption. An element in the formation of smog.

photosynthesis In plants light plus water plus carbon dioxide, functioning with chlorophyll as a catalyst, produce sugar plus free oxygen.

phytoplankton Microscopic plant life in the sea.

placental Mammals whose reproductive scheme involves an extended gestation period with the young in its mother's womb.

planation Erosion cutting a landform down to a plain or near plain. See peneplain.

Planck's law The hotter a body the more electromagnetic energy it will radiate and the shorter the wave length of that radiation.

plankton Microscopic plant and animal life in the sea.

plasma Superheated, ionized gases in a fusion energy device.

plastic shading (map) Utilizing a presumed low-angle light from the northwest, steep southeastern slopes are shaded darkest.

plate tectonics The theory that envisions the crust of the earth as an assemblage of huge freely moving plates.

playa lake A shallow short-lived sheet of water with no permanent source. Usually occurs in a desert basin where rapid evaporation in the wake of a rare storm quickly reduces it to a salt flat.

plucking A valley glacier in contact with cirque or valley walls pulls away great chunks of rock by freezing them into the ice and then moving off downslope.

plug (volcanic) A volcanic remnant, only the hard, resistant, igneous rock vent pipe of an ancient volcano stands above the surrounding landscape. Also called neck.
But the term plug volcano refers to an active, often craterless, cone which fea-

tures thick, pasty dacite or rhyolite pushing up a dome or spire. Trapped gasses make this type of volcano subject to massive explosion.

plutonic rock Coarse-grained igneous rock that has cooled slowly underground.

podzol A leached forest soil of the high and middle latitudes featuring a greyish A horizon, high in silicon.

Polar continental, cP (air mass) Cold and dry reflecting its source region, the interior of high latitude continents.

Polar Front The line of contact, in the middle latitudes, between Polar and Tropical air masses.

Polar high A polar region featuring constant cold and descending air, hence a permanent high pressure.

Polaris The north star almost directly over the North Pole. From the earth it appears perpetually stationary unlike any other heavenly body.

Polar Maritime, mP (air mass) Cold and moist reflecting its source region, the high latitude oceans.

porosity The water-holding capacity of rocks or soil.

post meridiem (P.M.) After the sun has transited a meridian, hence afternoon.

precipitation Liquid water droplets or ice falling out of the atmosphere. Solid matter may also precipitate from oceans or atmosphere.

pressure (atmospheric) The weight of the total atmosphere at a given location on the earth's surface.

pressure gradient The pressure differential expressed by the spacing of isobars. The control of wind velocity.

Prime meridian See Greenwich.

profile, soil A cross section of soil to a depth of several feet, normally revealing distinctive horizons.

projection (map) Transference of the globe's network of latitude and longitude lines onto a flat piece of paper.

psychrometer (wet and dry bulb) An instrument for measuring humidity utilizing two thermometers, one with a piece

of wet muslin wrapped about its bulb. The different temperature readings are reflective of the evaporative capacity of the air.

pyroclastic A volcanic eruption made up exclusively of broken solid particles, from dust to jagged boulders.

Q

quarrying (ice) Sizable chunks of bedrock frozen into the underside of a glacier are pulled away and carried off.

quartzite The metamorphic rock formed from sandstone.

R *Genji's Dog*

radar An active remote-sensing system which emits radio waves. As these bounce off the earth, both surface character and distance may be perceived.

radiation The transfer of energy through space.

radioactive dating Determining the age of rocks based on the known rate of decay of radioactive parent elements into stable daughter elements.

rain shadow The dry lee of a mountain whose windward slope is experiencing orographic precipitation.

range, annual (temperature) The number of degrees differential between the coldest and warmest month's averages in one year.

range, diurnal (temperature) The number of degrees differential between the maximum and minimum temperatures over a 24 hour period.

recumbent folds Involuted folding of rock strata wherein the flexures collapse.

reef, barrier A long coral reef formation parallel to but well offshore. Sea waves are thus intercepted and there is a body of quiet water behind the reef.

reef, fringing A coral reef formation along the immediate shoreline.

reflection (solar energy) A mirrorlike reversal of selected wave lengths by atmospheric dust, water, gas, or the earth's surface into outer space.

refraction, wave The tendency for waves, as they approach the beach, at an angle, to align themselves parallel to that beach.

reg Pebble desert, including desert pavement.

relative humidity The amount of water vapor in the air relative to that which it is capable of holding at a given temperature.

remote sensing High altitude platforms carry sensors capable of discerning diverse radiation and reflection from the earth's surface. Such information can be rendered into various useful forms via computer manipulation.

rendzina An intrazonal soil developed from a limestone parent material; less acid than would be expected in a humid region.

representative fraction A fraction or ratio indicating how many times a map has been reduced.

residual ores The nonsoluble end-products of massive soil leaching, for example, iron or aluminum. Most common in the wet tropics.

revolution (earth) The movement of the earth about the sun once each year.

rhumb line A true compass heading on a map.

rhyolite A fine-grained igneous rock of the same chemical composition as granite, differing only in size of crystal because of rapid cooling. Silica-rich.

rift valley A fault valley or graben.

rip currents As the water thrown up on the beach by each wave retreats to sea, it is frequently channeled into swiftly running discrete currents.

rocks Naturally occurring combinations of minerals.

rotation (earth) The movement of the earth about its axis once each day.

S

sandbar A variation on sandspit. Bar means barrier, hence a sandy deposition obstructing a river or bay mouth.

sandspit A curved sandy deposition anchored at one end to the shore, often a headland. It is a product of continued longshore drift.

sandstone A sedimentary rock consisting of loose sand particles cemented together by any of a variety of materials. Normally high in silica and often permeable.

Sargasso Sea The calm center of the North Atlantic ocean current gyral, characterized especially by the wide occurrence of sargassum weed.

saturation (humidity) The dew point—100% relative humidity. At a given temperature the air is incapable of holding any additional water vapor.

savanna A tropical grassland made up of tall, rank individual plants developing tussocks rather than a cohesive turf.

scale (map) The index of reduction on a map.

scattering The dispersal or minor reduction of selective wave lengths of the solar energy stream by dust, water, or gas molecules in the atmosphere.

schist A metamorphic rock featuring mica crystals aligned in a rough foliation and giving the rock a shiny, almost metallic appearance.

scoriacious (vesicular) Igneous rock that congealed as gases continued to escape from the lava. The result is a permeable stratum often acting as an aquifer.

scree See talus.

sea-bottom trenches Linear deeps [up to 36,000 feet (10,973 m)] around the margins of most oceans and inevitably adjacent to active seismic/volcanic mountain chains. They are interpreted as marking lines of direct conflict between moving crustal plates and hence subduction and diastrophism.

seafloor spreading Crustal mantle convection causes midocean volcanic ridges. Here new igneous rocks are formed and the seafloor moves away.

seamount A submarine volcanic peak. It may be a flat-topped guyot.

sedimentary rock A rock formed from the recombination of an accumulation of loose materials via pressure or cementation. It may also include rocks of organic origin (coal or coral) as well as evaporative salt deposits.

seiche A destructive wave resulting from large-scale earth slides into a confined body of water, e.g., a fjord or a lake.

seismic Literally means "to shake." Pertaining to earthquakes.

selva Dense tropical rain forest.

sensible temperature The temperature that the human body senses. A result of both actual temperature and humidity.

7th Approximation A new soil classification system officially adopted by the U.S. government in 1965.

sextant A navigational instrument capable of observing the horizon and a heavenly body simultaneously. It is used for measuring altitude of sun or stars to determine latitude.

shale A sedimentary rock of accumulated fine muds compressed into a solid—normally displays foliation at right angles to exerted pressure.

shield, volcanic cone A low-angle volcanic cone built by continued quiet flows of highly liquid lava.

shields The ancient hard-rock cores of every continent.

sill A relatively small intrusive plutonic mass. It is fed from below by a conduit cutting across strata, but the bulk inserts itself between layers and spreads out as a sheet.

silver iodide A substance pumped into a potential rain cloud to provide condensation and freezing nuclei, thereby enhancing precipitation.

sink A deep surface depression caused by collapse into a subterranean cavern.

slate The metamorphic rock formed from shale. Normally displays even more definitive foliation than the parent shale.

slipoff slope Shelving inner bend of a stream meander, the result of channel deposition.

solar constant The energy received at a plane placed at right angles to the solar radiation stream at the outer edge of the atmosphere.

solfatara See fumerole.

solstice, summer The maximum exposure of the Northern Hemisphere to the sun. The midday sun is overhead at the Tropic of Cancer and the Arctic Circle experiences 24 hours of light. On or about June 21.

solstice, winter The minimum exposure of the Northern Hemisphere to the sun. The midday sun is overhead at the Tropic of Capricorn and the Arctic Circle experiences 24 hours of darkness. On or about December 21.

solution load (streams) Dissolved minerals transported by surface stream flow.

sonic sounding Determining the depth of the sea by timing the speed of sound waves. The ship emits sonar "beeps" and times their return as they rebound off the bottom.

source area (air mass) That large earthly region over which an air mass is normally resident and from which the air mass derives its moisture and temperature characteristics.

South Pole The southern end of the earth's axis—90°S. lat.

spalling Rock surfaces peeling away in conchoidal layers as a result of weathering.

specific heat Almost five times as much energy is expended to raise a gram of water 1°C as opposed to a gram of dry earth.

spissatus (clouds) Greyish.

squall line The Cold Front in a standard Middle Latitude cyclone.

stacks Small islands or offshore rocks temporarily left behind as ocean waves selectively wear away the softer shoreline materials.

stalactite An iciclelike deposit formed by the evaporation of dripping lime-rich groundwater from the roof of a cave.

stalagmite A limestone pinnacle on the floor of a cave. Lime-rich groundwater evaporates as it drips from the roof.

standard atmosphere (air pressure) The average air pressure at sea level in the middle latitudes, about 15 pounds per square inch (1 km per cm²) or a 29.9 inch (1013.2 mb) barometric reading.

stated scale (map) A prose sentence indicating the inch to mile (or centimeter to kilometer) relationship on a map.

steppe Middle latitude short grass prairie.

stock A relatively small intrusive plutonic mass—less than 40 square miles (104 km²) in size.

storm Generally, any low pressure center. A cyclone.

storm surge Excessively high tides resulting from the passage of a hurricane.

stratopause The line separating stratosphere from mesosphere—35 miles (56 km) above the earth.

stratosphere That atmospheric layer extending outward from the tropopause to about 35 miles (56 km).

strato volcanic cone See composite volcanic cone.

stratus clouds Flat, tabular clouds covering the entire sky and appearing grey from underneath.

striations Parallel scratches and grooves on bedrock gouged out by sharp boulders frozen into passing glaciers.

structure, soil The arrangement of soil particles from granular to blocky.

subduction The steeply angled descent of a crustal plate as, along it advancing margin, it underruns another plate. Frequently marked by a linear oceanic trench.

sublimation The change from water vapor to ice crystals as the temperature of the air is decreased.

submerged shoreline One in which the most recent relative movement of land and water has resulted in drowning. Frequently a broken serrated coastline.

Subpolar Low A subpolar region of constant low pressure. Encountered, especially at sea, in both hemispheres at 60° latitude.

subsidence inversion A sinking air mass, heated via compression, results in a warm layer aloft. An increased temperature with altitude gain results.

succulents Soft bodied plants, frequently capable of storing significant water in their cells.

suspension load (streams) Fine-grained sediments and muds easily transported by surface stream flow.

symbolism (map) Map data represented by a pictorial or other non-digital character.

syncline Any downbowed rock strata.

synodic month The time required for the apparent position of moon to sun to repeat itself. About 29.5 days.

T

taiga High latitude coniferous forest.

talus (scree) An accumulation of detritus at the base of a weathered cliff. Delivered by gravity, it arranges itself according to size into an apron.

tarn lake The deepest part of an ice-abandoned cirque is often occupied by a lake.

tectonic forces Those forces that buckle, distort, and fracture the earth's crust folding, faulting, vulcanism.

Tethys Basin An ancient east/west trending sea in the general location of the current Mediterranean and Indo/Gangetic plain.

texture, soil The size of soil particles, from fine clay to coarse sand.

thermosphere (ionosphere) That atmospheric layer extending outward from the mesopause to the far periphery of the atmosphere—perhaps 6000 miles (9656 km).

thunder The instantaneous expansion of the air as it is heated by a flash of lightning.

tidal bore Channeled incoming tidal flow rushing up the lower course of a river as a wall of water a foot or more in height.

tidal rips Along a broken coastline, retreating tides from opposite shores meet in frothing eddies.

tidal waves (tsunamis) Earthquakes or volcanic explosions on the sea-bottom transmit their shock waves through the entire liquid medium. The deep-seated waves move rapidly outward from the epicenter, overwhelming coasts even thousands of miles distant.

tide, neap The lowest tides of each month when the gravitational forces of moon and sun are at right angles and act to cancel each other.

tide, spring The highest tides of each month when the moon and sun are in exact alignment and their combined gravitational forces activate the seas.

tiering, vegetative Vegetative communities arrange themselves in stories, with the taller plants requiring more light than those growing in their shade.

till Unsorted ice-deposited debris beneath the rotten ice at the forward edge of a glacier. Because it was laid down under the ice, it is compressed.

Tokamak The Russian device producing a controlled magnetic field to contain superheated plasma. This is necessary to produce energy from fusion.

tombolo A small offshore island connected to the mainland by a sandspit.

tornado The most violent storm in nature. It is small, swiftly moving, lethal, and almost exclusively American—normally a product of the Polar Front.

Trade Winds Constant brisk breezes, encountered especially at sea in both hemispheres (northern NE winds, southern SE winds) at 10° to 30° lat.

transpiration The evaporation via leaf surface of the cellular moisture in plants.

transverse faults (sea-bottom) Thousands of right angle fractures along the flanks of midocean ridges, in some fashion a product of seafloor spreading.

Tropical continental, cT (air mass) Hot and dry, reflective of its source region, the tropical deserts.

Tropical maritime, mT (air mass) Hot and moist, reflective of its source region, the tropical seas.

Tropic of Cancer The northernmost advance of the overhead sun—23.5° N. lat.

Tropic of Capricorn The southernmost advance of the overhead sun—23.5° S. lat.

tropopause The line separating the troposphere from the stratosphere, 8 to 9 miles (13 to 15 km) above the earth.

troposphere That atmospheric layer closest to the earth extending outward about 8 to 9 miles (13 to 15 km).

tsunami See tidal waves.

tundra Low ground-hugging vegetation, often amphibious in character, beyond the high latitude "tree line".

typhoon See hurricane.

U

uncinus (clouds) Comma shaped.

undertow The general rapid retreat of water brought forward by each wave, back down the beach. See rip currents.

unloading Abrupt melting of massive ice sheets, or the removal by erosion of overlying rock strata, can result in a measurable uplift of compressed materials below.

upwelling Cold water from the oceanic depths ascends as the surface-waters are pulled away by currents or winds.

V

vadose water The shallow surface zone of transitory groundwater, fluctuating radically in response to rain showers and evaporation.

valley (Alpine) glacier A tongue of ice flowing down a mountain valley.

valley train A glaciofluvial depositional apron downstream from the terminal moraine of an Alpine glacier. The material is sorted sequentially according to size. Lesser in scope but otherwise very similar to an outwash plain.

value symbols (map) A symbol that shows not only location but indicates some value as well, for example, dot maps.

Van Allen radiation belts Concentrations of highly charged particles from the sun trapped by the earth's magnetic field. One is about 10,000 miles (16,000 km) above the earth, a second at about 23,000 miles (37,000 km).

variation (compass) See declination.

varved clays Ancient lake bottoms displaying great depths of stratified sediments. These are useful as dating tools.

vesicular See scoriacious

Virgin Lands (USSR) Newly plowed wheat fields east of the Volga and on into the semiarid steppes of northern Kazakhstan.

volcano A cone-shaped mountain built from liquid lava or solid ejecta emanating from its vent(s).

vulcanism Any invasion of the earth's crust by magma from below.

vulcanism, extrusive Lava welling out onto the surface of the earth via volcano or fissure.

vulcanism, intrusive An invasion of the earth's crust by magma from below, but failing to reach the surface.

W

Warm Front A segment of the Polar Front, on the eastern side of Standard Middle Latitude cyclones, where local circulation causes warm air to advance against cold.

water table The top of saturated rock or soil.

water vapor The gaseous form of water.

wave-built terrace An extension of the submarine wave-cut platform by the deposition of fine sediments drifting offshore.

wave-cut bench Sea-bottom erosion just outside of the breakers line, resulting in a shallow rock platform.

waves refraction The tendency for waves, as they "feel bottom," to pull around roughly parallel to the shore.

weather The day-to-day or even hour-to-hour observation of the elements.

weathering, chemical The rotting, mouldering, or decay of rocks with water as the chief active agent.

weathering, mechanical The disintegration of rock via non-chemical means, such as ice wedging in cracks or shattering on impact.

Westerlies A middle latitude region of squally weather but featuring a dominant westerly wind drift. Encountered in both hemispheres at latitudes 35° to 60°.

West Wind drift Eastward flowing warm ocean currents in the middle latitudes of both hemispheres, impelled in part by the prevailing Westerlies.

X

xerophyte A plant with special adaptation of root or leaf to withstand a chronic absence of moisture.

Y

youthful landform One in which the original form and shape is scarcely altered by erosion.

youthful stream One in which high velocity imparts a much greater sedimentary load-carrying capacity than is required. An underloaded stream. As a result, there is active bed erosion.

Z

zircon A hard, durable, crystal compounded of zirconium, silicon and oxygen. Widely occurring.

zonal (mature) soils Those soils in equilibrium with their environment, especially climate and vegetation.

zone time Zones roughly 15° wide, keeping the time of a central meridian, hence each zone is one hour different from its immediate neighbor.

zooplankton Microscopic animal life in the sea.

Bibliography

PART I THE EARTH AS A GLOBE

*Beiser, A. and the editors of Time-Life, *The Earth*, Time-Life, New York, 1970.

Carter, D. B., T. H. Schmudde and D. M. Sharpe, *The Interface as a Working Environment: A Purpose for Physical Geography*, Comm. on College Geog. Tech Paper No. 7, Assoc. of American Geogs., Washington, 1972.

*Dasmann, R. F., *Environmental Conservation*, 5th ed. Wiley, New York, 1984.

*Gabler, Robert E. et al, *Essentials of Physical Geography*, 2nd ed., Holt, Rinehart and Winston, New York, 1982.

Knapp, B. J., ed., *Practical Foundations of Physical Geography*, Allen & Unwin, Winchester, MA, 1981.

*Navarra, G., *Contemporary Physical Geography*, Saunders, New York, 1981.

Ridley, B. K., *The Physical Environment*, Halstead, New York, 1979.

Sheffield, C. ed., *Man on Earth*, Macmillan, New York, 1983.

*Strahler, A. N. and A. H. Strahler, *Elements of Physical Geography*, 3rd ed., Wiley, New York, 1984.

*Tarbuck, E. J. and F. K. Lutgens, *Earth Science*, Merrill, Columbus, OH, 1982.

---

*A moderately elementary treatment recommended for students using this text.

CHAPTER ONE THE EARTH IN SPACE

*Barnett, L. and the editors of *Life*, *The World We Live In*, Time, New York, 1965.

*Bergamini, D. and the editors of *Life*, *The Universe*, Time, New York, 1962.

Bowditch, N., *American Practical Navigator*, U.S. Gov't. Print. Office, Washington, (issued irregularly as U.S. Hydrographic Office Pub. No. 9 since 1802).

Canadian Royal Astrological Society, *The Observer's Handbook*, Univ. of Toronto Press, Toronto, Annual.

*Harrison, L. C., *Sun, Earth, Time and Man*, Rand McNally, Chicago, 1960.

*Howse, D., *Greenwich Time and the Discovery of Longitude*, Oxford Univ. Press, 1983.

Munk, W. H., *The Rotation of the Earth*, Cambridge Univ. Press, Cambridge, England, 1960.

Nicks, O. W., ed., *This Island Earth*, NASA publication No. SP-2500, Office of Technical Utilization, U.S. Gov't. Print. Office, Washington, 1970.

Pyne, T.E., *Standard Time*, Interstate Commerce Commission, U.S. Gov't. Print. Office, Washington, 1958.

*Sheffield, C., *Earth Watch: A Survey of the World from Space*, Macmillan, New York, 1981.

Whipple, F. L., *Earth, Moon and Planets*, 3rd ed., Harvard Univ. Press, Cambridge, MA, 1968.

CHAPTER TWO MAPS

Bertin, J., *Semiology of Graphics: Diagrams, Networks, Maps*, Univ. of Wisconsin Press, Madison, 1984.

Bies, J. D. and R. A. Long, *Mapping and Topographic Drafting*, South-Western, Cincinnati, 1983.

Blakemore, M.J. and J.B. Harley, *Concepts in History of Cartography: A Review and Perspectives*, Univ. of Toronto Press, Toronto, 1980.

*Brown, L. A., *The Story of Maps*, Little, Brown, Boston, 1950.

*Campbell, J., *Introductory Cartography*, Prentice-Hall, Englewood Cliffs, NJ, 1984.

Dickenson, G.C., *Maps and Air Photographs*, Wiley, New York, 1979.

*Greenhood, D., *Mapping*, 3rd ed., Univ. of Chicago Press, Chicago, 1964.

*Keates, J.S., *Understanding Maps*, Wiley, New York, 1982.

*Lobeck, A.K., *Things Maps Don't Tell Us*, Macmillan, New York, 1957.

Monmonies, M. S., *Computer Assisted Cartography: Principles and Prospects*, Prentice-Hall, Englewood Cliffs, NJ, 1982.

Robinson, A. H. et al, *Elements of Cartography*, 5th ed., Wiley, New York, 1984.

Taylor, G. and R. Fraser, *Graphic Communication and Design in Contemporary Cartography*, Wiley, New York, 1983.

*Thrower, N.J.W., *Maps and Man*, Prentice-Hall, Englewood Cliffs, NJ, 1972.

*U.S. Department of the Army, *Map Reading*, Technical Manual FM 21–26, U.S. Gov't. Printing Office, Washington, 1969.

*Wilford, J. N., *The Mapmakers*, Vintage, New York, 1982.

PART II WEATHER

Breuer, G., *Weather Modification*, Cambridge Univ. Press, New York, 1980.

*Browning, K. A., *Nowcasting*, Academic Press, New York, 1982.

*Clayton, H. H., *World Weather Records*, Smithsonian Misc. Collection Vol. 79, Smithsonian Inst., Washington, 1944.

*Cole, F. W., *Introduction to Meteorology*, 3rd ed., Wiley, New York, 1980.

Hoskins, B. and R. Pearce, eds., *Large-Scale Dynamical Processes in the Atmosphere*, Academic Press, New York, 1983.

Lutgens, F. K. and E. J. Tarbuck, *The Atmosphere: An Introduction to Meteorology*, Prentice-Hall, Englewood Cliffs, NJ, 1982.

*Pryde, P. R., *Non-Conventional Energy Sources*, Wiley, New York, 1983.

Riehl, H., *Tropical Meteorology*, McGraw-Hill, New York, 1954.

*Thompson, P., R. O. O'Brien and the editors of Life, *Weather*, Time, New York, 1965.

Whittow, J., *Disasters: The Anatomy of Environmental Hazards*, Univ. of Georgia Press, Athens, 1980.

CHAPTER THREE THE ATMOSPHERE

*Allen, O. E. and the editors of Time-Life Books, *Atmosphere*, Time-Life Books, Alexandria, VA, 1983.

Anthes, R.A. et al, *The Atmosphere*, 2nd ed., Merrill, Columbus, OH, 1981.

*Battan, L.J., *The Unclear Sky*, Greenwood, New York, 1980.

*Beiser, A. and G. Beiser, *The Story of Cosmic Rays*, Dutton, New York, 1962.

Biswas, A.K., ed., *The Ozone Layer*, Proceedings of the Washington, D.C. Conference, March, 1977, Environmental Sciences and Applications (Volume 4), Pergamon, New York, 1980.

Hanwell, J.D. *Atmospheric Processes*, Allen & Unwin, Winchester, MA, 1980.

*Idso, S.B., *Carbon Dioxide: Friend or Foe; An Inquiry Into the Climatic and Agricultural Consequences of the Rapidly Rising CO_2 Content of the Earth's Atmosphere*, IBR Press, Tempe, AZ, 1982.

Meetham, A.R. et al., *Atmospheric Pollution*, 4th ed., Pergamon, Oxford, England, 1981.

National Research Council, *The Upper Atmosphere and Magnetosphere*, Nat. Acad. of Sciences, U.S. Gov't. Print. Office, Washington, 1977.

Oppenheim, A., *Impact of Aerospace Technology on Studies of the Earth's Atmosphere*, Pergamon, New York, 1974.

Pasquill, F., *Atmospheric Diffusion: The Dispersal of Windborne Material from Industrial and Other Sources*, 3rd Ed., Wiley, New York, 1983.

Ramage, C.S., *Monsoon Meteorology*, Academic Press, New York, 1971.

*Torrey, V., *Wind-catchers: American Windmills of Yesterday and Tomorrow*, S. Greene, Brattleboro, VT, 1977.

CHAPTER FOUR TEMPERATURE AND THE ENERGY BUDGET

*Behrman, D., *Solar Energy: The Awakening Science*, Little, Brown, Boston, 1976.

Eddy, J.A., *A New Sun: The Solar Results from Skylab*, NASA, Houston, 1979.

Gribbin, J., *Future Weather and the Greenhouse Effect*, Delacorte Press, New York, 1982.

Kondratev, K.Y., *Heat Exchange in the Atmosphere*, Pergamon, New York, 1965.

Miller, D.H., *Energy at the Surface of the Earth*, Academic Press, New York, 1981.

*Ponte, L., *The Cooling*, Prentice-Hall, Englewood Cliffs, NJ, 1974.

Robinson, N., *Solar Radiation*, Elsevier, New York, 1966.

*Secreteriat for Future Studies, Stockholm, *Solar Versus Nuclear: Choosing Energy Futures*, Pergamon, London, 1980.

Stambolis, C., ed., *Solar Energy in the 80's*, Pergamon, London, 1980.

CHAPTER FIVE PRESSURE AND WINDS

Corby, G.A., ed., *The Global Circulation of the Atmosphere*, Royal Meteorological Society, London, 1970.

*Eldridge, F.S., *Wind Machines*, Van Nostrand Reinhold, New York, 1980.

Franklin Institute, *Wind Energy*, Franklin Institute Press, Philadelphia, 1978.

Palmen, E. and C.W. Newton, *Atmospheric Circulation Systems: Their Structure and Physical Interpretation*, Academic Press, New York, 1969.

Panofsky, H.A. and J. Dutton, *Atmospheric Turbulence*, Wiley, New York, 1983.

*Park, J., *The Wind Power Book*, Chesire, Palo Alto, CA, 1981.

CHAPTER SIX MOISTURE

Battan, L.J., *Cloud Physics and Cloud Seeding*, Doubleday, New York, 1962.

Bentley, W.A. and W.J. Humphreys, *Snow Crystals*, Dover, New York, 1962.

Gokhale, N.R., *Hailstorms and Hailstone Growth*, State Univ. of New York Press, Albany, 1975.

*Kirk, R., *Snow*, Morrow, New York, 1978.

Ludlam, F.H., *Clouds and Storms: The Behavior and Effect of Water in the Atmosphere*, Penn. St. Univ. Press, University Park, 1980.

*Mason, B.J., *Clouds, Rain and Rainmaking*, 2nd ed., Cambridge Univ. Press, Cambridge, England, 1975.

Miller, D.H., *Water at the Surface of the Earth: An Introduction to Ecosystem Hydrodynamics*, Academic Press, New York, 1982.

Perry, A.H. and J.M. Walker, *The Ocean-Atmosphere System*, Longmans, London, 1977.

Thornthwaite, C.W. and J.R. Mather, *The Moisture Balance*, Princeton Univ. Press, Princeton, 1949.

*World Meteorological Organization, *International Cloud Atlas*, Geneva, Switzerland, 1956.

CHAPTER SEVEN AIR MASSES, FRONTS AND STORMS

*Battan, L.J., *The Thunderstorm*, Signet Science Library, New York, 1964.

*Eagleman, J.R., *Severe and Unusual Weather*, Van Nostrand Reinhold, New York, 1982.

Flora, S.D., *Hailstorms of the U.S.*, Univ. of Oklahoma Press, Norman, 1956.

Flora, S.D., *Tornadoes of the United States*, rev. ed., Univ. of Oklahoma Press, Norman, 1973.

Harman, J. R., *Tropospheric Waves, Jet Streams, and United States Weather Patterns*, Comm. on College Geog. Resource Paper No. 11, Assoc. of American Geogs., Washington, 1971.

* Jennings, G., *The Killer Storms, Hurricanes, Typhoons and Tornadoes*, Lippincott, New York, 1970.

Reiter, E. R., *Jet Streams*, Doubleday, Garden City, N.Y., 1967.

Simpson, R. H. and H. Riehl, *The Hurricane and its Impact*, Louisiana St. Univ. Press, Baton Rouge, 1981.

* Whipple, A. B. C. and the editors of Time-Life Books, *Storm*, Time-Life Books, Alexandria, VA, 1982.

Climate and Global Survival, Plenum, London, 1976.

* Skinner, B.J., ed., *Climates Past and Present*, Wm. Kaufmann, Los Altos, CA, 1981.

* Trewartha, G. and J. Horn, *An Introduction to Climate*, 5th ed., McGraw-Hill, New York, 1980.

* Trewartha, G.T., *The Earth's Problem Climates*, 2nd ed., Univ. of Wisconsin Press, Madison, 1981.

U.S. Oceanic and Atmospheric Adm., *Climates of the States*, 2 vols., U.S. Gov't. Print. Office, Washington, 1974.

Walls, J., *Desertification*, Pergamon, New York, 1980.

Williams, P.J., *Pipelines and Permafrost*, Longman, London, 1980.

PART III CLIMATE

Ayoade, J.O., *Introduction to Climatology for the Tropics*, Wiley, New York, 1983.

Barrett, E.C., *Climatology from Satellites*, Methuen, London, 1979.

Boucher, L., *Global Climate*, Wiley, New York, 1975.

Budyko, M.I., *The Earth's Climate*, Academic Press, New York, 1982.

* Critchfield, H.J., *General Climatology*, 4th ed., Prentice-Hall, Englewood Cliffs, NJ, 1983.

Golany, G.S., *Design for Arid Regions*, Van Nostrand Reinhold, New York, 1983.

Jaeger, J., *Climate and Energy Systems*, Wiley, New York, 1983.

Kendrew, W.G., *The Climates of the Continents*, Oxford Univ. Press, London, 1953.

Köppen, W. and R. Geiger, *Handbuch der Klimatologie*, 5 vols., Verlagsbuchhandlung Gebrüder Borntaeger, Berlin, 1930.

Lamb, H.H., *Climate, History and the Modern World*, Methuen, New York, 1982.

Landsberg, H.E., *The Urban Climate*, Academic Press, New York, 1981.

* Ley, W. and the editors of Time-Life, *The Poles*, Time-Life, New York, 1962.

Matthews, W.H., W.W. Kellogg and G.D. Robinson, eds., *Man's Impact on Climate*, M.I.T. Press, Cambridge, 1971.

* Schneider, S.H. and L.E. Mesirow, *The Genesis Strategy:*

CHAPTER EIGHT CLASSIFICATION OF CLIMATE

CHAPTER NINE THE TROPICAL CLIMATES

CHAPTER TEN THE SUBTROPICAL CLIMATES

CHAPTER ELEVEN THE MID-LATITUDE CLIMATES

CHAPTER TWELVE THE HIGH LATITUDE CLIMATES

PART IV NATURAL VEGETATION AND SOILS

Cox, C.B. and P.D. Moore, *Biogeography: An Ecological and Evolutionary Approach*, 3rd ed., Halstead, New York, 1980.

Eyre, S.R., *Vegetation and Soils, a World Picture*, Aldine, Chicago, 1968.

* Farb, P. and the editors of Life, *Ecology*, Time, New York, 1963.

Furley, P.A. and W.W. Newey, *Geography of the Biosphere: An Introduction to the Nature, Distribution and Evolution of the World's Life Zones*, Butterworths, Boston, 1983.

Money, D.C., *Climate, Soils and Vegetation*, University Tutorial Press, London, 1965.

Simmons, I., *Biographical Processes*, Allen and Unwin, Winchester, MA, 1982.

Sims, R.W., J.H. Price and P.E.S. Whalley, eds., *Evolution, Time and Space: The Emergence of the Biosphere*, Academic Press, London, 1983.

Tivy, J., *Biogeography, A Study of Plants in the Ecosphere*, Longman, London, 1982.

*U.S. Department of Agriculture, *Grass*, The Yearbook of Agriculture, U.S. Gov't. Print. Office, Washington, 1948.

*U.S. Department of Agriculture, *Trees*, The Yearbook of Agriculture, U.S. Gov't. Print. Office, Washington, 1949.

University of Arizona, *Desert Plants*, Vol. 1, No. 1, Boyce Thompson S.W. Arboretum, Superior, AZ, 1979.

Vankat, J.L., *The Natural Vegetation of North America*, Wiley, New York, 1979.

*Went, F.W. and the editors of Life, *The Plants*, Time, New York, 1963.

CHAPTER THIRTEEN INTRODUCTION TO NATURAL VEGETATION

Beazley, M., ed., *The International Book of the Forest*, Mitchell Beazley, 1981.

Emsley, M., *Rain Forests and Cloud Forests*, Harry N. Abrams, New York, 1979.

*Farb, P. and the editors of Time-Life Books, *The Forest*, Time-Life Books, New York, 1969.

Grime, J.P., *Plant Strategies and Vegetation Processes*, Wiley, Chichester, England, 1979.

Humphrey, R.R., *The Desert Grassland*, Univ. of Arizona Press, Tucson, 1968.

Kellman, M.C., *Plant Geography*, 2nd ed., St. Martins, New York, 1980.

Kozlowski, T.T., ed., *Fire and Ecosystems*, Academic Press, New York, 1974.

Larsen, J.A., *The Boreal Ecosystem*, Academic Press, New York, 1980.

Larsen, J.A., *Ecology of the Northern Lowland Bogs and Conifer Forests*, Academic Press, New York, 1982.

*Page, J. and the editors of Time-Life Books, *Forest*, Time-Life Books, Alexandria, VA, 1983.

*Perl, P. and the editors of Time-Life Books, *Cacti and Succulents*, Time-Life Books, New York, 1978.

*Richards, P.W., *The Life of the Jungle*, McGraw-Hill, New York, 1970.

Spurr, S.H. and B.V. Barnes, *Forest Ecology*, Wiley, New York, 1980.

*Thirgood, J.V., *Man and the Mediterranean Forest*, Academic Press, New York, 1981.

CHAPTER FOURTEEN CLASSIFICATION AND NATURAL VEGETATION REGIONS

CHAPTER FIFTEEN INTRODUCTION TO SOILS

*Basile, R.M., *A Geography of Soils*, Brown, Dubuque, 1971.

Finkle, C.W., Jr., ed., *Soil Classification*, Hutchinson Ross, New York, 1982.

FitzPatrick, E.A., *Soils, Their Formation, Classification and Distribution*, Longman, London, 1980.

*Foth, H.D., *Soil Geography and Land Use*, Wiley, New York, 1980.

*Gardner, B.L., *The Geography of Agriculture*, Regents, Lawrence, KS, 1981.

Glinka, K.D., *The Great Soil Groups of the World and Their Development*, Edwards, Ann Arbor, 1927.

Gerrard, J., *Soils and Landforms*, Allen and Unwin, Winchester, MA, 1981.

Jumikis, A., *Soil Mechanics*, Krieger, Melbourne, FL, 1983.

*Kirkby, M.J., *Soil Erosion*, Halstead, New York, 1981.

Lugan, A.L., *The Origin and Source of Loess*, Univ. of Nebraska Press, Lincoln, 1962.

McFarlane, M.J., *Laterite and Landscape*, Academic, New York, 1976.

Prakash, S., *Soil Dynamics*, McGraw-Hill, New York, 1981.

U.S. Soil Conservation Services, Soil Survey Staff, U.S. Dept. of Agriculture, *Soil Classification, a Comprehensive System: 7th Approximation*, U.S. Gov't. Print. Office, Washington, 1960.

Wallwork, J.A., *Desert Soil Fauna*, Praeger, New York, 1982.

Young, A., *Tropical Soils and Soil Survey*, Cambridge Univ. Press, Cambridge, 1976.

CHAPTER SIXTEEN CLASSIFICATION AND SOIL REGIONS

PART V LANDFORMS

Craig, R.G. and J.L. Croft, *Applied Geomorphology*, Allen & Unwin, Winchester, MA, 1982.

Embleton, C. and J. Thornes, eds., *Processes in Geomorphology*, Halstead, New York, 1979.

*Faul, H. and C. Faul, *It Began With Stone: A History of Geology From the Stone Age to the Age of Plate Tectonics*, Wiley, New York, 1983.

King, C.A.M., *Landforms and Geomorphology: Concepts and History*, Halstead, New York, 1976

*Lobeck, A., *Geomorphology*, McGraw-Hill, New York, 1939.

Ludman, A., *Physical Geology*, McGraw-Hill, Hightstown, NJ, 1982.

*Rothwell, S.C., *A Geography of Earth Form; Preface to Physical Geography*, Brown, Dubuque, IA, 1968.

*Snead, R., *World Atlas of Geomorphic Features*, Litton, New York, 1981.

Tank, R.W., *Environmental Geology*, Oxford Univ. Press, New York, 1983.

Thorn, C.E., *Space and Time in Geomorphology*, Allen and Unwin, Winchester, MA, 1982.

Williams, P.J., *The Surface of the Earth: An Introduction to Geotechnical Science*, Longman, New York, 1982.

CHAPTER SEVENTEEN INTRODUCTION TO LANDFORMS

Adams, G. and J. Wyckoff, *Landforms*, Golden Press, New York, 1972.

*Brown, G.C. and A.E. Mussett, *The Inaccessible Earth*, Allen and Unwin, Winchester, MA, 1981.

Davis, W.M., *Geographical Essays*, Ginn, Boston, 1909, (reprinted Dover, NY, 1954).

Hsu, K.J., *Mountain Building Processes*, Academic Press, New York, 1983.

*Lyttleton, R.A., *The Earth and its Mountains*, Wiley, Somerset, NH, 1982.

Mabbutt, J.A., *Desert Landforms*, M.I.T. Press, Cambridge, 1977.

Melhorn, W.N. and R.C. Flemal, *Theories of Landform Development*, Allen and Unwin, Winchester, MA, 1981.

*Milne, L.J., M. Milne and the editors of Life, *The Mountains*, Time, New York, 1962.

Nir, D., *Man, A Geomorphological Agent: An Introduction to Anthropic Geomorphology*, Reidel, Dordrecht, W. Germany, 1983.

Miyashiro, A., K. Aki, and C. Sengor, *Orogony*, Wiley, Somerset, NJ, 1982.

Ollier, C.D., *Tectonics and Landforms*, Longman, New York, 1981.

Weyman, D., *Tectonic Processes*, Allen and Unwin, Winchester, MA, 1981.

CHAPTER EIGHTEEN THE EARTH'S CRUST

Ager, D.V., *Nature of the Stratigraphical Record*, 2nd ed., Wiley, Somerset, NJ, 1981.

*Bascom, W., *A Hole in the Bottom of the Sea*, Doubleday, N.Y., 1961.

Condie, K.C., *Plate Tectonics and Crustal Evolution*, 2nd ed., Pergamon, New York, 1982.

Cox, K.G., J.D. Bell and R.J. Pankhurst, *The Interpretation of Igneous Rocks*, Allen & Unwin, Winchester, MA, 1979.

Cullingford, R.A., D.A. Davidson and J. Lewin, eds., *Timescales in Geomorphology*, Wiley, New York, 1980.

*Eicher, D.L., *History of the Earth*, Prentice-Hall, Englewood Cliffs, NJ, 1980.

*Elder, J., *The Bowels of the Earth*, Oxford Univ. Press, Oxford, 1978.

Fry, N., *The Field Description of Metamorphic Rocks*, Wiley, New York, 1983.

*Hallam, A., *A Revolution in the Earth Sciences: From Continental Drift to Plate Tectonics*, Clarendon Press, Oxford, England, 1973.

Leeder, M.R., *Sedimentology: Process and Product*, Allen and Unwin, Winchester, MA, 1982.

MacKenzie, W.S., C.H. Donaldson and C. Guilford, *Atlas of Igneous Rocks and Their Texture*, Wiley, Somerset, NJ, 1982.

*Miller, R. and the editors of Time-Life Books, *Continents in Collision*, Time-Life Books, Alexandria, Va, 1983.

*Raymo, C., *The Crust of Our Earth: An Armchair Traveler's Guide to the New Geology*, Prentice-Hall, Englewood Cliffs, NJ, 1984.

Tarling, D.T. ed., *Paleomagnetism: Principles and Applications in Geology, Geophysics and Archaeology*, Methuen, New York, 1983.

Tucker, M.E., *The Field Description of Sedimentary Rocks*, Wiley, Somerset, NJ, 1982.

Wegener, A., *The Origins of the Continents and Oceans*, (translated by John Byram), Dover, New York, 1966.

*Wilson, J.T., ed., *Continents Adrift and Continents Aground*, Freeman, San Francisco, 1976.

CHAPTER NINETEEN FOLDING

Collett, L.W., *The Structure of the Alps*, R.E. Krieger, Huntington, NY, 1974.

Colwell, Robert N., *Manual of Remote Sensing*, 2 Vols., 2nd ed., American Society of Photogrammetry, Sheridan Press, Falls Church, VA, 1983.

Holz, R.K., ed., *The Surveillant Science: Remote Sensing of the Environment*, 2nd ed., Houghton Mifflin, Wiley, N.Y., 1985.

Measures, R.M, *Laser Remote Sensing*, Wiley, New York, 1983.

Ramsay, J.G., *Folding and Fracturing of Rocks*, McGraw-Hill, New York, 1967.

Townshend, J.R.G., *Terrain Analysis and Remote Sensing*, Allen & Unwin, Winchester, MA, 1981.

*Wallace, R. and the editors of Time-Life Book, *The Grand Canyon*, Time-Life, Alexandria, VA, 1973.

Whitten, E.H.T., *Structural Geology of Folded Rocks*, Rand McNally, Chicago, 1966.

CHAPTER TWENTY FAULTING

Anderson, E.M., *The Dynamics of Faulting and Dyke Formation*, Oliver and Boyd, Edinburgh, 1951.

*Burton, I., R.W. Kates and G. White, *The Environment as Hazard*, Oxford Univ. Press, New York, 1978.

Cresswell, M.M., ed., *Tsunami Research Symposium*, Unesco Press, Paris, 1976.

Eiby, G.A., *Earthquakes*, Van Nostrand, New York, 1980.

*Iacopi, R., *Earthquake Country: How, Why, and When Earthquakes Strike In California*, rev. ed., Lane, Menlo Park, CA, 1976.

Richter, C.E. *Elementary Seismology*, W.H. Freeman, San Francisco, 1958.

U.S. National Academy of Sciences, Panel on Earthquake Prediction, *A Scientific and Technical Evaluation With Implications for Society*, N.A.S., Washington, 1976.

*Walker, B. and the editors of Time-Life Books, *Earthquake*, Time-Life, Alexandria, VA, 1982.

CHAPTER TWENTY-ONE VULCANISM

Bolt, B.A., *Earthquakes and Volcanoes*, W.H Freeman, San Francisco, 1980.

Civetta, L., ed., *Physical Volcanology*, Elsevier, New York, 1974.

Cotton, C.A., *Volcanoes as Landscape Forners*, Whitcombe and Tombs, Christchurch, 1944.

*Editors of Time-Life Books, *Volcano*, Time-Life, Alexandria, VA, 1982.

Green, J. and N.M. Short, *Volcanic Landforms and Surface Features*, Springer-Verlag, New York, 1972.

*Harris, S.L., *Fire and Ice: The Cascade Volcanoes*, 2nd ed., Pacific Search Press, Seattle, 1980.

Herbert, D. and F. Bardossi, *Kilauea: Case History of a Volcano*, Harper & Row, New York, 1968.

Lambert, M.B. *Volcanoes*, Univ. of Washington, Press, Seattle, 1980.

Nokes, R., ed., *Mount St. Helens: The Volcano*, Oregonian Pub. Co., Portland, OR, 1980.

*Rosenfeld, C. and R. Cooke, *Earthfire: The Eruption of Mount St. Helens*, M.I.T. Press, Cambridge, 1982.

Sheets, P.D. and D. Grayson, *Volcanic Activity and Human Ecology*, Academic Press, New York, 1979.

*Simkin, T., et al, *Volcanoes of the World*, Academic Press, New York, 1981.

CHAPTER TWENTY-TWO WEATHERING AND MASS WASTING

deBoodt, M. and D. Gabriels, eds., *Assessment of Erosion*, Wiley, New York, 1980.

Carroll, D., *Rock Weathering*, Plenum, New York, 1970.

*Holy, M., *Erosion and Environment*, Pergamon, New York, 1980.

Ollier, C., *Weathering*, 2nd ed., Longman, London, 1976.

Selby, M.J., *Hillslope Materials and Processes*, Wiley, New York, 1983.

Small, R.J., and M.J. Clark, *Slopes and Weathering*, Cambridge Univ. Press, New York, 1982.

Trudgill, S.J., *Weathering and Erosion*, Butterworths, Boston, 1983.

Zarubia, Q., and V. Mendel, *Landslides and Their Control*, 2nd rev. ed. Elsevier, New York, 1982.

CHAPTER TWENTY-THREE RUNNING WATER

Chorley, R.C., *Introduction to Geographical Hydrology*, Barnes and Noble, New York, 1969.

*Clark, C. and the editors of Time-Life Books, *Flood*, Time-Life, Alexandria, VA, 1982.

Hey, R.D., J.C. Bathurst and C.R. Thorne, *Gravel Bed Rivers: Fluvial Processes, Engineering and Management*, Wiley, Somerset, NJ, 1982.

Morisawa, M., *Fluvial Geomorphology*, Allen & Unwin, Winchester, MA, 1981.

*Petts, G.E., *Rivers*, Butterworths, Woburn, MA, 1983.

Pitty, A. F., ed., *Geographical Approaches to Fluvial Processes*, Univ. of East Anglia, Norwich, 1979.

Rachocki, A.H. *Alluvial Fans*, Wiley, New York, 1981.

Rhodes, D.D. and G.P. Williams, *Adjustments of the Fluvial System*, Allen & Unwin, Winchester, MA, 1982.

CHAPTER TWENTY-FOUR GLACIATION

Azassiz, L., *Geological Sketches*, 2 series, Houghton Mifflin, New York, 1886 and 1903.

*Bailey, R.H. and the editors of Time-Life Books, *Glacier*, Time-Life, Alexandria, VA, 1982.

Banks, M., *Greenland*, Rowman & Littlefield, Totowa, NJ, 1975.

*Bond, C., and R. Siegfried, *Antarctica: No Single Country, No Single Sea*, Mayflower, New York, 1979.

*Chorlton, W. and the editors of Time-Life Books, *Ice Ages*, Time-Life, Alexandria, VA, 1983.

Coats, D.R., *Glacial Geomorphology*, Allen & Unwin, Winchester, MA, 1981.

Colbeck, S.C., ed., *Dynamics of Snow and Ice Masses*, Academic Press, New York, 1980.

Davidson-Arnott, R., W. Nickling and B.D. Fahey, eds., *Research in Glacial, Glaciofluvial and Glaciolacustrine Systems*, Geo Abstracts, Norwich, England, 1982.

*Denton, G.H. and T. Hughes, eds., *The Last Great Ice Sheets*, Wiley, New York, 1981.

Embleton, C. and C.A.M. King, *Glacial and Periglacial Geomorphology*, 2nd ed., Wiley, New York, 1975.

Flint, R.F., *Glacial and Quarternary Geology*, Wiley, New York, 1971.

Frasier, Colin, *The Avalanche Enigma*, Rand McNally, Chicago, 1966.

Goldthwaite, R.P., ed., *Glacial Deposits*, Dowden, Hutchinson and Ross, Stroudsburg, PA, 1975.

*Mathes, F.E., F. Fryxell, eds., *The Incomparable Valley: A Geological Interpretation of Yosemite*, Univ. of California Press, Berkeley, 1950.

*Muir, J.A., and William E. Colby, ed., *Studies in the Sierra*, rev. ed., Sierra Club, San Francisco, 1960.

Paterson, W.S.B., *The Physics of Glaciers*, 2nd ed., Pergamon, Toronto, 1981.

CHAPTER TWENTY-FIVE WIND

Johnson, V., *Heaven's Tableland; The Dust Bowl Story*, DeCapo Press, New York, 1974.

Lai, R.J., *Wind Erosion and Deposition Along a Coastal Sand Dune*, Univ. of Delaware Press, Newark, 1978.

* Leopold, A.S. and the editors of Time-Life Books, *The Desert*, Time-Life, New York, 1967.

Peure, T.L., *Desert Dust: Origin, Characteristics and Effect on Man*, Geol. Soc. of America, Boulder, CO, 1981.

* Sears, P.B., *Deserts on the March*, 3rd ed., Univ. of Oklahoma Press, Norman, 1959.

Smalley, I.J., *Loess: Lithology and Genesis*, Halstead, New York, 1975.

CHAPTER TWENTY-SIX WAVES

Bascom, W., *Waves and Beaches: The Dynamics of the Ocean Surface*, rev. ed., Doubleday, Garden City, NY, 1980.

Coates, D.R., *Coastal Geomorphology*, Allen & Unwin, Winchester, MA, 1981.

Davies, J.L., *Geographical Variation in Coastal Development*, Longman, London, 1977.

* Fox, W.T., *At the Sea's Edge: An Introduction to Coastal Oceanography for the Amateur Naturalist*, Prentice-Hall, Englewood Cliffs, NJ, 1983.

Gresswell, R.K., *The Physical Geography of Beaches and Coastlines*, Hulton Educational Pubs., London, 1957.

Johnson, D.W., *Shoreline Processes and Shoreline Development*, Hafner, New York, 1965.

* Kauffman, W. and O. Pilkey, *The Beaches Are Moving: The Drowning of America's Shoreline*, Doubleday, New York, 1979.

King, C.A.M., *Beaches and Coasts*, 2nd ed., St. Martin's, New York, 1972.

Komar, P.D., *Beach Processes and Sedimentation*, Prentice-Hall, Englewood Cliffs, NJ, 1976.

Mei, C.C., *The Applied Dynamics of Ocean Surface Waves*, Wiley, New York, 1982.

* Sackett, R. and the editors of Time-Life Books, *Edge of the Sea*, Time-Life, Alexandria, VA, 1983.

Schwartz, M.L., *Spits and Bars*, Dowden, Hutchinson and Ross, Stroudsburg, PA, 1972.

Smith, D.E. and A.G. Dawson, *Shorelines and Isostasy*, Academic Press, New York, 1984.

Snead, R.D., *Coastal Landforms and Surface Features*, Hutchinson and Ross, Stroudsburg, PA, 1982.

CHAPTER TWENTY-SEVEN GROUNDWATER RELATED WATER PROBLEMS

* Bitton, G. and C.P. Gerbs, *Groundwater Pollution Microbiology*, Wiley, New York, 1984.

Blair, P.B., T.A.V. Cassel and R.H. Edelstein, *Geothermal Energy*, Wiley, New York, 1982.

Bögli, A., *Karst Hydrology and Physical Speleology* (translated by June C. Schmid), Springer-Verlag, New York, 1980.

Bowen, R., *Geothermal Resources*, Wiley, New York, 1980.

Bowen, R., *Groundwater*, Halstead, New York, 1980.

Edwards, L.M. et al, eds., *Handbook of Geothermal Energy*, Gulf Pub. Co., Houston, 1982.

* Howe, C.W. and K.W. Easter, *Interbasin Transfers of Water: Economic Issues and Impacts*, Johns Hopkins Press, Baltimore, MD, 1971.

* Jackson, D.D. and the editors of Time-Life Books, *Underground Worlds*, Time-Life, Alexandria, VA, 1982.

* Leopold, L.B., K.S. Davis and the editors of Time-Life, *Water*, Time-Life, New York, 1969.

Matthess, G., *The Properties of Groundwater* (translated by John C. Harvey), Wiley, Somerset, NJ, 1981.

McCaul, J. and J. Crossland, *Water Pollution*, Harcourt Brace Jovanovich, New York, 1974.

Middlebrooks, E.J., *Water Reuse*, Ann Arbor Science, Ann Arbor, MI, 1982.

Raghunath, H.M., *GroundWater Hydrology*, Wiley, New York, 1982.

Shainberg, I. and J.D. Oster, *Quality of Irrigation Water*, Pergamon, New York, 1979.

Shuval, H.I., *Water Quality and Management Under Conditions of Scarcity*, Academic Press, New York, 1980.

Sweeting, M., *Karst Landforms*, Columbia Univ. Press, New York, 1973.

Todd, D.K., *Groundwater Hydrology*, 2nd ed., Wiley, New York, 1980.

PART VI THE OCEANS

Brewer, P.G., ed., *Oceanography: The Present and Future*, Springer-Verlag, New York, 1983.

*Carson, R., *The Sea Around Us*, Oxford Univ. Press, Fairlawn, NJ, 1951.

Deacon, G.E.R. and M.B. Deacon, eds., *Modern Concepts of Oceanography*, Hutchinson and Ross, Stroudsburg, PA, 1982.

*Earle, S.A. and A. Giddings, *Exploring the Deep Frontier: The Adventure of Man in the Sea*, National Geog. Soc., Washington, DC, 1980.

*Engel, L. and the editors of Life, *The Sea*, Time, New York, 1961.

*Gross, M.G., *Oceanography: A View of the Earth*, Prentice-Hall, Englewood Cliffs, NJ, 1982.

King, C.A.M., *Oceanography for Geographers*, Arnold, London, 1975

Pickard, G.L. and W.J. Emery, *Physical Oceanography: An Introduction*, Pergamon, New York, 1982.

Rand McNally Atlas of the Oceans, Rand McNally, Chicago, 1977.

Rey, L., *The Arctic Ocean*, Wiley, New York, 1983.

Robinson, A.R., ed., *Eddies in Marine Science*, Springer-Verlag, New York, 1983.

Shepard, F.P., *Geological Oceanography: Evolution of Coasts, Continental Margins and the Sea Floor*, Crane, Russak, New York, 1977.

Stowe, K., *Ocean Science*, 2nd ed., Wiley, New York, 1983.

Svendrup, H.V., M.W. Johnson and R.H. Fleming, *The Oceans*, Prentice-Hall, Englewood Cliffs, NJ, 1942.

*Whipple, A.B.C. and the editors of Time-Life Books, *The Restless Oceans*, Time-Life, Alexandria, VA, 1984.

CHAPTER TWENTY-EIGHT THE BOTTOM OF THE SEA

Darwin, C., *The Structure and Distribution of Coral Reefs*, 3rd ed., Appleton-Century-Crofts, New York, 1898.

Fridriksson, S., *Surtsey: Evolution of Life on a Volcanic Island*, Halstead Press, New York, 1975.

Heezen, B.C., M. Tharp and M. Ewing, *The Floors of the Oceans*, Geolo. Soc. of America, Special Paper No. 65, New York, 1959.

Hopley, D., *The Geomorphology of the Great Barrier Reef*, Wiley, New York, 1982.

Johnson, D., *The Origin of Submarine Canyons*, Columbia Univ. Press, New York, 1939.

Kennett, J.P., *Marine Geology*, Prentice-Hall, Englewood Cliffs, NJ, 1982.

King, C.A.M., *Introduction to Marine Geology*, Arnold, London, 1975.

*Linklater, E., *The Voyage of the Challenger*, Sphere Books, London, 1974.

*Macdonald, G.A., A.T. Abbott and F.L. Peterson, *Volcanoes in the Sea: The Geology of Hawaii*, 2nd ed., Univ. of Hawaii Press, Honolulu, 1983.

*Scherman, K., *Daughter of Fire: A Portrait of Iceland*, Little, Brown, Boston, 1976.

Scrutton, R.A. and M. Talwani, eds., *The Ocean Floor*, Wiley, New York, 1982.

Vinogradov, A.P. and G.B. Udinstev, *Rift Zones of the World Oceans*, Wiley, New York, 1975.

CHAPTER TWENTY-NINE CURRENTS AND TIDES

Charlies, R.H., *Tidal Energy*, Van Nostrand Reinhold, New York, 1982.

*Gaskell, T.F., *The Gulf Stream*, John Day, New York, 1973.

Marchuck, G.I. and B.A. Kagan, *Ocean Tides*, Pergamon, New York, 1982.

*Maury, M.F., J. Leighly, ed., *The Physical Geography of the Sea and its Meterology*, Harvard Univ. Press, Cambridge, 1963.

Melchior, P.J., *The Tides of the Planet Earth*, Pergamon, New York, 1978.

Neumann, G., *Ocean Currents*, Elsevier, New York, 1968.

Roberts, J., *Internal Gravity Waves in the Ocean*, Dekker, New York, 1975.

Strommel, H., *The Gulf Stream: A Physical and Dynamical Description*, 2nd ed., Univ. of California Press, Berkeley, 1976.

Wylie, Francis E., *Tides and the Pull of the Moon*, Stephen Greene Press, Brattleboro, VT, 1979.

CHAPTER THIRTY RESOURCES FROM THE SEA

Armstrong, J.M. and P.C. Rynes, *Ocean Management: A New Perspective*, Ann Arbor Science, Ann Arbor, MI, 1981.

*Borgese, E.M., *Seafarm: The Story of Aquiculture*, Harry N. Abrams, New York, 1980.

Duedall, I.W. et al, eds., *Wastes in the Ocean*, Vol. 1: *Industrial and Sewage*, Vol. 2: *Dredged Material Disposal*, Vol. 3: *Radioactive Wastes*, Wiley, New York, 1983.

*Earney, F.C.F., *Ocean Mining: Geographic Perspectives*, Meddeleser Fra Geografisk Institutt, Bergen, 1982.

Earney, F.C.F., *Petroleum and Hard Minerals from the Sea*, Halstead, New York, 1980.

Hepher, B. and Y. Pruginin, *Commercial Fish Farming*, Wiley, New York, 1981.

Hollander, A., ed., *The Biosaline Concept: An Approach to the Utilization of Unexploited Resources*, Plenum, London, 1982.

Kent, P., *Minerals from the Marine Environment*, Halstead, New York, 1981.

*Limburg, P.R., *Farming the Waters*, Beaufort, New York, 1980.

Malik, M.A.S., et al, *Solar Distillation*, Pergamon, New York, 1982.

*Marx, W., *The Oceans, Our Last Resource*, Sierra Club, San Francisco, 1981.

Moghissi, A.A., ed., *Oil Spills*, Pergamon, New York, 1980.

Oxman, B.H. and D.D. Caron, *Law of the Sea: U.S. Policy Dilemma*, Chas. L.O. Buderi, New York, 1983.

Palmer, H.D. and M. Grant Gross, eds., *Ocean Dumping and Marine Pollution*, Dowden, Hutchinson and Ross, Stroudsburg, PA, 1979.

Polking, K., *Oceans of the World: Our Essential Resource*, Philomel, New York, 1983.

Post, R.G. and R.L. Seale, eds., *Water Production Using Nuclear Energy*, Univ. of Arizona Press, Tucson, AZ, 1966.

Scott, J., *Desalinization of Seawater by Reverse Osmosis*, Noyes Data Corp., Park Ridge, NJ, 1981.

*Simon, A.W., *The Thin Edge: Coast and Man in Crisis*, Avon Books, New York, 1979.

Spiegler, K.S. and A.D.K. Laird, eds., *Principles of Desalinization*, 2 vols., Academic Press, New York, 1980.

Tippie, K. and D.R. Kester, eds., *Impact of Marine Pollution on Society*, I.F. Bergin, South Hadley, MA, 1982.

PHOTO CREDIT LIST

Weyerhaeuser Company. Fig. 13–6: Grant Heilman. Fig. 13–7: Harold Hungerford. Box: Grant Heilman.

Chapter 14
Opener: Courtesy New Zealand Forest Products Co. Ltd. Fig. 14–1: Aero Exploration. Fig. 14–2a: Jack E. Boucher/National Park Service, Fig. 14–2b: Australian News & Information Bureau, 14–2c: National Park Service. Fig. 14–3: South African Information Services." Fig. 14–4: Ira Kirshenbaum/Stock Boston. Fig. 14–5: Daniel S. Brody/Stock Boston. Fig. 14–6: G.R. Roberts. Fig. 14–7: Courtesy Spanish Ministry of Information & Tourism. Fig. 14–8: Courtesy U.S. Forest Service. Fig. 14–9: Courtesy Redwood Empire Association. Fig. 14–10: Courtesy Weyerhauser Company. Fig. 14–11: Pierre Berger/NAS—Photo Researchers. Fig. 14–12: Grant Heilman. Fig. 14–13: Courtesy Environment Canada. Fig. 14–14: D. Wilkinson/National Film Board of Canada. Box: Grant Heilman.

Chapter 15
Opener: Almasy. Fig. 15–2: Peter Menzel/Stock Boston. Fig. 15–3: Daniel S. Brody/Stock Boston. Fig. 15–5: Grant Heilman. Fig. 15–7: Grant Heilman. Fig. 15–10: New Zealand National Publicity Studios. Fig. 15–11: Grant Heilman. Fig. 15–12: U.S.D.A. Box: Grant Heilman.

Chapter 16
Opener: Owen Franken/Stock Boston. Fig. 16–1: G.R. Roberts. Fig. 16–2: Roy Simonson/U.S.D.A. Fig. 16–3: Roy Simonson/U.S.D.A. Fig. 16–4: Grant Heilman. Fig. 16–5a: Ned Haines/Rapho—Photo Researchers. Fig. 16–5b: Victor Englebert /Photo Researchers. Fig. 16–6: Georg Gerster/Rapho-Photo Researchers. Fig. 16–7: Courtesy Shell Oil Company Box: Dr. John S. Shelton.

Chapter 17
Opener: Aero Exploration. Fig. 17–1: Pierre Berger/Photo Researchers. Fig. 17–3a: Courtesy Japan National Tourist Organi-

zation. Fig. 17–3b: Courtesy Spanish Ministry of Information & Tourism. Fig. 17–5: Dr. John S. Shelton. Fig. 17–6: Grant Heilman. Box: Harvard University Library.

Chapter 18
Opener: National Air Photo Library of Canada. Fig. 18–5: Frank Wing/Stock Boston. Fig. 18–8: W. J. Lee/U.S. Geological Survey. Fig. 18–10: The American Museum of Natural History. Fig. 18–11: Grant Heilman. Fig. 18–12: Harry Wilks/Stock Boston. Fig. 18–13: National Park Service. Fig. 18–15: G.R. Roberts. Fig. 18–17: Jerome Wyckoff. Box 1: Courtesy Australian Tourist Commission. Box 2: Wide World Photos.

Chapter 19
Opener: Dr. John S. Shelton. Fig. 19–2: Jerome Wyckoff. Fig. 19–4: National Park Service. Fig. 19–5: NASA. Fig. 19–6: Grant Heilman. Fig. 19–8: Swiss National Tourist Office. Box: NASA.

Chapter 20
Opener: Georg Gerster/Photo Researchers. Fig. 20–1: Mary Hill/California Division of Mines & Geology. Fig. 20–4: NASA. Fig. 20–5: Dr. John S. Shelton. Fig. 20–7: Dr. John S. Shelton. Fig. 20–9: Owen Franken/Sygma. Fig. 20–10: Grant Heilman. Box: Sovfoto.

Chapter 21
Opener: Georg Gerster/Photo Researchers. Fig. 21–1: Courtesy French Government Tourist Office. Fig. 21–4: Courtesy Oregon State Highway Travel Division. Fig. 21–6: U.S. Geological Survey. Fig. 21–7: Owen Franklin/Stock Boston. Fig. 21–8: Grant Heilman. Fig. 21–9: Dr. John S. Shelton. Fig. 21–10A: Dr. John S. Shelton. Fig. 21–10B: Alan Pitcairn/Grant Heilman. Box: Ira Kirschenbaum/Stock Boston.

Chapter 22
Opener: Courtesy Oregon State Highway Travel Division. Fig. 22–1: National Park

Service. Fig. 22–2: G.K. Gilbert/National Park Service. Fig. 22–3: John Christopher/Photo Researchers. Fig. 22–4: G.R. Roberts. Fig. 22–5: H.E. Stork/National Park Service. Fig. 22–6: Read D. Brugger/Picture Cube. Fig. 22–7: U.P.I. Fig. 22–9a: Los Angeles County Fire Department. 22–9b: Los Angeles County Fire Department. Fig. 22–10: W.B. Finch/Stock Boston. Box: Lily Solmssen/Photo Researchers.

Chapter 23
Opener: Coast & Geodetic Survey. Fig. 23–1: Jen & Des Bartlett/Photo Researchers. Fig. 23–4: Oregon State Highway Travel Division. Fig. 23–5: Dr. John S. Shelton. Fig. 23–7: Grant Heilman. Fig. 22–10: Dr. John S. Shelton. Fig. 23–12: Grant Heilman. Fig. 23–14: Grant Heilman. Box: An Keren/Eastfoto.

Chapter 24
Opener: Dr. John S. Shelton. Fig. 24–1: Courtesy Swissair. Fig. 24–2: Courtesy Danish Information Office, New York. Fig. 24–3: Grant Heilman. Fig. 24–5: French Government Tourist Office. Fig. 24–6: National Park Service. Fig. 24–7: G.K. Gilbert/U.S. Geological Survey. Fig. 24–8: Grant Heilman. Fig. 24–9: New Zealand National Publicity Studios. Fig. 24–11: Ira Kirschenbaum/Stock Boston. Fig. 24–13: Courtesy Danish Information Office, New York. Fig. 24–14: Canadian National Railways. Fig. 24–15: E.E. Hertzog/Bureau of Reclamation. Box: Culver Pictures.

Chapter 25
Opener: Peter Menzel/Stock Boston. Fig. 25–2: Hubertus Kanus/Rapho-Photo Researchers. Fig. 25–2: National Park Service. Fig. 25–3: National Park Service. Fig. 25–4: Georg Gerster/Photo Researchers. Fig. 25–5: New Zealand Forest Service Library. Fig. 25–6: M. Woodbridge Williams/National Park Service. Box 2: New Zealand Forest Service Library.

Chapter 26
Opener: Oregon State Highway Department. Fig. 26–1: Bill Bachman/Photo Researchers. Fig. 26–2: German Information Center. Fig. 26–3: Verna R. Johnson/NAS-Photo Researchers. Fig. 26–7: Washington State Department of Commerce and Economic Development. Fig. 26–8: New Zealand National Publicity Studios. 26–9: New Zealand National Publicity Studios. Fig. 26–10: G.R. Roberts. Fig. 26–11a: U.S. Army Engineer District, San Francisco. Fig. 26–11b: Port Angeles Chamber of Commerce. Fig. 26–12: Mark Leet/Sygma. Fig. 26–14: NASA Photo Research by Grant Heilman.

Chapter 27
Opener: Aero Exploration. Fig. 27–2: Harold R. Hungerford. Fig. 27–5: U.S. Army Corps of Engineers, San Francisco District. Fig. 27–6: Santa Clara County Water Conservation District. Fig. 27–7: The Permuitt Company. Fig. 27–8: State of California Department of Water Resources. Fig. 27–11: New Zealand National Publicity Studios. Fig. 27–12: New Zealand Consulate General, New York. Fig. 27–14: Courtesy Luray Caverns. Fig. 27–15: Jen & Des Bartlett/Photo Researchers. Box: Edvard Kotlyakov/Sovfoto.

Chapter 28
Opener: Maurice & Sally Landre/Photo Researchers. Fig. 28–1: U.S. Navy. Fig. 28–3: Horace Bristol, Jr./Photo Researchers. Fig. 28–6: Casa de Portugal. Fig. 28–7: U.S. Coast and Geodetic Survey. Fig. 28–8: G.R. Roberts. Fig. 28–11; New Zealand Information Service. Box: John Donnelly/Woods Hole Oceanographic Institution.

Chapter 29
Opener: Serge de Sazo/Rapho-Photo Researchers. Fig. 29–2: George Bellrose/Stock Boston. Fig. 29–4a: Fredrik D. Bodin/Stock Boston. Fig. 29–4b: Fredrik D. Bodin/Stock Boston. Fig. 29–5: Phototeque EDF, Michel Brigaud, French Embassy Press & Information Division. Fig. 29–6: G.R. Roberts. Box: Woods Hole Oceanographic Institution.

Chapter 30
Opener: Leslie Salt Company. Fig. 30–2: International Salt Company. Fig. 30–3: Mobile Oil Corporation. Fig. 30–6: Johnnie Walker/Picture Cube. Fig. 30–7: Georg Gerster/Photo Researchers. Fig. 30–8: Anestis Diakopoulos/Stock Boston. Fig. 30–9: N.O.A.A. Box 1: Lamont-Doherty Geological Observatory of Columbia University. Box 2: Tass from Sovfoto.

INDEX

Numbers in **boldface** type indicate most important references.